Advances in Electrochemical Science and Engineering

Volume 13
Bioelectrochemistry

Advances in Electrochemical Science and Engineering

In collaboration with the International Society of Electrochemistry

Advances in Electrochemical Science and Engineering

Volume 13
Bioelectrochemistry

Edited by
Richard C. Alkire, Dieter M. Kolb, and Jacek Lipkowski

WILEY-VCH Verlag GmbH & Co. KGaA

The Editors

Prof. Richard C. Alkire
University of Illinois
600 South Mathews Avenue
Urbana, IL 61801
USA

Prof. Dieter M. Kolb
University of Ulm
Institute of Electrochemistry
Albert-Einstein-Allee 47
89081 Ulm
Germany

Prof. Jacek Lipkowski
University of Guelph
Department of Chemistry
N1G 2W1 Guelph, Ontario
Canada

Library of Congress Card No.: applied for

British Library Cataloguing-in-Publication Data
A catalogue record for this book is available from the British Library.

Bibliographic information published by the Deutsche Nationalbibliothek
The Deutsche Nationalbibliothek lists this publication in the Deutsche Nationalbibliografie; detailed bibliographic data are available on the Internet at http://dnb.d-nb.de.

Typesetting Toppan Best-set Premedia Limited, Hong Kong
Printing and Binding betz-druck GmbH, Darmstadt
Cover Design Grafik-Design Schulz, Fußgöheim

Printed in the Federal Republic of Germany
Printed on acid-free paper

Print ISBN: 978-3-527-32885-7
ISSN: 0938-5193

Contents

Preface

The intent of this edition is to provide an up-to-date account of the recent development in the fast-growing field of bioelectrochemistry. Significant methodological advances in studies of model biomembranes supported at electrode surfaces provided new tools for drug screening, detection of proteins, and electrochemical DNA screening. Significant progress in the understanding of the enzymatic reactions at electrode surfaces lead to development of biofuel cells. The knowledge of membrane and cell electroporation by high electric fields finds application in tumor therapy. This volume reviews the progress in bioelectrochemical science with a particular emphasis on the recent methodological developments and new applications of biochemistry for analytical detection, in medicine, and for energy conversion in biofuel cells. The volume should be of interest to students and researchers working in several fields such as electrochemistry, biochemistry, and analytical and medicinal chemistry. Each chapter provides sufficient background material so that it can be read by a non-specialist and specialist alike.

August 2011 *Jacek Lipkowski*
University of Guelph, Guelph, Ontario, Canada

List of Contributors

Philip Bartlett
University of Southampton
School of Chemistry
Southampton SO17 1BJ
UK

Lucia Becucci
Florence University
Department of Chemistry
Via della Lastruccia 3
50019 Sesto Fiorentino Firenze
Italy

Sabine Borgmann
Ruhr-Universität Bochum
Analytische Chemie – Elektroanalytik & Sensorik
Universitätsstrasse 150
D-44780 Bochum
Germany

Ian J. Burgess
University of Saskatchewan
Department of Chemistry
Room 256, Thorvaldson, 110 Science Place
Saskatoon, SK, S7N 5C9
Canada

Qijin Chi
Technical University of Denmark
DTU Chemistry
Department of Chemistry
Kemitorvet, Building 207
2800 Kongens Lyngby
Denmark

Rumiana Dimova
Max Planck Institute of Colloids and Interfaces
Science Park Golm
14424 Potsdam
Germany

Julie Gehl
Copenhagen University Hospital Herlev
Center for Experimental Drug and Gene Electrotransfer (C*EDGE)
Department of Oncology
Herlev Ringvej 75
2730 Herlev
Denmark

Rolando Guidelli
Florence University
Department of Chemistry
Via della Lastruccia 3
50019 Sesto Fiorentino, Firenze
Italy

Palle Skovhus Jensen
Technical University of Denmark
DTU Chemistry
Department of Chemistry
Kemitorvet, Building 207
2800 Kongens Lyngby
Denmark

Paul Kavanagh
National University of Ireland Galway
School of Chemistry
University Road, Galway
Republic of Ireland

Dónal Leech
National University of Ireland Galway
School of Chemistry
University Road, Galway
Republic of Ireland

Sumeet Mahajan
University of Cambridge
Centre for Physics of Medicine
Cavendish Laboratory
Cambridge CB3 0HE
UK

Sebastian Neugebauer
Ruhr-Universität Bochum
Analytische Chemie – Elektroanalytik & Sensorik
Universitätsstrasse 150
D-44780 Bochum
Germany

Wolfgang Schuhmann
Ruhr-Universität Bochum
Analytische Chemie – Elektroanalytik & Sensorik
Universitätsstrasse 150
D-44780 Bochum
Germany

Albert Schulte
Suranaree University of Technology
Institute of Science
Schools of Chemistry and Biochemistry
Biochemistry – Electrochemistry Research Unit
111 University Avenue
Nakhon Ratchasima 30000
Thailand

Jens Ulstrup
Technical University of Denmark
DTU Chemistry
Department of Chemistry
Kemitorvet, Building 207
2800 Kongens Lyngby
Denmark

Jingdong Zhang
Technical University of Denmark
DTU Chemistry
Department of Chemistry
Kemitorvet, Building 207
2800 Kongens Lyngby
Denmark

1
Amperometric Biosensors

Sabine Borgmann, Albert Schulte, Sebastian Neugebauer, and Wolfgang Schuhmann

1.1
Introduction

The scope of this chapter is to review the advancements made in the area of amperometric biosensors. It is intended to provide general background about biosensor technology and to discuss important aspects for developing and optimizing biosensors. A major concern of this chapter is also to critically review the benefits, limitations, and potential of the different approaches to biosensor research and its applications. An introduction to biosensor research is given (Section 1.1) before criteria of "good to excellent" biosensor research are outlined (Section 1.2), and a standard for characterizing biosensor performance is defined (Section 1.3). Endeavor has been made to define what "good to excellent" biosensor research represents.

Because of the volume of the literature regarding amperometric biosensors as well as space limitations it is not possible to cite any substantial contribution to the field. We selected – to the best of our knowledge – representative work that can be of use not only for beginners but also for advanced researchers in the field as a basis for discussion. Examples of success stories accomplished in biosensor research are given as case studies in Section 1.4. General milestones and achievements relevant to biosensor research and development are listed in Table 1.2. The final conclusions are given in Section 1.5.

A way to address the current impact of biosensor research on analytical chemistry, biochemistry, biology, and medicine is to have a look at the number of publications. Table 1.1 contains the number of articles and reviews with the keyword "biosensor" and related keywords published between 2005 and 2010. About 11 345 papers and 549 reviews have been published containing the keyword "biosensor."

Almost 2000 papers dealing with glucose or employing glucose oxidase as biological recognition element have been published during the last five years. Glucose sensing is one of the success stories of biosensing. The health and the quality of life of diabetes patients depend on the accurate monitoring of their blood glucose levels by means of glucose biosensors. [59–65] The widespread use of glucose

Advances in Electrochemical Science and Engineering. Edited by Richard C. Alkire, Dieter M. Kolb, and Jacek Lipkowski

ISBN: 978-3-527-32885-7

Table 1.1 Numbers of papers published in important fields of biosensor research between 2005 and 2010.

Keywords	Number of papers published	Number of reviews published
"Biosensor"	11 345	549
"Biosensor" and "glucose"	1 974	96
"Biosensor" and "glucose oxidase"	1 331	37
"Biosensor" and "laccase"	109	7
"Biosensor" and "cellobiose dehydrogenase"	11	0
"Biosensor" and "DNA"	2 166	156
"Biosensor" and "disposable"	287	6
"Biosensor" and "amperometric"	1 931	86
"Biosensor" and "electrochemistry"	1 715	63
"Biosensor" and "reagentless"	177	3
"Biosensor" and "direct electron transfer"	570	25
"Biosensor" and "mediated electron transfer"	53	2
"Biosensor" and self-assembled monolayer"	389	12
"Biosensor" and "conducting polymer"	220	20
"Biosensor" and "osmium"	40	0
"Biosensor" and "PQQ"	26	6
"Biosensor" and "NADH"	190	5
"Biosensor" and "biofuel cell"	73	8
"Biosensor" and "microsensor"	45	2
"Biosensor" and "microelectrode"	185	16
"Biosensor" and "microarray"	257	26
"Biosensor" and "biochip"	155	12
"Biosensor" and "protein chip"	304	14
"Biosensor" and "microfabrication"	57	3
"Biosensor" and "microfluidics"	184	15
"Biosensor" and "scanning electrochemical microscope"	51	
"Biosensor" and "nano"	476	29
"Biosensor" and "nanobiosensor"	33	4
"Biosensor" and "nanomaterial"	56	16

Database search for publications of the latest five years was performed on 10 May 2010 with Web of Science (Thomson Reuters).

oxidase (GOx, EC 1.1.3.4) as analytical reagent has been reviewed in detail. [63, 66, 67] The success of GOx as biological recognition element for biosensors is not only due to the importance of its substrate glucose and its enzymatic performance but also to its outstanding high stability and relatively low price.

Thus, it is not surprising that GOx has also evolved into an initial testing tool for the primary evaluation of new biosensor architectures. It seems to be almost the indestructible "working horse" as a model system. However, one needs to be careful with the general applicability for transferring the findings from initial studies to other more challenging biological recognition elements without providing substantial experimental evidence. This highlights the importance of designing smart electron transfer (ET) pathways allowing the use of a general biosensor design for more than one biological recognition element (and analyte).

The number of papers for the different keywords from Table 1.1 gives hints on the current trends in the field of biosensor research. As mentioned above, glucose sensing is an ongoing trend. Whereas "biosensing and DNA" (2166 papers, 156 reviews) is still a hot topic, especially for the area of low-cost diagnostic devices. The use of biosensor approaches to biofuel cells (about 80 papers in the last five years) is increasingly of interest, and the use of nanomaterials is just evolving and becoming a hot topic. Although, to the best of our knowledge, nanobiosensors not only fabricated out of nanomaterials but also with a transducer surface confined to nanometric dimensions have not yet been realized.

The number of publications, however, does not address the level of quality of the presented research. What, in fact, represents the outcome of all these publications? What represents the resulting scientific advancement? This may not be so easy to answer as it as first seems. Thus, we will first give a general introduction to amperometric biosensors in the rest of this section, before we address the quality issue of biosensor research in Sections 1.2 and 1.3 and before we present some of the success stories of biosensor research (Section 1.4) in order to address the questions mentioned at the beginning of this paragraph.

1.1.1
Definition of the Term "Biosensor"

The use of enzyme electrodes was reported for the first time in 1962 [68]. The term "biosensor" was introduced by Cammann in 1977 [6]. The IUPAC definition of a biosensor, however, was introduced as recently as 1999 to 2001 [3–5]. Figure 1.1 schematically summarizes the set-up of a biosensor. A biosensor is a device that enables the identification and quantification of an analyte of interest from a sample matrix, for example, water, food, blood, or urine. As a key feature of the biosensor architecture, biological recognition elements that selectively react with the analyte of interest (e.g., antibody–antigen or enzymatic reactions) are employed. It is important to note that the biological recognition element is either integrated within or in close proximity to the transducer. The transducer enables the transformation of the analyte recognition and/or catalytic conversion event into a quantifiable physical signal, for example, a current in an amperometric biosensor.

As outlined in Figure 1.1, a biosensor consists of different components. Examples of these components are given in Figure 1.2. It is obvious that there are many ways to design a biosensor architecture. A variety of biological recognition elements ranging from enzymes to antibodies can be employed. The compilation given in Figure 1.2 helps one to understand which parameters change during a biological recognition event in a biosensor. This knowledge is fundamental for developing and optimizing biosensors. The choice of the transduction process and transduction material is dependent on this knowledge as well as the chemical approach to construct the sensing layer on the transducer surface.

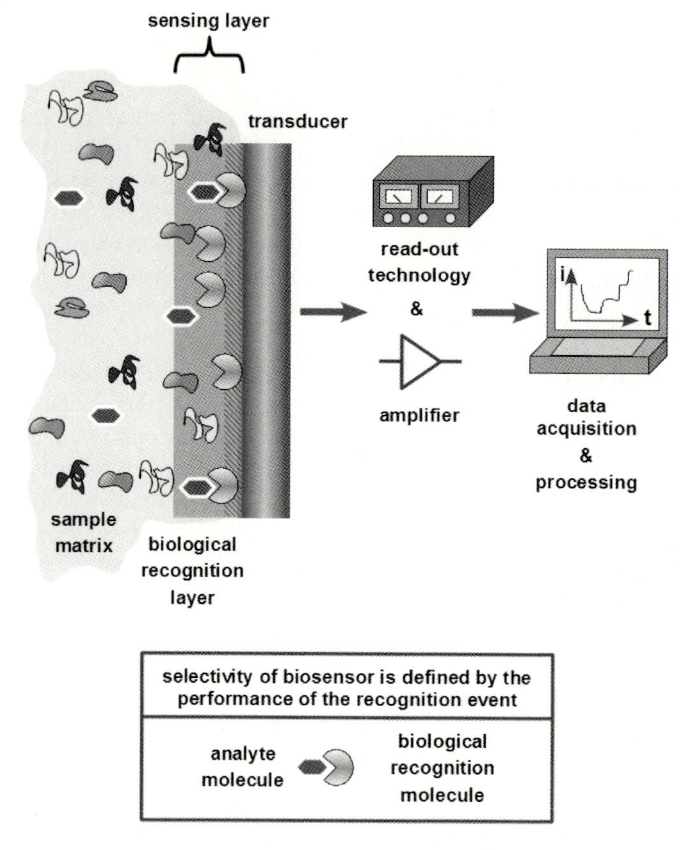

Figure 1.1 Typical biosensor set-up.

The choice of the biological recognition element is the crucial decision that is taken when developing a novel biosensor design. It is important to define criteria for, for example, a suitable redox enzyme for a specific biosensor. Most importantly, the enzyme needs to selectively react with the analyte of interest. The redox potential of the primary redox center needs to be within a suitable potential window (usually between –0.6 and 0.9 V vs. Ag/AgCl). The enzyme needs to be stable under the operation and storage conditions of the biosensor and should provide a reasonable long-term stability. It is advantageous if the chemical structure of the enzyme allows the introduction of additional functionalities for chemical modification with redox mediators, binding, or crosslinking with the immobilization matrix. In addition, the potential for tuning the properties of the redox enzyme by means of genetic or chemical techniques can be helpful for biosensor optimization. An important factor, especially with respect to potential commercialization, is that the redox enzyme is available at reasonable costs and effort.

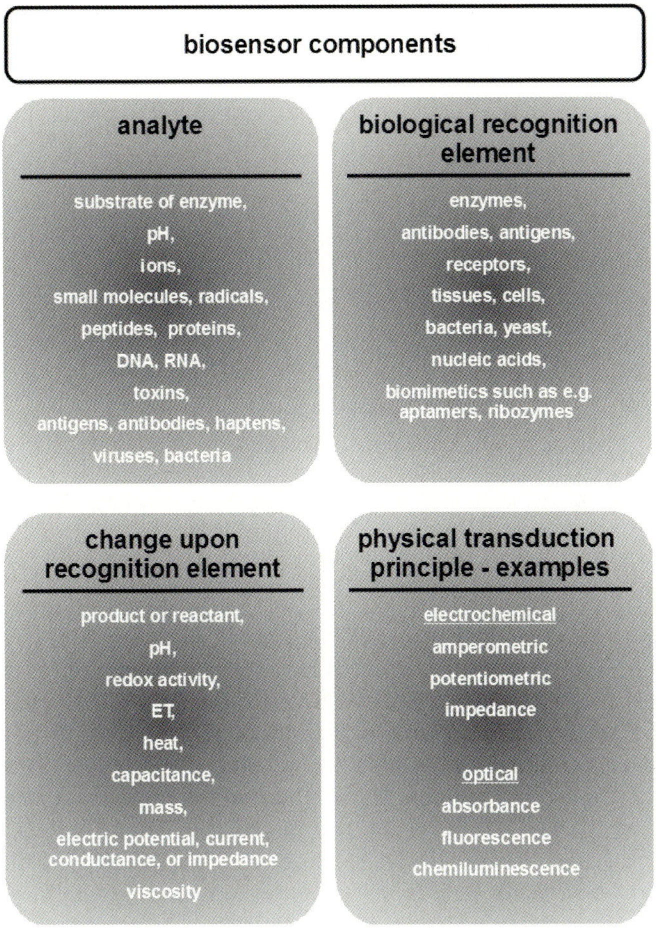

biosensor components

analyte
substrate of enzyme,
pH,
ions,
small molecules, radicals,
peptides, proteins,
DNA, RNA,
toxins,
antigens, antibodies, haptens,
viruses, bacteria

biological recognition element
enzymes,
antibodies, antigens,
receptors,
tissues, cells,
bacteria, yeast,
nucleic acids,
biomimetics such as e.g. aptamers, ribozymes

change upon recognition element
product or reactant,
pH,
redox activity,
ET,
heat,
capacitance,
mass,
electric potential, current, conductance, or impedance
viscosity

physical transduction principle - examples
electrochemical
amperometric
potentiometric
impedance

optical
absorbance
fluorescence
chemiluminescence

Figure 1.2 Examples for biosensor components.

The advantages of employing enzymes in biosensor architectures are the following:

i) They exhibit a very high catalytic activity with a turnover on a per mole basis which makes them not only exceptional bioelectrocatalysts for effective signal amplification in biosensors but also for biofuel cells. Good turnover frequencies k_{cat} are in the range of up to at least $100\,s^{-1}$.

ii) Typically, enzymes have a high selectivity for their substrates.

iii) In addition, the driving force, the redox potential that is needed to achieve enzymatic biocatalysis, is often very close to that of the substrate of the enzyme. Therefore, biosensors can operate at moderate potentials.

iv) In several cases, an improvement of the enzyme stability was found when enzymes were immobilized on transducer surfaces [25, 69].

The disadvantages of using enzymes in bioelectrochemical devices are the following:

i) Enzymes are rather large molecules. Thus, despite the high catalytic turnover at the active site of the enzyme, the overall catalytic (volume) density is low. As an example, at most about a few picomoles of enzyme molecules per square centimeter are contained in a monolayer of enzymes. Barton and coworkers calculated that the theoretical current density in such a monolayer is about $80\,\mu A\,cm^{-2}$ under the assumption that the "footprint" of the enzyme is about $100\,nm^2$ and the turnover frequency is about $500\,s^{-1}$ [70].

ii) Often the active site of the enzyme is deeply buried within the surrounding protein shell. Thus, direct ET is often not possible and artificial redox mediators are required.

iii) Enzymes have a limited lifetime and, therefore, biosensors exhibit only a limited long-term stability. So far, operational lifetimes of biosensors have been realized to up to 30 to 60 days [71, 72].

Mainly oxidoreductases have been employed for biosensors [73]. However, especially in the context of biofuel cell development, the spectrum of enzymes employed as bioelectrocatalysts is increasing [25]. For biosensor applications, it is important that the catalytic activity strongly depends on the substrate concentration which corresponds to an operating range of about the K_M value or below. This is important for obtaining a suitable dynamic range of the envisaged biosensor. In the case of blood glucose, for example, normal glucose levels are between 4 and 8 mM [74]. Typically, sugar-oxidizing enzymes have rather high K_M values (about 10 mM). Thus, if such enzymes are employed, the resulting biosensor can operate below substrate saturation. In contrast, in the case of biofuel cells the substrates are often present at concentrations well above the K_M value.

Electroanalytical techniques (also in combination with other techniques, e.g., optical techniques such as photometry and Raman spectrometry) can be employed to investigate many functional aspects of proteins and enzymes in particular. It is possible to study the biocatalytic process with respect to the chemistry of the active site, the interfacial and intramolecular ET, slow enzyme activators or inhibitors, the pH dependence, the transport of the substrate, and even more parameters. For example, slow scan voltammetry can be used to determine the relation of ET rates or of protonation and ligand binding. In contrast, fast scan voltammetry allows the determination of rates of interfacial ET. In addition, it is also possible to investigate chemical reactions that are coupled to the ET process, such as protonation. The use of direct ET for mechanistic studies of redox enzymes was recently reviewed by Léger and Bertrand [27]. Mathematical models help to elucidate the impact of different variables on the entire current signal [27, 75, 76].

1.1.2
Milestones and Achievements Relevant to Biosensor Research and Development

Biosensors have been studied extensively during the last fifty years. Hence, a number of milestones mark the progress made in biosensor research. Table 1.2 summarizes the main scientific milestones that are relevant to biosensor discovery and further development of this technology.

1.1.3
"First-Generation" Biosensors

Though many highly complex detection schemes can be found in biosensor designs, the simplest approach to a biosensor is the direct detection of either the increase of an enzymatically generated product or the decrease of a substrate of the redox enzyme. Additionally, a natural redox mediator that is participating in the enzymatic reaction can be monitored. In all three cases it is necessary that the compound monitored is electrochemically active. The use of GOx as biological recognition element for a "first-generation" biosensor design is the typical case and has been employed numerous times (Figure 1.3). Here, the increasing concentration of the product H_2O_2 or the decrease in O_2 concentration as natural co-substrate can be electrochemically detected in order to monitor glucose concentration [68, 103, 110, 150, 151].

The major drawbacks of the first-generation biosensor approach are the following: (i) if the O_2 concentration is monitored, it is challenging to maintain a reasonable reproducibility due to varying O_2 concentrations within the sample and (ii) working electrode potentials for either the oxidation of H_2O_2 or the reduction of O_2 are not optimal because these potentials are prone to the impact of interferences present in biological samples, such as ascorbic acid or dopamine.

1.1.4
"Second-Generation" Biosensors

In order to achieve biosensors which operate at moderate redox potentials the use of artificial redox mediators was introduced for the "second-generation" biosensors [135, 152–157]. Following the pioneering work by Kulys and Svirmickas [124, 125], Cass *et al.* were the first to show that an artificial redox mediator, ferrocene, could be employed for an amperometric glucose biosensor [135]. Figure 1.4 schematically explains how such a redox mediator can be used to read out the analyte concentration within a sample. The employed redox enzyme for the analyte of interest is able to donate or accept electrons to or from an electrochemically active redox mediator. It is important that the redox potential of this mediator is in tune with the cofactor(s) of the enzyme. Preferably, the redox mediator is highly specific for the selected ET pathway between the biological recognition element and the electrode surface. Note that the difference in potential between the different cofactors and the introduced artificial redox mediator should not be less than $\Delta E \sim 50\,\text{mV}$

Table 1.2 Milestones and achievements relevant to biosensor research and development.

Year	Contribution
1800s	Alessandro Giuseppe Anastasio Volta (1745–1827) introduced modern electrochemistry, and found out at that the frog legs employed in the 1791 experiments of Luigi Galvani (1737–1798) to generate currents were not the true source for the stimulation. Actually, it was the contact between two dissimilar metals. He termed this type of electricity "metallic electricity" and demonstrated the first electrochemical battery using his voltaic piles [77].
1839	The principle of the fuel cell was discovered by Christian Friedrich Schönbein (1799–1868) presenting a hydrogen–oxygen fuel cell [78, 79]. Sir William Robert Grove (1811–1896) created one of the first fuel cells which he called a "gas battery" [80]. He also wrote one of the first books that stated the principle of conservation of energy in 1846. Grove is known as the "father of the fuel cell." Friedrich Wilhelm Ostwald (1853–1932), a founder of the field of physical chemistry, contributed significantly to the operation principles of fuel cells [81]. The term "fuel cell" became fashionable around 1889.
1889	Walther Nernst (1864–1941) introduced the Nernst equation [82].
1894	Emil Fischer (1852–1919) introduced the key-lock-principle (specific binding between enzyme and substrate) [83].
1913	Leonor Michaelis (1875–1949) and Maud Leonora Menten (1879–1960) developed the basis for enzyme kinetics and defined a mathematical model, the Michaelis–Menten kinetics [84].
1916	Immobilization of proteins (adsorption of invertase on activated charcoal) reported for the first time by Nelson and Griffin [85].
1922	Jaroslav Heyrovský (1890–1967) invented polarography and the use of the dropping mercury electrode for electroanalysis [86, 87]. Heyrovský and Masuro Shikata (1895–1965) developed a polarograph that was able to automatically record cyclic voltammograms and that was the first automated analytical instrument [88]. In 1959, Heyrovský received a Nobel prize for the development of polarography [89].
1925	George E. Briggs and John B.S. Haldane re-evaluated the Michaelis–Menten equation and contributed to the modern view on the steady-state treatment of enzyme-catalyzed reactions [90].
1926	Otto Warburg (1859–1938) discovered cytochrome c oxidase ("Warburg ferment"). This represents the basis for the description of the mechanism of cellular respiration (Nobel prize in 1931) [91]. Later, Warburg discovered the cofactors (NADH) and the mechanism of dehydrogenases. This leads to optical tests for NADH and NADPH which allows for testing the activity of dehydrogenases. This indicator reaction can be coupled with other enzyme reactions. These advancements were the basis of the work of Hans-Ulrich Bergmeyer (Boehringer Mannheim) promoting enzymatic analysis in the 1960s [92].
1950	Erwin Chargaff (1905–2002) discovered that the ratio of adenine to thymine and guanosine to cytosine is in all living creatures about 1 (Chargaff's rules) [93].
1953	James Dewey Watson (born 1928) and Francis Harry Compton Crick (1916–2002) developed a model for the structure of the double helix of DNA [94].
1955	Frederick Sanger (born 1918) determined the complete amino acid sequence of the two polypeptide chains of insulin. He received a Nobel prize in 1958 for his work on the structure of proteins, especially insulin [95].

Table 1.2 (*Continued*)

Year	Contribution
1956	Rudolph A. Marcus (born 1923) introduced a theory of electron transfer, named Marcus theory. He received the Nobel Prize in Chemistry for this achievement in 1992 [19, 29, 30].
1956	Leland C. Clark Jr. (1918–2005) presented his first paper about the oxygen electrode, later named the Clark electrode, on 15 April 1956, at a meeting of the American Society for Artificial Organs during the annual meetings of the Federated Societies for Experimental Biology [96].
	In 1962, Clark and Ann Lyons from the Cincinnati Children's Hospital developed the first glucose enzyme electrode. This biosensor was based on a thin layer of glucose oxidase (GOx) on an oxygen electrode. Thus, the readout was the amount of oxygen consumed by GOx during the enzymatic reaction with the substrate glucose [68]. This publication became one of the most often cited papers in life sciences. Due to this work he is considered the "father of biosensors," especially with respect to the glucose sensing for diabetes patients.
1957	The first crystal structures of proteins were resolved [97].
1959	Rosalyn Sussman Yalow (born 1921) and Solomon Aaron Berson (1918–1972) developed the radioimmunoassay (RIA) which allows the very sensitive determination of hormones such as insulin based on an antigen–antibody reaction [98, 99]. In 1997, Yalow received the Nobel Prize in Medicine for developing RIA. Today the RIA technology is surpassed by enzyme-linked immunosorbent assay (ELISA) because the colorimetric or fluorescent detection principles are favored over radioactive-based technologies.
1960s	General Electric (GE) developed a fuel cell-based electrical power system employing the so-called "Bacon cell" in order to maintain the Gemini and Apollo space capsules of NASA.
1963	Garry A. Rechnitz together with S. Katz introduced one of the first papers in the field of biosensors with the direct potentiometric determination of urea after urease hydrolysis. At that time the term "biosensor" had not yet been coined. Thus, these types of devices were called enzyme electrodes or biocatalytic membrane electrodes [100].
1964	For the first time, enzymes were used as fuel cell catalysts by Yahiro *et al.* in a glucose/O_2 biofuel cell [101].
1967	G.P. Hicks und S.J. Updike introduced the first practical enzyme electrode immobilizing the enzyme within a gel [102, 103].
1969	George Guilbault introduced the potentiometric urea electrode [104].
1970	Bergveld introduced the ion selective field effect transistor (ISFET) [105].
1970s	ELISA was introduced by Stratis Avrameas (Institut Pasteur, France) und G. Barry Peiers (University of Michigan, USA) and others [106, 107].
1972	Betso *et al.* showed for the first time that direct electron transfer (ET) of cytochrome c could be realized at mercury electrodes. This breakthrough suffers from nonreversible electrochemistry due to protein denaturation on this electrode material [108].
1973	Ph. Racinee and W. Mindt (Hoffmann La Roche) developed a lactate electrode [109].
1973	G.G. Guilbault and G.J. Lubrano introduced an amperometric glucose enzyme electrode that was based on the detection of the product of the enzymatic reaction, hydrogen peroxide [110].
1975	The first commercial biosensor (YSI analyzer) was introduced [60, 61]. A review by Newman and Turner summarized the commercial development of blood glucose biosensors used at home by diabetes patients [59].

(*Continued*)

Table 1.2 (Continued)

Year	Contribution
1976	First microbe-based biosensors [111–113].
1976	The first bedside artificial pancreas was introduced. The glucose analyzer allows one to control an insulin infusion system (the Biostator) [114–116].
1977	Karl Cammann introduced the term "biosensor" [6].
1977	First realization of reversible ET of cytochrome c employing tin-doped indium oxide electrodes [117] and 4,4′-bipyridiyl as a promoting monolayer on gold electrodes [118, 119].
1979	First steps towards biofuel cells were realized [120–123].
1979	Pioneering work by J. Kulys using artificial redox mediators [124, 125].
1980s	Self-assembled monolayers (SAMs) start to receive considerable attention in the scientific community and are employed in biosensor research [49–52].
1981	Oxidation of NADH at graphite electrodes is described for the first time [126, 127].
1982	First needle-type enzyme electrode for subcutaneous implantation by Shichiri [128].
1982	First biologically engineered proteins using site-directed mutagenesis, enabling work on specific mutants of enzymes [129–131].
1983	First surface plasmon resonance (SPR) immunosensor [132–134].
1984	First ferrocene-mediated amperometric glucose biosensor by Cass *et al.* [135]. The work led to the development of the first electronic blood glucose measuring system which was commercialized by MediSense Inc. (later bought by Abbott Diagnostics) in 1987.
1988	Adam Heller and Yinon Degani introduced the electrical connection ("wiring") of redox centers of enzymes to electrodes through electron-conducting redox hydrogels [47, 136]. This work was the basis for continuous glucose monitoring employing subcutaneously implanted miniaturized glucose biosensors [137–139].
1988	Direct ET by means of immobilized enzymes was introduced [22, 120, 122, 123, 140–142].
1990	Bartlett *et al.* introduce mediator-modified enzymes [143].
1980s to 1990s	Nanostructured carbon materials such as C_{60} and nanotubes were discovered [144, 145].
1997	IUPAC introduced for the first time a definition for biosensors in analogy to the definition of chemosensors [3–5].
2002	Schuhmann *et al.* introduced the use of electrodeposition paints (EDPs) as immobilization matrices for biosensors [17]. Following work enabled the incorporation of redox mediators into the polymer structure of EDPs [18, 146].
2003	An enzymatic glucose/O_2 fuel cell which was implanted in a living plant was presented by Heller and coworkers [147].
2006	The first H_2/O_2 biofuel cell based on the oxidation of low levels of H_2 in air was introduced by Armstrong and coworkers [148].
2007	An implanted glucose biosensor (Freestyle Navigator System) operated for five days [149].

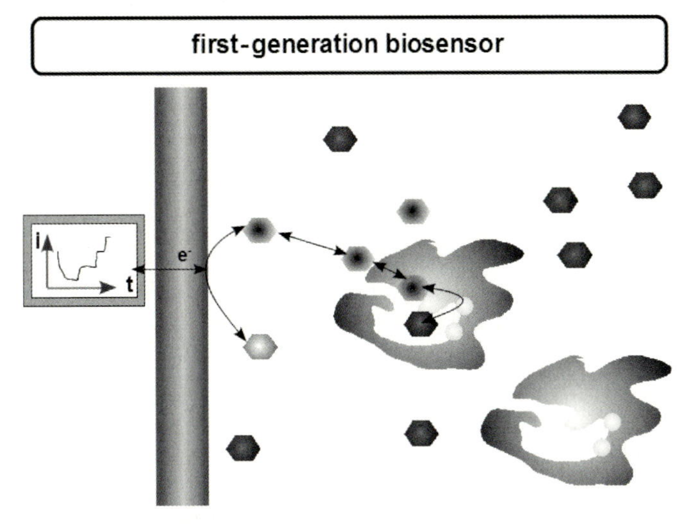

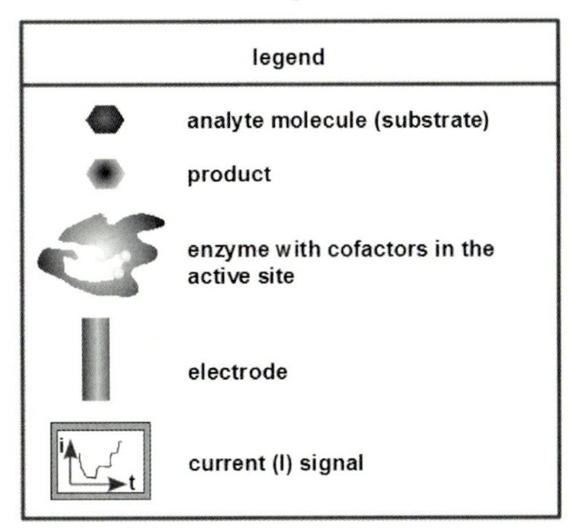

Figure 1.3 Schematic representation of the architecture of a "first-generation" biosensor.

in order to provide a reasonable driving force of the reaction. As free-diffusing mediators, a large variety of compounds such as ferrocene derivatives, organic dyes, ferricyanide, ruthenium complexes and osmium complexes have been used [158].

What are the most important properties of redox mediators suitable for biosensors? First of all, the electrochemistry has to be reversible and they need to be stable in the oxidized and reduced forms. No side reactions should occur. The redox potential needs to be compatible with the enzymatic reaction. It is helpful if the basic structure of the redox mediator also allows for chemical modifications

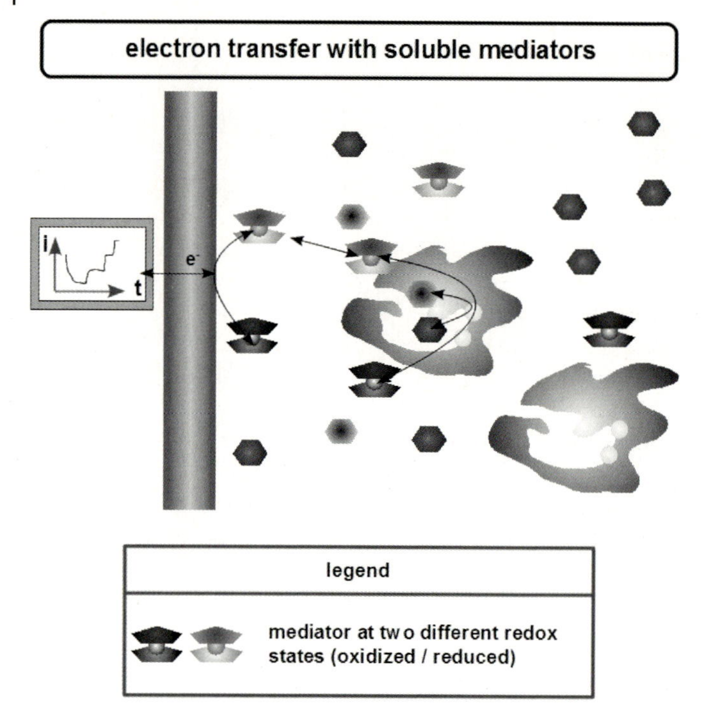

Figure 1.4 Schematic representation of a biosensor operating with soluble mediators.

enabling tuning of the desired redox potential. The immobilization of the redox mediator on the electrode surface and/or the redox enzyme needs to be possible. For instance, functional side chains for, for example, covalent binding to the polymer backbone, redox enzyme, or electrode surface need to be available. The redox mediator should not be toxic and available at reasonable cost and experimental effort. Note that the K_M value of the enzyme for a specific redox mediator also impacts the sensor response.

The major drawback of using either a natural or an artificial free-diffusing redox mediator in a biosensor design as illustrated in Figures 1.3 and 1.4 is that sufficient natural (e.g., O_2) or artificial mediator needs to be available to the active site of the enzyme and, subsequently, at the electrode surface for generating a detectable current signal. In addition, and of more importance to the accuracy and long-term stability as well as product safety, artificial mediator molecules that are not securely fixed within the sensing film can leak from the electrode surface [159]. This will change the sensor performance over time. In addition, not all redox mediators are biocompatible. The described problems with the use of free-diffusing redox mediators are not critical for single-use devices. For example, self-monitoring devices for monitoring blood glucose levels are very successfully used by diabetes patients at home [59, 61, 65, 160–163].

1.1.5
"Third-Generation" Biosensors

A different approach to realize biosensor architectures is the immobilization of a redox enzyme on the electrode surface in such a manner that direct ET is possible between the active side of the enzyme and the transducer [22, 140]. Thus, free-diffusing redox mediators are not necessary for these types of biosensors [164–169]. Biosensor designs based on direct ET have been investigated thoroughly and comprehensively reviewed [22–26, 170–182]. Figure 1.5 schematically illustrates how such direct ET can be realized within "third-generation" biosensors.

Proteins can spontaneously adsorb on many electrode materials [176] as schematically shown in Figure 1.5a. The interaction is mainly governed by hydrogen bonds as well as electrostatic, dipole–dipole, or hydrophobic interactions. It is important to take into account spontaneous adsorption on the electrode surface because it might also contribute to the overall current signal of a biosensor based on a more complex architecture. The impact of this effect can be evaluated by performing suitable control experiments.

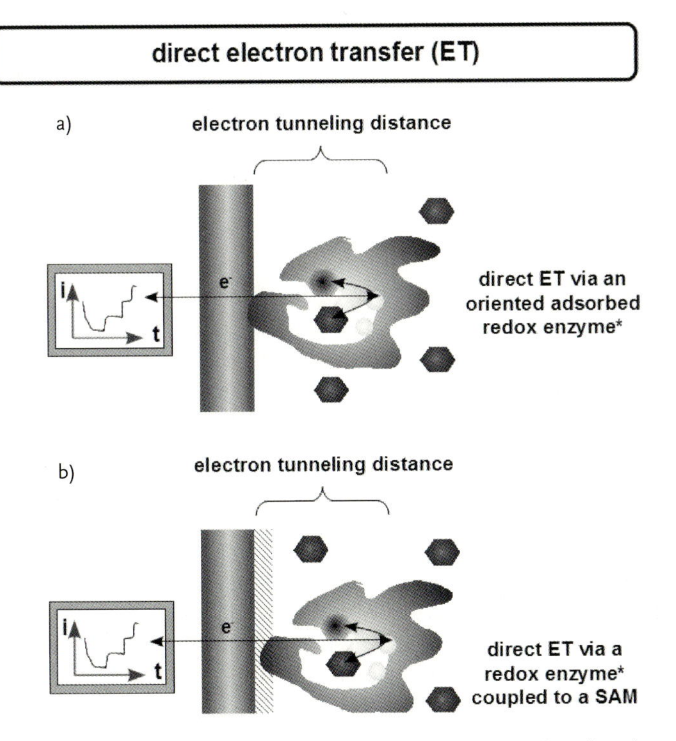

Figure 1.5 Schematic representation of biosensor architectures based on direct ET: (a) via an oriented adsorbed redox enzyme; (b) via a redox enzyme coupled to a self-assembled monolayer (SAM).

It is important to note that proteins tend to denature during such an adsorption process on noble metals or carbon electrodes. In addition, the stability of the adsorbed sensing layer is highly dependent on the pH value and ionic strength of the solution as well as the temperature, the electrode material, and other additional factors. For instance, as early as 1972 direct ET was observed on mercury electrodes employing cytochrome c as redox protein [108]. Reversible electrochemical behavior of cytochrome c was not observed because the protein denatured on the surface.

Therefore, it was a milestone when reversible ET of cytochrome c was achieved for the first time by employing tin-doped indium oxide electrodes [117] and 4,4′-bipyridiyl as a promoting monolayer on gold electrodes [118, 119]. Starting from the 1980s self-assembled monolayers (SAMs) have received considerable attention in the scientific community and have been successfully employed in biosensors [49–52]. The advantages of SAM-based biosensor architectures are the following: (i) the technology is easy and straightforward to use because the formation of SAMs is relatively fast [52]; (ii) enzymes can be adsorbed on SAMs providing the enzyme has an overall surface charge opposite to that of the SAM [176, 183]; (iii) the covalent attachment of redox enzymes and mediators is possible [184–186]; (iv) the design of alkanethiols can be tailored to particular needs (e.g., length of spacer, type of functional groups, mixture of alkanethiols with different properties for integrating different chemical functionalities, generation of multilayers); and (v) SAMs are sufficiently stable with respect to temperature and pH and can be operated in a rather broad potential range (about −1.4 to 0.8 V vs. SCE). This stability window with respect to applied redox potential can also be utilized to generate structured biosensor designs by, for example, stripping of confined areas within the SAM using a scanning electrochemical microscope (SECM) [187–189]. Thus, complex biosensor architectures even employing multiple redox enzymes or mediators can be realized.

It needs to be taken into account that direct ET between a redox enzyme in very close vicinity to the electrode surface (e.g., first monolayer) is normally very slow. It might even be impossible due to the shielding of the active side and/or redox-active cofactors of the enzyme by the surrounding insulating protein shell. Therefore, observation of direct ET has so far been restricted to either small redox proteins or redox enzymes that are characterized by the location of their cofactor(s) close to the protein shell. If the distance is significantly longer than 10 to 15 Å the chance for efficient direct ET is also significantly reduced according to Marcus theory.

Biology has solved this problem by introducing multi-cofactor enzymes in which the overall distance between two redox sites is divided into a number of shorter distances (multiple cofactors with different redox potentials instead of just one cofactor) or by introducing small redox shuttle proteins such as cytrochromes in the respiratory chain. Thus, ET cascades have been proven to be very efficient and useful. This principle was borrowed from nature not only for direct ET-based biosensors but also for mediated ET-based biosensors. Note that the overall efficiency of ET cascades depends on the entire architecture of a biosensor. A striking option is to reduce the distance between the active site of the redox enzyme and the transducer surface. Figuratively speaking, one could bring the enzyme closer to the electrode surface or vice versa. For example, the enzyme could be genetically or chemically modified to reduce the impact of the protein shell on the ET distance between the electrode and the active site, or the orientation of the active site towards the electrode surface is controlled by chemically (or genetically) modifying the enzyme or the electrode surface. Strategies for such modifications have been extensively used and evaluated [186]. An example of bringing the electrode surface closer to the enzyme could be the introduction of conducting nanoparticles or nanostructures into the sensing layer in order to increase the probability of ET taking place [190].

However, ET efficiency is not only dependent on the distance of the involved redox relays but also on the properties of the electrode material, the nature of the enzyme, the properties of the immobilization matrix, and the redox mediator (if any) in a complex manner.

For the evaluation of a biosensor design based on direct ET, one needs to take into account that not all the enzymes immobilized on the transducer surface are at productive ET distance. A portion of the immobilized enzymes may be oriented in such a way that direct ET is not possible due to a longer distance between the catalytic side of the enzyme and the transducer in comparison to the ideal oriented distance. A certain percentage of the enzymes immobilized may lose or change their catalytic activity during the immobilization procedure. Therefore, one approach to estimate the ratio of enzyme molecules that communicate via direct ET and enzyme molecules that are fully functional but do not contribute to the overall current response of the biosensor is to measure the current in the absence and presence of a suitable free-diffusing redox mediator. Another approach would be to estimate the catalytically active enzyme concentration on the electrode surface by means of a standard optical enzyme activity test. This is also helpful in case no direct ET can be detected and it is not clear if the enzyme undergoes denaturation during the sensor fabrication and operating process.

1.1.6 Reagentless Biosensor Architectures

For a variety of applications it is useful to employ artificial redox mediators for biosensor architectures. In the case of mediated ET, redox mediators shuttle electrons between the active side of the enzyme and the electrode. The main advantages of employing mediated ET within a biosensor device are that the ET process is independent of the presence of natural electron acceptors or donors. For instance, oxygen is ubiquitously present in biological systems. Given that the redox mediator is appropriately selected, the influence of possible interfering compounds can be reduced because the working potential of the biosensor is defined by the formal potential of the redox mediator. In addition, the pH dependence of the sensor response can be better controlled. Furthermore, the sensitivity and overall current response can be increased. Multicomponent ET cascades can be designed.

Due to the drawbacks of free-diffusing redox mediators, especially with respect to continuous monitoring of the analyte of interest, the development of reagentless biosensors has become of importance over the last 15 years [191]. The outstanding feature of a reagentless biosensor is that all components required for the electroanalytical reaction are securely immobilized on a transducer surface. This is a nontrivial task because the immobilization approach needs to ensure a microenvironment within the biosensor film that facilitates the biological recognition reaction and an efficient ET cascade. The only free-diffusing component of the overall assay reaction is the analyte which is provided by the sample solution. The most appropriate approaches towards reagentless biosensor architectures are the use of an appropriate redox enzyme and an ET pathway that either uses direct ET or mediated ET via securely immobilized redox mediators within the biosensor film. There are several technologies available for immobilizing biological recognition elements on transducer surfaces: adsorption, microencapsulation, entrapment, covalent attachment, and crosslinking. These techniques have been comprehensively reviewed [173, 192–197].

What does a suitable immobilization matrix for a reagentless biosensor need to provide? First of all, the selected biological recognition element needs to be securely fixed at the electrode surface. Furthermore, it needs to provide a microenvironment that either maintains or tunes the enzyme activity at a desired level. This is important not only for the efficiency of the reaction with the analyte and the ET but also for the operation, storage, and long-term stability of the biosensor. Matching charges and hydrophobicity/hydrophilicity as well as hydrogen bonds, complexation, or covalent binding sites are important. It is advantageous if the nature of the immobilization matrix is intrinsically open for optimization, for example, by means of adapting the chemical structure, tuning the chemical and/or physical properties as well as the immobilization technique. With respect to miniaturization and automation of biosensor production as well as the reproducibility within the production process, it becomes of importance that the immobilization approach can be done in a way that enables exclusive addressing of the electrode surface. A certain flexibility of the backbone of the immobilization matrix has proven to be advantageous, especially with respect to the aspect of diffusion limitations. Note that one limiting case for biosensor performance can be the diffusion limit. In addition, suitable binding sites for redox mediators, spacers, or (multiple) redox enzymes are useful. Redox mediators also need to be securely immobilized within the biosensor film. If *in vivo* or *in vitro* use of the device is intended the film needs to be biocompatible and compatible with special needs such as sterilization.

Figure 1.6 describes what reagentless biosensor structures based on mediated ET can look like.

Designing appropriate and efficient ET pathways within a biosensor intrinsically requires that the generated environment is suitable for the chosen biological recognition element. Its catalytic activity and stability might be tuned to the desired performance by the immobilization matrix. The immobilization procedure itself needs to be compatible with the ET pathway strategy. For instance, the stress on

mediated electron transfer (ET)

a) **ET via a conductive polymer chain**

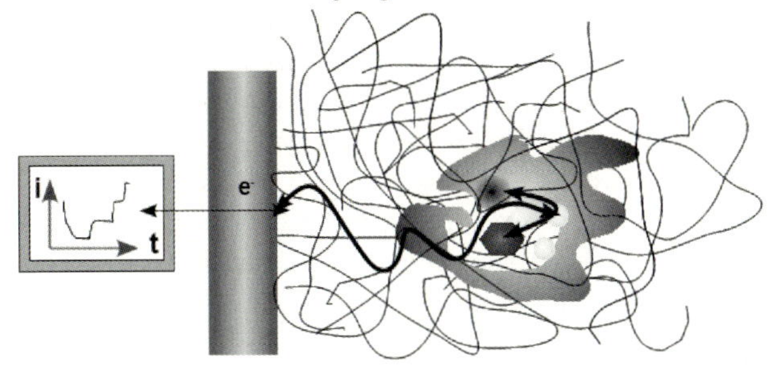

b) **ET via a redox-relay modified polymer**

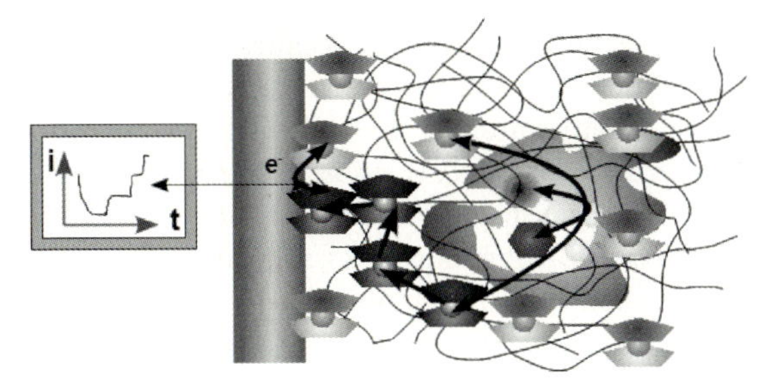

Figure 1.6 Schematic representation of reagentless biosensors: (a) ET via conductive polymer chains; (b) ET via redox-relay modified polymer chains.

an enzyme by either the chemical nature of the immobilization reaction or the applied potential for electrodeposition on the transducer surface may lead to a significant loss of enzyme activity (enzymatic sensor) or binding reaction (affinity sensor). Efficient ET between a redox enzyme and the redox relays in the immobilization matrix to the electrode surface is mainly controlled by the distance between the individual redox couples [198] participating in the overall ET reaction. Thus, for further optimization, it is crucial to elucidate the rate-determining steps of all involved processes.

For instance, if one would like to design a suitable redox polymer for a certain enzymatic reaction, it helps to think about the following factors. First, the redox polymer needs to create a three-dimensional network that allows secure immobilization of the enzyme with a reasonable pore size. In addition, fast diffusion of the analyte, products, or counter-ions and fast ET kinetics need to be ensured. The polymer film deposited as sensing layer creates a diffusion barrier which often prolongs the response time, shifts the linear measuring range, and decreases the sensitivity of the sensor. Second, the redox polymer should create a local microenvironment that is beneficial for enzyme immobilization, functionality, selectivity, kinetics, and stability. Good interaction between the active site of the enzyme and the redox relays, especially for the first ET step, is vital. For example, ET transfer distances have to be reasonably short (<10 Å).

The general advantages of reagentless biosensor structures can be summarized as follows. Since all components of the assay are securely immobilized on the electrode surface, there is no or just a negligible loss of redox mediators, cofactors, and/or enzymes over the time of operation. This is of importance for the performance and safety of a device because the impact of free-diffusing possibly toxic substances is minimized. Therefore, reagentless biosensor architectures are often used for *in vitro* and *in vivo* measurements as outlined in Section 1.4.5.

Which advantage of a certain generation of biosensors outweighs the advantages of the other generations will, however, depend on the analytical task. The specifications of requirements (e.g., type and concentration range of analyte, composition of sample matrix and occurrence of possible interferences, official regulations for the final application, overall cost limitation for the device) need to be defined. A comprehensive review of the state of the art for the specific analytical task helps to develop a suitable strategy. Sampling and sample preparation are additional important aspects. After a preliminary testing of the envisaged biosensor design, it has to be thoroughly tested to determine if it is capable of identifying and quantifying the analyte of interest. For applications with high throughput and/or commercial interest, approaches have to be evaluated to properly deal with data collection, processing, interpretation, documentation, and reporting. It might be necessary to provide suitable instrumentation to operate the sensing device, and features such as a self-referencing system for calibration, temperature control, etc., might become assets to be considered.

1.1.7
Parameters with a Major Impact on Overall Biosensor Response

The choice of biosensor architecture depends to a major extent on the (bio)chemical processes involved in the biorecognition process. The processes in close proximity to the electrode surface that are involved in a typical biosensor reaction are rather complex. It is important to have an overview about the variables that affect the performance of a biosensor (Figure 1.7) and which of these parameters may have a major impact on the signal response. As a matter of fact, the design of appropriate sensor architecture depends on the specific demands arising from the particular analytical task. Thus, a sound understanding of the challenges of the analytical task with respect of the main reactions involved in the overall sensing process is mandatory. It is important to note that due to the complexity of a biosensor architecture one deals with a multiparameter space of which, most likely, only a limited number of parameters can be controlled. This implies that it may be helpful to visualize the analytical task and the envisaged biosensor design at molecular dimensions in order to better understand the processes taking place, thus being able to identify at an early stage potential pitfalls with

variables affecting the electroanalytical performance of a biosensor

electrode variables
- material
- surface area
- geometry
- surface condition

mass transfer variables
- convection
- diffusion
- surface concentration
- adsorption

electrical variables
- potential
- current
- charge
- impedance

electrolyte variables
- bulk concentration of electroactive species
- pH
- solvent
- trace impurities

external variables
- temperature
- pressure
- time
- composition of sample matrix
- occurrence of interferences and/or electrode fouling processes

reaction variables
- kinetic and thermodynamic parameters of ET & reactions(s) (bulk, surface, biosensor film)
- rate-determining step (rds) of the reaction
- stability of reaction partners/ products

potential noise variables
- electrochemical noise
- thermal noise
- environmental noise (e.g. electromagnetic)

Figure 1.7 Parameters influencing the overall response of a specific biosensor architecture.

respect to general device layout as well as the interpretation of the obtained data. It is indispensable to always have in mind the impact of diffusion, enzyme, and ET kinetics. In addition, the impact of temperature is not negligible. Diffusion, kinetics, selectivity, and overall (bio)sensor performance are highly temperature dependent [199, 200].

A detailed list of variables affecting the electroanalytical performance of a biosensor (Figure 1.7) enables an awareness of the main "adjusting screws" of a particular system for further optimization.

Knowing the most influential parameters of a specific biosensor architecture is the basis to understand and fine tune the performance of these devices in a rational manner. Figure 1.8 summarizes the key features of typical biosensors and lists several that are of additional importance for commercial devices. Among these, selectivity, sensitivity, accuracy, response, and recovery time as well as operating lifetime are some of the most important key factors. Keeping in mind the needs of the specific analytical task of interest, it seems to be necessary to characterize at least the key parameters mentioned in Figure 1.8 in order to specify the analytical performance of a biosensor design.

It is indispensable to elucidate the rate-limiting steps of the overall reaction sequence in order to develop an appropriate optimization strategy. Thus, mathematical and chemometric approaches are expected to promote a deeper understanding of the processes involved. As a useful source of information for modeling biosensor responses, a related book chapter by Bartlett *et al.* [201] is recommended.

An important aspect of biosensor optimization is the elimination of interferences, or at least a reduction of the impact of interferences. In many samples there are components that either directly react at the electrode surface or the involved redox centers or interfere with the biological recognition reaction (e.g., inhibitors or other substrates for the enzyme). In addition, leakage from the sensing layer, loss in enzyme activity or electrode fouling may occur. Thus, changes in sensitivity and baseline drifts may occur during biosensor operation. Therefore, suitable strategies for calibration are needed to ensure reproducible and quantitative results. For real-world applications it is imperative to characterize and optimize the biosensor architecture under actual measuring conditions. A useful review by Phillips and Wightman [202] discusses critical guidelines for the validation of *in vivo* microsensors.

Electroanalytical methods and biosensor architectures offer various strategies to tune selectivity for the analyte of interest [203]. Among the diverse approaches to tackle problems arising from interferences and electrode fouling are the following. (i) Size exclusion and/or charge repulsion are utilized when additional films or membranes are placed as upper layers on top of the actual sensing layer. Typical membranes are polymethylcellulose, Nafion, hydrogels, polypyrrole, *o*-phenylenediamine, polyeugenol, and other electrodepositable films (conducting or nonconducting). However, this approach is used at the expense of biosensor response time. (ii) Use of suitable redox mediators allows operation at moderate potentials below the potential of abundant interferences such as ascorbic acid. (iii) The applied potential is a useful tool to discriminate between different electroactive species under the assumption that the redox waves are distinguishable. Electroanalytical techniques such as cyclic voltammetry (CV), fast-scan CV, differential pulse voltammetry (DPV), square wave voltammetry (SWV) or differential pulse amperometry (DPA) as well as multiple pulse amperometry [204, 205] are useful for determining several species in parallel and discriminating between them.

key features of a biosensor

key parameters of biosensor performance

- analyte
- employed biocomponent for recognition of analyte
- type & composition of sample matrices
- selectivity
- sensitivity
- dynamic range
- detection limit
- response time
- reproducibility
- precision
- stability (in use, long-term, storage)
- calibration approach (requirements, intervals, drift)
- duration to reach baseline

additional parameters of biosensor performance, e.g. relevant for commercialization

- cost per measurement
- duration of measurement
- turnover of measurements (time resolution)
- warm-up time
- sample volume
- measuring temperature
- size, weight & price of the device
- costs, size & weight of required instrumentation
- delivery time
- development status of the device/procedure (commercial product, standard procedure (in house or commercial, established research procedure, basic research, proof-of-principle)
- time between maintenance checks
- potential to be coupled with other analytical techniques
- data storage and processing

Figure 1.8 List of key characteristics of a biosensor.

(iv) The choice of the immobilization matrix can be important for the susceptibility to interferences. (v) Mathematical models and chemometrics can be employed. One approach utilized the impact of temperature on biosensor performance to improve selectivity [199, 200]. A recent review summarizes for example some of the strategies towards the elimination of interference of glucose biosensors [206].

1.1.8
Application Areas of Biosensors

Today biosensors are mainly used for healthcare applications, controlling industrial processes, and environmental monitoring, as outlined in Figure 1.9. In all cases the biosensor design, packaging, and instrumentation required are dependent on the purpose of the measuring approach. In several cases, the type of sample dictates the biosensor design. For example, if potentially harmful samples such as blood or contaminated waste water are of interest, disposable sensor formats are preferred. Samples can be analyzed off-line in a laboratory, such as glucose testing of patients' blood samples in a hospital laboratory or water samples from rivers. In addition, off-line analysis can also be performed close to the operation side of an industrial plant or process or glucose monitoring can be performed at home by patients themselves. For several applications, however, there is a need for online analysis in real time, such as quality control in the food or drug industries or metabolite monitoring at the bedside or during surgery. The requirements for

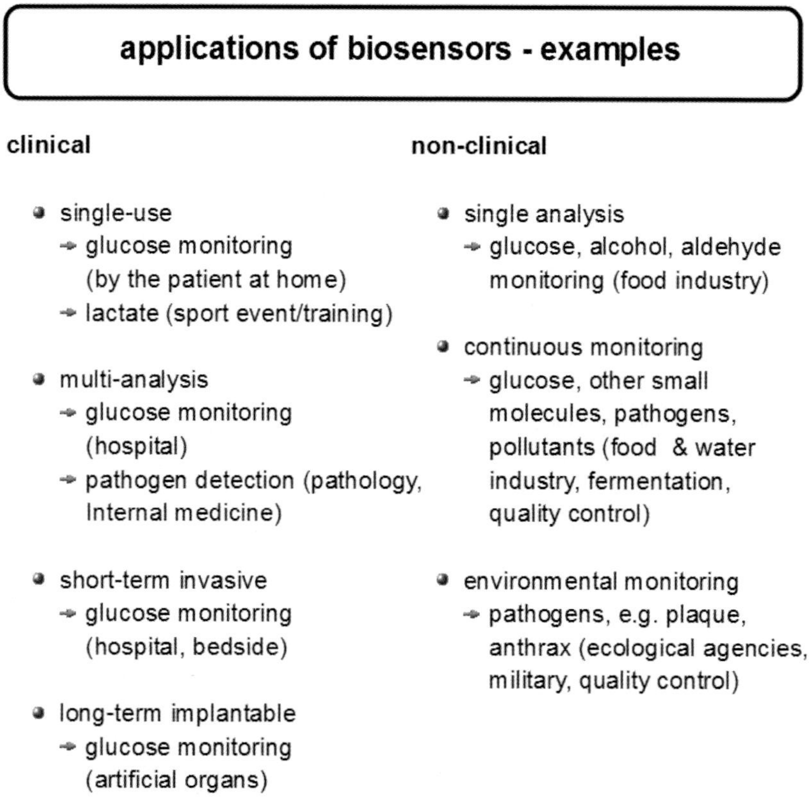

applications of biosensors - examples

clinical

- single-use
  - glucose monitoring (by the patient at home)
  - lactate (sport event/training)
- multi-analysis
  - glucose monitoring (hospital)
  - pathogen detection (pathology, Internal medicine)
- short-term invasive
  - glucose monitoring (hospital, bedside)
- long-term implantable
  - glucose monitoring (artificial organs)

non-clinical

- single analysis
  - glucose, alcohol, aldehyde monitoring (food industry)
- continuous monitoring
  - glucose, other small molecules, pathogens, pollutants (food & water industry, fermentation, quality control)
- environmental monitoring
  - pathogens, e.g. plaque, anthrax (ecological agencies, military, quality control)

Figure 1.9 Areas of application for biosensors.

a single-use device differ from those for multi-analysis and continuous monitoring and have to be taken into account when considering overall biosensor architecture.

1.2 Criteria for "Good" Biosensor Research

It is obvious that science does not always lead to ground-breaking advancements that are worth publishing. This is also very much true for publications from biosensor-related research. The area of biosensor research is even more susceptible to publications that do not significantly contribute to the present state of the art, since the basic equipment for doing high-level biosensor research is comparatively cheap. Moreover, nearly any modified electrode with an immobilized biorecognition element will show a certain response upon the addition of a specific analyte, leading to a calibration graph, the possibility of determining the pH optimum, etc. This is not intrinsically a sign of low quality if the contribution is otherwise scientifically sound. An example would be a publication that only slightly changes an existing biosensor design, but otherwise is unambiguously supported by technically sound data and interpretation. However, there are also a large number of publications that suffer from technically wrong or biased data acquisition, processing, and interpretation or an insufficient amount of data for the hypothesis proposed.

To do research is basically to generate knowledge which is made available to the scientific community via publication. The main aim is that other scientists will be convinced by the scientific approach and they can adopt the strategy or scientific principle for answering their own research questions. Thus, criteria for "good to excellent" biosensor research have to be measured in terms of the following questions:

i) Does the research work introduce a novel sensing principle, a novel signal amplification strategy, a novel specifically adapted redox mediator with improved properties, a novel immobilization scheme, a novel sensor architecture with tunable parameters? Does the proposed research work contribute to an increase in fundamental knowledge, an in-depth evaluation of the signal transduction mechanism, or an in-depth physicochemical evaluation of the rate-determining steps and the interplay of the parameters in the complex parameter space?

ii) Does the research work introduce novel aspects to an already known sensing architecture, to an already known application, or does it extend a sensing principle to be more general? Does the work include the discovery of surprising results by combining a specific biological recognition element with an already known sensing principle and is there a rational way to understand this surprising result? Does the adaptation of an already known sensing principle to a specific application require innovative features?

iii) Is the proposed research work just a variation of an existing principle by varying the biological recognition element, the electrode material, the size and integration of the electrode? Is there any predictable contribution of the elements of which the sensing layer is composed to the expected signal generation, interference elimination, improvement of long-term or operational stability, etc.?

iv) Does the contribution solve a previously unsolved scientific question? Can the principle be the basis for improvements in sensitivity, selectivity, applicability?

v) Is the selection of the compounds used for creating the sensor architecture based on buzzwords such as nanomaterial, nanosensor, etc.? Is the effect of the material used correlated with the meaning of the buzzword or is it just used because of the buzzword?

vi) Does the complexity of the sensor architecture allow a rational investigation of the complex influence of all compounds used on the final sensor output? Is it an effect or a scientific result? If a novel effect is discovered, can it be explained by a scientifically sound argumentation chain? Is it possible to design control experiments to provide evidence for the hypothesis concerning the sensing mechanism?

vii) Is there any possibility of reproducing the measurements in the same laboratory at another time or even in a different laboratory? Are all results derived from one sensor? Is there any statistical evaluation of the repeatability of the sensor fabrication protocol, and of the obtained signals?

viii) If the results are sound and justified, do the authors try to benchmark the results with existing sensing strategies for the same analyte and in the same application?

Taking the above into account it seems to be straightforward to distinguish between fundamental biosensor research and biosensor development. For fundamental biosensor research, the discovery of novel sensing strategies or bioelectrochemical signal transduction schemes, the elucidation of the fundamental processes, and the understanding of the complex parameter interplay that finally leads to the observed sensor signal are the focus of the research. For biosensor development, an existing sensor principle has to be adapted to a specific application taking into account costs, storage time, reproducibility, calibration, validation, legal consequences, etc. Presently, many biosensor papers that predominantly deal with fundamental biosensor design try to include some application aspects by showing that some standard samples can be measured at a required quality. However, these results are most often obtained in the research laboratory using the standard addition method and well-trained personnel. On the other hand, papers on application-oriented research often try to include basic mechanistic studies at a limited depth. This is also reflected by the editorial policy of international journals accepting work on biosensors. Recently, fundamental studies are more often published in physical chemistry journals while the biosensor journals have shifted to be more application oriented.

Ideally, a biosensor design needs to be adaptable to a certain application or analytical task. If the design is a general principle, the biosensor performance needs to be tunable to the needs of a specific analytical task and be open for modifications leading to a broader range of analytes or applications. If the mechanism behind the biosensor function is at least partially understood, a fine tuning of the sensing layer or a rational adaptation of the sensor design may become possible. As pointed out above, the parameters influencing and limiting the overall sensing process of a biosensor are often not fully understood. In a classical approach, it was assumed that the parameters are linearly independent. Thus, it was assumed that one may independently vary one parameter while the others are kept constant. For example, parameters such as transducer type and pretreatment, enzyme and mediator concentration and their ratios within the film, type of immobilization matrix, immobilization parameters, and film thickness are varied. A rational optimization approach includes that the main parameters affecting the overall ET pathway and hence the final sensor response have to be investigated in order to find reasonable tools for tuning the performance of the selected biosensor design. As a matter of fact, it is well known that it is impossible to keep parameters constant while changing others. For example, if the enzyme loading is increased, the film thickness, the diffusional properties for the substrate and the products, the counter-ion movement, possible ET reactions, etc., may be altered simultaneously. Thus, a "pseudo" rational approach has to be complemented by combinatorial approaches in which the overall parameter space is addressed by means of a large number of measurements after permutation of all possible influencing parameters. The knowledge gained should be comprehensively summarized in a related publication. The final consideration should be whether the sensor fabrication process described in a paper can be directly repeated successfully in another laboratory leading to similar sensor responses.

1.3
Defining a Standard for Characterizing Biosensor Performances

Considering the workflow as proposed in Figure 1.10 it becomes obvious that it is essential to characterize a specific set of key parameters mainly determining biosensor performance. A selection of suitable key parameters is given in Figure 1.8. With respect to the particular needs of a specific analytical task one needs to decide which performance parameters make sense to be evaluated at a certain stage of the process. For example, Phillips and Wightman evaluated guidelines for the validation of *in vivo* microsensors [202]. In the following, the most common and most important characteristics of biosensors are discussed.

Though a biological recognition reaction is typically very selective, interferences may occur due to substances other than the analyte of interest. Such interferences can be converted by the biorecognition element or at the transducer surface and

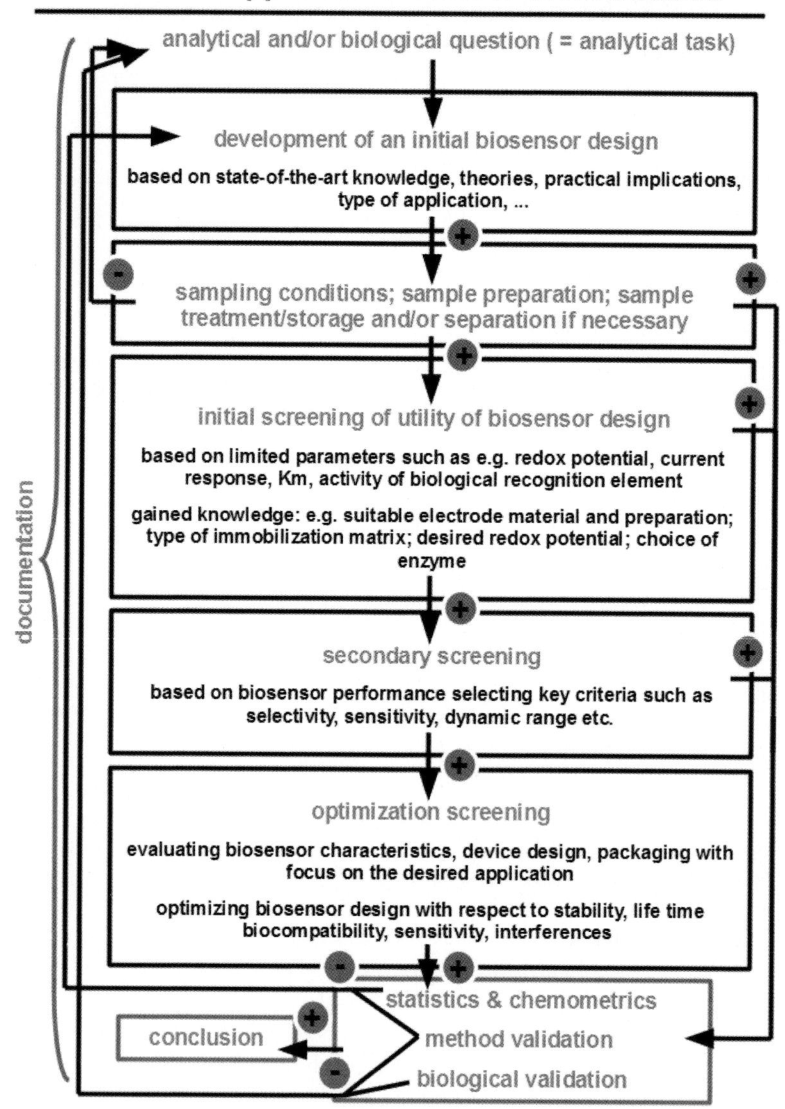

Figure 1.10 Workflow for successful biosensor research.

thus create false-positive results. The *selectivity* of a biosensor is characterized by the selectivity coefficient. The selectivity coefficient is defined as the quotient of the respective binding constants of the analyte of interest A and a potential interference I with the biorecognition element B:

$$B + A \rightarrow BA;\ K_{BA}$$

$$B + I \rightarrow BI;\ K_{BI}$$

$$K_{AI} = \frac{K_{BA}}{K_{BI}} K_{AI}$$

In addition to selectivity, *sensitivity* (S) is a vital parameter of the performance of a biosensor. Sensitivity is defined as the slope of change in signal with change in concentration:

$$S = \frac{d(\text{signal})}{d(\text{concentration})}$$

This is, however, only straightforward if the sensor response is linear.

The *linear range* of a sensor is defined as the range in which the sensor signal is proportional to a change in concentration. Linear range should not be confused with dynamic range. *Dynamic range* describes the range in which a change in concentration will lead to any sort of noticeable change in signal. Typically, the whole dynamic range will not yield a signal suitable to determine the analyte of interest in a reliable way. In most cases the *working range* of a sensor corresponds to the linear range.

The *limit of detection* (LOD) of a biosensor is one of the most important parameters to be determined. That holds especially true when disease markers have to be determined. The LOD is typically defined as

$$\text{LOD} = k \times \text{std}_{\text{background}}$$

where k is the *signal-to-noise ratio* and $\text{std}_{\text{background}}$ is the standard deviation of the background signal. The value of k can be chosen deliberately depending on the desired accuracy of the LOD but is typically 3. Another definition describes the smallest detectable concentration of an analyte c_{LOD} as

$$c_{\text{LOD}} = \frac{\text{std}_{\text{background}}}{S}$$

where S is the sensitivity.

It has to be pointed out that the LOD cannot be properly discussed without any knowledge of the binding constant of the primary biorecognition process. If the binding constant of thc biorccognition process is, for example, in the nanomolar range, a detection limit far below seems to be thermodynamically impossible. Thus, it is very important to gain a solid understanding about the difference in which signal can be measured and amplified and which limit of detection can be achieved based on a distinct biological recognition process. Hence, it would be very helpful if together with a LOD an estimation of the binding constant was given.

Terms rarely mentioned in the biosensor literature are *accuracy* and *precision*. Accuracy describes the agreement between the average of the measured value and a reference value. To determine accuracy is relatively straightforward. Normally sensors are tested using solutions with well-known concentrations of the analyte ("true value"). Obtained values can easily be compared to the true value.

Precision describes the scatter of measured values around the average of the measured values. Precision is much more important for biosensor performance than accuracy. Accuracy can be influenced by systematic errors that can be corrected. However, a sensor producing values that are scattered will not be regarded as very reliable. Some confusion can be found in the use of terms that are measures for the precision of a sensor, *repeatability* and *reproducibility*. Many papers report work as highly reproducible even though reproducibility is defined as the *between-laboratory precision*. That simply means that a sensor's measurements are reproducible if the same results are obtained in different laboratories with the same sensor architecture [207]. This is, however, rarely tested. Most authors really test the repeatability of a sensor. Repeatability is the *in-laboratory precision*. In-laboratory precision means that the sensor yields the same value of, for example, concentration in repeated measurements. Repeatability also means that the same sensor architecture will yield the same result if manufactured in the same laboratory. It would hence make more sense to speak about the standard deviation of a single sensor and to analyze the repeatability of the respective sensor architecture than to speak about reproducibility if the latter has not been tested.

Another measure for sensor performance that is a potential source of confusion is the *stability* of a sensor. Sensor stability can mean different things including but not limited to *working stability*, *storage stability*, and *long-term stability*. Working stability (sometimes also called usage stability) describes the stability of the sensor during continuous operation. Storage stability obviously describes the stability of the sensor upon storage, while long-term stability describes the sensor stability during operation in a sample solution but not necessarily continuous operation. It already becomes clear that the "stability" of a sensor will rarely mean the same thing for different sensor architectures.

Sensor performance is also characterized by commercial means, most importantly *time per measurement* and *cost per measurement*. Again, there is no one-size-fits-all definition for these terms and one should emphasize on being transparent in the way in which these numbers are determined.

In conclusion, a number of parameters are helpful when characterizing biosensor performance. It is, however, extremely important to be precise in the use of these terms. A clear definition of the measured variable is mandatory in any case and needs to be reported in a transparent fashion.

1.4 Success Stories in Biosensor Research

This section aims at discussing how "good" biosensor research inspires the scientific community to achieve advancements that reach from novel basic concepts to real-world applications. To do so, selected examples with a significant scientific impact are presented. This selection is not exhaustive, and there are many other possible success stories, but is just some examples selected by a very personal view.

1.4.1 Direct ET Employed for Biosensors and Biofuel Cells

The fundamentals of biosensors that exhibit direct ET between biological recognition element and electrode have been discussed thoroughly in the section on third-generation biosensors (Section 1.1.5). This section concentrates on highlighting the major contributions made in the area of direct ET in biosensors and biofuel cells. Though the latter is not a sensing application it draws from the same concepts as third-generation biosensors, and biofuel cells are a striking example of the continued development of redox-enzyme electrodes.

The realization of direct ET poses some challenges that already have been outlined above. Figure 1.11 summarizes the key features and challenges of this approach. The protein shell may prevent ET processes due to a large distance between active site and electrode if the active site is deeply buried within the protein. Proteins with suitable characteristics for direct ET have to be securely fixed to the electrode surface in an orientation facilitating direct ET. The orientation of the redox enzyme towards the electrode surface is the main challenge in designing direct ET pathways. Basically, two cases can be distinguished: direct ET between the active site of the enzyme or direct ET via an internal electron pathway within the protein. The first case is relevant for rather small redox proteins or for redox enzymes exhibiting an active site closely located at the outer protein shell. The second case mainly applies to multi-cofactor enzymes. However, the orientation is critical in both cases since either the redox-active center or the cofactor closest to the protein shell has to be located within a productive ET distance [27, 208–211]. Recently, a study investigated the different orientations of recombinant horseradish peroxidases to gold surfaces [212].

Early work on the direct electrochemistry of redox proteins suffered from the poor stability of those proteins at electrode surfaces. Though a significant body of work demonstrates the direct electrochemistry of redox proteins at graphite electrodes [213], the major breakthrough came with the use of surface-modified electrodes that provided a substrate for the stable orientation and immobilization of redox proteins at the electrode surface [208, 214, 215]. Surface-modified electrodes allowed for the study of the direct electrochemistry of cytochrome c on 4,4′-bipyridyl-modified gold electrodes [216]. As previously mentioned, SAMs on electrode surfaces are useful tools to realize biosensors suitable for direct ET [52]. The creation of monolayers on electrode surfaces with immobilized recognition elements is not limited to SAMs; a review by Willner and Katz summarizes the different approaches to realize covalent binding of enzymes to functionalized electrode surfaces as well as strategies to employ modified enzymes (e.g., protein conjugates) [186].

Enzymes that have been much studied in direct ET configuration include peroxidases [217, 218], especially horseradish peroxidase [219, 220], laccase [120, 121],

analytical task

direct electron transfer (ET)

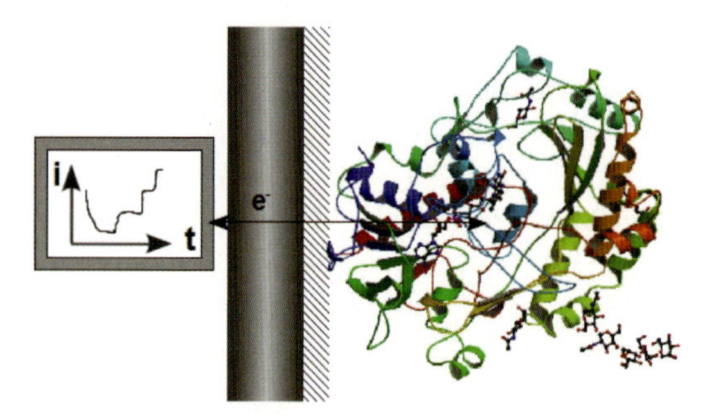

key features & challenges for direct ET

- immobilization of the enzyme on the electrode surface
  (impact on the structure of the enzyme, on the enzyme activity, on the diffusion of the substrate, on the ET kinetics, ...)
- orientation of the active site of the enzyme to the electrode surface
- orientation of the cofactor(s) of the enzyme to the electrode surface
- distance(s) between the redox relays taking part in the direct ET pathway
- selection of a suitable redox potential for the measurement
  (driving force of the reaction, catalytic current should be observed at the redox potential of the redox relay (e.g. a cofactor) communicating with the electrode surface)
- efficiency of the overall ET process

Figure 1.11 Biosensors based on direct ET.

and dehydrogenases [221] including fructose dehydrogenase [222], cellobiose dehydrogenase [223–225], and quinohemoprotein alcohol dehydrogenase [226]. It is important to keep in mind that for characterizing biosensor responses it is important to check if the enzyme employed is still able to efficiently catalyze the physiological reaction at a rational range of redox potential.

It is very important to define criteria to unequivocally proof a direct ET pathway between an immobilized redox protein and an electrode surface. The first important prerequisite is the occurrence of the direct electrochemistry of the redox cofactor inside the protein in the absence of the substrate. Hence, a reversible redox wave in a cyclic voltammogram of the protein-integrated cofactor has to be visible with a formal potential which clearly shows that the protein structure is not disturbed during the immobilization process. The second and most important prerequisite is that upon addition of the substrate the catalytic current increases at the redox potential of the protein-integrated cofactor without any significant overpotential at least at slow scan rates. If these two features in the cyclic voltammogram are not seen, a direct ET pathway can be excluded.

The development of enzyme electrodes with immobilized redox enzymes in direct ET communication was the prerequisite for the design of enzyme-based biofuel cells. For representative recent reviews see [70, 227, 228]. A fuel cell generally converts chemical energy into electrical energy in a continuous process as long as fuel is supplied. Biofuel cells convert chemical energy by means of a biocatalytic process as can be seen in Figure 1.12. Typically, enzyme-based biofuel cells consist of at least one enzyme electrode on either the cathode or anode side of the fuel cell or enzyme electrodes on both the cathode and the anode sides. The high specificity of fuel conversion by enzyme electrodes allows for membrane-free

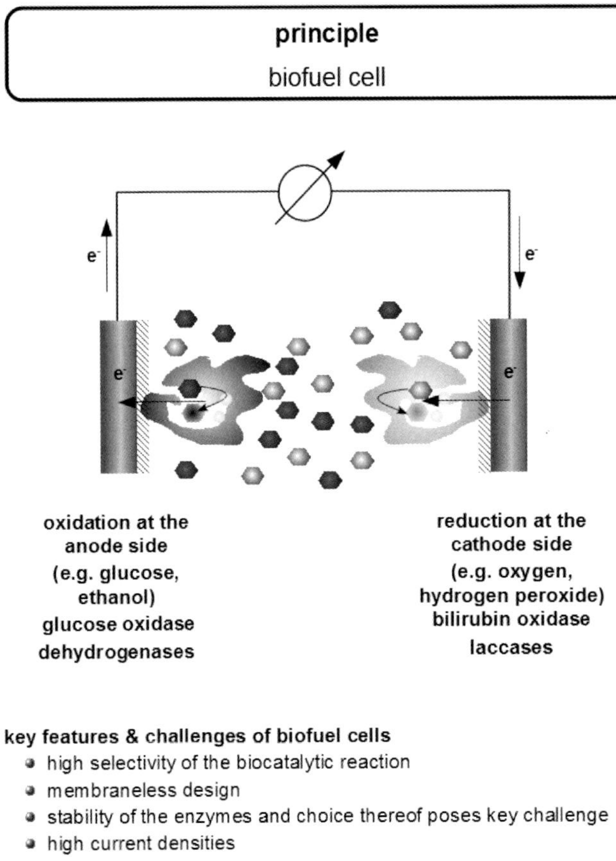

Figure 1.12 Principle of biofuel cells.

designs in which the cathode and anode reactions proceed in a single compartment.

The typical reaction on the cathode side is the reduction of oxygen at either a platinum catalyst or an electrode. The ET pathway at the enzyme electrode can be mediated (Section 1.4.3) or direct. Enzymes that have been employed in biofuel cells relying on direct ET include laccases [229, 230] which, however, suffer from a pH optimum in the acidic range and inhibition by halide ions. Thus, despite their favorable high potential for oxygen reduction they show poor stability in human tissue and fluids. Alternatively, bilirubin oxidase has been used as oxygen reduction biocatalyst in biofuel cell cathodes due to the better pH optimum and the insensitivity towards chloride ions. However, bilirubin oxidase has an about 200 mV lower reduction potential for molecular oxygen [231]. Alternatively, reduction of peroxide can be the cathodic reaction in biofuel cells. Relying on the experience with peroxidase-modified electrodes in biosensor research, electrodes modified with peroxidases have been shown to be highly efficient biocatalysts in biofuel cells [232, 233]. Microperoxidases that are truncated forms of cytochrome c have also been employed in biofuel cells [234] in which they convert hydrogen peroxide that is supplied by the enzymatic reaction of GOx with glucose and oxygen.

The most common reaction at the anodic side of biofuel cells is the oxidation of sugars which relies on the catalytic properties of oxidases. This class of enzymes has, however, usually poor potential for direct ET. Direct ET on the anodic site was, however, described for a number of hydrogenases [235, 236] and cellobiose dehydrogenase [225, 237, 238]. Enzymatic catalysis by means of direct ET was also realized on conducting graphite or TiO_2 particles [239, 240].

In conclusion, biofuel cells have a tremendous potential to be applied in, for example, implantable sensors or similar functional devices. They are a striking example of the continued development and application of the principles of biosensors employing direct ET.

1.4.2
Direct ET with Glucose Oxidase

Glucose sensors are *the* success story with respect to biosensor research and application. Today, diabetes patients are able to monitor their blood glucose levels on their own at home with commercial devices [59–65]. All these successful devices use a mediated ET pathway with natural or artificial redox mediators irrespective of whether GOx or other glucose-converting enzymes such as pyrroloquinoline quinone (PQQ)-dependent glucose dehydrogenase or nicotinamide adenine dinucleotide (NAD^+)-dependent glucose dehydrogenases are used. Also, there have been continuing attempts to demonstrate direct ET between the active-site integrated flavin adenine dinucleotide (FAD) cofactor of GOx and an electrode surface. However, after solving the crystal structure of GOx [241] it becomes clear that the ET distance from the protein-integrated FAD is large and hence fast ET kinetics are unlikely.

Despite the knowledge about the large ET distance, there is an ongoing attempt to propose direct ET between GOx and specifically prepared electrode surfaces (Figure 1.13). However, the specific nature of GOx, namely its reaction with its natural electron acceptor O_2 and the probably unwanted generation of H_2O_2, may lead to glucose-proportional current changes which are incorrectly attributed to direct ET reaction. Even if all traces of molecular oxygen are removed the source for the obtained current changes is often not clear. For example, if carbon nanotube-modified sensor surfaces are used, it is difficult to unequivocally confirm that no traces of the metal catalyst used for the growth of the nanotubes are left providing free-diffusing metal complexes which may serve as redox mediator for shuttling electrons between the active site of GOx and the electrode surface. As a

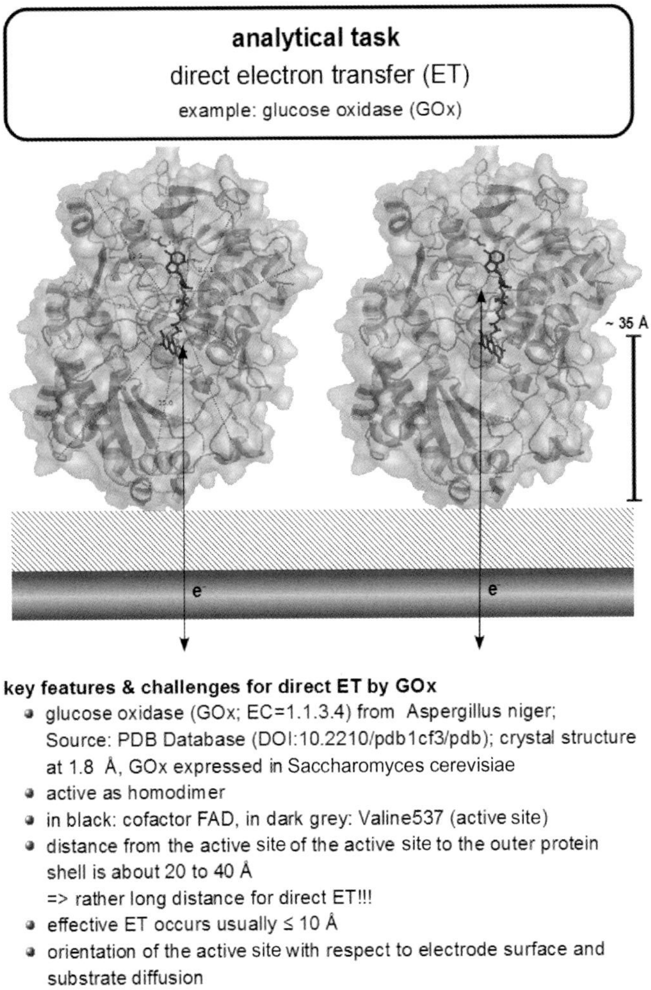

analytical task
direct electron transfer (ET)
example: glucose oxidase (GOx)

~ 35 A

e⁻

e⁻

key features & challenges for direct ET by GOx

- glucose oxidase (GOx; EC=1.1.3.4) from Aspergillus niger; Source: PDB Database (DOI:10.2210/pdb1cf3/pdb); crystal structure at 1.8 Å, GOx expressed in Saccharomyces cerevisiae
- active as homodimer
- in black: cofactor FAD, in dark grey: Valine537 (active site)
- distance from the active site of the active site to the outer protein shell is about 20 to 40 Å
  => rather long distance for direct ET!!!
- effective ET occurs usually ≤ 10 Å
- orientation of the active site with respect to electrode surface and substrate diffusion

Figure 1.13 Direct ET between GOx and an electrode surface.

matter of fact, these potential sources for free-diffusing redox species become increasingly unpredictable the more different components are used to fabricate the sensor.

Thus, as already pointed out, there are clear presuppositions which have to be met before a potential direct ET pathway may be discussed. First, the $FAD/FADH_2$ redox wave in a cyclic voltammogram has to be visible at the potential which is characteristic for the cofactor bound within the active enzyme. Moreover, upon addition of glucose a clear oxidation current has to commence at this redox potential without any significant overpotential. In our opinion, this is one of the main sources for falsely assuming direct ET. It is known that FAD is not covalently bound to the enzyme and hence can become dissolved in the electrolyte during denaturation of the protein. The redox potential of free-diffusing or surface-adsorbed FAD differs from that of FAD located at the active site of the enzyme. Thus, due to the high surface area of the often-used sensor architectures and the comparatively large amount of GOx adsorbed on the electrode surface, the $FAD/FADH_2$ redox couples may often be due to released FAD. Upon addition of glucose, the catalytic current is then not closely related to the observed redox wave and hence is no criterion for a potential direct ET between GOx and the electrode surface. In the following paragraphs a number of recent publications are briefly mentioned in which the source of the observed glucose-proportional current is not completely clear and there may or may not be alternative possibilities to a direct ET process to explain the observed effects.

In the following, a number of recent papers proposing direct ET of GOx will be discussed. A uniformly porous TiO_2 material was synthesized using a carbon nanotube template-assisted hydrothermal method and GOx was adsorbed leading to glucose-proportional currents [242]. Similarly, three-dimensional macroporous inverse TiO_2 opals were synthesized from a sol–gel procedure using polystyrene colloidal crystals as templates. Glucose oxidase was successfully immobilized on the surface of an indium tin oxide electrode modified using inverse TiO_2 opals. Cyclic voltammetry showed stable and well-defined redox peaks for the direct ET of GOx in the absence of glucose. This redox peak increased upon addition of glucose [243]. Along the same lines, direct electrochemistry of GOx adsorbed on boron-doped carbon nanotubes/glassy carbon surfaces [244] or an oxidized boron-doped diamond electrode [245], nitrogen-doped carbon nanotubes [246], exfoliated graphite nanosheets [247], single-wall carbon nanotubes in combination with an amine-terminated ionic liquid [248], and GOx incorporated into polyaniline nanowires on carbon cloth [249] was proposed. Entrapping GOx at the inner wall of highly ordered polyaniline nanotubes [250] or chemically synthesized multi-walled carbon nanotube–SnO_2–Au composites [251], co-deposited GOx–NiO nanoparticles [252], and immobilization of GOx in a natural nanostructural attapulgite clay film-modified glassy carbon electrode [253] have been investigated. Biologically synthesized silica–carbon nanotube–enzyme composites displayed stable redox peaks at a potential close to that of the $FAD/FADH_2$ cofactor of immobilized GOx. The immobilized enzyme was stable for one month and retained catalytic activity for the oxidation of glucose [254].

Direct electrochemistry of GOx immobilized on a hexagonal mesoporous silica-modified glassy carbon electrode was investigated. A pair of redox peaks at a potential of −417 mV was obtained and a diffusion-controlled electrode process with a two-electron transfer coupled with a two-proton transfer reaction process was postulated [255]. However, despite of the well-defined $FAD/FADH_2$ redox process, biocatalytic oxidation of glucose was only possible in the presence of a free-diffusing redox mediator such as ferrocene monocarboxylic acid. This behavior is quite common and supports the assumption that the FAD causing the redox process may be free or surface-adsorbed FAD, which is no longer bound to the enzyme. There are a number of similar studies in which the first criterion, namely the visible voltammogram of the cofactor, seems to be met; however, no electrocatalytic current could be obtained upon addition of glucose. Another series of publications propose direct ET based on a decrease of the electrocatalytic response of the reduced form of GOx to dissolved oxygen [256, 257] or using complex multicomponent immobilization layers with integrated nanomaterials and binders such as GOx–graphene–chitosan [258], dispersed multiwalled carbon nanotubes in a gold nanoparticle colloid stabilized by chitosan and an ionic liquid [259], a carbon nanotube-modified glassy carbon electrode with GOx immobilized within a chitosan film containing gold nanoparticles [260, 261], CdTe quantum dot–carbon nanotube–Nafion films [262], a conductive cellulose–multiwalled carbon nanotube matrix with a porous structure using a room temperature ionic liquid as solvent and encapsulating GOx within this matrix [263], or carbon nanotubes in combination with platinum nanoparticles and chitosan [264].

In all the attempts mentioned above the enzyme was not modified, and hence its size and the large ET distance from GOx to the (nanostructured and high-surface-area) electrode remained constant. Despite the $FAD/FADH_2$ redox wave often being visible in the related cyclic voltammogram, the measured redox potentials varied largely between about −0.49 and −0.41 V which remained without large changes upon addition of glucose. Due to the limitations for direct ET as derived from Marcus theory, these observations are most likely not caused by a true direct ET process but alternative explanations have to be considered despite the observed and repeatedly obtained effects. Alternatively, the ET distance may be decreased by the formation of enzyme–nanoparticle hybrids in which the nanoparticle penetrates into the protein shell [265]. However, in these cases the catalytic current for glucose oxidation is often obtained at high overpotentials. Recently, a more rational approach aimed at decreasing the size of GOx either by preparing genetically modified GOx [266] or by wrapping off the glycosylation shell of the enzyme [267–269]. However, even then it is very hard to distinguish if the catalytic reaction is at the potential of the functional enzyme-integrated FAD or of FAD which may have been released from enzyme molecules.

Thus, despite the large number of publications and the steep increase in the number of publications about direct ET between GOx and modified electrode surfaces, one has to be extremely careful with the possible over-interpretation of the observed effects. The proposed sensors may work fine in dedicated applications; however, it is a fundamental difference if a sensor concept can be applied and glucose concentrations can be reliably determined or if a basic physicochemical claim about a potential direct ET pathway is suggested.

1.4.3 Mediated ET Employed for Biosensors and Biofuel Cells

As already described in Sections 1.1.4 and 1.1.6 (mobile or immobilized) mediators and/or conducting polymers can also be employed to shuttle electrons between a redox enzyme and an electrode surface [11, 12, 16, 20, 21, 24, 25, 28, 270–273]. This approach is called mediated ET (see also Figure 1.6). Efficient ET throughout the entire sensing layer is envisaged in order to avoid only the enzyme layer in close vicinity to the electrode surface contributing to the overall current signal.

For many applications, soluble mediators are not suitable. Thus, redox hydrogels (hydrogels covalently modified with a redox-active mediator) are increasingly being used for reagentless biosensor structures and more recently also for biofuel cells. Heller and coworkers introduced osmium complex-modified redox hydrogels as matrices for biosensors [47, 137–139]. It was determined that the linker length between the osmium complex and the polymer backbone has an impact on the sensor response [274]. Most likely, the mobility of the osmium complex is affected by the length and, hence, flexibility of the polymer backbone and results in a higher efficiency of ET if optimized.

The polymer backbone of typical redox hydrogels is highly hydrophilic and is based on, for example, poly(vinyl pyridine) [48, 275–279], poly(vinyl imidazole) [280, 281], poly(acrylic acid) [282], or poly(allyl amine) [283]. Onto these backbones redox mediators, for example, osmium complexes or ferrocene derivatives, are covalently attached. The biosensors are typically realized by dropping a mixture of the redox hydrogel, a bifunctional linker, and the biological recognition element on the electrode surface. The obtained sensing film adheres well on the electrode surface in most cases and swells in aqueous solutions. Thus, the polymer is rather flexible which promotes the ET rate, the mobility of the counter-ions, and the diffusion of the substrate of the enzyme and the resulting reaction products within the sensing layer [284, 285]. The properties of a hydrogel may also provide an enzyme-friendly microenvironment, and even extend the lifetime of the involved biological recognition elements.

Electron hopping between redox relays covalently incorporated at the polymer backbone dominates the ET. Note, however, that often the first ET between the active site of the redox enzyme and the polymer-bound redox relay represents the rate-limiting step of the entire ET reaction. Biosensors have been miniaturized on the basis of redox hydrogels by employing manual dropping or dipping procedures and, for example, needle-type implantable glucose sensors have been fabricated [137, 286–289]. Properties of electron-conducting redox hydrogels were reviewed most recently in 2006 [272]. Figure 1.14 highlights the analytical task and the challenges involved for mediated ET-based devices.

The approach of employing redox hydrogels helps one to obtain higher current densities which are not only advantageous for, for example, long-term glucose

analytical task
mediated electron transfer (ET)

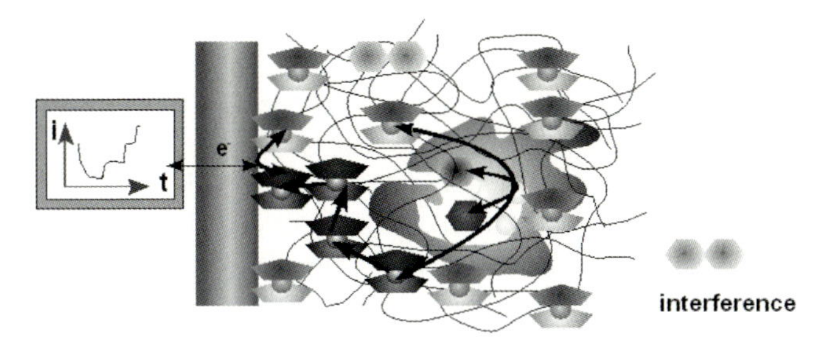

key features & challenges for mediated ET

- efficiency of the overall ET process
- enzyme-friendly formation of the sensing layer
- enzyme-friendly microenvironment (important for biocatalysis and stability)
- pore size of the polymer film should securely incorporate the enzyme
- mobility of substrate and products of the biocatalytic reaction, and counter ions
- incorporation of a suitable amount of biological recognition elements for amplification
- selection of a suitable redox relay (refer to ideal characteristics of a redox mediator) that is preferably covalently attached to the polymer
- The number of redox relays within the polymer chains should facilitate fast ET.
- The contribution of current signal obtained by plain adsorption of the enzyme and/or polymer directly on the electrode surface needs to be significantly smaller than the current signal generated by mediated ET
- reduction of the influence of potential interferences or biofouling processes by selecting suitable film compositions and redox mediators

Figure 1.14 Analytical task of developing and optimizing biosensors based on mediated ET.

determinations but also for biofuel cell applications. Higher current densities compared to those of conventional biosensors are a prerequisite for bringing fundamental studies on biofuel cells closer to real-world applications. Osmium complexes exhibit many properties of an ideal mediator as outlined in Section 1.1.4. For example, their coordination structure is not very much impacted by the oxidation or reduction of the complexes. By modifying the ligand structure the redox potential can be fine tuned to the desired range [272]. The same principle is true for redox-modified electrodeposition paints (EDPs) [146], which were introduced by our group in 2002 [17, 290, 291].

For a rational design of biosensor devices, it is advantageous to aim for non-manual fabrication processes. Electrochemical techniques provide advantages as many polymers can be electrochemically formed or deposited such as conducting polymers and EDPs, for example. The sensing layer is formed by applying potential cycles or sequences of suitable potential pulses while the biological recognition element is present in the solution [17, 292]. The advantage of this approach is that the films are formed exclusively on the electrode surfaces due to the electrochemical initiation of the deposition process. Thus, miniaturization of model biosensor architectures is straightforward and mass production of devices even at small dimensions is feasible. In addition, by automating the fabrication process, the reproducibility of the obtained biosensors should be improved.

One successful strategy to improve ET rates between enzyme and electrode is the modification of conducting polymers with redox mediators in order to obtain reagentless biosensors [11, 270, 271, 292–299]. The drawback of electropolymerization of conducting polymers is that the reaction is sensitive to oxygen, which complicates fabrication at the industrial scale.

Mediated enzyme electrodes were also realized on combined microscale and nanoscale supports [300]. Bioelectrocatalytic hydrogels have also been realized by co-assembling electron-conducting metallopolypeptides with bifunctional building blocks [301]. More recently, redox-modified polymers have been employed to build biofuel cells [25, 70, 302, 303]. In 2003, an enzymatic glucose/O_2 fuel cell which was implanted in a living plant was introduced [147].

The main potential of mediated ET lies in the increase of current densities, as the essential challenge of designing biofuel cells is to increase the biocatalytic power of these devices. Biofuel cells presently reach a power output in the range of about 10^{-6} to $10^{-3}\,W\,cm^{-2}$. Practical conventional fuel cells operate in the range of about 1 to $10^{8}\,W\,cm^{-2}$ [303]. Taking the calculations from Barton and coworkers into consideration [70], in which, as mentioned above, the theoretical current density of a monolayer was estimated to be about $80\,\mu A\,cm^{-2}$, one would require thousands of layers to obtain a current density above $10\,mA\,cm^{-2}$.

To summarize, the advent of redox-relay modified polymers, such as redox hydrogels, conducting polymers, or EDPs, enabled the development of biosensors that even made it to commercial applications such as implantable glucose sensors. In addition, this approach is now increasingly used for the development of biofuel cells.

1.4.4
Nanomaterials and Biosensors

Without any doubt, nanotechnology has had and is still having an enormous impact on science. When speaking of nanotechnology one typically assumes that structures are used with at least one dimension being in the sub-100 nm range. The advantages of and new possibilities offered by nanotechnology are manifold. Materials exhibit new properties when scaled down from bulk material to nanometric dimensions. These properties can be precisely fine tuned, thus allowing for the fabrication of defined structures and materials optimized for a certain purpose. Consequently, nanomaterials and concepts from nanotechnology have been much employed in biosensor development. Several reviews on the topic [182,

304–306] provide a detailed overview of the possibilities of nanotechnology in the field of biosensor research. The following summarizes the most important trends.

The main challenges in the application of nanomaterials for biosensor designs are the definition of the material properties, the reproducible synthesis of materials with suitable properties, and the meaningful application of nanotechnological concepts to biosensors. Definition of material properties and, thus, the choice of materials are common to other areas of biosensor research and have been discussed earlier in more general terms. The question of how to synthesize or otherwise access these materials will not be answered exclusively by the biosensor expert. Instead, multidisciplinary effort will be necessary to obtain nanomaterials with properties as required for a novel biosensor design. The seemingly most challenging task of applying nanotechnology to biosensors is to really make use of "nano features" and not simply using nanomaterials without them adding value to the biosensor architecture. In the area of biosensor research some features of nanostructures become important in addition to pure material properties. For instance, in nanometric structures diffusion lengths become very short and hence mass transport is highly efficient. Since mass transport is crucial in many biosensor designs, an increase or at least a change in sensor performance can be expected from using nanometric structures.

There are basically three broad categories of approaches towards nanobiosensors and in particular in electrochemical nanobiosensor development. The modification of a (macroscopic) transducer with nanomaterials is the first of these approaches. In electrochemical biosensors, this would translate into large electrodes modified with nanomaterials. The second approach is the miniaturization of the transducer, namely the use of nanoelectrodes [307] or other miniaturized circuitry of nanometric dimensions. The modification of biomolecules with nanomaterials or coupling of biomolecules and nanomaterials is the third category of approach towards nanobiosensors. Of course the lines between these approaches are blurred and some sensor designs may draw from more than one of these concepts.

1.4.4.1 Modification of Macroscopic Transducers with Nanomaterials

There is an enormous variety of nanomaterials that can potentially be employed in biosensor architectures. The most prominent among them are metal nanoparticles [304], quantum dots [308], and carbon nanotubes [309–311]. All of them have been employed in biosensors though not necessarily exclusively electrochemical biosensors. Quantum dots (QDs) offer unique absorption properties making them highly suitable for the construction of biosensors with optical readout. The most diverse electrochemical nanobiosensors are, however, obtained from carbon nanotubes (CNTs) which offer a wide range of different applications.

CNTs were discovered in the early 1990s [312]. CNTs have a tubular structure of closed topology and consist of hexagonal honeycomb lattices made up of sp^2 carbon units. A schematic of the structure of CNTs is shown in Figure 1.15. The diameters of CNTs are typically several nanometers. The length of CNTs can be up to several micrometers. Two basic forms are distinguished, single-walled

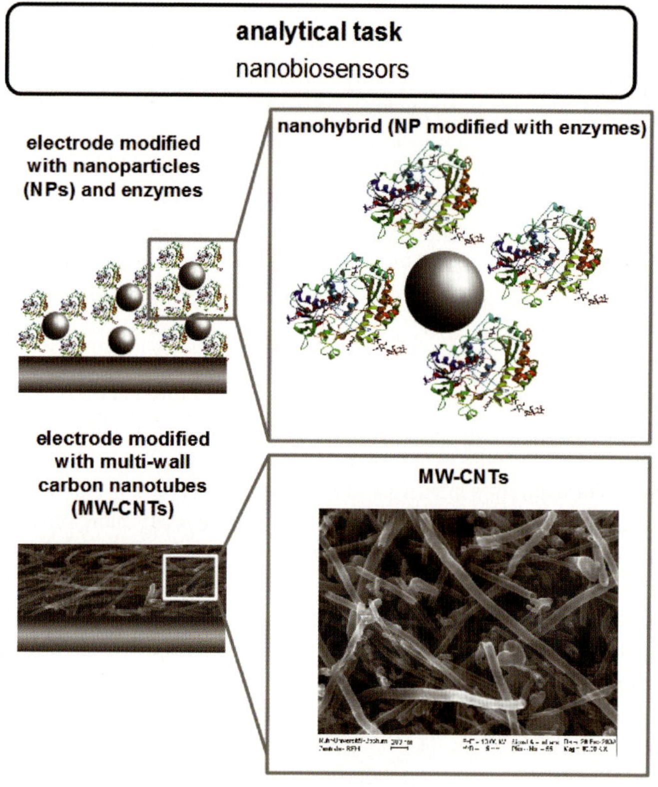

key features & challenges for nanobiosensors

- novel material properties due to nanometric dimensions
- tunable properties of nanomaterials
- cost and time effectiveness of nanobiosensors (small amount of material, small dimensions)
- comparable sizes of nanomaterials and biological molecules
- potential harm of nanomaterials

Figure 1.15 Analytical task of nanobiosensors.

carbon nanotubes (SWCNTs) and multiwalled carbon nanotubes (MWCNTs). Besides their chemical stability [313], one of the most interesting characteristics of CNTs for electrochemical biosensors are their ET properties [314]. The ET properties of CNTs can be modified by surface groups such as oxygen, NO_2, or amino groups. Electrodeposition and other forms of growth of metal nanoparticles on CNTs result in another class of nanomaterials with high application potential in electrochemical biosensors [315]. The suitability of CNTs as immobilization matrices retaining or even enhancing the activity of the respective biomolecule has been discussed [316]. In addition, the large surface area of CNTs results in a large active electrode area and CNTs can prevent electrode fouling such as caused by NADH oxidation [317].

The fabrication of electrodes (often glassy carbon or gold electrodes) modified with CNTs typically suffers from the low solubility of CNTs in most commonly used solvents. Hence, CNTs are in many cases dispersed within solvents or polyelectrolytes and drop-coated onto the electrode to be modified. Alternatively, CNTs are incorporated within composite binding materials such as Teflon [318]. Another route to CNT-modified electrodes is the direct growth of CNTs on the electrode material [319]. Electrochemical biosensors based on CNTs have been used in the determination of a wide variety of analytes including glucose, fructose, cholesterol, lactate, catechols, hydrogen peroxide, alcohols, cholines, and organophosphates, as recently reviewed in [310], as well as DNA and proteins [320]. With the level of pioneering work left behind, the powerful combination of biorecognition and extraordinary ET properties and material properties of CNTs can be expected to yield even more high-performance electrochemical biosensors in the near future.

1.4.4.2 **Nanometric Transducers**

This section highlights two trends in nanobiosensing. First, the use of nanofluidics [321, 322] in biosensing and, second, the use of nanoelectrodes [307] and nanoelectrode arrays [323] will be briefly discussed.

Nanofluidics is part of the field of so-called lab-on-a-chip analytical devices which integrate all essential tasks of an analytical problem into a chip-based format [45]. Lab-on-a-chip devices originate from microsystems technology and have several of advantages over conventional instrumental analysis, such as cost-effectiveness due to small material amounts used, time-effectiveness due to small diffusion lengths and therefore extremely efficient mixing of reagents, and other transport phenomena that can be employed to efficiently separate reagents. Furthermore, lab-on-a-chip devices are ideally suited for automated analysis allowing for high-throughput screening. These advantages become even more pronounced when the devices are of nanometric dimensions. In nanofluidic devices, at least one dimension of the device is close to the Debye length and hence transport phenomena not known at the macroscale and even microscale predominate. Biosensing employing nanofluidics includes immunoassays [324] among other analytical schemes such as reviewed in [322].

The electrochemical properties of nanoelectrodes differ significantly from those of macroelectrodes. Like microelectrodes [325], nanoelectrodes are characterized by a hemispherical diffusion field (whereas at macroelectrodes, linear diffusion dominates in amperometric measurements and voltammetric experiments at slow and moderate sweep rates). Consequences of the hemispherical diffusion field are the fast establishment of a stationary diffusion current, high current densities, and a favorable signal-to-noise ratio. Hence, nanoelectrodes have proven to be highly sensitive probes of biorecognition reactions. Often nanoelectrodes are employed as nanoelectrode arrays. Interdigitated nanoelectrode arrays allowed for the label-free detection of DNA using redox mediators [326]. Just as microelectrodes were much employed as probes in scanning electrochemical microscopy [327, 328] to study the immobilization processes of biomolecules on surfaces [329, 330], nanoelectrodes have the potential to allow for an even more detailed mapping of biological activity.

1.4.4.3 Modification of Biomolecules with Nanomaterials

The direct modification of biomolecules with nanomaterials resulting in biomolecule–nanomaterial hybrids offers interesting possibilities for biosensing. Gold nanoparticles can be used to immobilize enzymes or other biorecognition elements on electrodes or other supports. However, in this case the nanoparticles often just function as a linker and the sensor architecture does not benefit from a unique property due to nanometric dimensions. In contrast, the already mentioned unique optical properties of QDs make these materials well suited as fluorescent labels in optical sensors [331, 332] really taking advantage of a nanofeature.

The use of nanoparticle–enzyme hybrids has been recently reviewed [333, 334] as has been the use of nanotechnology in the manipulation of redox systems at an earlier stage [335]. The wiring of enzymes by redox hydrogels or osmium-modified EDPs is the subject to another part of this chapter. Such an establishment of electrical contact between a redox enzyme and an electrode can also be achieved by nanoparticles. Standing out in this field of research is the wiring of redox enzymes by gold nanoparticles [265]. In this work, GOx was reconstituted with a gold nanoparticle (1.4 nm in diameter, corresponding to the size of the redox center of the enzyme) that was functionalized with the enzyme's cofactor FAD. Such enzyme–nanoparticle hybrids were assembled on gold electrodes leading to exceptionally good electrical contact between the enzyme redox center and the macroscopic electrode.

In conclusion, nanotechnology has contributed significantly to recent developments in biosensor research. Modification of macroelectrodes with nanomaterials has resulted in new exciting ET properties and biocompatibility. Nanometric transducers have been used to obtain new classes of biosensor devices. Finally, biomolecule–nanoparticle conjugates show a promising application potential in biosensor development. An aspect of nanotechnology that is rarely mentioned is the potential harm of nanomaterials towards health. Though a significant effort has been put into the research of this field [265, 336, 337], the consequences of the use of nanomaterials in everyday life are not yet fully understood. It seems, however, that the potential risks of nanomaterials are by far outweighed by the possibilities offered by nanotechnology.

1.4.5 Implanted Biosensors for Medical Research and Health Check Applications

As depicted in Figure 1.16, (electrochemical) biosensors are either placed in laboratory animals for fundamental (patho-)physiological and neurochemical *in vivo* measurements or implanted in the human body for health check purposes and metabolite monitoring. In the field of *in vivo* medical research, enzyme-based

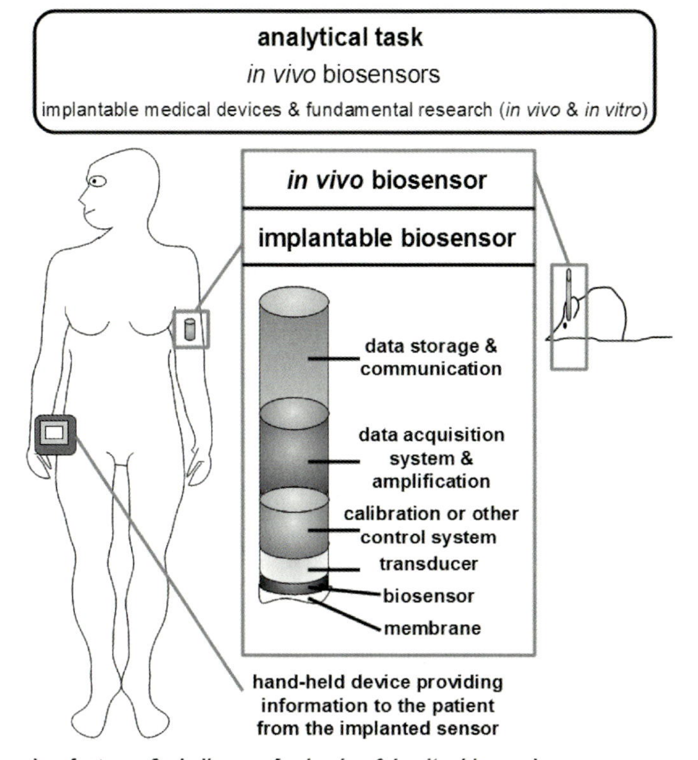

Figure 1.16 Analytical task of developing and optimizing *in vivo* biosensors.

analytical tools are often used for spatially confined measurements of their corresponding target species in preselected regions of living test subjects.

The main challenges in the field of implantable sensors are the stability of the sensor, the selectivity of the sensor, and the biocompatibility of the sensor. First of all, the sensor must not be rejected by the body. When implanted, the sensor should operate for a prolonged time to justify any surgical procedure necessary for the introduction of the sensor into the body. Even when these two challenges are met, the sensor has typically to deal with a very complex sample matrix, most commonly blood. *In vitro* sensors have to cope with the same demands in terms of their selectivity while long-term stability is usually not such a critical issue. However, *in vitro* sensors also have to be biocompatible in such a way that their presence should not influence the biological environment in which they operate.

At present, the highly heterogeneous rat or mouse brain environment is probably most prominently addressed by *in vivo* and *in vitro* biosensors. Nevertheless, other parts of the rodent central nervous system, the many secretory glands of the regulatory endocrine system, or the tissue of muscles are sites of interest for implantable biosensors. Fixed in a particular brain region for fundamental cognitive, pathological, and pharmacological investigations, the sensing tips of, for instance, tapered voltammetric enzyme microbiosensors have demonstrated their ability to directly record up- and down-regulations of neurochemicals that may appear in response to premeditated external stimuli such as feeding, drug administration, or gratification at the local level with high time resolution. Biosensors that are implanted in humans, on the other hand, are supposed to report the dynamics of the levels of metabolites related to mental or physical disease states. To reach this challenging goal they may be placed subcutaneously, just beneath the carrier's skin, or at deeper body locations close to target organs such as the kidney, liver, and pancreas or the muscles of the extremities. Specific applications of biosensors in the human body include clinical point-of-care testing in hospital settings and personal diabetes management using handheld monitoring devices. However, the main focus in this success story of biosensor development is not on the description of a number of examples from *in vivo* and healthcare measurements but advancements that were reported in the last five years in terms of the design and quality of electrochemical biosensors for successful analysis in a firmly implanted configuration, be it in animals or humans. For in-depth information on specific examples of both classical *in vivo* (neurochemical) studies and standard human metabolite monitoring with implanted glucose, glutamate, lactate, acetylcholine, or peroxide biosensors the reader is referred to several recently published comprehensive review articles on the two subjects [65, 163, 289, 338–351].

The clever involvement of new enzymes or adapted enzyme blends in the design of implantable biosensors was used to detect physiologically or pathologically relevant biochemical compounds other than the five conventional ones already mentioned. The release of the well-known purine ATP as potent extracellular signaling molecule was, for instance, demonstrated *in vivo* for the *Xenopus* tadpole spinal cord during motor activity with implanted biosensors that had co-immobilized glycerol kinase, glycerol-3-phosphate oxidase, and phosphocreatine kinase [352]. Miniaturized carbon fiber-based biosensors for *in vivo* measurements of acetylcholine and choline have been prepared by means of a co-immobilization of acetylcholine esterase and choline oxidase [353]. The gliotransmitter D-serine, well-known for a long time to modulate neurotransmission at the glutamatergic synapse, has been monitored in the rat brain striatal extracellular fluid with implanted biosensors employing mammalian D-amino acid oxidase as the indicating biological recognition element [354]. The common neurotransmitter dopamine is typically measured *in vivo* in particular brain sections with direct fast CV at the solid graphite disc of polished glass–epoxy insulated carbon-fiber microelectrodes

[355, 356]. In an attempt to improve the selectivity of local dopamine measurements in the complex extracellular matrix of brain fluid, an implantable enzyme-based dopamine microbiosensor has been constructed based on the immobilization of tyrosinase in a thin-film chitosan coating of carbon-fiber disc microelectrodes [357]. *o*-Dopaquinone, which is the product of the tyrosinase reaction with dopamine, was monitored via its reduction at the modified microelectrode surface. The application of these cathodic tyrosinase dopamine microbiosensors was reported for the continuous real-time *in vivo* visualization of electrically stimulated dopamine release in the brain of anesthetized laboratory rats. Remarkably, due to the cathodic potential the sensor response was not significantly disturbed by the presence of typical interferences such as ascorbic and uric acid, serotonin, norepinephrine, and epinephrine.

As with any conventional electrochemical biosensor, an implanted biosensor should also have an exceptional selectivity and sensitivity for the target compound, a low detection limit, and a fast response time that is well tailored to the time course of the expected dynamic changes in the concentration of the target analyte in the surrounding tissue. There are, however, important additional properties to look for when the ambition is for long-term stable electrochemical biosensor performance in the complex matrix of the bodies of animals or humans. It is very important for *in vivo* brain biosensor analysis, but also valid for other situations, to obtain a sensitive acquisition of a strongly localized signal from the molecule in question. In this case, a sensor design is needed that offers a positionable tapered sensor of small total tip dimension which often equals the diameter of an electroactive disc plus twice the thickness of its insulating sheath. Small sensor tip size will of course also be beneficial for placement with minimal (brain) cell and surrounding tissue damage. The exploration of glass- or polymer-insulated needle-type carbon or metal microelectrodes as diminutive precursor structures for biosensors offered an appropriate solution for this problem and no real innovation in this aspect arose in the period under consideration.

The second relevant issue for success with biosensors in the chronically implanted configuration is sufficient sensor stability over the extended time of data acquisition throughout a trial. For a lot of significant behavioral studies but basically in the general implantation case the desired period is days if not weeks of measuring time. The long-term quality of the sensor performance is of course impeded by the gradual loss of proper signal generation caused by the foreign body response and contaminating contact of functional sensor entities on the electrode surface, the immobilization matrix, and the immobilized biological recognition element with protein and lipid contents of the immediate physiological measuring environment [358–362]. Among the issues that can be adverse to long-lasting sensor functioning are (i) the fouling of the immobilization layer in the form of a delamination or loss of porosity which is essential for substrate (analyte) diffusion, (ii) the degradation or denaturation of the biological recognition element, (iii) the passivation of the electrode surface by nonspecific adsorption of proteins and lipids, and (iv) the slow formation of a barrier for substrate diffusion through an ongoing fibrous encapsulation of the biosensor tip. In view of these considerations, both the optimization of immobilizing top coat and an advanced morphological and chemical design of the transducing electrode surface of enzyme-based *in vivo* biosensors set the scope for the development of new concepts that offer a well-thought-out prevention against the listed set of detrimental effects of sensor tip implantation and the preservation of the analytical response. Several reports in this context have dealt with the adaptation of redox hydrogels employed for the entrapment of the enzymes used via either a mild nondegrading biocompatible environment for the active macromolecule or the creation of hydrogel surfaces that are less prone to the adsorption of contaminating (protein) species. Suggestions include a self-cleaning nanocomposite hydrogel membrane [363], biomimetic hydrogels [364], the involvement of surfactants in the formation procedure of redox hydrogels [365], and hydrogels with optimized type and ratios of individual polymerizing components [366, 367]. Taking advantage of the fact that nitric oxide effectively inhibits platelet and bacterial surface adhesion, Shin and Schoenfisch proposed advanced biosensor interfaces with a high potential to resist biofouling via the implementation of an additional nitric oxide-releasing top coating made of *N*-diazeniumdiolate-modified polymers [368]. Self-assembling polyelectrolyte–poly(ethylene glycol)-based nanofilm multilayers have been demonstrated on porous alumina supports as effective diffusion-controlling and protein adsorption-resistant coatings and were reported as optimized dual-function immobilization matrices for implanted biosensors [369]. Also recommended as surface modifications with promising biocompatibility properties were apparently low-fouling zwitterionic carboxybetaine methacrylate coatings [370], microporous collagen scaffolds that minimized unfavorable tissue reactions while stimulating angiogenesis in the vicinity of biosensor tips [371], porous poly(L-lactic acid) coatings to reduce fibrosis and promote new blood microvessel formation in the tissue surrounding the implanted biosensor surface [372], intentionally pre-adsorbed coatings of constructive proteins capable of inhibiting bad foreign body responses [373], new hydrophilic poly(ethylene glycol)-based redox copolymers bearing electrochemically active ferrocene and thiol/disulfide functionalities for anchoring to a gold electrode surface [374], and special nanoporous membranes [375, 376].

At present, roughly a quarter of a billion (!!!), a still steeply increasing number, worldwide cases of diabetes are reported. Situations of hypo- and hyperglycemia in patients have to be avoided and thus effective blood glucose measurement and control is a top analytical task in medical diagnostics and healthcare, respectively. Already prior to the period covered by this section, personal self-monitoring of internal glucose levels became routine in small-volume blood samples obtained, for instance, by piercing the fingertip or arm. The required commercial tools and information on both their technology and on the glucose meter marketplace is available, for instance, in [59, 65, 163, 343]. Glucose meters typically take advantage of sophisticated single-use screen-printed arrays of electrodes one of which is designed as the glucose sensor via specific immobilization of mostly GOx as the biological recognition element and the involvement of artificial free-diffusing redox mediators. Upon placement of a microliter droplet of whole blood, the electronics of the glucose measuring device assesses and digitally displays a glucose equivalent in reasonably short time. Dependent on the quality of the measured value in relation to the accepted normal level, insulin injection or dietary carbohydrate uptake should be performed. Even if carried out several times a day, timed glucose monitoring with external sensors activated at user-chosen intervals obviously has the shortcomings that it fails to report irregular up and downs in between assessments and cannot utilize trends associated with daily habits of diabetics for an instant therapeutic action. In this context a better, albeit more challenging, route of blood glucose analysis and management is the operation of permanently implanted glucose sensors for a continuous direct detection of the analyte either in the bloodstream or the interstitial fluid of the subcutaneous tissue. The advantages and disadvantages of continuous glucose sensing as well as the difficulties in and steps forward to the establishment of the approach have been discussed in depth elsewhere [65, 338, 340, 341, 344, 346, 348, 377, 378].

Worth mentioning here as an excellent example of the remarkable achievements of focused joint academic and industrial glucose biosensor R&D is the appearance of the Freestyle Navigator® continuous glucose monitoring system from Abbot Diabetes Care/TheraSense, which recently got approval by the US Department of Health and Human Service, Section Food and Drug Administration (FDA) and became commercially available for diabetics in 2008 [65, 341, 379]. Other similar systems seem to be on their way, and include, as an example, a device that has already been tested successfully for stable long-term glucose monitoring in diabetic and nondiabetic animal models [380] and currently is awaiting FDA approval for GlySens Incorporation.

Electrochemical enzyme biosensors for *in vivo* studies and human body metabolite monitoring have in recent years been brought to quite an advanced level. A clear proof of the achievements is the good number of successful biosensor recordings of brain activity and the enterprise of marketable continuous blood glucose monitoring. Further improvements in the spatial and time resolution of *in vivo* measurements would need further sensor miniaturization and tapered nanobiosensors that should be similar to their microelectrode analogues in terms of the proper conductor embedment and resistance against sensor fouling. However, they should be equipped with a reduced total tip dimension for better positioning and nanometric sensing areas for fast and highly localized recording. *In vivo* biosensor measurements at the single-cell level or at least a very small cell cluster level could then add novel information to the fundamental insights that were and still are gained through *in vitro* single-cell electrochemistry with isolated cells out of their native matrix [381]. Though tapered enzyme-based nanobiosensors with small total tip radii have already been reported [382], these fragile needle-like versions of biosensors have, to the best knowledge of the authors, not yet been successfully operated *in vivo*. Here, there is definite room for future innovative research activity. Another area worth working on is the further extension of the lifetime of sensors for continuous blood glucose monitoring and the transfer of the principles of well-working GOx-based implanted biosensors to those incorporating other enzyme systems for broadening the scope of target analytes. The related possible enhancements and expansion of vital health and disease marker monitoring could open up the individualized and portable medication and care plan that is envisioned by clinicians and patients around the globe.

1.4.6
Nucleic Acid-Based Biosensors: Nucleic Acid Chips, Arrays, and Microarrays

Biosensors and high-throughput electrochemical screening devices based on DNA, ribonucleic acid (RNA), and peptide nucleic acid (PNA) gain their outstanding sensitivity and selectivity from the very strong base pair affinity between complementary sections of lined-up nucleotide strands, which are the evolutionary genetic code maps of living beings. In fact, all of the genetic information required for body development and functioning as well as details on disease prevalence and states of an organism are smartly made available in the programmed sequence of the nitrogenous bases adenine (A), cytosine (C), guanine (G), thymine (T; in DNA), and uracil (U; in RNA). To create so-called nucleic acid (NA; or gene) chips (or (micro-)arrays), physicochemical transducers (e.g., the surface of an electrode, a microscope glass slide, or a quartz crystal microbalance) are decorated with regular patterns of spots of synthetic single-stranded oligonucleotides, each of them being associated with an intentionally designed A, C, G, and T/U order. Subsequent to the exposure of the sensor surface to a (clinical) sample the remaining analytical task is the identification of all NA spots that underwent hybridization. Definite localization of the collection of immobilized "probe" NA strands that undoubtedly found their matching binding partner (the "target") in a pre-prepared complex blend of genetic material is at the center of the analysis of an apparent gene expression.

The major challenge for an (electrochemical) NA biosensor is the full exploration of the massive amount of information that is buried in the totality of the genomes of mammalian species. Currently, only the very tip of the "genetic iceberg" is revealed and a vast amount of effort has to be invested to finally make the best out of the technology for health science and clinical diagnosis and medication. Figure 1.17 visualizes the analytical task for electrochemical NA sensors and assays.

The main challenge for designing NA-based biosensors is that the hybridization event needs to be monitored correctly. Mismatches need to be distinguished from matches. The assay procedure needs to be compatible with the conditions of hybridization and to allow tuning of the binding specificity. For this, assay strategies employing labels such as redox-active dyes or intercalators as well as enzymes or nonlabeled detection schemes can be used. Each approach has its own benefits and limitations. The design of the capture probe is critical for overall assay performance. If one is employing electrodes, one always has to take into account the impact of the electric field on the orientation of the NA molecules, because NAs are highly negatively charged molecules. In cases where the biosensor design should be compatible with commercialization of the device, the NA chip fabrication procedure needs to take production effort and costs into account.

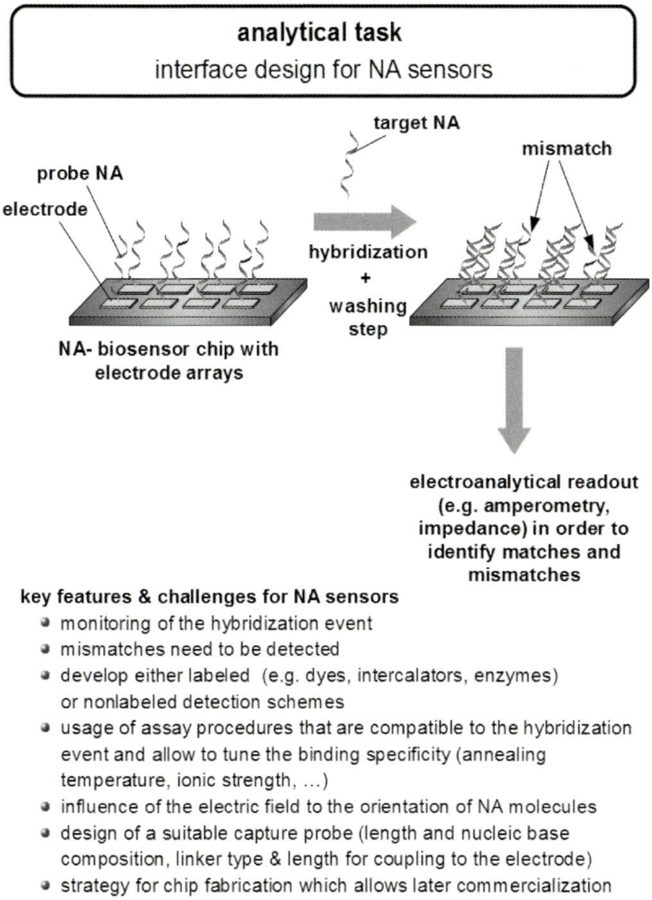

analytical task
interface design for NA sensors

target NA
mismatch
probe NA
electrode
hybridization
+
washing step
NA- biosensor chip with electrode arrays

electroanalytical readout (e.g. amperometry, impedance) in order to identify matches and mismatches

key features & challenges for NA sensors

- monitoring of the hybridization event
- mismatches need to be detected
- develop either labeled (e.g. dyes, intercalators, enzymes) or nonlabeled detection schemes
- usage of assay procedures that are compatible to the hybridization event and allow to tune the binding specificity (annealing temperature, ionic strength, ...)
- influence of the electric field to the orientation of NA molecules
- design of a suitable capture probe (length and nucleic base composition, linker type & length for coupling to the electrode)
- strategy for chip fabrication which allows later commercialization

Figure 1.17 Analytical task of NA-based biosensors.

The most common electrode material for electrochemical NA chips is gold; however, other metals, carbon, and certain semiconductors have been used as well. The immobilization of NA probe strands can be achieved, for instance, via simple physisorption, the chemisorption of thiol-modified NA (when gold is the transducer material), the covalent binding of, for example, biotinylated NA to (strept)avidin-modified surfaces, and NA fixation into ultrathin polymeric surface coatings. For ultrasensitive electrochemical recognition of hybridization on NA chips, reagentless, labeled, and label-free schemes have been reported. Enzymes, for instance, may be attached to the endings of target NA fragments and trace detection of double-strand formation can be established via adapted electroreduction or electrooxidation of a product of an enzyme interaction with intentionally added substrate. Other successful schemes exploit redox-active intercalating or groove-binding mediator molecules that can enter the tubular and twisted structure of the double helix of fused probe and target NA strands to become available for the generation of electrical current. Hybridization detection strategies may use redox mediators that are tethered to either probe or target strands for the creation of configurations favorable for the induction of a Faradaic sensor response, or utilize the distinguished electrostatic properties of immobilized single- and double-stranded NA for impedimetric or voltammetric hybridization detection. An already quite matured NA chip technology in its various facets has entered almost all fields of biology and medicine and a thorough analysis of gene screening data is currently an extensively explored specialty for genotyping, pharmacogenomics, pathogen classification, gene expression profiling, drug discovery, and molecular medical diagnostics. The very obvious indication of the prominent role of state-of-the-art NA biosensing is the noticeable explosion in the number of publications on the subject. A scientific literature screening with the search terms "DNA chip, DNA array, DNA microarray, gene chip, gene array, gene microarray genosensor, DNA biosensor," all combined with the Boolean operator OR and tested for topic or title appearance gave 19 227 and 1148 hits, respectively, for the five-year period 2006–2010. A search refinement with the phrase combination "electrochemical OR voltammetry" reveals close to 1000 topical publications, which, of course, are still far too many to have all been summarized in the constraint of the NA subsection of an overview article on electrochemical biosensors. Hence, the focus was placed on a selection of issues that – from the viewpoint of the authors – provide promising threads for practical advancements of NA-based diagnostics and reflect emerging trends in the field. Several recently published methodical review articles are, however, recommended as excellent additional sources of information about the basic concepts of NA immobilization on single or arrayed electrodes, for the fundamental details of the many existing analytical schemes for electrochemical detection of hybridization, and for the particulars of the variety of published designs of (electrochemical) NA chips [36, 38, 80, 95, 109, 128, 140, 145, 153, 207, 302, 383–386]. Good summaries on the state of the art of the specialties of aptamer-based NA sensing [70, 96, 106, 144, 300, 387] and on impedance/capacitance-based hybridization detection [388–390] are also available in the literature.

Aiming at an improvement of the analytical performance of NA chips/arrays in general and targeting a more sensitive detection of clinically relevant point mutations in particular, immobilized probes made of synthetic PNA have been brought into play as promising alternative biological key components of gene assays. In contrast to "normal" biological NA strands with negatively charged phosphate groups contained in their filamentary polymeric structure, PNA has an entirely uncharged backbone. In contact with single-stranded DNA or RNA target strands of a sample, PNA probe strands obey the rules of NA hybridization. However, as there is no negative polarity in PNA backbones, electrostatic repelling forces between hybridizing PNA and complementary DNA or RNA pieces are absent and the strand bonds in PNA–DNA/RNA hybrids are thus stronger than in the conventional case [391–393]. Benefits related to this effect are improved hybridization properties in terms of affinity, specificity, and sensitivity against single-base mismatches, a better chemical stability of the obtained duplex structures and resistance against enzyme cleavage by, for example, nucleases and proteases, and, last but not least, a reduced impact of the ionic strength of the measuring buffer on the outcome of a hybridization screening. An early demonstration of the potential of PNA probes was their successful immobilization onto a quartz crystal microbalance transducer and the subsequent application of the construct in hybridization experiments for the discrimination of perfect matches and single-base mismatches [394]. Sensitive single nucleotide mismatch detection with PNA-modified electrode transducers was the subject of a number of studies in the past five years. Representative examples are the exploitation of the impedance characteristics of a PNA–metal ion interaction for the identification of an individual C–T mismatch in a 15-mer PNA–DNA hybrid [40], the discrimination of completely complementary from mismatched double helices via specifically acting redox-active diviologen indicator molecules [49], the application of osmium mediator end-labeled PNA and stripping voltammetry for single-base mismatch detection [110], the use of ferrocene-labeled PNA for full match/mismatch identification [94], a single nucleotide polymorphism detection via joint employment of electroactive chitosan nanoparticles and PNA strings [97], and hybridization and mismatch recognition via the diverse electrostatic interaction of a cationic ruthenium(III) mediator with neutral PNA capture probes and anionic target DNA backbones [301]. Involvement of PNA probes in work on samples with clinical relevance includes, for instance, the detection of short sequences of the hepatitis C 3a virus [35], implementation in silicon nanowire biosensors for highly sensitive and rapid detection of Dengue virus [395], the application with a PNA-modified electrode for *Mycobacterium tuberculosis* pathogen detection [93], and the development of a PNA array for the identification and quantification of the cancer gene c-Ki-ras [396]. Another form of uncharged synthetic DNA is the morpholinos. With a comparable motivation as for PNA they have been explored as components of (label-free) surface hybridization assays [34, 50, 101].

Common electrochemical DNA chips/arrays are fabricated either with automated spotting procedures and controlled dispensing of small-volume droplets of NA solutions into a regular pattern or via a microlithography-controlled patterned (bio-)chemical on-chip synthesis. Both methodologies work well but they are associated with expensive apparatus and have their practical limits when the desired characteristic array dimensions such as probe spot diameters and distances approach the nanometric level. In view of that, cheap and spatially more accurate NA patterning techniques are sought after. In this context NA origami, the folding of longer filamentary NA structures into nanoscale two- and three-dimensional surface features [37, 51, 104, 397–400], has recently been suggested in the field of molecular NA sensing technology as a tactic for manufacturing nanoscopic NA chips [39, 98, 169, 401]. The potential of a combination of NA origami with existing nanoelectrode array fabrication technology has not yet been explored; however, this is an attractive theme that could lead to the development of powerful novel electrochemical tools for scaled-down NA hybridization screening in ultrasmall sample volumes.

Usually the electrode surface in an electrochemical NA biosensor acts as the physicochemical transducer that responds to exposure to target NA-containing sample solution and the formation of probe–target hybrids with a change in current flow, capacitance, or impedance. An interesting atypical configuration of a NA probe-carrying electrode is to use alternating electrode potential variations for controlled modulation of the structural conformation of the surface-tethered probe and fused probe–target strands and establish what is named a switchable DNA interface [402–404]. Hybridization with this method is not determined as usual via the acquisition of electrode properties but by means of an extra optical scheme that is susceptible to situations in which NA strands extend away from or fold onto the electrified sensor surface. At frequencies in the kilohertz region, the combination of electrode potential variation and an optical readout revealed distinct switching kinetics for changes from the upright to the flat surface position of NA strands, and the sensitivity was reported good enough to allow single-mismatch detection.

Without a doubt, NA-based biosensing and the application of the related arrayed gene chips for studying mRNA levels and examining gene expression profiles in human, animal, and plant samples are among the hot topics in current analytical chemistry. The prominent role is of course strongly related to the accumulated success of the many running genome projects. In particular the previous disclosure of the human genome [405] and the expectations from specific gene identification for disease diagnostics are fostering the exclusive position of hybridization assays. Electrochemical hybridization detection has been proven competent and competitive enough to be an attractive alternative to the more expensive and often technically more complex optical options in certain circumstances. However, handling the massive genetic data material of the human genome is a huge challenge for NA chip technology no matter what detection scheme is employed. So far, only an extremely small percentage of the wide range of possibilities has been accomplished. Significant advancements in both NA array fabrication and electrochemical, mass, or optical readout strategies and equipment are indispensable to enable personalized medical care based on individual gene profiles become part of daily life.

1.4.7 Immunosensors

Immunosensors rely on the extremely high binding affinity of antibodies towards their respective antigens. Apart from the very specific biorecognition reaction, antibodies and antigens can be produced to obtain a specific binding partner for a target of interest. A comprehensive overview of antibody production is given in [406]. However, the challenge of producing antibodies should not be underestimated. Apart from the production of the actual biorecognition element, the challenge in the design of (electrochemical) immunosensors lies in the development of a suitable detection scheme. Basically two types of immunosensor approaches can be distinguished: labeled and nonlabeled approaches.

1.4.7.1 Labeled Approaches

Labeled immunosensors typically make use of a reporter molecule attached to the respective antibody. Apart from the electrochemical detection schemes discussed below, immunosensors employ fluorescent labels, radioactive labels, or nanoparticles, among other reporter systems [407–410]. Labeled immunosensors normally operate using either a direct or indirect sandwich procedure or a competitive format as depicted in Figure 1.18.

Sandwich assays rely on secondary antibodies binding to the target antigen or antibody after the primary biorecognition reaction. In a competitive assay format, labeled and nonlabeled antibodies (or antigens) compete for the binding sites of

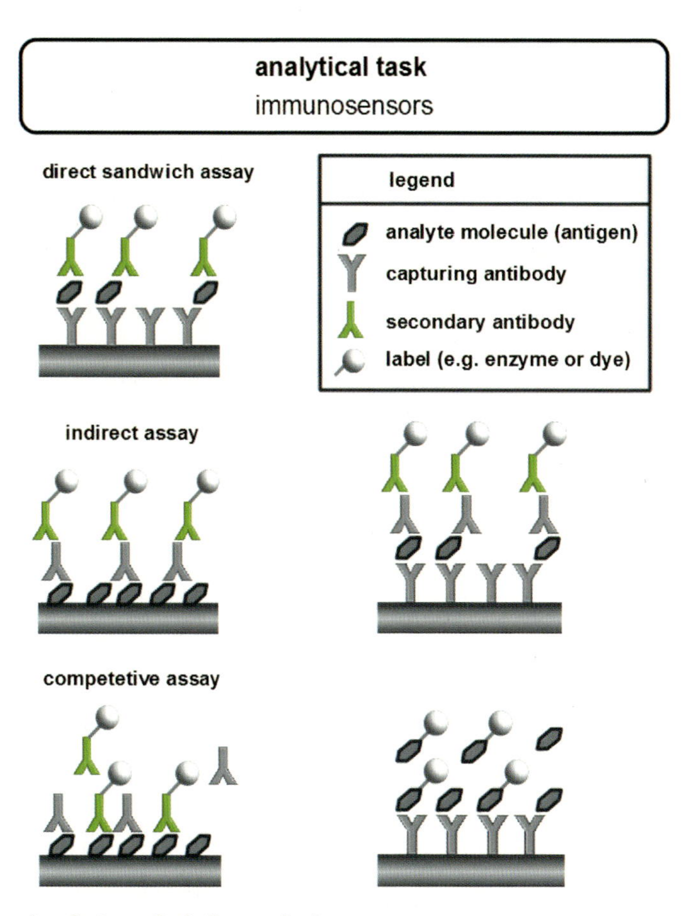

Figure 1.18 Analytical task of immunosensors.

the biorecognition layer. In direct assay formats basically all binding events of the target analyte can be addressed by the secondary antibody leading to typically higher sensor signals than in competitive assays. Competitive assays, however, require less working steps leading to potentially more cost- and time-effective sensors.

Electrochemical immunosensors have been the subject of a large number of reviews [409, 411–413]. By far the most prominent electrochemical immunosensor is the enzyme-linked immunosorbent assay (ELISA) [98, 99]. This class of sensor uses enzyme labels that produce or consume an electroactive substrate or cofactor which can be monitored at an electrode interface. The most common enzyme labels are horseradish peroxidase, GOx, and alkaline phosphatase. The latter opens up a tremendous potential for powerful immunosensors. Alkaline phosphatase cleaves the non-electroactive *p*-aminophenylphosphate into p-aminophenol. *p*-Aminophenol undergoes a reversible oxidation at moderate working potentials of 150 mV (vs. Ag/AgCl) [414]. This feature does allow for the amplification of the obtained signal. If an electrode system is used comprising two closely spaced electrodes such as interdigitated electrodes (IDEs) [415, 416], *p*-aminophenol can be recycled (reduction of *p*-benzoquinonimine) at the second electrode after primary detection at the first electrode (oxidation of *p*-aminophenol). The reporter molecule is cycled between the two electrodes leading to a significant amplification of the sensor signal. This detection scheme has hence not only introduced an alternative enzyme as reporter system but has inspired a whole class of amplified electrochemical sensors including but not limited to immunosensors. Analytes that have been detected using this detection scheme as reviewed in [409, 411–413] include a wide range of bacteria, viruses, tumor markers, and others.

1.4.7.2 **Nonlabeled Approaches**

Detection of biological recognition reactions between antibodies and antigens omitting labels and reporter systems typically relies on sophisticated instrumental analysis. Among others, mass spectrometry and chromatographic approaches can be used to detect immunoreactions. Among optical methods, surface plasmon resonance (SPR) spectroscopy is an interesting way to detect antibody–antigen binding events [58]. Mass-sensitive biosensors based on vibrating cantilevers [417] or the quartz crystal microbalance [418, 419] offer a straightforward way to detect antibody–antigen binding.

Instrumental electroanalysis offers some alternatives for the detection of biological recognition events in immunosensors such as capacitive immunosensors and sensors based on electrochemical impedance spectroscopy. Both techniques are universal platforms for the detection of immunoreactions and hence bacteria, viruses, tumor markers, and more, as described above, have been detected with these methods.

Capacitive biosensors [390] detect changes in the capacitance of an electrode upon the occurrence of a binding event. The capacitive structure comprises a series of components such as the electrochemical double layer including the diffuse layer from ions in solution, the grafting layer, and the biorecognition layer. Since the contribution of the biorecognition layer to the overall capacitance is typically large compared to that of the other components, changes in the biorecognition layer upon the binding of antibodies or antigens can be probed by measuring the changes in the capacitance of the biosensor. This is often accomplished by potential-step experiments that require relatively cost-effective electrochemical equipment.

Electrochemical impedance spectroscopy (EIS) [420, 421] measures the complex resistance of an electrochemical system. An electrochemical system in equilibrium is perturbed by a low-amplitude sinusoidal perturbation, typically in the range 5–10 mV around the equilibrium potential. The different components of the electrochemical system will react at different speeds resulting in a frequency-dependent shift in magnitude and phase between perturbation and response of the electrochemical system. A detailed analysis of the response of the electrochemical system and its components such as double-layer capacitance, charge-transfer resistance, or even diffusion coefficients of the molecules involved can be quantified separately. The components of an electrochemical system including a biorecognition layer are altered upon a biorecognition event and can thus be probed by EIS. The occurrence of an immune recognition reaction typically results in an increase of the charge-transfer resistance of an electrochemical system. Consequently, EIS detection schemes have been widely employed in biosensors and in immunosensors in particular, for the quantification of target analytes [422, 423].

In conclusion, electrochemical immunosensors are a useful class of biosensors that have taken advantage of some major developments during the past decades. The use of enzyme labels in ELISA-type immunosensors and simple amperometric detection schemes resulted in simple and cost-effective alternatives to fluorescence immunosensors. In particular, the use of alkaline phosphatase as enzyme label allowed for the fabrication of advanced immunosensors with signal amplification by means of redox cycling, which has been a success story of its own. This detection scheme has been used in immunosensors and other biosensors and has stimulated significant developments in electrode fabrication. Instrumental electroanalysis, namely capacitance measurements and EIS allow for label-free detection of immunoreactions.

1.5 Conclusion

The full potential of amperometric biosensors has not yet been tapped, especially with respect to the versatile and broad range of applications for which biosensors can be used. Many contributions to the field of biosensors and biofuel cells still are at the "proof-of-concept" stage. Thus, the authors hope that this chapter will promote lively and valuable discussions in order to generate new ideas and approaches towards the development and optimization of biosensor architectures.

This field of research has a true multidisciplinary, "boundary-crossing" nature which is actually one of the driving forces in science in general and a great impetus for the biosensor field in particular. From the perspective of the later application of a biosensor design, for example, physicians, biologists, or engineers are involved in order to provide the specifications for a certain analyte of interest and to define the analytical challenge. Chemists and engineers are involved in the packaging of the biosensor device and additional instrumentation for readout of the signals. With respect to the actual sensing layer on the transducer, not only analytical chemists such as electrochemists but also material chemists, organic chemists, polymer chemists, biochemists, and biologists may well work hand in hand in order to achieve the desired performance of the biosensor.

There has been more than four decades of developing amperometric biosensors. Of the many approaches towards biosensor architectures reported in the literature, a vast majority of the possibilities introduced still remain restricted to the realm of academic papers. This is mainly due to the still unsolved problem of unspecific adsorption which affects biosensor performance in many cases, degradation of the sensing layer in *in vivo* applications, limited long-term stability of biological recognition elements, etc. In addition, in complex biosensor architectures comprising a large number of different components for immobilization, for a designed ET pathway, anti-interference layers, amplification systems, etc., the extremely complex multiparameter optimization procedure still does not provide sufficient information for a rational design of sensors. Thus, many papers provide recipes leading to a concentration-dependent change in a sensor signal without a full understanding of the underlying physicochemical properties of the sensor components and the interplay of the most significant influencing parameters. Thus, especially the knowledge gained for a particular sensor design cannot be easily transferred to other biological recognition elements, other transducer materials, etc. Having this in mind, we feel that cooking-book-type mixing of many components on a transducer surface, recording of a calibration graph, and using the same sensor for the quantification of an artificial sample do not provide substantial progress in the field. Therefore, we would like to propose, on the one hand, an application-driven approach in which a real analytical task is finally accomplished taking also into account the repeatable and reproducible fabrication of the developed sensor concepts and statistical evidence for the applicability of the biosensor. On the other hand, a more in-depth understanding of the complex influence of the different parameters has to be acquired and a general understanding about the functioning of sensors at a molecular level is required to allow for a rational optimization of the sensors. Many routes are still open to novel approaches in biosensor design and to bridge the gap between academic model studies and real-world applications.

Acknowledgments

The many valuable discussions with current and former colleagues are highly appreciated. A.S. expresses his thanks to the Suranaree University of Technology for its kind financial support of the Biochemistry–Electrochemistry Research Unit through a supplementary grant. S.B. and W.S. are grateful for the financial support of the BMBF project "Das Taschentuchlabor" (03IS2201F).

Abbreviations

CV	cyclic voltammetry
DNA	deoxyribonucleic acid
DPV	differential pulse voltammetry
E	electrode potential
ET	electron transfer
FAD	flavin adenine dinucleotide
GOx	glucose oxidase
k_{cat}	turnover rate
K_M	Michaelis constant
MA	microarray
NA	nucleic acid
NAD^+	nicotinamide adenine dinucleotide
NP	nanoparticle
PQQ	pyrroloquinoline quinone
SAM	self-assembled monolayer
SCE	standard calomel electrode
SECM	scanning electrochemical microscope

Glossary

This glossary explains many of the technical terms relevant to amperometric biosensors. For additional terms not listed here, the reader is referred to, for example, the *Electrochemical Dictionary* [1].

bioelectrochemistry	a scientific discipline describing ET reactions between biological redox-active entities such as enzymes, redox-labeled proteins, living cells, etc., and suitable modified electrode surfaces.
biofuel cell	a device that converts chemical energy to electrical energy by means of biocatalysis.
biosensor	an analytical device that consists of a biological recognition element for the analyte of interest either integrated within or in close proximity to a transducer. The transducer enables the transformation of selective and quantitative information on the presence of the analyte into a quantifiable signal [2–7].
– first generation	direct oxidation or reduction of natural and freely diffusing electroactive reactants, cofactors, or products of a biological recognition reaction at an underlying electrode.
– second generation	oxidation or reduction of an artificial electroactive mediator that transfers electrons between the enzyme and the electrode surface.

– third generation	direct ET between the enzyme and the electrode surface excluding any intermediate ET reactions with redox mediators.
chemical sensor	a device that identifies and/or quantifies an analyte-of-interest as a result of a chemical reaction or interaction of the analyte with a sensor components. This chemical information is transduced into a physical signal which can be amplified.
cofactor	the terms "prosthetic group" and "coenzyme" are used for tightly bound, specific non-polypeptide units required for the biological activities of proteins. The term "cofactor" is often used synonymously with "prosthetic group" or "coenzyme." However, it is important to note that enzymes containing more than one prosthetic group are usually called multi-cofactor enzymes. Here, the term "cofactor" is exclusively used for redox-active non-polypeptide substructures in an enzyme which are tightly but not necessarily covalently bound within the protein. Free-diffusing compounds such as NAD^+/NADH are not included in this meaning of the term "cofactor." They are called "coenzymes" or "mediators."
conducting polymer	polymers or "synthetic metals" that are intrinsically conducting or semiconducting. Examples are, among many others, polypyrrole, polyaniline, and polythiophenes [8–16].
diffusion	random movement of a species under the influence of a concentration gradient.
– linear	diffusion that is found at electrodes with an electroactive surface that is large in comparison to the thickness of the diffusion layer at the electrode surface. Commonly associated with macroelectrodes.
– hemispherical	diffusion that is found at electrodes with an electroactive surface that has a similar dimension to the thickness of the diffusion layer. Commonly associated with disc-shaped microelectrodes.
electrodeposition	a process in which a film of metal, polymer, oxide, or another composite is formed on an electrode surface by electrochemically induced oxidation or reduction of a precursor reagent.
electrodeposition polymer	a polymer that can be electrodeposited. For example, protons or hydroxide ions can be electrochemically generated within the diffusion zone of an electrode. This enables a pH-induced modulation of the solubility of the electrodeposited polymer, leading to a precipitation of the polymer and the formation of a polymer film on the electrode surface [17, 18].

electron transfer (ET)	a process in which an electron is transferred from one redox center to a second. ET according to Marcus theory is mainly dependent on the distance between the electron donor and acceptor, the driving force for the overall process (i.e., the potential difference), the reorganization energy of the involved redox centers, and the intervening medium (i.e., the nature of the protein matrix) [19, 20].
– direct ET in biosensors	ET from the active center of the biological recognition element to an electrode without the involvement of a mediator or cofactor [20–27].
– mediated ET in biosensors	ET transfer from the active center of the biological recognition element to an electrode via intermediate ET processes with a redox mediator [20, 21, 24, 25, 28].
enzyme electrodes	electrodes modified with an immobilized enzyme.
flow injection analysis	an analysis technique which is based on the injection of a defined volume of a liquid sample into a continuous flowing carrier stream that at one point passes a detector.
immobilization, electrochemically induced	a process in which a composite or reagent is altered from a free-moving state to a fixed state by means of an electrochemically induced modulation of molecular properties of the compound that has to be deposited.
impedance	complex resistance of an electrochemical system.
interdigitated array	interdigitated arrays normally consist of two electrodes with finger-like structures which intertwine like a zipper structure.
interferences	substances that can electrochemically react with transducer surfaces or redox centers, which are not the substrate of the biological recognition element in a biosensor. In addition, compounds can interfere with the biological recognition reaction (e.g., inhibitors) or substrates other than the analyte of interest if the enzyme does not exhibit narrow substrate selectivity.
ion-sensitive field effect transistor (ISFET)	a semiconductor device in which the current between two electrodes (source and drain) is controlled by a third electrode (gate) which is ion-sensitive. Changes in ion activity in the surrounding electrolyte result in a change of the potential at the gate and hence in a modulation of the current between source and drain.
label-free	an assay strategy or a biosensor architecture in which no artificial reporter molecules have to be attached to one of the assay components.
Marcus theory	theory for single-electron transfer reactions [19, 29, 30].
microarray (MA)	electrochemical MAs are electrode structures that consist of ensembles of at least four or more microelectrodes that are typically arranged in an orthogonal grid. Electrodes can be either interconnected or individually addressable [31–33].

microfabrication	techniques that lead to well-defined structures and patterns in the micrometer range or below. Includes techniques such as physical or chemical vapor deposition and photolithography [34–40].
microfluidics	discipline that deals with devices in which fluids are confined and moved in micrometer-sized spaces and channels [41–46].
multi-electrode array	electrode structure that consists of four or more electrodes that are arranged in an orthogonal grid. Electrodes are typically individually addressable and miniaturized (see microarray).
multiplexing	the signals from individual sensors are sequentially read out by a single measuring device.
nanomaterials	materials in which at least one of the spatial dimensions of the material is in the sub-100 nm range.
nanosensors	a sensor that makes use of either nanomaterials or transducers with at least one spatial dimension in the sub-100 nm range.
protein conjugate	a synthetic molecule linking a protein with a partner molecule. The linked molecule can be either a small molecule such as a linker, a dye, or biotin or larger molecules such as antibodies or enzymes or other materials such as nanoparticles or quantum dots. The resulting hybrid molecules combine properties of both of the linked molecules.
protein chips	analytical devices to probe protein–protein interactions. A protein chip typically comprises an array of sites capable of carrying out an analytical task.
protein modification	chemical or genetic modification of a protein to adapt its properties to a certain task.
reagentless biosensor	a biosensor which comprises all components required for the analytical reaction securely fixed on the transducer surface. The only free-diffusing component of the sensing process is the analyte.
redox hydrogel	a hydrogel consists of a polymer that is highly dispersed in water and hence provides fast diffusion for molecules through the polymer network. In addition, a redox hydrogel is modified with redox-active groups such as redox mediators [47, 48].
redox mediator	a redox mediator shuttles electrons between two molecules or between a molecule and an electrode (homogeneous or heterogeneous charge transfer).
redox relay	see redox mediator.
selectivity	the ability to discriminate between different substances.
self-assembled monolayer	molecular monolayers spontaneously formed at interfaces (e.g., electrode–solution) by self-assembly. The monolayer is

	characterized by a high degree of orientation, molecular order, and packing [49–53].
scanning electrochemical microscope (SECM)	a tool for imaging of local electrochemical activity and topography of a sample by means of an ultramicroelectrode (tip) with the tip scanning either at constant height or constant distance. A variety of detection modes, such as feedback, generator–collector, direct, alternating current, and redox competition modes, are available [54–57].
surface plasmon resonance (SPR)	surface-sensitive optical technique in which the changes in the oscillation of surface electromagnetic waves are employed to detect changes on the surface of a metallic substrate [58].
thin-film electrode	an electrode that is modified with a thin layer of a substance that is not the basic electrode material. Includes metal films, polymer films, and self-assembled monolayers.

References

1 Bard, A.J., Inzelt, G., and Scholz, F. (2008) *Electrochemical Dictionary*, Springer-Verlag, Berlin, Heidelberg.

2 Hall, E.A.H. (1992) Overview of biosensors. *ACS Symposium Series*, **487**, 1–14.

3 Thevenot, D.R., Toth, K., Durst, R.A., and Wilson, G.S. (1999) Electrochemical biosensors: recommended definitions and classification. *Pure and Applied Chemistry*, **71** (12), 2333–2348.

4 Thevenot, D.R., Toth, K., Durst, R.A., and Wilson, G.S. (2001) Electrochemical biosensors: recommended definitions and classification. *Biosensors & Bioelectronics*, **16** (1–2), 121–131.

5 Thevenot, D.R., Toth, K., Durst, R.A., and Wilson, G.S. (2001) Electrochemical biosensors: recommended definitions and classification. *Analytical Letters*, **34** (5), 635–659.

6 Cammann, K. (1977) Biosensors based on ion-selective electrodes. *Fresenius Zeitschrift für Analytische Chemie*, **287** (1), 1–9.

7 Schuhmann, W. (2002) Amperometric enzyme biosensors based on optimised electron-transfer pathways and non-manual immobilisation procedures. *Reviews in Molecular Biotechnology*, **82**, 425–441.

8 Heeger, A.J. (2001) Semiconducting and metallic polymers: the fourth generation of polymeric materials (Nobel lecture). *Angewandte Chemie International Edition*, **40** (14), 2591–2611.

9 MacDiarmid, A.G. (2001) "Synthetic metals": a novel role for organic polymers (Nobel lecture). *Angewandte Chemie International Edition*, **40** (14), 2581–2590.

10 Shirakawa, H. (2001) The discovery of polyacetylene film: the dawning of an era of conducting polymers (Nobel lecture). *Angewandte Chemie International Edition*, **40** (14), 2575–2580.

11 Schuhmann, W. (1995) Conducting polymer based amperometric enzyme electrodes. *Microchimica Acta*, **121** (1–4), 1–29.

12 Cosnier, S. (1999) Biomolecule immobilization on electrode surfaces by entrapment or attachment to electrochemically polymerized films. A review. *Biosensors & Bioelectronics*, **14** (5), 443–456.

13 Cosnier, S. (2003) Biosensors based on electropolymerized films: new trends. *Analytical and Bioanalytical Chemistry*, **377** (3), 507–520.

14 Trojanowicz, M. (2003) Application of conducting polymers in chemical

analysis. *Microchimica Acta*, **143** (2–3), 75–91.

15 Cosnier, S. (2007) Recent advances in biological sensors based on electrogenerated polymers: a review. *Analytical Letters*, **40** (7), 1260–1279.

16 Trojanowicz, M. and Krawczyk, T.K.V. (1995) Electrochemical biosensors based on enzymes immobilized in electropolymerized films. *Microchimica Acta*, **121** (1–4), 167–181.

17 Kurzawa, C., Hengstenberg, A., and Schuhmann, W. (2002) Immobilization method for the preparation of biosensors based on pH shift-induced deposition of biomolecule-containing polymer films. *Analytical Chemistry*, **74** (2), 355–361.

18 Guschin, D.A., Shkil, H., and Schuhmann, W. (2009) Electrodeposition polymers as immobilization matrices in amperometric biosensors: improved polymer synthesis and biosensor fabrication. *Analytical and Bioanalytical Chemistry*, **395** (6), 1693–1706.

19 Marcus, R.A. (1993) Electron-transfer reactions in chemistry–theory and experiment (Nobel lecture). *Angewandte Chemie (International Edition in English)*, **32** (8), 1111–1121.

20 Borgmann, S., Hartwich, G., Schulte, A., and Schuhmann, W. (2006) Amperometric enzyme sensors based on direct and mediated electron transfer, in *Electrochemistry of Nucleic Acids and Proteins. Towards Electrochemical Sensors for Genomics and Proteomics* (eds E. Palecek, F. Scheller, and J. Wang), Elsevier, Amsterdam, pp. 599–655.

21 Ikeda, T., Miyaoka, S., Ozawa, S., Matsushita, F., Kobayashi, D., and Senda, M. (1991) Amperometric biosensors based on biocatalyst electrodes. Mediated and mediatorless enzyme electrodes. *Analytical Sciences*, **7**, 1443–1446.

22 Frew, J.E. and Hill, H.A.O. (1988) Review: direct and indirect electron transfer between electrodes and redox proteins. *European Journal of Biochemistry*, **172**, 261–269.

23 Ghindilis, A.L., Atanasov, P., and Wilkins, E. (1997) Enzyme-catalyzed direct electron transfer: fundamentals and analytical applications. *Electroanalysis*, **9** (9), 661–674.

24 Habermüller, K., Mosbach, M., and Schuhmann, W. (2000) Electron-transfer mechanisms in amperometric biosensors. *Fresenius Journal of Analytical Chemistry*, **366** (6–7), 560–568.

25 Cracknell, J.A., Vincent, K.A., and Armstrong, F.A. (2008) Enzymes as working or inspirational electrocatalysts for fuel cells and electrolysis. *Chemical Reviews*, **108** (7), 2439–2461.

26 Armstrong, F.A. and Wilson, G. (2000) Recent developments in faradaic bioelectrochemistry. *Electrochimica Acta*, **45** (15–16), 2623–2645.

27 Léger, C. and Bertrand, P. (2008) Direct electrochemistry of redox enzymes as a tool for mechanistic studies. *Chemical Reviews*, **108** (7), 2379–2438.

28 Chaubey, A. and Malhotra, B.D. (2002) Mediated biosensors. *Biosensors & Bioelectronics*, **17** (6–7), 441–456.

29 Marcus, R.A. (1956) On the theory of oxidation–reduction reactions involving electron transfer. *Journal of Chemical Physics*, **24** (5), 966–978.

30 Beratan, D.N. and Skourtis, S.S. (1998) Electron transfer mechanisms. *Current Opinion in Chemical Biology*, **2** (2), 235–243.

31 Matsue, T. (1993) Electrochemical sensors using microarray electrodes. *Trends in Analytical Chemistry: TRAC*, **12** (3), 100–108.

32 Albers, J., Grunwald, T., Nebling, E., Piechotta, G., and Hintsche, R. (2003) Electrical biochip technology–a tool for microarrays and continuous monitoring. *Analytical and Bioanalytical Chemistry*, **377**, 521–527.

33 Dill, K., Ghindilis, A., and Schwarzkopf, K. (2006) Multiplexed analyte and oligonucleotide detection on microarrays using several redox enzymes in conjunction with electrochemical detection. *Lab on a Chip*, **6** (8), 1052–1055.

34 Roy, S. (2007) Fabrication of micro- and nano-structured materials using mask-less processes. *Journal of Physics D: Applied Physics*, **40** (22), R413–R426.

35 Weibel, D.B., DiLuzio, W.R., and Whitesides, G.M. (2007) Microfabrication meets microbiology. *Nature Reviews Microbiology*, **5** (3), 209–218.

36 Bratton, D., Yang, D., Dai, J.Y., and Ober, C.K. (2006) Recent progress in high resolution lithography. *Polymers for Advanced Technologies*, **17** (2), 94–103.

37 Tseng, A.A. (2004) Recent developments in micromilling using focused ion beam technology. *Journal of Micromechanics and Microengineering*, **14** (4), R15–R34.

38 Ziaie, B., Baldi, A., Lei, M., Gu, Y.D., and Siegel, R.A. (2004) Hard and soft micromachining for BioMEMS: review of techniques and examples of applications in microfluidics and drug delivery. *Advanced Drug Delivery Reviews*, **56** (2), 145–172.

39 Suzuki, H. (2000) Advances in the microfabrication of electrochemical sensors and systems. *Electroanalysis*, **12** (9), 703–715.

40 Spearing, S.M. (2000) Materials issues in microelectromechanical systems (MEMS). *Acta Materialia*, **48** (1), 179–196.

41 Sanders, G.H.W. and Manz, A. (2000) Chip-based microsystems for genomic and proteomic analysis. *Trends in Analytical Chemistry: TRAC*, **19** (6), 364–378.

42 deMello, A.J. (2006) Control and detection of chemical reactions in microfluidic systems. *Nature*, **442** (7101), 394–402.

43 Blow, N. (2009) Microfluidics: the great divide. *Nature Methods*, **6** (9), 683–686.

44 Kricka, L.J. (1998) Miniaturization of analytical systems. *Clinical Chemistry*, **44** (9), 2008–2014.

45 Arora, A., Simone, G., Salieb-Beugelaar, G.B., Kim, J.T., and Manz, A. (2010) Latest developments in micro total analysis systems. *Analytical Chemistry*, **82** (12), 4830–4847.

46 Young, E.W.K. and Beebe, D.J. (2010) Fundamentals of microfluidic cell culture in controlled microenvironments. *Chemical Society Reviews*, **39** (3), 1036–1048.

47 Degani, Y. and Heller, A. (1988) Direct electrical communication between chemically modified enzymes and metal electrodes. 2. Methods for bonding electron-transfer relays to glucose oxidase and D-amino-acid oxidase. *Journal of the American Chemical Society*, **110** (8), 2615–2620.

48 Degani, Y. and Heller, A. (1989) Electrical communication between redox centers of glucose oxidase and electrodes via electrostatically and covalently bound redox polymers. *Journal of the American Chemical Society*, **111**, 2357–2358.

49 Nuzzo, R.G. and Allara, D.L. (1983) Adsorption of bifunctional organic disulfides on gold surfaces. *Journal of the American Chemical Society*, **105** (13), 4481–4483.

50 Finklea, H.O., Avery, S., Lynch, M., and Furtsch, T. (1987) Blocking oriented monolayers of alkyl mercaptans on gold electrodes. *Langmuir*, **3** (3), 409–413.

51 Sabatini, E., Rubinstein, I., Maoz, R., and Sagiv, J. (1987) Organized self-assembling monolayers on electrodes. 1. Octadecyl derivatives on gold. *Journal of Electroanalytical Chemistry*, **219** (1–2), 365–371.

52 Finklea, H.O. (2000) Self-assembled monolayers on electrodes, in *Encyclopedia of Analytical Chemistry* (ed. R.A. Meyers), John Wiley & Sons, Inc., New York, pp. 10090–10115.

53 Wink, T., van Zuilen, S.J., Bult, A., and Van Bennekom, W.P. (1997) Self-assembled monolayers for biosensors. *The Analyst*, **122** (4), r43–r50.

54 Mirkin, M.V. and Horrocks, B.R. (2000) Electroanalytical measurements using the scanning electrochemical microscope. *Analytica Chimica Acta*, **406** (2), 119–146.

55 Sun, P., Laforge, F.O., and Mirkin, M.V. (2007) Scanning electrochemical microscopy in the 21st century. *Physical Chemistry Chemical Physics*, **9** (7), 802–823.

56 Amemiya, S., Bard, A.J., Fan, F.R., Mirkin, M.V., and Unwin, P.R. (2008) Scanning electrochemical microscopy. *Annual Review of Analytical Chemistry*, **1**, 95–131.

57 Nakamura, H. (2010) Recent organic pollution and its biosensing methods. *Analytical Methods*, **2** (5), 430–444.

58 Scarano, S., Mascini, M., Turner, A.P.F., and Minunni, M. (2010) Surface plasmon resonance imaging for affinity-based biosensors. *Biosensors & Bioelectronics*, **25** (5), 957–966.

59 Newman, J.D., and Turner, A.P.F. (2005) Home blood glucose biosensors: a commercial perspective. *Biosensors & Bioelectronics*, **20** (12), 2435–2453.

60 Wang, J. (2001) Glucose biosensors: 40 years of advances and challenges. *Electroanalysis*, **13** (12), 983–988.

61 Wang, J. (2008) Electrochemical glucose biosensors. *Chemical Reviews*, **108** (2), 814–825.

62 Turner, A.P.F. (1997) Biosensors–realities and aspirations. *Annali di Chimica*, **87** (3–4), 255–260.

63 Wilson, R. and Turner, A.P.F. (1992) Glucose oxidase: an ideal enzyme. *Biosensors & Bioelectronics*, **7**, 165–185.

64 Wilson, G.S. and Gifford, R. (2005) Biosensors for real-time *in vivo* measurements. *Biosensors & Bioelectronics*, **20** (12), 2388–2403.

65 Heller, A. and Feldman, B. (2010) Electrochemistry in diabetes management. *Accounts of Chemical Research*, **43** (7), 963–973.

66 Raba, J. and Mottola, H.A. (1995) Glucose oxidase as an analytical reagent. *Critical Reviews in Analytical Chemistry*, **25** (1), 1–42.

67 Bankar, S.B., Bule, M.V., Singhal, R.S., and Ananthanarayan, L. (2009) Glucose oxidase–an overview. *Biotechnology Advances*, **27** (4), 489–501.

68 Clark, L.C. and Lyons, C. (1962) Electrode systems for continuous monitoring in cardiovascular surgery. *Annals of the New York Academy of Sciences*, **102** (1), 29–45.

69 Bowers, L.D. (1982) Immobilized enzymes in chemical analysis. *Trends in Analytical Chemistry: TRAC*, **1**, 191–198.

70 Barton, S.C., Gallaway, J., and Atanassov, P. (2004) Enzymatic biofuel cells for implantable and microscale devices. *Chemical Reviews*, **104** (10), 4867–4886.

71 Ruediger, O., Abad, J.M., Hatchikian, E.C., Fernandez, V.M., and de Lacey, A.L. (2005) Oriented immobilization of Desulfovibrio gigas hydrogenase onto carbon electrodes by covalent bonds for nonmediated oxidation of H-2. *Journal of the American Chemical Society*, **127** (46), 16008–16009.

72 Blanford, C.F., Heath, R.S., and Armstrong, F.A. (2007) A stable electrode for high-potential, electrocatalytic O-2 reduction based on rational attachment of a blue copper oxidase to a graphite surface. *Chemical Communications*, **17**, 1710–1712.

73 May, S.W. (1999) Applications of oxidoreductases. *Current Opinion in Biotechnology*, **10** (4), 370–375.

74 Berger, A.J., Itzkan, I., and Feld, M.S. (1997) Feasibility of measuring blood glucose concentration by near-infrared Raman spectroscopy. *Spectrochimica Acta. Part A, Molecular and Biomolecular Spectroscopy*, **53** (2), 287–292.

75 Leger, C., Jones, A.K., Albracht, S.P., and Armstrong, F.A. (2002) Effect of a dispersion of interfacial electron transfer rates on steady state catalytic electron transport in [NiFe]-hydrogenase and other enzymes. *Journal of Physical Chemistry B*, **106** (50), 13058–13063.

76 Yankovskaya, V., Horsefield, R., Tornroth, S., Luna-Chavez, C., Miyoshi, H., Leger, C., Byrne, B., Cecchini, G., and Iwata, S. (2003) Architecture of succinate dehydrogenase and reactive oxygen species generation. *Science*, **299** (5607), 700–704.

77 Volta, A. (1800) On the electricity excited by the mere contact of conducting substances of different kinds (in French). *Philosophical Transactions of the Royal Society of London*, **90**, 403–431.

78 Schoenbein, C.F. (1839) On the theory of the gaseous voltaic battery. *Philosophical Magazine*, **14**, 43–45.

79 Bossel, U. (2000) The Birth of the Fuel Cell (1835–1845). Complete Correspondence between Christian Friedrich Schoenbein and William Robert Grove, Oberrohrdorf.

80 Grove, W.R. (1839) On voltaic series and the combination of gases by platinum. *Philosophical Magazine*, **14**, 127–130.

81 Kunze, J. and Stimming, U. (2009) Electrochemical versus heat-engine energy technology: a tribute to Wilhelm Ostwald's visionary statements.

Angewandte Chemie International Edition, **48** (49), 9230–9237.
82 Nernst, W.H. (1889) Die elektromotorische Wirksamkeit der Ionen. *Zeitschrift für Physikalische Chemie, Stöchiometrie und Verwandtschaftslehre*, **IV**, 129–181.
83 Fischer, E. (1894) Einfluss der Configuration auf die Wirkung der Enzyme. *Berichte der Deutschen Chemischen Gesellschaft*, **27**, 2985–2993.
84 Michaelis, L. and Menten, M.L. (1913) Die Kinetik der Invertinwirkung. *Biochemische Zeitschrift*, **49**, 333–369.
85 Nelson, J.M. and Griffin, E.G. (1916) Adsorption of invertase. *Journal of the American Chemical Society*, **38** (5), 1109–1115.
86 Heyrovský, J. (1922) Electrolysis with a dropping mercury cathode. *Chemické Listy*, **16**, 256.
87 Heyrovský, J. (1923) Electrolysis with a dropping mercury cathode. Part I. Deposition of alkali and alkaline earth metals. *Philosophical Magazine*, **45**, 303–314.
88 Heyrovský, J. and Shikata, M. (1925) Researches with the dropping mercury cathode. Part II. *The Polarograph, Recueil des Travaux Chimiques des Pays-Bas*, **44**, 496–498.
89 Heyrovský, J. (1964) The trends of polarography, in *Nobel Lectures Chemistry 1942–1962*, Elsevier, Amsterdam, London, New York, pp. 564–584.
90 Briggs, G.E. and Haldane, J.B.S. (1925) A note on the kinetics of enzyme action. *Biochemical Journal*, **19**, 338–339.
91 Warburg, O.H. (2010) The classic: the chemical constitution of respiration ferment. *Clinical Orthopaedics and Related Research*, **468** (11), 2833–2839.
92 Bergmeyer, H.U. (1984) Enzymatic analysis of the new generation. *Fresenius Journal of Analytical Chemistry*, **319** (8), 883–889.
93 Chargaff, E., Zamenhof, S. and Green, C. (1950) Composition of human desoxypentose nucleic acid. *Nature*, **165** (4202), 756–757.
94 Watson, J.D. and Crick, F.H.C. (1953) Molecular structure of nucleic acids–a structure for deoxyribose nucleic acid. *Nature*, **171** (4356), 737–738.
95 Sanger, F. (1955) La structure de l'insuline. *Bulletin de la Societe de Chimie Biologique*, **37** (1), 23–35.
96 Clark, L.C. (1956) Monitor and control of blood and tissue oxygen tensions. *Transactions of the American Society for Artificial Internal Organs*, **2**, 41–48.
97 Kendrew, J.C. and Perutz, M.F. (1957) X-ray studies of compounds of biological interest. *Annual Review of Biochemistry*, **26** (1), 327–372.
98 Yalow, R.S. and Berson, S.A. (1960) Immunoassay of endogenous plasma insulin in man. *Journal of Clinical Investigation*, **39** (7), 1157–1175.
99 Berson, S.A. and Yalow, R.S. (2006) General principles of radioimmunoassay (reprinted from *Clinica Chimica Acta*, **22**, 51–69, 1968). *Clinica Chimica Acta*, **369** (2), 125–143.
100 Katz, S.A. and Rechnitz, G.A. (1963) Direct potentiometric determination of urea after urease hydrolysis. *Fresenius Zeitschrift für Analytische Chemie*, **196** (4), 248–251.
101 Yahiro, A.T., Lee, S.M., and Kimble, D.O. (1964) Biolectrochemistry I. Enzyme utilizing bio-fuel cell studies. *Biochimica et Biophysica Acta*, **88** (2), 375–383.
102 Updike, S.J. and Hicks, G.P. (1967) Reagentless substrate analysis with immobilized enzymes. *Science*, **158**, 270–272.
103 Updike, S.J. and Hicks, G.P. (1967) The enzyme electrode. *Nature*, **214**, 986–988.
104 Guilbault, G.G. and Montalvo, J.G. (1969) A urea-specific enzyme electrode. *Journal of the American Chemical Society*, **91** (8), 2164–2165.
105 Bergveld, P. (1970) Development of an ion-sensitive solid-state device for neurophysiological measurements. *IEEE Transactions on Biomedical Engineering*, **BM17** (1), 70–71.
106 Guilien, P., Avrameas, S., and Burtin, P. (1970) Specificity of antibodies in single cells after immunization with antigens bearing several antigenic determinants–study with a new paired staining technique. *Immunology*, **18** (4), 483–491.
107 Ternynck, T. and Avrameas, S. (1976) A new method using parabenzoquinone

for coupling antigens and antibodies to marker substances. *Annales de l'Institut Pasteur Immunologie*, **C127** (2), 197–208.

108 Betso, S.R., Klapper, M.H., and Anderson, L.B. (1972) Electrochemical studies of heme proteins. Coulometric, polarographic, and combined spectroelectrochemical methods for reduction of the heme prosthetic group in cytochrome c. *Journal of the American Chemical Society*, **94** (23), 8197–8204.

109 Racine, P., Mindt, W., and Schlaepf, P. (1973) Eelctrochemical sensor for lactate. *Journal of the Electrochemical Society*, **120** (3), C115.

110 Guilbault, G.G. and Lubrano, G.J. (1973) Enzyme electrode for amperometric determination of glucose. *Analytica Chimica Acta*, **64** (3), 439–455.

111 Karube, I., Matsunaga, T., Tsuru, S., and Suzuki, S. (1976) Continuous hydrogen production by immobilized whole cells of clostridium-butyricum. *Biochimica et Biophysica Acta*, **444** (2), 338–343.

112 Hikuma, M., Obana, H., Yasuda, T., Karube, I., and Suzuki, S. (1980) A potentiometric microbial sensor based on immobilized Escherichia coli for glutamic acid. *Analytica Chimica Acta*, **116** (1), 61–67.

113 Riedel, K., Neumann, B., and Scheller, F. (1992) Microbial sensors based on respiration measurements. *Chemie Ingenieur Technik*, **64**, 518–528.

114 Pfeiffer, E.F., Thum, C., Beischer, W., Meissner, C., Tamas, G., and Clemens, A.H. (1975) Artificial pancreas–application to experimental and clinical research. *Diabetologia*, **11** (4), 369–370.

115 Fogt, E.J., Dodd, L.M., Jenning, E.M., and Clemens, A.H. (1978) Development and evaluation of a glucose analyzer for a glucose-controlled insulin infusion system (Biostator). *Clinical Chemistry*, **24** (8), 1366–1372.

116 Nishida, K., Shimoda, S., Ichinose, K., Araki, E., and Shichiri, M. (2009) What is artificial endocrine pancreas? Mechanism and history. *World Journal of Gastroenterology*, **15** (33), 4105–4110.

117 Yeh, P. and Kuwana, T. (1977) Reversible electrode reaction of cytochrome c. *Chemistry Letters*, **6** (10), 1145–1148.

118 Eddowes, M.J. and Hill, H.A.O. (1977) Novel method for the investigation of the electrochemistry of metalloproteins: cytochrome c. *Journal of the Chemical Society. Chemical Communications*, 771–772.

119 Eddowes, M.J., Hill, H.A.O., and Uosaki, K. (1980) The electrochemistry of cytochrome c. Investigation of the mechanism of the 4,4′-bipyridyl surface modified gold electrode. *Bioelectrochemistry and Bioenergetics*, **7** (3), 527–537.

120 Tarasevich, M.R., Yaropolov, A.I., Bogdanovskaya, V.A., and Varfolomeev, S.D. (1979) Electrocatalysis of a cathodic oxygen reduction by laccase. *Bioelectrochemistry and Bioenergetics*, **6** (3), 393–403.

121 Berezin, I.V., Bogdanovskaia, V.A., Varfolomeev, S.D., Tarasevich, M.R., and Yaraoplov, A.I. (1978) Bioelectrocatalysis–equilibrium oxygen potential in presence of laccase. *Doklady Akademii Nauk SSSR*, **240** (3), 615–618.

122 Ikeda, T., Fushimi, F., Miki, K., and Senda, M. (1988) Direct bioelectrocatalysis at electrodes modified with D-gluconate dehydrogenase. *Agricultural and Biological Chemistry*, **52** (10), 2655–2658.

123 Ikeda, T., Miyaoka, S., and Miki, K. (1993) Enzyme-catalyzed electrochemical oxidation of D-gluconate at electrodes coated with D-gluconate dehydrogenase, a membrande-bound flavohemoprotein. *Journal of Electroanalytical Chemistry*, **352** (1–2), 267–278.

124 Kulys, J.J. and Svirmickas, G.J.S. (1979) Biochemical cell for the determination of lactate. *Analytica Chimica Acta*, **109** (1), 55–60.

125 Kulys, J.J. and Svirmickas, G.J.S. (1980) Reagentless lactate sensor based on cytochrome-b2. *Analytica Chimica Acta*, **117**, 115–120.

126 Huck, H. and Schmidt, H.L. (1981) Chloranil as a catalyst for the electrochemical oxidation of NADH to NAD. *Angewandte Chemie International Edition*, **20** (4), 402–403.

127 Jaegfeldt, H., Torstensson, A.B., Gorton, L.G., and Johansson, G. (1981) Catalytic-oxidation of reduced nicotonamide adenine-dinucleotide by graphite electrodes modified with adsorbed aromatics containing catechol functionalities. *Analytical Chemistry*, **53** (13), 1979–1982.

128 Shichiri, M., Kawamori, R., Yamasaki, Y., Hakui, N., and Abe, H. (1982) Wearable artificial endocrine pancreas with needle-type glucose sensor. *Lancet*, **2** (8308), 1129–1131.

129 Smith, M. (1982) Site-directed mutagenesis. *Trends in Biochemical Sciences*, **7** (12), 440–442.

130 Itakura, K., Rossi, J.J., and Wallace, R.B. (1984) Synthesis and use of synthetic oligonucleotides. *Annual Review of Biochemistry*, **53**, 323–356.

131 Carter, P. (1986) Site-directed mutagenesis. *Biochemical Journal*, **237** (1), 1–7.

132 Liedberg, B., Nylander, C., and Lundstrom, I. (1983) Surface-plasmon resonance for gas-detection and biosensing. *Sensors and Actuators*, **4** (2), 299–304.

133 Hutchinson, A.M. (1995) Evanescent-wave biosensors – real-time analysis of biomelocular interactions. *Molecular Biotechnology*, **3** (1), 47–54.

134 Schuck, P. (1997) Use of surface plasmon resonance to probe the equilibrium and dynamic aspects of interactions between biological macromolecules. *Annual Review of Biophysics and Biomolecular Structure*, **26**, 541–566.

135 Cass, A.G., Davis, G., Francis, G.D., Hill, H.A.O., Aston, W.J., Higgins, I.J., Plotkin, E.V., Scott, L.D.L., and Turner, A.P.F. (1984) Ferrocene-mediated enzyme electrode for amperometric determination of glucose. *Analytical Chemistry*, **56** (4), 667–671.

136 Forster, R.J. and Vos, J.G. (1990) Synthesis, characterization, and properties of a series of osmium-containing and ruthenium-containing metallopolymers. *Macromolecules*, **23** (20), 4372–4377.

137 Csöregi, E., Schmidtke, D.W., and Heller, A. (1995) Design and optimization of a selective subcutaneously implantable glucose electrode based on wired glucose-oxidase. *Analytical Chemistry*, **67** (7), 1240–1244.

138 Schmidtke, D.W., Freeland, A.C., Heller, A., and Bonnecaze, R.T. (1998) Measurement and modeling of the transient difference between blood and subcutaneous glucose concentrations in the rat after injection of insulin. *Proceedings of the National Academy of Sciences of the United States of America*, **95** (1), 294–299.

139 Wagner, J.G., Schmidtke, D.W., Quinn, C.P., Fleming, T.F., Bernacky, B., and Heller, A. (1998) Continuous amperometric monitoring of glucose in a brittle diabetic chimpanzee with a miniature subcutaneous electrode. *Proceedings of the National Academy of Sciences of the United States of America*, **95** (11), 6379–6382.

140 Armstrong, F.A., George, S.J., Thomson, A.J., and Yates, M.G. (1988) Direct electrochemistry in the characterization of redox proteins – novel properties of *Azobacter* 7Fe ferredoxin. *FEBS Letters*, **234** (1), 107–110.

141 Armstrong, F.A. (1990) Probing metalloproteins by voltammetry. *Bioinorganic Chemistry*, **72**, 137–230.

142 Armstrong, F.A., Heering, H.A., and Hirst, J. (1997) Reactions of complex metalloproteins studied by protein-film voltammetry. *Chemical Society Reviews*, **26** (3), 169–179.

143 Bartlett, P.N., Bradford, V.Q., and Whitaker, R.G. (1991) Enzyme electrode studies of glucose oxidase modified with a redox mediator. *Talanta*, **38** (1), 57–63.

144 Kroto, H.W., Heath, J.R., Obrien, S.C., Curl, R.F., and Smalley, R.E. (1985) C-60 – Buckminsterfullerene. *Nature*, **318** (6042), 162–163.

145 Monthioux, M. and Kuznetsov, V.L. (2006) Who should be given the credit for the discovery of carbon nanotubes? *Carbon*, **44** (9), 1621–1623.

146 Guschin, D.A., Castillo, J., Dimcheva, N., and Schuhmann, W. (2010) Redox electrodeposition polymers: adaptation of the redox potential of polymer-bound Os complexes for bioanalytical

applications. *Analytical and Bioanalytical Chemistry*, **398** (4), 1661–1673.

147 Mano, N., Mao, F., and Heller, A. (2003) Characteristics of a miniature compartment-less glucose-O-2 biofuel cell and its operation in a living plant. *Journal of the American Chemical Society*, **125** (21), 6588–6594.

148 Vincent, K.A., Cracknell, J.A., Clark, J.R., Ludwig, M., Lenz, O., Friedrich, B., and Armstrong, F.A. (2006) Electricity from low-level H2 in still air–an ultimate test for an oxygen tolerant hydrogenase. *Chemical Communications*, **48**, 5033–5035.

149 Weinstein, R.L., Bugler, J.R., Schwartz, S.L., Peyser, T.A., Brazg, R.L., and McGarraugh, G.V. (2007) Accuracy of the 5-day freestyle navigator continuous glucose monitoring system–comparison with frequent laboratory reference measurements. *Diabetes Care*, **30** (5), 1125–1130.

150 Harrison, D.J., Turner, R.B.F., and Baltes, H.P. (1988) Characterization of perfluorosulfonic acid polymer coated enzyme electrodes and a miniaturized integrated potentiostat for glucose analysis in whole blood. *Analytical Chemistry*, **60**, 2002–2007.

151 Shimizu, Y. and Morita, K. (1990) Microhole array electrode as a glucose sensor. *Analytical Chemistry*, **62** (14), 1498.

152 Kulys, J.T. and Cenas, N.K. (1983) Oxidation of glucose oxidase from *Penicillium vitale* by one- and two-electron acceptors. *Biochimica et Biophysica Acta*, **744**, 57–63.

153 Ianniello, R.M., Lindsay, T.J., and Yacynych, A.M. (1982) Immobilized xanthine-oxidase chemically modified electrode as a dual analytical sensor. *Analytical Chemistry*, **54** (12), 1980–1984.

154 Umana, M. and Waller, J. (1986) Protein-modified electrodes. The glucose oxidase/polypyrrole system. *Analytical Chemistry*, **58**, 2979–2983.

155 Foulds, N.C. and Lowe, C.R. (1988) Immobilization of glucose oxidase in ferrocene-modified pyrrole polymers. *Analytical Chemistry*, **60**, 2473–2478.

156 Kajiya, Y., Tsuda, R., and Yoneyama, H. (1991) Conferment of cholesterol sensitivity on polypyrrole films by immobilization of cholesterol oxidase and ferrocenecarboxylate ions. *Journal of Electroanalytical Chemistry*, **301** (1–2), 155–164.

157 Hill, H.A.O. and Sanghera, G.S. (1990) Mediated amperometric enzyme electrodes, in *Biosensors: A Practical Approach* (ed. A.E.G. Cass), Oxford University Press, London, pp. 19–46.

158 Bartlett, P.N., Tebbutt, P., and Whitaker, R.G. (1991) Kinetic aspects of the use of modified electrodes and mediators in bioelectrochemistry. *Progress in Reaction Kinetics*, **16** (2), 55–155.

159 Schuhmann, W., Wohlschläger, H., Lammert, R., Schmidt, H.-L., Löffler, U., Wiemhöfer, H.-D., and Göpel, W. (1990) Leaching of dimethylferrocene, a redox mediator in amperometric enzyme electrodes. *Sensors and Actuators B*, **B1**, 571–574.

160 Tang, Z.P., Louie, R.F., Lee, J.H., Lee, D.M., Miller, E.E., and Kost, G.J. (2001) Oxygen effects on glucose meter measurements with glucose dehydrogenase- and oxidase-based test strips for point-of-care testing. *Critical Care Medicine*, **29** (5), 1062–1070.

161 Louie, R.F., Tang, Z.P., Sutton, D.V., Lee, J.H., and Kost, G.J. (2000) Point-of-care glucose testing–effects of critical care variables, influence of reference instruments, and a modular glucose meter design. *Archives of Pathology & Laboratory Medicine*, **124** (2), 257–266.

162 Kost, G.J., Nguyen, T.H., and Tang, Z.P. (2000) Whole-blood glucose and lactate–trilayer biosensors, drug interference, metabolism, and practice guidelines. *Archives of Pathology & Laboratory Medicine*, **124** (8), 1128–1134.

163 Yoo, E.H., and Lee, S.Y. (2010) Glucose biosensors: an overview of use in clinical practice. *Sensors*, **10** (5), 4558–4576.

164 Kulys, J.J., Samalius, A.S., and Svirmickas, G.J.S. (1980) Electron exchange between the enzyme active center and organic metals. *FEBS Letters*, **114**, 7–10.

165 Lötzbeyer, T., Schuhmann, W., and Schmidt, H.L. (1996) Electron transfer principles in amperometric biosensors:

direct electron transfer between enzymes and electrode surface. *Sensors and Actuators B*, **33** (1–3), 50–54.

166 Contractor, A.Q., Sureshkumar, T.N., Narayanan, R., Sukeerthi, S., Lal, R., and Srinivasa, R. (1994) Conducting polymer-based biosensors. *Electrochimica Acta*, **39** (8–9), 1321–1324.

167 Schuhmann, W. (1995) Electron-transfer pathways in amperometric biosensors–ferrocene-modified enzymes entrapped in conducting-polymer layers. *Biosensors & Bioelectronics*, **10** (1–2), 181–193.

168 Khan, G.F. (1996) Organic charge transfer complex based printable biosensor. *Biosensors & Bioelectronics*, **11** (12), 1221–1227.

169 Chaubey, A., Pande, K.K., Singh, V.S., and Malhotra, B.D. (2000) Co-immobilization of lactate oxidase and lactate dehydrogenase on conducting polyaniline films. *Analytica Chimica Acta*, **407** (1–2), 97–103.

170 Gorton, L., Lindgren, A., Larsson, T., Munteanu, F.D., Ruzgas, T., and Gazaryan, I. (1999) Direct electron transfer between heme-containing enzymes and electrodes as basis for third generation biosensors. *Analytica Chimica Acta*, **400**, 91–108.

171 Ghindilis, A. (2000) Direct electron transfer catalysed by enzymes: application for biosensor development. *Biochemical Society Transactions*, **28**, 84–89.

172 Varfolomeev, S.D., Kurochkin, I.N., and Yaropolov, A.I. (1996) Direct electron transfer effect biosensors. *Biosensors & Bioelectronics*, **11** (9), 863–871.

173 Zhang, S., Wright, G., and Yang, Y. (2000) Materials and techniques for electrochemical biosensor design and construction. *Biosensors & Bioelectronics*, **15** (5–6), 273–282.

174 Shleev, S., Tkac, J., Christenson, A., Ruzgas, T., Yaropolov, A.I., Whittaker, J.W., and Gorton, L. (2005) Direct electron transfer between copper-containing proteins and electrodes. *Biosensors & Bioelectronics*, **20** (12), 2517–2554.

175 Wu, Y.H. and Hu, S.S. (2007) Biosensors based on direct electron transfer in redox proteins. *Microchimica Acta*, **159** (1–2), 1–17.

176 Armstrong, F.A. (2005) Recent developments in dynamic electrochemical studies of adsorbed enzymes and their active sites. *Current Opinion in Chemical Biology*, **9** (2), 110–117.

177 Freire, R.S., Pessoa, C.A., Mello, L.D., and Kubota, L.T. (2003) Direct electron transfer: an approach for electrochemical biosensors with higher selectivity and sensitivity. *Journal of Brazilian Chemical Society*, **14** (2), 230–243.

178 Bistolas, N., Wollenberger, U., Jung, C., and Scheller, F.W. (2005) Cytochrome P450 biosensors–a review. *Biosensors & Bioelectronics*, **20** (12), 2408–2423.

179 Gooding, J.J., Mearns, F., Yang, W.R., and Liu, J.Q. (2003) Self-assembled monolayers into the 21st century: recent advances and applications. *Electroanalysis*, **15** (2), 81–96.

180 Borgmann, S., Hartwich, G., Schulte, A., and Schuhmann, W. (2005) Amperometric enzyme sensors based on direct and mediated electron transfer, in *Electrochemistry of Nucleic Acids and Proteins*, 1st edn (eds E. Palecek, F. Scheller, and J. Wang), Elsevier, Amsterdam, Boston, pp. 599–656.

181 Polohova, V. and Snejdarkova, M. (2008) Electron transfer in amperometric biosensors. *Chemické Listy*, **102** (3), 173–182.

182 Pumera, M., Sanchez, S., Ichinose, I., and Tang, J. (2007) Electrochemical nanobiosensors. *Sensors and Actuators B*, **123** (2), 1195–1205.

183 Armstrong, F.A. (1992) Dynamic electrochemistry of iron-sulfur proteins. *Advances in Inorganic Chemistry*, **38**, 117–163.

184 Richardson, J.N., Peck, S.R., Curtin, L.S., Tender, L.M., Terril, R.H., Carter, M.T., Murray, R.W., RoweE, G.K., and Creager, S.E. (1995) Electron-transfer kinetics of self-assembled ferrocene octanethiol monolayers on gold and silver electrodes from 115K to 170K. *Journal of Physical Chemistry*, **99** (2), 766–772.

185 Gooding, J.J. and Hibbert, D.B. (1999) The application of alkanethiol self-assembled monolayers to enzyme electrodes. *Trends in Analytical Chemistry: TRAC*, **18** (8), 525–533.

186 Willner, I. and Katz, E. (2000) Integration of layered redox proteins and conductive supports for bioelectronic applications. *Angewandte Chemie International Edition*, **39** (7), 1180–1218.

187 Wittstock, G. and Schuhmann, W. (1997) Formation and imaging of microscopic enzymically active spots on an alkanethiolate-covered gold electrode by scanning electrochemical microscopy. *Analytical Chemistry*, **69** (24), 5059–5066.

188 Wilhelm, T. and Wittstock, G. (2000) Localised electrochemical desorption of gold alkanethiolate monolayers by means of scanning electrochemical microscopy (SECM). *Microchimica Acta*, **133** (1–4), 1–9.

189 Wilhelm, T. and Wittstock, G. (2001) Patterns of functional proteins formed by local electrochemical desorption of self-assembled monolayers. *Electrochimica Acta*, **47** (1–2), 275–281.

190 Katz, E. and Willner, I. (2004) Integrated nanoparticle–biomolecule hybrid systems: synthesis, properties, and applications. *Angewandte Chemie International Edition*, **43** (45), 6042–6108.

191 Schmidt, H.L. and Schuhmann, W. (1996) Reagentless oxidoreductase sensors. *Biosensors & Bioelectronics*, **11** (1–2), 127–135.

192 Murray, R.W. (1980) Chemically modified electrodes. *Accounts of Chemical Research*, **13** (5), 135–141.

193 Bard, A.J. (1983) Chemical modification of electrodes. *Journal of Chemical Education*, **60** (4), 302–304.

194 White, S.F. and Turner, A.P.F. (1997) Part 1: mediated amperometric biosensors, in *Handbook of Biosensors and Electronic Noses: Medicine, Food and Environment* (ed. E. Kress-Rogers), CRC Press, Boca Raton, FL, USA, pp. 227–244.

195 Barendrecht, E. (1990) Chemically and physically modified electrodes – some new developments. *Journal of Applied Electrochemistry*, **20** (2), 175–185.

196 Stoecker, P.W. and Yacynych, A.M. (1990) Chemically modified electrodes as biosensors. *Selective Electrode Reviews*, **12**, 137–160.

197 Tombelli, S. and Mascini, M. (2010) Aptamers biosensors for pharmaceutical compounds. *Combinatorial Chemistry & High Throughput Screening*, **13** (7), 641–649.

198 Marcus, R.A. and Sutin, N. (1985) Electron transfers in chemistry and biology. *Biochimica et Biophysica Acta – Bioenergetics*, **811** (3), 265–322.

199 Li, N., Assmann, J., Schuhmann, W., and Muhler, M. (2007) Spatially resolved characterization of catalyst-coated membranes by distance-controlled scanning mass spectrometry utilizing catalytic methanol oxidation as gas-solid probe reaction. *Analytical Chemistry*, **79** (15), 5674–5681.

200 Lau, C., Borgmann, S., Maciejewska, M., Ngounou, B., Gruendler, P., and Schuhmann, W. (2007) Improved specificity of reagentless amperometric PQQ-sGDH glucose biosensors by using indirectly heated electrodes. *Biosensors & Bioelectronics*, **22** (12), 3014–3020.

201 Bartlett, P.N., Toh, C.S., Calvo, E.J., and Flexer, V. (2008) Modelling biosensor responses, in *Bioelectrochemistry: Fundamentals, Experimental Techniques and Applications* (ed. P.N. Bartlett), John Wiley & Sons, Ltd, Chichester, pp. 267–325.

202 Phillips, P.E. and Wightman, R.M. (2003) Critical guidelines for validation of the selectivity of in-vivo chemical microsensors. *Trends in Analytical Chemistry: TRAC*, **22** (9), 509–514.

203 Vessman, J., Stefan, R.I., van Staden, J.F., Danzer, K., Lindner, W., Burns, D.T., Fajgelj, A., and Muller, H. (2001) Selectivity in analytical chemistry (IUPAC Recommendations 2001). *Pure and Applied Chemistry*, **73** (8), 1381–1386.

204 Bieniasz, L.K. (2008) Cyclic voltammetric current functions determined with a prescribed accuracy by the adaptive Huber method for Abel integral equations. *Analytical Chemistry*, **80** (24), 9659–9665.

205 Borgmann, S. (2009) Electrochemical quantification of reactive oxygen and nitrogen: challenges and opportunities.

Analytical and Bioanalytical Chemistry, **394** (1), 95–105.

206 Jia, W.Z., Wang, K., and Xia, X.H. (2010) Elimination of electrochemical interferences in glucose biosensors. *Trends in Analytical Chemistry: TRAC*, **29** (4), 306–318.

207 Svoboda, V., Cooney, M., Liaw, B.Y., Minteer, S., Piles, E., Lehnert, D., Barton, S.C., Rincon, R., and Atanassov, P. (2008) Standardized characterization of electrocatalytic electrodes. *Electroanalysis*, **20** (10), 1099–1109.

208 Zimmermann, H., Lindgren, A., Schuhmann, W., and Gorton, L. (2000) Anisotropic orientation of horseradish peroxidase by reconstitution on a thiol-modified gold electrode. *Chemistry–A European Journal*, **6** (4), 592–599.

209 Presnova, G., Grigorenko, V., Egorov, A., Ruzgas, T., Lindgren, A., Gorton, L., and Borchers, T. (2000) Direct heterogeneous electron transfer of recombinant horseradish peroxidases on gold. *Faraday Discussions*, **116** (116), 281–289.

210 Ferapontova, E.E., Grigorenko, V.G., Egorov, A., Borchers, T., Ruzgas, T., and Gorton, L. (2001) Direct electron transfer in the system gold electrode–recombinant horseradish peroxidases. *Journal of Electroanalytical Chemistry*, **509** (1), 19–26.

211 Ferapontova, E.E., Grigorenko, V.G., Egorov, A., Borchers, T., Ruzgas, T., and Gorton, L. (2001) Mediatorless biosensor for H_2O_2 based on recombinant forms of horseradish peroxidase directly adsorbed on polycrystalline gold. *Biosensors & Bioelectronics*, **16** (3), 147–157.

212 Kartashov, A.V., Serafini, G., Dong, M.D., Shipovskov, S., Gazaryan, I., Besenbacher, F., and Ferapontova, E.E. (2010) Long-range electron transfer in recombinant peroxidases anisotropically orientated on gold electrodes. *Physical Chemistry Chemical Physics*, **12** (34), 10098–10107.

213 Armstrong, F.A., Hill, H.A.O., Oliver, B.N., and Walton, N.J. (1984) Direct electrochemistry of redox proteins at pyrolytic graphite electrodes. *Journal of the American Chemical Society*, **106**, 921–923.

214 Armstrong, F.A. and Brown, K.J. (1987) Studies of protein–electrode interactions by organosilyl derivatisation of pyrolytic graphite electrodes. *Journal of Electroanalytical Chemistry*, **219**, 319–325.

215 Schuhmann, W., Zimmermann, H., Habermüller, K., and Laurinavicius, V. (2000) Electron-transfer pathways between redox enzymes and electrode surfaces: reagentless biosensors based on thiol-monolayer-bound and polypyrrole-entrapped enzymes. *Faraday Discussions*, **116**, 245–255.

216 Eddowes, M.J. and Hill, H.A.O. (1977) Novel method for investigation of electrochemistry of metalloproteins – cytochrome c. *Journal of the Chemical Society. Chemical Communications*, 771–772.

217 Yaraoplov, A.I., Malovik, V., Varfolomeev, S.D., and Berezin, I.V. (1979) Electroreduction of hydrogen peroxides on the electrode with immobilized peroxidase. *Doklady Akademii Nauk SSSR*, **249** (6), 1399–1401.

218 Mondal, M.S., Fuller, H.A., and Armstrong, F.A. (1996) Direct measurement of the reduction potential of catalytically active cytochrome c peroxidase compound. I: Voltammetric detection of a reversible, cooperative two-electron transfer reaction. *Journal of the American Chemical Society*, **118** (1), 263–264.

219 Jonsson, G. and Gorton, L. (1989) An electrochemical sensor for hydrogen-peroxide based on peroxide adsorbed on a spectrographic graphite electrode. *Electroanalysis*, **1** (5), 465–468.

220 Jia, W.Z., Schwamborn, S., Jin, C., Xia, W., Muhler, M., Schuhmann, W., and Stoica, L. (2010) Towards a high potential biocathode based on direct bioelectrochemistry between horseradish peroxidase and hierarchically structured carbon nanotubes. *Physical Chemistry Chemical Physics*, **12** (34), 10088–10092.

221 Ikeda, T., Kobayashi, D., Matsushita, F., Sagara, T., and Niki, K. (1993) Bioelectrocatalysis at electrodes coated with alcohol-dehydrogenase, a

quinohemoprotein with heme-c serving as a built-in mediator. *Journal of Electroanalytical Chemistry*, **361** (1–2), 221–228.

222 Ikeda, T., Matsushita, F., and Senda, M. (1991) Amperometric fructose sensor based on direct bioelectrocatalysis. *Biosensors & Bioelectronics*, **6**, 299–304.

223 Lindgren, A., Larsson, T., Ruzgas, T., and Gorton, L. (2000) Direct electron transfer between the heme of cellobiose dehydrogenase and thiol modified gold electrodes. *Journal of Electroanalytical Chemistry*, **494** (2), 105–113.

224 Stoica, L., Ruzgas, T., Ludwig, R., Haltrich, D., and Gorton, L. (2006) Direct electron transfer – a favorite electron route for cellobiose dehydrogenase (CDH) from *Trametes villosa*. Comparison with CDH from *Phanerochaete chrysosporium*. *Langmuir*, **22** (25), 10801–10806.

225 Ludwig, R., Harreither, W., Tasca, F., and Gorton, L. (2010) Cellobiose dehydrogenase: a versatile catalyst for electrochemical applications. *ChemPhysChem*, **11** (13), 2674–2697.

226 Schuhmann, W., Zimmermann, H., Habermüller, K., and Laurinavicius, V. (2000) Electron-transfer pathways between redox enzymes and electrode surfaces: reagentless biosensors based on thiol-monolayer-bound and polypyrrole-entrapped enzymes. *Faraday Discussions*, **116**, 245–255.

227 Heller, A. (2004) Miniature biofuel cells. *Physical Chemistry Chemical Physics*, **6** (2), 209–216.

228 Frew, J.E., Foulds, N.C., Wilshere, J.M., Forrow, N.J., and Green, M.J. (1989) Measurement of alkaline phosphatase activity by electrochemical detection of phosphate esters. Application to amperometric enzyme immunoassay. *Journal of Electroanalytical Chemistry*, **266**, 309–316.

229 Johnson, D.L., Thompson, J.L., Brinkmann, S.M., Schuller, K.A., and Martin, L.L. (2003) Electrochemical characterization of purified Rhus vernicifera laccase: voltammetric evidence for a sequential four-electron transfer. *Biochemistry*, **42** (34), 10229–10237.

230 Blanford, C.F., Foster, C.E., Heath, R.S., and Armstrong, F.A. (2008) Efficient electrocatalytic oxygen reduction by the "blue" copper oxidase, laccase, directly attached to chemically modified carbons. *Faraday Discussions*, **140**, 319–335.

231 Kang, C., Shin, H., and Heller, A. (2006) On the stability of the "wired" bilirubin oxidase oxygen cathode in serum. *Bioelectrochemistry*, **68** (1), 22–26.

232 Pizzariello, A., Stredansky, M., and Miertus, S. (2002) A glucose/hydrogen peroxide biofuel cell that uses oxidase and peroxidase as catalysts by composite bulk-modified bioelectrodes based on a solid binding matrix. *Bioelectrochemistry*, **56** (1–2), 99–105.

233 Mondal, M.S., Goodin, D.B., and Armstrong, F.A. (1998) Simultaneous voltammetric comparisons of reduction potentials, reactivities, and stabilities of the high-potential catalytic states of wild-type and distal-pocket mutant (W51F) yeast cytochrome c peroxidase. *Journal of the American Chemical Society*, **120** (25), 6270–6276.

234 Ramanavicius, A., Kausaite, A., and Ramanaviciene, A. (2005) Biofuel cell based on direct bioelectrocatalysis. *Biosensors & Bioelectronics*, **20** (10), 1962–1967.

235 Vincent, K.A., Cracknell, J.A., Parkin, A., and Armstrong, F.A. (2005) Hydrogen cycling by enzymes: electrocatalysis and implications for future energy technology. *Dalton Transactions*, 3397–3403.

236 Coman, V., Ludwig, R., Harreither, W., Haltrich, D., Gorton, L., Ruzgas, T., and Shleev, S. (2010) A direct electron transfer-based glucose/oxygen biofuel cell operating in human serum. *Fuel Cells*, **10** (1), 9–16.

237 Lindgren, A., Stoica, L., Ruzgas, T., Ciucu, A., and Gorton, L. (1999) Development of a cellobiose dehydrogenase modified electrode for amperometric detection of diphenols. *The Analyst*, **124** (4), 527–532.

238 Stoica, L., Ludwig, R., Haltrich, D., and Gorton, L. (2006) Third-generation biosensor for lactose based on newly discovered cellobiose dehydrogenase. *Analytical Chemistry*, **78** (2), 393–398.

239 Vincent, K.A., Li, X., Blanford, C.F., Belsey, N.A., Weiner, J.H., and Armstrong, F.A. (2007) Enzymatic catalysis on conducting graphite particles. *Nature Chemical Biology*, **3** (12), 760–761.

240 Reisner, E., Powell, D.J., Cavazza, C., Fontecilla-Camps, J.C., and Armstrong, F.A. (2009) Visible light-driven H_2 production by hydrogenases attached to dye-sensitized TiO_2 nanoparticles. *Journal of the American Chemical Society*, **131** (51), 18457–18466.

241 Hecht, H.J., Schomburg, D., Kalisz, H., and Schmid, R.D. (1993) The 3D structure of glucose oxidase from *Aspergillus niger*. Implications for the use of GOD as a biosensor enzyme. *Biosensors & Bioelectronics*, **8** (3–4), 197–203.

242 Li, L.M., Xu, S.J., Du, Z.F., Gao, Y.F., Li, J.H., and Wang, T.H. (2010) Electrografted poly(N-mercaptoethyl acrylamide) and Au nanoparticles-based organic/inorganic film: a platform for the high-performance electrochemical biosensors. *Chemistry–An Asian Journal*, **5** (4), 919–924.

243 Shan, C.S., Yang, H.F., Han, D.X., Zhang, Q.X., Ivaska, A., and Niu, L. (2010) Electrochemical determination of NADH and ethanol based on ionic liquid-functionalized graphene. *Biosensors & Bioelectronics*, **25** (6), 1504–1508.

244 Chen, S.H., Yuan, R., Chai, Y.Q., Yin, B., Li, W.J., and Min, L.G. (2009) Amperometric hydrogen peroxide biosensor based on the immobilization of horseradish peroxidase on core-shell organosilica@chitosan nanospheres and multiwall carbon nanotubes composite. *Electrochimica Acta*, **54** (11), 3039–3046.

245 Li, F.H., Song, J.X., Li, F., Wang, X.D., Zhang, Q.X., Han, D.X., Ivaska, A., and Niu, L. (2009) Direct electrochemistry of glucose oxidase and biosensing for glucose based on carbon nanotubes@ SnO_2–Au composite. *Biosensors & Bioelectronics*, **25** (4), 883–888.

246 Chen, X.H., Guo, H., Yi, J., and Hu, J.Q. (2009) Fabrication of AucoreCo_3O_4shell/PAA/HRP composite film for direct electrochemistry and hydrogen peroxide sensor applications. *Sensors & Materials*, **21** (8), 433–444.

247 Fu, C.L., Yang, W.S., Chen, X., and Evans, D.G. (2009) Direct electrochemistry of glucose oxidase on a graphite nanosheet–Nafion composite film modified electrode. *Electrochemistry Communications*, **11** (5), 997–1000.

248 Gao, R.F. and Zheng, J.B. (2009) Amine-terminated ionic liquid functionalized carbon nanotube–gold nanoparticles for investigating the direct electron transfer of glucose oxidase. *Electrochemistry Communications*, **11** (3), 608–611.

249 Horozova, E., Dodevska, T., and Dimcheva, N. (2009) Modified graphites: application to the development of enzyme-based amperometric biosensors. *Bioelectrochemistry*, **74** (2), 260–264.

250 Zhang, Y., Zhang, H.F., and Zheng, J.B. (2008) Direct electrochemistry of horseradish peroxidase modified electrode in ionic liquid [EMIM]BF_4. *Acta Chimica Sinica*, **66** (19), 2124–2130.

251 Sarma, A.K., Vatsyayan, P., Goswami, P., and Minteer, S.D. (2009) Recent advances in material science for developing enzyme electrodes. *Biosensors & Bioelectronics*, **24** (8), 2313–2322.

252 Liu, Y., Geng, T.M., and Gao, J. (2008) Layer-by-layer immobilization of horseradish peroxidase on a gold electrode modified with colloidal gold nanoparticles. *Microchimica Acta*, **161** (1–2), 241–248.

253 Tang, L.H., Zhu, Y.H., Xu, L.H., Yang, X.L., and Li, C.Z. (2007) Amperometric glutamate biosensor based on self-assembling glutamate dehydrogenase and dendrimer-encapsulated platinum nanoparticles onto carbon nanotubes. *Talanta*, **73**, 438–443.

254 Kang, X.H., Wang, J., Wu, H., Aksay, I.A., Liu, J., and Lin, Y.H. (2009) Glucose oxidase-graphene-chitosan modified electrode for direct electrochemistry and glucose sensing. *Biosensors & Bioelectronics*, **25** (4), 901–905.

255 Ansari, A.A., Kaushik, A., Solanki, P.R., and Malhotra, B.D. (2009) Electrochemical cholesterol sensor based

on tin oxide–chitosan nanobiocomposite film. *Electroanalysis*, **21** (8), 965–972.

256 Branzoi, V., Pilan, L., and Branzoi, F. (2009) Amperometric glucose biosensor based on electropolymerized carbon nanotube/polypyrrole composite film. *Revue Roumaine de Chimie*, **54** (10), 783–789.

257 Liu, S.Q., Lin, B.P., Yang, X.D., and Zhang, Q.Q. (2007) Carbon-nanotube-enhanced direct electron-transfer reactivity of hemoglobin immobilized on polyurethane elastomer film. *Journal of Physical Chemistry B*, **111** (5), 1182–1188.

258 Lin, J.H., He, C.Y., Zhang, L.J., and Zhang, S.S. (2009) Sensitive amperometric immunosensor for alpha-fetoprotein based on carbon nanotube/gold nanoparticle doped chitosan film. *Analytical Biochemistry*, **384** (1), 130–135.

259 Wan, J., Bi, J.L., Du, P., and Zhang, S.S. (2009) Biosensor based on the biocatalysis of microperoxidase-11 in nanocomposite material of multiwalled carbon nanotubes/room temperature ionic liquid for amperometric determination of hydrogen peroxide. *Analytical Biochemistry*, **386** (2), 256–261.

260 You, C.P., Li, X., Zhang, S., Kong, J.L., Zhao, D.Y., and Liu, B.H. (2009) Electrochemistry and biosensing of glucose oxidase immobilized on Pt-dispersed mesoporous carbon. *Microchimica Acta*, **167** (1–2), 109–116.

261 Rahman, M.A., Noh, H.B., and Shim, Y.B. (2008) Direct electrochemistry of laccase immobilized on au nanoparticles encapsulated-dendrimer bonded conducting polymer: application for a catechin sensor. *Analytical Chemistry*, **80** (21), 8020–8027.

262 Wang, Z.Y., Liu, S.N., Wu, P., and Cai, C.X. (2009) Detection of glucose based on direct electron transfer reaction of glucose oxidase immobilized on highly ordered polyaniline nanotubes. *Analytical Chemistry*, **81** (4), 1638–1645.

263 Mackey, D., Killard, A.J., and Ambrosi, A. (2007) Optimizing the ratio of horseradish peroxidase and glucose oxidase on a bienzyme electrode: comparison of a theoretical and experimental approach. *Sensors and Actuators B*, **122** (2), 395–402.

264 Withey, G.D., Kim, J.H., and Xu, J. (2007) Wiring efficiency of a metallizable DNA linker for site-addressable nanobioelectronic assembly. *Nanotechnology*, **18** (42), 424025.

265 Xiao, Y., Patolsky, F., Katz, E., Hainfeld, J.F., and Willner, I. (2003) "Plugging into enzymes": nanowiring of redox enzymes by a gold nanoparticle. *Science*, **299** (5614), 1877–1881.

266 Demin, S. and Hall, E.A. (2009) Breaking the barrier to fast electron transfer. *Bioelectrochemistry*, **76** (1–2), 19–27.

267 Courjean, O., Gao, F., and Mano, N. (2009) Deglycosylation of glucose oxidase for direct and efficient glucose electrooxidation on a glassy carbon electrode. *Angewandte Chemie International Edition*, **48** (32), 5897–5899.

268 Prevoteau, A., Courjean, O., and Mano, N. (2010) Deglycosylation of glucose oxidase to improve biosensors and biofuel cells. *Electrochemistry Communications*, **12** (2), 213–215.

269 Courjean, O., Flexer, V., Prevoteau, A., Suraniti, E., and Mano, N. (2010) Effect of degree of glycosylation on charge of glucose oxidase and redox hydrogel catalytic efficiency. *ChemPhysChem*, **11** (13), 2795–2797.

270 Bartlett, P.N. and Cooper, J.M. (1993) A review of the immobilization of enzymes in electropolymerized films. *Journal of Electroanalytical Chemistry*, **362** (1–2), 1–12.

271 Emr, S.A. and Yacynych, A.M. (1995) Use of polymer films in amperometric biosensors. *Electroanalysis*, **7** (10), 913–923.

272 Heller, A. (2006) Electron-conducting redox hydrogels: design, characteristics and synthesis. *Current Opinion in Chemical Biology*, **10** (6), 664–672.

273 Gerard, M., Chaubey, A., and Malhotra, B.D. (2002) Application of conducting polymers to biosensors. *Biosensors & Bioelectronics*, **17** (5), 345–359.

274 Mano, N., Soukharev, V., and Heller, A. (2006) A laccase-wiring redox hydrogel for efficient catalysis of O_2

electroreduction. *Journal of Physical Chemistry B*, **110** (23), 11180–11187.

275 Gregg, B.A. and Heller, A. (1990) Crosslinked redox gels containing glucose oxidase for amperometric biosensor applications. *Analytical Chemistry*, **62**, 258–263.

276 Leech, D., Forster, R.J., and Smyth and Vos, J.G. (1991) Effect of composition of polymer backbone on spectroscopic and electrochemical properties of ruthenium(II)bis(2,2′-bipyridyl)-containing 4-vinylpyridine styrene copolymers. *Journal of Materials Chemistry*, **1** (4), 629–635.

277 Doherty, A.P., Forster, R.J., Smyth, M.R. and Vos, J.G. (1991) Development of a sensor for the detection of nitrite using a glassy-carbon electrode modified with the electrocatalyst [Os(bipy)2(PVP)10Cl] Cl. *Analytica Chimica Acta*, **255** (7), 45–52.

278 Calvo, E.J., Etchenique, R., Danilowicz, C., and Diaz, L. (1996) Electrical communication between electrodes and enzymes mediated by redox hydrogels. *Analytical Chemistry*, **68** (23), 4186–4193.

279 Mao, F., Mano, N., and Heller, A. (2003) Long tethers binding redox centers to polymer backbones enhance electron transport in enzyme "wiring" hydrogels. *Journal of the American Chemical Society*, **125** (16), 4951–4957.

280 Taylor, C., Kenausis, G., Katakis, I., and Heller, A. (1995) Wiring of glucose oxidase within a hydrogel made with polyvinyl imidazole complexed with [(Os-4,4′-dimethoxy-2,2′-bipyridine) Cl]$^{(+/2+)}$. *Journal of Electroanalytical Chemistry*, **396** (1–2), 511–515.

281 Larson, N., Ruzgas, T., Gorton, L., Kokaia, M., Kissinger, P., and Csöregi, E. (1998) Design and development of an amperometric biosensor for acetylcholine determination in brain microdialysates. *Electrochimica Acta*, **43** (23), 3541–3554.

282 Kashiwagi, Y., Pan, Q.H., Yanagisawa, Y., Sibayama, N., and Osa, T. (1994) The effects of chain-length of ferrocene moiety on electrical communication of mediators-modified and enzyme-modified electrodes. *Denki Kagaku*, **62** (12), 1240–1246.

283 Danilowicz, C., Corton, E., Battaglini, F., and Calvo, E.J. (1998) An Os(byp)(2) ClPyCH(2)NHpoly(allylamine) hydrogel mediator for enzyme wiring at electrodes. *Electrochimica Acta*, **43** (23), 3525–3531.

284 Aoki, A. and Heller, A. (1993) Electron-diffusion coefficients in hydrogels formed of cross-linked redox polymers. *Journal of Physical Chemistry*, **97** (42), 11014–11019.

285 Aoki, A., Rajagopalan, R., and Heller, A. (1995) Effect of quaternization on electron diffusion coefficients for redox hydrogels based on poly(4-vinylpyridine). *Journal of Physical Chemistry*, **99** (14), 5102–5110.

286 Csöregi, E., Quinn, C.P., Schmidtke, D.W., Lindquist, S.E., Pishko, M.V., Ye, L., Katakis, I., Hubbell, J.A., and Heller, A. (1994) Design, characterization and one point *in vivo* calibration of a subcutaneously implanted glucose electrode. *Analytical Chemistry*, **66** (19), 3131–3138.

287 Yang, Q.L., Atanasov, P., and Wilkins, E. (1997) A novel amperometric transducer design for needle-type implantable biosensor applications. *Electroanalysis*, **9** (16), 1252–1256.

288 Abel, P.U. and von Woedtke, T. (2002) Biosensors for *in vivo* glucose measurement: can we cross the experimental stage. *Biosensors & Bioelectronics*, **17** (11–12), 1059–1070.

289 D'Orazio, P. (2003) Biosensors in clinical chemistry. *Clinica Chimica Acta*, **334** (1–2), 41–69.

290 Ngounou, B., Aliyev, E.H., Guschin, D.A., Sultanov, Y.M., Efendiev, A.A., and Schuhmann, W. (2007) Parallel synthesis of libraries of anodic and cathodic functionalized electrodeposition paints as immobilization matrix for amperometric biosensors. *Bioelectrochemistry*, **71** (1), 81–90.

291 Ngounou, B., Guschin, D.A., Castillo, J., and Schuhmann, W. (2009) Combinatorial polymer synthesis as a tool in biosensor and biofuel cell development and optimization. *ECS Transactions*, **19** (6), 119–128.

292 Schuhmann, W., Kranz, C., Wohlschlager, H., and Strohmeier, J.

(1997) Pulse technique for the electrochemical deposition of polymer films on electrode surfaces. *Biosensors & Bioelectronics*, **12** (12), 1157–1167.

293 Ryabova, V., Schulte, A., Wolfgang, T.E., and Schuhmann, W. (2005) Robotic sequential analysis of a library of metalloporphyrins as electrocatalysts for voltammetric nitric oxide sensors. *The Analyst*, **130** (9), 1245–1252.

294 Schuhmann, W., Kranz, C., Huber, J., and Wohlschlager, H. (1993) Conducting polymer-based amperometric enzyme electrodes: towards the development of miniaturized reagentless biosensors. *Synthetic Metals*, **61** (1–2), 31–35.

295 Reiter, S., Habermüller, K., and Schuhmann, W. (2001) A reagentless glucose biosensor based on glucose oxidase entrapped into osmium-complex modified polypyrrole films. *Sensors and Actuators B*, **79** (2–3), 150–156.

296 Habermüller, K., Reiter, S., Buck, H., Meier, T., Staepels, J., and Schuhmann, W. (2003) Conducting redox polymer-based reagentless biosensors using modified PQQ-dependent glucose dehydrogenase. *Microchimica Acta*, **143** (2–3), 113–121.

297 Devries, E.F.A., Schasfoort, R.B.M., Vanderplas, J., and Greve, J. (1994) Nucleic acid detection with surface plasmon resonance using cationic latex. *Biosensors & Bioelectronics*, **9** (7), 509–514.

298 Cosnier, S. (1997) Electropolymerization of amphiphilic monomers for designing amperometric biosensors. *Electroanalysis*, **9** (12), 894–902.

299 Gerlache, M., Kauffmann, J.M., Quarin, G., Vire, J.C., Bryant, G.A., and Talbot, J.M. (1996) Electrochemical analysis of surfactants–an overview. *Talanta*, **43** (4), 507–519.

300 Barton, S.C., Sun, Y.H., Chandra, B., White, S., and Hone, J. (2007) Mediated enzyme electrodes with combined micro- and nanoscale supports. *Electrochemical and Solid-State Letters*, **10** (5), B96–B100.

301 Wheeldon, I.R., Gallaway, J.W., Barton, S.C., and Banta, S. (2008) Bioelectrocatalytic hydrogels from electron-conducting metallopolypeptides coassembled with bifunctional enzymatic building blocks. *Proceedings of the National Academy of Sciences of the United States of America*, **105** (40), 15275–15280.

302 Hudak, N.S., Gallaway, J.W., and Barton, S.C. (2009) Formation of mediated biocatalytic cathodes by electrodeposition of a redox polymer and laccase. *Journal of Electroanalytical Chemistry*, **629** (1–2), 57–62.

303 Bullen, R.A., Arnot, T.C., Lakeman, J.B., and Walsh, F.C. (2006) Biofuel cells and their development. *Biosensors & Bioelectronics*, **21** (11), 2015–2045.

304 Pandey, P., Datta, M., and Malhotra, B.D. (2008) Prospects of nanomaterials in biosensors. *Analytical Letters*, **41** (2), 159–209.

305 Zhang, X.Q., Guo, Q., and Cui, D.X. (2009) Recent advances in nanotechnology applied to biosensors. *Sensors*, **9** (2), 1033–1053.

306 Urban, G.A. (2009) Micro- and nanobiosensors: state of the art and trends. *Measurement Science and Technology*, **20** (1), 12001.

307 Arrigan, D.W.M. (2004) Nanoelectrodes, nanoelectrode arrays and their applications. *The Analyst*, **129** (12), 1157–1165.

308 Frasco, M.F. and Chaniotakis, N. (2009) Semiconductor quantum dots in chemical sensors and biosensors. *Sensors*, **9** (9), 7266–7286.

309 Gooding, J.J. (2005) Nanostructuring electrodes with carbon nanotubes: a review on electrochemistry and applications for sensing. *Electrochimica Acta*, **50** (15), 3049–3060.

310 Rivas, G.A., Rubianes, M.D., Rodriguez, M.C., Ferreyra, N.E., Luque, G.L., Pedano, M.L., Miscoria, S.A., and Parrado, C. (2007) Carbon nanotubes for electrochemical biosensing. *Talanta*, **74** (3), 291–307.

311 Qureshi, A., Kang, W.P., Davidson, J.L., and Gurbuz, Y. (2009) Review on carbon-derived, solid-state, micro and nano sensors for electrochemical sensing applications. *Diamond and Related Materials*, **18** (12), 1401–1420.

312 Iijima, S. (1991) Helical microtubules of graphitic carbon. *Nature*, **354** (6348), 56–58.

313 Ajayan, P.M. (1999) Nanotubes from carbon. *Chemical Reviews*, **99** (7), 1787–1799.

314 Che, G.L., Lakshmi, B.B., Fisher, E.R., and Martin, C.R. (1998) Carbon nanotubule membranes for electrochemical energy storage and production. *Nature*, **393** (6683), 346–349.

315 Agui, L., Yanez-Sedeno, P., and Pingarron, J.M. (2008) Role of carbon nanotubes in electroanalytical chemistry – a review. *Analytica Chimica Acta*, **622** (1–2), 11–47.

316 Britto, P.J., Santhanam, K.S., Rubio, A., Alonso, J.A., and Ajayan, P.M. (1999) Improved charge transfer at carbon nanotube electrodes. *Advanced Materials*, **11** (2), 154–157.

317 Musameh, M., Wang, J., Merkoci, A., and Lin, Y.H. (2002) Low-potential stable NADH detection at carbon-nanotube-modified glassy carbon electrodes. *Electrochemistry Communications*, **4** (10), 743–746.

318 Wang, J. and Musameh, M. (2003) Carbon nanotube/teflon composite electrochemical sensors and biosensors. *Analytical Chemistry*, **75** (9), 2075–2079.

319 Nguyen, C.V., Delzeit, L., Cassell, A.M., Li, J., Han, J., and Meyyappan, M. (2002) Preparation of nucleic acid functionalized carbon nanotube arrays. *Nano Letters*, **2** (10), 1079–1081.

320 Ahammad, A.J., Lee, J.J., and Rahman, M.A. (2009) Electrochemical sensors based on carbon nanotubes. *Sensors*, **9** (4), 2289–2319.

321 Eijkel, J.C. and van den Berg, A. (2005) Nanofluidics: what is it and what can we expect from it? *Microfluidics and Nanofluidics*, **1** (3), 249–267.

322 Abgrall, P. and Nguyen, N.T. (2008) Nanofluidic devices and their applications. *Analytical Chemistry*, **80** (7), 2326–2341.

323 Compton, R.G., Wildgoose, G.G., Rees, N.V., Streeter, I., and Baron, R. (2008) Design, fabrication, characterisation and application of nanoelectrode arrays. *Chemical Physics Letters*, **459** (1–6), 1–17.

324 Magliulo, M., Michelini, E., Simoni, P., Guardigli, M., and Roda, A. (2006) Ultrasensitive and rapid nanodevices for analytical immunoaassays *Analytical and Bioanalytical Chemistry*, **384** (1), 27–30.

325 Stulik, K., Amatore, C., Holub, K., Marecek, V., and Kutner, W. (2000) Microelectrodes. Definitions, characterization, and applications. *Pure and Applied Chemistry*, **72** (8), 1483–1492.

326 Finot, E., Bourillot, E., Meunier-prest, R., Lacroute, Y., Legay, G., Cherkaoui-malki, M., Latruffe, N., Siri, O., Braunstein, P., and Dereux, A. (2003) Performance of interdigitated nanoelectrodes for electrochemical DNA biosensor. *Ultramicroscopy*, **97** (1–4), 441–449.

327 Bard, A.J., Fan, F.R., Kwak, J., and Lev, O. (1989) Scanning electrochemical microscopy – introduction and principles. *Analytical Chemistry*, **61** (2), 132–138.

328 Wittstock, G., Burchardt, M., Pust, S.E., Shen, Y., and Zhao, C. (2007) Scanning electrochemical microscopy for direct imaging of reaction rates. *Angewandte Chemie International Edition*, **46** (10), 1584–1617.

329 Turyan, I., Matsue, T., and Mandler, D. (2000) Patterning and characterization of surfaces with organic and biological molecules by the scanning electrochemical microscope. *Analytical Chemistry*, **72** (15), 3431–3435.

330 Kueng, A., Kranz, C., Lugstein, A., Bertagnolli, E., and Mizaikoff, B. (2003) Integrated AFM-SECM in tapping mode: simultaneous topographical and electrochemical imaging of enzyme activity. *Angewandte Chemie International Edition*, **42**, 3238–3240.

331 Medintz, I.L., Uyeda, H.T., Goldman, E.R., and Mattoussi, H. (2005) Quantum dot bioconjugates for imaging, labelling and sensing. *Nature Materials*, **4** (6), 435–446.

332 Costa-Fernandez, J.M., Pereiro, R., and Sanz-Medel, A. (2006) The use of luminescent quantum dots for optical sensing. *Trends in Analytical Chemistry: TRAC*, **25** (3), 207–218.

333 Willner, I., Basnar, B., and Willner, B. (2007) Nanoparticle–enzyme hybrid

systems for nanobiotechnology. *FEBS Journal*, **274** (2), 302–309.

334 Guo, S.J. and Dong, S.J. (2009) Biomolecule–nanoparticle hybrids for electrochemical biosensors. *Trends in Analytical Chemistry: TRAC*, **28** (1), 96–109.

335 Gilardi, G. and Fantuzzi, A. (2001) Manipulating redox systems: application to nanotechnology. *Trends in Biotechnology*, **19** (11), 468–476.

336 Wiesner, M.R., Lowry, G.V., Jones, K.L., Hochella, M.F., Di Giulio, R.T., Casman, E., and Bernhardt, E.S. (2009) Decreasing uncertainties in assessing environmental exposure, risk, and ecological implications of nanomaterials. *Environmental Science & Technology*, **43** (17), 6458–6462.

337 Shvedova, A.A. and Kagan, V.E. (2010) The role of nanotoxicology in realizing the "helping without harm" paradigm of nanomedicine: lessons from studies of pulmonary effects of single-walled carbon nanotubes. *Journal of Internal Medicine*, **267** (1), 106–118.

338 Vaddiraju, S., Tomazos, I., Burgess, D.J., Jain, F.C., and Papadimitrakopoulos, F. (2010) Emerging synergy between nanotechnology and implantable biosensors: a review. *Biosensors & Bioelectronics*, **25** (7), 1553–1565.

339 Ronkainen, N.J., Halsall, H.B., and Heineman, W.R. (2010) Electrochemical biosensors. *Chemical Society Reviews*, **39** (5), 1747–1763.

340 Skyler, J.S. (2009) Continuous glucose monitoring: an overview of its development. *Diabetes Technology & Therapeutics*, **11**, S5–S10.

341 Heller, A. and Feldman, B. (2008) Electrochemical glucose sensors and their applications in diabetes management. *Chemical Reviews*, **108** (7), 2482–2505.

342 Wilson, G.S. and Johnson, M.A. (2008) In-vivo electrochemistry: what can we learn about living systems? *Chemical Reviews*, **108** (7), 2462–2481.

343 Lee, T.M. (2008) Over-the-counter biosensors: past, present, and future. *Sensors*, **8** (9), 5535–5559.

344 Wang, J. (2008) *In vivo* glucose monitoring: towards "sense and act" feedback-loop individualized medical systems. *Talanta*, **75** (3), 636–641.

345 Qin, S., van der Zeyden, M., Oldenziel, W.H., Cremers, T., and Westerink, B.H. (2008) Microsensors for *in vivo* measurement of glutamate in brain tissue. *Sensors*, **8** (11), 6860–6884.

346 Koschwanez, H.E. and Reichert, W.M. (2007) *In vitro, in vivo* and post explantation testing of glucose-detecting biosensors: current methods and recommendations. *Biomaterials*, **28** (25), 3687–3703.

347 Wilson, G.S. and Ammam, M. (2007) *In vivo* biosensors. *FEBS Journal*, **274** (21), 5452–5461.

348 Pickup, J.C., Hussain, F., Evans, N.D., and Sachedina, N. (2005) *In vivo* glucose monitoring: the clinical reality and the promise. *Biosensors & Bioelectronics*, **20** (10), 1897–1902.

349 Wilson, G.S. and Gifford, R. (2005) Biosensors for real-time *in vivo* measurements. *Biosensors & Bioelectronics*, **20** (12), 2388–2403.

350 Wilson, G.S. and Hu, Y. (2000) Enzyme based biosensors for *in vivo* measurements. *Chemical Reviews*, **100** (7), 2693–2704.

351 O'Neill, R.D., Lowry, J.P., and Mas, M. (1998) Monitoring brain chemistry *in vivo*: voltammetric techniques, sensors, and behavioral applications. *Critical Reviews in Neurobiology*, **12** (1–2), 69–127.

352 Llaudet, E., Hatz, S., Droniou, M., and Dale, N. (2005) Microelectrode biosensor for real-time measurement of ATP in biological tissue. *Analytical Chemistry*, **77** (10), 3267–3273.

353 Schuvailo, O.N., Dzyadevych, S.V., El'skaya, A., Gautier-Sauvigne, S., Csoregi, E., Cespuglio, R., and Soldatkin, A.P. (2005) Carbon fibre-based microbiosensors for *in vivo* measurements of acetylcholine and choline. *Biosensors & Bioelectronics*, **21** (1), 87–94.

354 Zain, Z.M., O'Neill, R.D., Lowry, J.P., Pierce, K.W., Tricklebank, M., Dewa, A., and Ab Ghani, S. (2010) Development of an implantable D-serine biosensor for *in* *vivo* monitoring using mammalian D-amino acid oxidase on a poly (*o*-phenylenediamine) and Nafion-modified platinum-iridium disk electrode. *Biosensors & Bioelectronics*, **25** (6), 1454–1459.

355 Kita, J.M. and Wightman, R.M. (2008) Microelectrodes for studying neurobiology. *Current Opinion in Chemical Biology*, **12** (5), 491–496.

356 Robinson, D.L., Hermans, A., Seipel, A.T., and Wightman, R.M. (2008) Monitoring rapid chemical communication in the brain. *Chemical Reviews*, **108** (7), 2554–2584.

357 Njagi, J., Chernov, M.M., Leiter, J.C., and Andreescu, S. (2010) Amperometric detection of dopamine *in vivo* with an enzyme based carbon fiber microbiosensor. *Analytical Chemistry*, **82** (3), 989–996.

358 Onuki, Y., Bhardwaj, U., Papadimitrakopoulos, F., and Burgess, D.J. (2008) A review of the biocompatibility of implantable devices: current challenges to overcome foreign body response. *Journal of Diabetes Science and Technology*, **2** (6), 1003–1015.

359 Dungel, P., Long, N., Yu, B., Moussy, Y., and Moussy, F. (2008) Study of the effects of tissue reactions on the function of implanted glucose sensors. *Journal of Biomedical Materials Research Part A*, **85A** (3), 699–706.

360 Gifford, R., Kehoe, J.J., Barnes, S.L., Kornilayev, B.A., Alterman, M.A., and Wilson, G.S. (2006) Protein interactions with subcutaneously implanted biosensors. *Biomaterials*, **27** (12), 2587–2598.

361 Khan, A.S. and Michael, A.C. (2003) Invasive consequences of using micro-electrodes and microdialysis probes in the brain. *Trends in Analytical Chemistry: TRAC*, **22** (9), 503–508.

362 Wisniewski, N., Moussy, F., and Reichert, W. (2000) Characterization of implantable biosensor membrane biofouling. *Fresenius Journal of Analytical Chemistry*, **366** (6–7), 611–621.

363 Gant, R.M., Abraham, A.A., Hou, Y., Cummins, B.M., Grunlan, M.A., and Cote, G.L. (2010) Design of a self-cleaning thermoresponsive nanocomposite hydrogel membrane for implantable biosensors. *Acta Biomaterialia*, **6** (8), 2903–2910.

364 Justin, G., Finley, S., Rahman, A.R., and Guiseppi-Elie, A. (2009) Biomimetic hydrogels for biosensor implant biocompatibility: electrochemical characterization using micro-disc electrode arrays (MDEAs). *Biomedical Microdevices*, **11** (1), 103–115.

365 Mitala, J.J. and Michael, A.C. (2006) Improving the performance of electrochemical microsensors based on enzymes entrapped in a redox hydrogel. *Analytica Chimica Acta*, **556** (2), 326–332.

366 Oldenziel, W.H., Dijkstra, G., Cremers, T., and Westerink, B.H. (2006) Evaluation of hydrogel-coated glutamate microsensors. *Analytical Chemistry*, **78** (10), 3366–3378.

367 Oldenziel, W.H. and Westerink, B.H. (2005) Improving glutamate microsensors by optimizing the composition of the redox hydrogel. *Analytical Chemistry*, **77** (17), 5520–5528.

368 Shin, J.H. and Schoenfisch, M.H. (2006) Improving the biocompatibility of *in vivo* sensors via nitric oxide release. *The Analyst*, **131** (5), 609–615.

369 Park, J. and McShane, M.J. (2010) Dual-function nanofilm coatings with diffusion control and protein resistance. *ACS Applied Materials & Interfaces*, **2** (4), 991–997.

370 Brault, N.D., Gao, C.L., Xue, H., Piliarik, M., Homola, J., Jiang, S.Y., and Yu, Q.M. (2010) Ultra-low fouling and functionalizable zwitterionic coatings grafted onto SiO_2 via a biomimetic adhesive group for sensing and detection in complex media. *Biosensors & Bioelectronics*, **25** (10), 2276–2282.

371 Ju, Y.M., Yu, B.Z., West, L., Moussy, Y., and Moussy, F. (2010) A novel porous collagen scaffold around an implantable biosensor for improving biocompatibility. II. Long-term *in vitro/in vivo* sensitivity characteristics of sensors with NDGA- or GA-crosslinked collagen scaffolds. *Journal of Biomedical Materials Research Part A*, **92A** (2), 650–658.

372 Koschwanez, H.E., Yap, F.Y., Klitzman, B., and Reichert, W.M. (2008) *In vitro* and *in vivo* characterization of porous poly-L-lactic acid coatings for subcutaneously implanted glucose sensors. *Journal of Biomedical Materials ResearchPart A*, **87A** (3), 792–807.

373 Geelhood, S.J., Horbett, T.A., Ward, W.K., Wood, M.D., and Quinn, M.J. (2007) Passivating protein coatings for implantable glucose sensors: evaluation of protein retention. *Journal of Biomedical Materials Research Part B*, **81B** (1), 251–260.

374 Nagel, B., Gajovic-Eichelmann, N., Scheller, F.W., and Katterle, M. (2010) Ionic topochemical tuned biosensor interface. *Langmuir*, **26** (11), 9088–9093.

375 Narayan, R.J., Jin, C.M., Menegazzo, N., Mizaikoff, B., Gerhardt, R.A., Andara, M., Agarwal, A., Shih, C.C., Shih, C.M., Lin, S.J., and Su, Y.Y. (2007) Nanoporous hard carbon membranes for medical applications. *Journal of Nanoscience and Nanotechnology*, **7** (4–5), 1486–1493.

376 Adiga, S.P., Jin, C.M., Curtiss, L.A., Monteiro-Riviere, N.A., and Narayan, R.J. (2009) Nanoporous membranes for medical and biological applications. *Wiley Interdisciplinary Reviews: Nanomedicine and Nanobiotechnology*, **1** (5), 568–581.

377 Renard, E., Costalat, G., Chevassus, H., and Bringer, J. (2006) Artificial beta-cell: clinical experience toward an implantable closed-loop insulin delivery system. *Diabetes & Metabolism*, **32** (5), 497–502.

378 Gough, D.A. and Bremer, T. (2000) Immobilized glucose oxidase in implantable glucose sensor technology. *Diabetes Technology & Therapeutics*, **2** (3), 377–380.

379 Heller, A. (2005) Integrated medical feedback systems for drug delivery. *AIChE Journal*, **51** (4), 1054–1066.

380 Gough, D.A., Kumosa, L.S., Routh, T.L., Lin, J.T., and Lucisano, J.Y. (2010) Function of an implanted tissue glucose sensor for more than 1 year in animals. *Science Translational Medicine*, **2** (42), 42ra53.

381 Schulte, A. and Schuhmann, W. (2007) Single-cell microelectrochemistry. *Angewandte Chemie International Edition*, **46** (46), 8760–8777.

382 Fei, J.J., Wu, K.B., Wang, F., and Hu, S.S. (2005) Glucose nanosensors based on redox polymer/glucose oxidase modified carbon fiber nanoelectrodes. *Talanta*, **65** (4), 918–924.

383 Gallaway, J., Wheeldon, I., Rincon, R., Atanassov, P., Banta, S., and Barton, S.C. (2008) Oxygen-reducing enzyme cathodes produced from SLAC, a small laccase from *Streptomyces coelicolor*. *Biosensors & Bioelectronics*, **23** (8), 1229–1235.

384 Kothandaraman, R., Deng, W.H., Sorkin, M., Kaufman, A., Gibbard, H.F., and Barton, S.C. (2008) Methanol anode modified by semipermeable membrane for mixed-feed direct methanol fuel cells. *Journal of the Electrochemical Society*, **155** (9), B865–B868.

385 Barton, S.C. (2005) Oxygen transport in composite biocathodes. *Proton Conducting Membrane Fuel Cells III, Proceedings*, **2002** (31), 324–335.

386 Hudak, N.S. and Barton, S.C. (2005) Mediated biocatalytic cathode for direct methanol membrane–electrode assemblies. *Journal of the Electrochemical Society*, **152** (5), A876–A881.

387 Tombelli, S. and Mascini, M. (2009) Aptamers as molecular tools for bioanalytical methods. *Current Opinion in Molecular Therapeutics*, **11** (2), 179–188.

388 Daniels, J.S. and Pourmand, N. (2007) Label-free impedance biosensors: opportunities and challenges. *Electroanalysis*, **19** (12), 1239–1257.

389 K'owino, I.O. and Sadik, O.A. (2005) Impedance spectroscopy: a powerful tool for rapid biomolecular screening and cell culture monitoring. *Electroanalysis*, **17** (23), 2101–2113.

390 Berggren, C., Bjarnason, B., and Johansson, G. (2001) Capacitive biosensors. *Electroanalysis*, **13** (3), 173–180.

391 Chandra, B., Kace, J.T., Sun, Y., Barton, S.C., and Hone, J. (2007) Growth of carbon nanotubes on carbon Toray paper for bio-fuel cell applications.

Proceedings of the 2nd Energy Nanotechnology International Conference 2007, pp. 69–71.

392 Sun, Y.H. and Barton, S.C. (2006) Methanol tolerance of a mediated, biocatalytic oxygen cathode. *Journal of Electroanalytical Chemistry*, **590** (1), 57–65.

393 Barton, S.C. and Hudak, N. (2004) Mediated gas diffusion biocathodes. *Abstracts of Papers of the American Chemical Society*, **228**, U484–U484.

394 Wang, J., Nielsen, P.E., Jiang, M., Cai, X.H., Fernandes, J.R., Grant, D.H., Ozsoz, M., Beglieter, A., and Mowat, M. (1997) Mismatch sensitive hybridization detection by peptide nucleic acids immobilized on a quartz crystal microbalance. *Analytical Chemistry*, **69** (24), 5200–5202.

395 Zhang, G.J., Zhang, L., Huang, M.J., Luo, Z.H., Tay, G.K., Lim, E.J., Kang, T.G., and Chen, Y. (2010) Silicon nanowire biosensor for highly sensitive and rapid detection of Dengue virus. *Sensors and Actuators B*, **146** (1), 138–144.

396 Barton, S.C. and West, A.C. (2001) Electrohydrodynamic impedance in the presence of nonuniform transport properties. *Journal of the Electrochemical Society*, **148** (4), A381–A387.

397 Wink, T., vanZuilen, S.J., Bult, A., and Van Bennekom, W.P. (1997) Self-assembled monolayers for biosensors. *The Analyst*, **122** (4), R43–R50.

398 Hudak, N.S., and Barton, S.C. (2005) Direct methanol fuel cells with mediated biocatalytic cathodes. *Abstracts of Papers of the American Chemical Society*, **230**, 139.

399 Wheeldon, I.R., Barton, S.C., and Banta, S. (2007) Bioactive proteinaceous hydrogels from designed bifunctional building blocks. *Biomacromolecules*, **8**, 2990–2994.

400 Barton, S.C. and West, A.C. (2001) Electrodissolution of zinc at the limiting current. *Journal of the Electrochemical Society*, **148** (5), A490–A495.

401 Barton, S.C., Patterson, T., Wang, E., Fuller, T.F., and West, A.C. (2001) Mixed-reactant, strip-cell direct methanol fuel cells. *Journal of Power Sources*, **96** (2), 329–336.

402 Rant, U., Pringsheim, E., Kaiser, W., Arinaga, K., Knezevic, J., Tornow, M., Fujita, S., Yokoyama, N., and Abstreiter, G. (2009) Detection and size analysis of proteins with switchable DNA layers. *Nano Letters*, **9** (4), 1290–1295.

403 Rant, U., Arinaga, K., Scherer, S., Pringsheim, E., Fujita, S., Yokoyama, N., Tornow, M., and Abstreiter, G. (2007) Switchable DNA interfaces for the highly sensitive detection of label-free DNA targets. *Proceedings of the National Academy of Sciences of the United States of America*, **104**, 17364–17369.

404 Rant, U., Arinaga, K., Fujita, S., Yokoyama, N., Abstreiter, G., and Tornow, M. (2004) Dynamic electrical switching of DNA layers on a metal surface. *Nano Letters*, **4** (12), 2441–2445.

405 Venter, J.C., Adams, M.D., Myers, E.W., Li, P.W., Mural, R.J., Sutton, G.G., Smith, H.O., Yandell, M., Evans, C.A., Holt, R.A., Gocayne, J.D., Amanatides, P., Ballew, R.M., Huson, D.H., Wortman, J.R., Zhang, Q., Kodira, C.D., Zheng, X.Q., Chen, L., Skupski, M., Subramanian, G., Thomas, P.D., Zhang, J.H., Miklos, G.L., Nelson, C., Broder, S., Clark, A.G., Nadeau, C., McKusick, V.A., Zinder, N., Levine, A.J., Roberts, R.J., Simon, M., Slayman, C., Hunkapiller, M., Bolanos, R., Delcher, A., Dew, I., Fasulo, D., Flanigan, M., Florea, L., Halpern, A., Hannenhalli, S., Kravitz, S., Levy, S., Mobarry, C., Reinert, K., Remington, K., Abu-Threideh, J., Beasley, E., Biddick, K., Bonazzi, V., Brandon, R., Cargill, M., Chandramouliswaran, I., Charlab, R., Chaturvedi, K., Deng, Z.M., Di Francesco, V., Dunn, P., Eilbeck, K., Evangelista, C., Gabrielian, A.E., Gan, W., Ge, W.M., Gong, F.C., Gu, Z.P., Guan, P., Heiman, T.J., Higgins, M.E., Ji, R.R., Ke, Z.X., Ketchum, K.A., Lai, Z.W., Lei, Y.D., Li, Z.Y., Li, J.Y., Liang, Y., Lin, X.Y., Lu, F., Merkulov, G.V., Milshina, N., Moore, H.M., Naik, A.K., Narayan, V.A., Neelam, B., Nusskern, D., Rusch, D.B., Salzberg, S., Shao, W., Shue, B.X., Sun, J.T., Wang, Z.Y., Wang, A.H., Wang, X., Wang, J., Wei, M.H., Wides, R., Xiao, C.L., Yan, C.H.,

Yao, A., Ye, J., Zhan, M., Zhang, W.Q., Zhang, H.Y., Zhao, Q., Zheng, L.S., Zhong, F., Zhong, W.Y., Zhu, S.P., Zhao, S.Y., Gilbert, D., Baumhueter, S., Spier, G., Carter, C., Cravchik, A., Woodage, T., Ali, F., An, H.J., Awe, A., Baldwin, D., Baden, H., Barnstead, M., Barrow, I., Beeson, K., Busam, D., Carver, A., Center, A., Cheng, M.L., Curry, L., Danaher, S., Davenport, L., Desilets, R., Dietz, S., Dodson, K., Doup, L., Ferriera, S., Garg, N., Gluecksmann, A., Hart, B., Haynes, J., Haynes, C., Heiner, C., Hladun, S., Hostin, D., Houck, J., Howland, T., Ibegwam, C., Johnson, J., Kalush, F., Kline, L., Koduru, S., Love, A., Mann, F., May, D., McCawley, S., McIntosh, T., McMullen, I., Moy, M., Moy, L., Murphy, B., Nelson, K., Pfannkoch, C., Pratts, E., Puri, V., Qureshi, H., Reardon, M., Rodriguez, R., Rogers, Y.H., Romblad, D., Ruhfel, B., Scott, R., Sitter, C., Smallwood, M., Stewart, E., Strong, R., Suh, E., Thomas, R., Tint, N.N., Tse, S., Vech, C., Wang, G., Wetter, J., Williams, S., Williams, M., Windsor, S., Winn-Deen, E., Wolfe, K., Zaveri, J., Zaveri, K., Abril, J.F., Guigo, R., Campbell, M.J., Sjolander, K.V., Karlak, B., Kejariwal, A., Mi, H.Y., Lazareva, B., Hatton, T., Narechania, A., Diemer, K., Muruganujan, A., Guo, N., Sato, S., Bafna, V., Istrail, S., Lippert, R., Schwartz, R., Walenz, B., Yooseph, S., Allen, D., Basu, A., Baxendale, J., Blick, L., Caminha, M., Carnes-Stine, J., Caulk, P., Chiang, Y.H., Coyne, M., Dahlke, C., Mays, A.D., Dombroski, M., Donnelly, M., Ely, D., Esparham, S., Fosler, C., Gire, H., Glanowski, S., Glasser, K., Glodek, A., Gorokhov, M., Graham, K., Gropman, B., Harris, M., Heil, J., Henderson, S., Hoover, J., Jennings, D., Jordan, C., Jordan, J., Kasha, J., Kagan, L., Kraft, C., Levitsky, A., Lewis, M., Liu, X.J., Lopez, J., Ma, D., Majoros, W., McDaniel, J., Murphy, S., Newman, M., Nguyen, T., Nguyen, N., Nodell, M., Pan, S., Peck, J., Peterson, M., Rowe, W., Sanders, R., Scott, J., Simpson, M., Smith, T., Sprague, A., Stockwell, T., Turner, R., Venter, E., Wang, M., Wen, M.Y., Wu, D., Wu, M., Xia, A., Zandieh, A., and Zhu, X.H. (2001) The sequence of the human genome. *Science*, **291** (5507), 1304–1351.

406 Pohanka, M. (2009) Monoclonal and polyclonal antibodies production – preparation of potent biorecognition element. *Journal of Applied Biomedicine*, **7** (3), 115–121.

407 Leca-Bouvier, B. and Blum, L.J. (2005) Biosensors for protein detection: a review. *Analytical Letters*, **38** (10), 1491–1517.

408 Seydack, M. (2005) Nanoparticle labels in immunosensing using optical detection methods. *Biosensors & Bioelectronics*, **20** (12), 2454–2469.

409 Marquette, C.A. and Blum, L.J. (2006) State of the art and recent advances in immunoanalytical systems. *Biosensors & Bioelectronics*, **21** (8), 1424–1433.

410 Hempen, C. and Karst, U. (2006) Labeling strategies for bioassays. *Analytical and Bioanalytical Chemistry*, **384** (3), 572–583.

411 Bange, A., Halsall, H.B., and Heineman, W.R. (2005) Microfluidic immunosensor systems. *Biosensors & Bioelectronics*, **20** (12), 2488–2503.

412 Diaz-gonzalez, M., Gonzalez-garcia, M.B., and Costa-garcia, A. (2005) Recent advances in electrochemical enzyme immunoassays. *Electroanalysis*, **17** (21), 1901–1918.

413 Privett, B.J., Shin, J.H., and Schoenfisch, M.H. (2010) Electrochemical sensors. *Analytical Chemistry*, **82** (12), 4723–4741.

414 Tang, H.T., Lunte, C.E., Halsall, H.B., and Heineman, W.R. (1988) *p*-Aminophenyl phosphate: an improved substrate for electrochemical enzyme immunoassay. *Analytica Chimica Acta*, **214**, 187–195.

415 Hintsche, R., Paeschke, M., Wollenberger, U., Schnakenberg, U., Wagner, B., and Lisec, T. (1994) Microelectrode arrays and application to biosensing devices. *Biosensors & Bioelectronics*, **9** (9–10), 697–705.

416 Morita, M., Niwa, O., and Horiuchi, T. (1997) Interdigitated array microelectrodes as electrochemical sensors. *Electrochimica Acta*, **42** (20–22), 3177–3183.

417 Ziegler, C. (2004) Cantilever-based biosensors. *Analytical and Bioanalytical Chemistry*, **379**, 946–959.

418 O'Sullivan, C.K. and Guilbault, G.G. (1999) Commercial quartz crystal microbalances – theory and applications. *Biosensors & Bioelectronics*, **14** (8–9), 663–670.

419 Cooper, M.A. and Singleton, V.T. (2007) A survey of the 2001 to 2005 quartz crystal microbalance biosensor literature: applications of acoustic physics to the analysis of biomolecular interactions. *Journal of Molecular Recognition*, **20**, 154–184.

420 Pejcic, B. and de Marco, R. (2006) Impedance spectroscopy: over 35 years of electrochemical sensor optimization. *Electrochimica Acta*, **51** (28), 6217–6229.

421 Park, S.M. and Yoo, J.S. (2003) Electrochemical impedance spectroscopy for better electrochemical measurements. *Analytical Chemistry*, **75** (21), 455A–461A.

422 Lillie, G., Payne, P., and Vadgama, P. (2001) Electrochemical impedance spectroscopy as a platform for reagentless bioaffinity sensing. *Sensors and Actuators B*, **78** (1–3), 249–256.

423 Lisdat, F. and Schafer, D. (2008) The use of electrochemical impedance spectroscopy for biosensing. *Analytical and Bioanalytical Chemistry*, **391** (5), 1555–1567.

2
Imaging of Single Biomolecules by Scanning Tunneling Microscopy

Jingdong Zhang, Qijin Chi, Palle Skovhus Jensen, and Jens Ulstrup

2.1
Introduction

Interfacial electrochemistry of biological molecules such as redox metalloproteins and their constituent amino acid building blocks, DNA components, biomimetic lipid membranes, and bioinorganic "hybrids" of metallic nanoparticles and metalloproteins is moving towards new levels of understanding. Key notions are structural imaging of biomolecules at the electrochemical interface, electrochemical adsorption, electron transfer, and even enzyme processes followed and controlled towards the resolution of the single molecule. Scanning tunneling microscopy (STM) and electrical conductivity of biological (macro)molecules directly in action in aqueous biological media under full electrochemical potential control (*in situ* STM) have been powerful tools in this exciting development. Atomic force microscopy (AFM) of biomolecules in aqueous biological environments has opened other routes to single-(macro)molecular biological function, particularly protein unfolding and DNA unzipping.

The remarkable evolution of physical electrochemistry from the late 1970s, almost a renaissance of the electrochemical sciences [1], prompted by novel close interaction between electrochemistry and both solid-state physics and surface science is well recognized. The introduction of single-crystal, atomically planar electrode surfaces [2–6] sometimes with quite simple preparation procedures [4–6] was a major breakthrough. This also laid the foundation for other crucial electrochemical technologies, not the least being the scanning probe microscopies. At the same time a range of other surface techniques and theories was introduced. These included spectroscopy (UV/visible [7, 8], infrared [9, 10], Raman [11, 12], and X-ray photoelectron spectroscopy [13]), quartz crystal microbalance [14], and other physical techniques. To these were added statistical mechanical [15, 16] and electronic structural theories and computations [17–19], warranted by the new electrochemistry. Only slightly later STM [20, 21] and AFM [22, 23] signaled a new lift of surface science and interfacial electrochemistry to an unprecedented level of structural resolution [24]. *Atomic* resolution of pure metal and semiconductor electrode surfaces and molecular or *sub-molecular* resolution of electrochemical

Advances in Electrochemical Science and Engineering. Edited by Richard C. Alkire, Dieter M. Kolb, and Jacek Lipkowski

ISBN: 978-3-527-32885-7

adsorbates could now be achieved. This opened a new world of microscopic structures and processes and a whole new understanding of electrochemical nanotechnology.

In addition to the new physical electrochemistry, imaging of single biomolecules has come to rest on the development of new biotechnology by use of mutant proteins, DNA-base variability, and *de novo* synthetic metalloproteins such as the 4α-helix heme proteins (synthetic biology). Combination of the new physical electrochemistry with new biotechnology has led interfacial bioelectrochemistry of redox metalloproteins and DNA-based molecules to a level where similar boundary-traversing efforts as in physical electrochemistry are being seen. This has been an essential prerequisite for structural mapping of the bioelectrochemical solid–liquid interface to the level of the *single* molecule.

Imaging of single biomolecules by *in situ* STM and mapping of the electrochemical interface have rested on pioneering efforts by Nichols and coworkers [25–27], Kolb [24], Wandlowski [28], Tao and coworkers [29], Itaya and coworkers [30, 31], and others. These efforts have been paralleled by other pioneering efforts in the electrochemistry of redox metalloproteins by Yeh and Kuwana [32], Niki *et al.* [33], Eddowes and Hill [34, 35], and others. Essential prerequisites in new protein film voltammetry and in data interpretation relating to single-biomolecule mapping are an underlying fundamental understanding of molecular electron transfer (ET) (and proton transfer) processes in homogeneous and interfacial environments. Molecular charge transport theory has continued to develop over the last few decades and more recently as a basis for *in situ* STM and other condensed matter single-molecule conductivity phenomena. Single-molecule mapping by (*in situ*) STM rests on molecular tunneling conductivity, rather than topographic shape. Theoretical support is therefore needed from the very beginning in order to translate, for example, STM contrasts into molecular structure and single-molecule function. The observation that molecules as large and as fragile as redox metalloproteins on atomically planar electrode surfaces can be mapped to single-molecule resolution in their natural functional state by a physical phenomenon as subtle as *the quantum mechanical tunneling effect* is in fact remarkable.

Theoretical support for single-molecule conductivity has assumed at least a twofold appearance. Electronic structure computations have disclosed, sometimes unexpected, details in the STM contrasts of "small" molecules such as single amino acids or DNA bases. Solvation may be of minor importance for nonpolar molecules such as straight and branched alkanethiols [36, 37]. On the other hand, static and dynamic solvation effects are crucial and computationally much more demanding for *in situ* STM of electrostatically charged molecules, for example functionalized alkanethiol self-assembled monolayers (SAMs), but can be accommodated within evolving computational schemes [38]. Other new challenges arise for redox molecules, not to say biological (macro)molecules such as redox metalloproteins. Molecular interfacial ET theory [39–41] has here emerged as a rigorous, powerful tool at first in a parameterized form due to the complexity of the systems but all parameters are now being addressed by large-scale computational schemes [42–52]. By their analytical form, the "phenomenological" theories have offered immediate insight into current/overpotential and other central *in situ* tunneling spectroscopic correlations and also revealed new phenomena specific to *in situ* STM [45, 51–53]. These include coherent multi-electron tunneling through single (biological) macromolecules, the role of the metal centers in single-molecule conductivity of transition metal complexes and metalloproteins, features specific to limited-size molecular assemblies (ultimately a single molecule), etc. [54–56].

The combination of protein and DNA biotechnology with well-defined (single-crystal) electrochemical interfaces offers other perspectives towards biosensing and bioelectrochemical communication for electrical signal transfer between target molecules and external electrochemical circuitry. Strategic surface preparation, functional linker molecules, and, last but not least, imaging of metalloproteins and metalloenzymes directly in electron transporting or enzyme action are all part of this.

2.2
Interfacial Electron Transfer in Molecular and Protein Film Voltammetry

Imaging of single biomolecules by (*in situ*) STM rests on interfacial ET phenomena and must be supported by notions from this broad class of processes. As a quantum mechanical tunneling phenomenon, *in situ* STM image interpretation and tunneling current/overpotential and bias voltage spectroscopy must also be based on theoretical support. Two kinds of electrochemical support are important. First, electrochemical determination of molecular monolayer coverage offers important "boundary conditions" for STM image contrast interpretation. The sharp reductive desorption signal of pure and functionalized alkanethiol SAMs is a key example. Second, protein film voltammetry (PFV) is crucial to characterize and control conditions for optimal immobilized protein function. The protein molecules must, thus, be immobilized firmly enough to stay on the surface but gently enough that the molecules retain their ET, enzyme, or other function, to be carried over to *in situ* STM.

Condensed matter molecular charge transfer theory, now developed to high sophistication [39–61], offers comprehensive support both for evolving bioelectrochemistry of redox proteins and DNA-based molecules and for *in situ* STM. The theory addresses, in these contexts, *two* fundamental features. One is the aspect of electron tunneling between the electrode and donor or acceptor groups in the molecules through intermediate protein or other molecular "matter." The other one is the nuclear environmental effects from local modes, collective protein and DNA nuclear dynamics, and external solvent. The strong interaction between the transferring electrons and the environment is thus crucial. Both need attention to be paid to the inhomogeneous, anisotropic environment of the electrode–solution interfacial region.

The high spatial *in situ* STM resolution of biological macromolecules offers new theoretical challenges. Even novel charge transfer phenomena have been revealed as noted. We therefore first overview a few conceptual notions of the fundamental electrochemical ET process, with emphasis on ET phenomena in nanogap electrode systems and *in situ* STM.

2.2.1
Theoretical Notions of Interfacial Chemical and Bioelectrochemical Electron Transfer

Views of the electrochemical ET process as an electronic transition composed of contributions from all the electronic levels of the enclosing electrodes carry over to (bio)electrochemical nanoscale systems. Notions in focus are illuminated by the following cathodic current density form (with an analogous form for the anodic process) [39–41]:

$$j(\eta)=\int d\varepsilon\, f(\varepsilon)\rho(\varepsilon)\, j(\varepsilon;\eta); \quad j(\varepsilon;\eta)=e\Gamma_{\mathrm{ox}}^{(1-\alpha)}\Gamma_{\mathrm{red}}^{\alpha}W(\varepsilon;\eta) \tag{2.1}$$

where $W(\varepsilon;\eta)$ is the rate constant, $j(\varepsilon;\eta)$ is the (infinitesimal) current density from a given electronic energy level in the metal electrode ε at the overpotential η, $f(\varepsilon)$ is the Fermi function, and $\rho(\varepsilon)$ is the electronic level density. Γ_{ox} and Γ_{red} are the populations of the oxidized and reduced state, respectively, of the redox (bio)molecule close to the electrode surface, e is the electronic charge, and α is the electrochemical transfer coefficient given by

$$\alpha=-k_{\mathrm{B}}T\frac{\mathrm{d}\ln j(\eta)}{\mathrm{d}(e\eta)} \tag{2.2}$$

$W(\varepsilon;\eta)$ holds all the information about electron tunneling, overpotential, and environmental nuclear reorganization. The following equations apply broadly [39–41, 60]:

$$\left.\begin{aligned} &j=j(\eta)=e\Gamma_{\mathrm{ox}}^{(1-\alpha)}\Gamma_{\mathrm{red}}^{\alpha}k^{\mathrm{o/r}}\\ &k^{\mathrm{o/r}}\approx\kappa_{\mathrm{eff}}\frac{\omega_{\mathrm{eff}}}{2\pi}\exp\left[-\frac{(E_{\mathrm{R}}+e\eta)^2}{4E_{\mathrm{R}}k_{\mathrm{B}}T}\right]\\ &\kappa_{\mathrm{eff}}=4\pi\kappa_{\mathrm{el}}(\varepsilon_{\mathrm{F}};\eta)\rho(\varepsilon_{\mathrm{F}})k_{\mathrm{B}}T\quad\text{if }\kappa_{\mathrm{eff}}\ll 1,\quad\text{otherwise }\kappa_{\mathrm{eff}}\rightarrow 1\end{aligned}\right\} \tag{2.3}$$

where E_{R} is the nuclear reorganization free energy, ω_{eff} the effective nuclear vibrational frequency of all the nuclear modes reorganized, ε_{F} the Fermi energy of the electrode, k_{B} Boltzmann's constant, and T the temperature. The parameter $\kappa_{\mathrm{el}}(\varepsilon_{\mathrm{F}};\eta)$ is the microscopic electronic transmission coefficient, the most important part of which is the electron exchange energy, $T_{\varepsilon\mathrm{A}}(\varepsilon_{\mathrm{F}};\eta)$, which couples the molecular acceptor level (A) with the metallic electronic level ε_{F}.

This reduction does not apply immediately to *in situ* STM of redox molecules. Equations (2.1)–(2.3), however, represent broadly interfacial (bio)electrochemical ET processes. The transparency of Eqs. (2.1)–(2.4) prompts some observations:

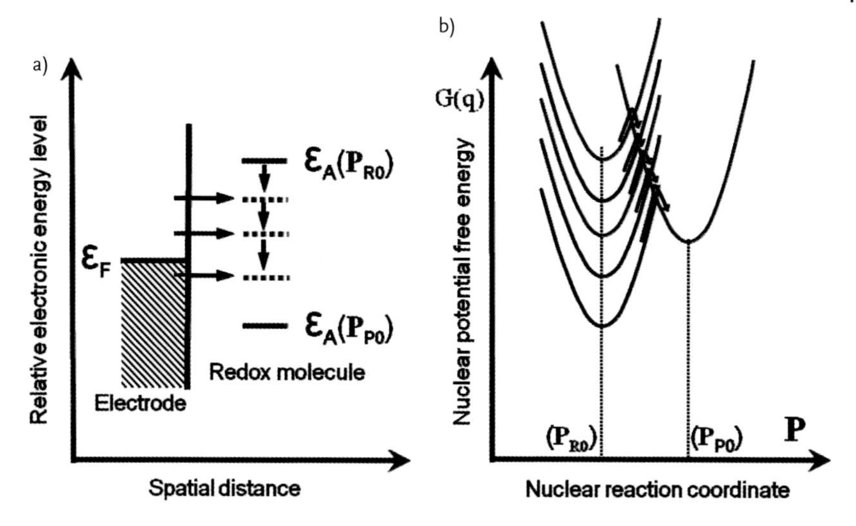

Figure 2.1 (a) Electronic energies of electrode and redox molecule in (cathodic) electrochemical ET at different solvent polarization **P**. $\mathbf{P}_{R0}$ and $\mathbf{P}_{P0}$ are the equilibrium values of **P** in the reactant state (electron in the electrode) and product state (electron on the molecule). (b) Corresponding potential free energy surfaces.

1) The current form contains an electronic tunneling factor, $\kappa_{eff}(\eta)$, and a nuclear activation factor. Nuclear activation in all the vibrational, protein conformational, and external solvent polarization modes, along with driving force effects, thus "precedes" the electronic transition, which occurs at the crossing between the potential surfaces of reactants and products (Figure 2.1).

2) The character of the electronic factor differs from ET in homogeneous solution, being composed of an infinitesimally small single-level transmission coefficient, $\kappa_{el}(\varepsilon;\eta)$, but a macroscopically large number of contributing levels, $4\pi\rho(\varepsilon_F)k_BT$. The electronic factor is therefore the "effective" transmission coefficient, $\kappa_{eff} = 4\pi\kappa_{e1}(\varepsilon_F;\eta)\rho(\varepsilon_F)k_BT$. The process belongs to the weak-coupling diabatic limit when $\kappa_{eff} \ll 1$, while the adiabatic limit $\kappa_{eff} \to 1$ prevails when the coupling is strong.

3) In addition to the driving force or overpotential, the activation free energy contains the nuclear reorganization free energy, E_R, addressed further in Section 2.2.2.

Molecular charge transfer theory has developed into much more powerful frameworks than implied by this simple formalism. Features that can be incorporated include [39, 40]:

1) Anharmonic nuclear motion, frequency changes, mode mixing, etc.
2) Dynamic effects in the electrochemical double layer.

3) Nuclear tunneling, important for proton and hydrogen atom transfer.
4) Details of the electronic structure of the metal electrode surface.
5) Interfacial electrochemical ET via adsorbate states and electrocatalysis.
6) Relaxation of common schemes of separating electronic and nuclear dynamics, important for long-range ET.
7) Stochastic theory and multistep ET via dynamically populated intermediate states.

We address briefly the two central quantities, the nuclear reorganization free energy and the electronic tunneling factor.

2.2.2 Nuclear Reorganization Free Energy

The reorganization free energy E_R contains an intramolecular and an environmental contribution. Modified forms of the simple quadratic form of Eq. (2.4) and other procedures are needed when vibrational frequency changes, nuclear mode anharmonicity, etc., are important [61]. This would apply when major local nuclear reorganization, proton or hydrogen atom transfer, or solvent dielectric saturation accompanies the ET process. Strong interaction between localized electronic states and the nuclear environment is crucial in chemical and biological charge transfer processes. At the same time the complexity of the inhomogeneous, anisotropic electrochemical interface is a theoretical challenge. Comprehensive descriptions of solvation phenomena are integrated in molecular charge transfer theory. These mostly rest on "structure-less" dielectric continuum theory [62–66], but extend to include vibrational and spatial dispersion [67, 68], dielectric interfaces [69], statistical mechanical views [70], and other reorganization phenomena, for example dynamic ionic strength features [71, 72].

The quadratic form, however, rests on the very general assumption that whatever structural features characterize the environment, conformational, polarization, density, or other physical properties respond *linearly* to the field changes, that is, changes in electric field, pressure, or other forces. The quadratic form of the activation free energy is thus broadly valid independently of the nature of the solvation and reorganization.

2.2.3 Electronic Tunneling Factor in Long-Range Interfacial (Bio)electrochemical Electron Transfer

Nuclear reorganization to a nonequilibrium configuration precedes electron tunneling, and is followed by electron (hole) tunneling along electronically facile route(s) in this configuration. The strong electronic–vibrational interaction thus leads to pre-organization prior to the electronic transition and nuclear relaxation subsequent to the transition.

Simple analytical forms illustrate the fundamental features of the (diabatic) tunneling factor. The tunneling process at the crossing of the potential surfaces (Figure 2.1) can be viewed as tunneling "percolation" through a network of intermediate groups between the electrode and the molecular redox group(s). LUMO or HOMO levels of amino acid residues, single DNA bases, or linker groups in molecular monolayers used to immobilize the proteins constitute the tunneling barrier [73–77]. The electron exchange factor, $T_{\varepsilon_F A}$ (Eq. (2.2)), can be given the following approximate form [40, 73–75, 78]:

$$T_{\varepsilon A} = \sum_j \frac{\beta_{\text{electr},1}\beta_{1,2}\ldots\beta_{j,A}}{(\varepsilon - \varepsilon_{1\text{mol}} - \Delta_1)(\varepsilon - \varepsilon_{2\text{mol}} - \Delta_2)\ldots(\varepsilon - \varepsilon_{j\text{mol}} - \Delta_j)} \tag{2.4}$$

which attaches closely to Green's function formalism broadly used to describe the conductivity of small molecules [79–82]. This form, however, also dates back to early approaches to long-range molecular ET [73, 83–89]. In Eq. (2.4), $\beta_{k,k+1}$ is the electron exchange factor for coupling between the kth and $(k + 1)th$ intermediate group, $\beta_{\text{electr},1}$ couples the electrode with the nearest molecular group, and $\beta_{j,A}$ the terminal bridge group with the acceptor group. The energy denominators are the energy gaps between the level ε in the electrode and the nearest intermediate molecular level, ε_k. Δ_k ($\approx \beta_{k-1,k} + \beta_{k,k+1}$) represents all other exchange couplings in which the kth group is engaged.

Equation (2.4) can be simplified and the tunneling feature directly disclosed if (i) only the shortest tunneling path through nearest-neighbor interactions is important and (ii) all the intermediate group exchange interactions and energy gaps take the same values, β and $\Delta\varepsilon$, respectively. Equation (2.4) then reduces to

$$T_{\varepsilon A} = \frac{\beta_{\text{electr},1}\beta_{\text{NA}}}{\beta}\left(\frac{\beta}{\Delta\varepsilon(\eta)}\right)^N = \frac{\beta_{\text{electr},1}\beta_{\text{NA}}}{\beta}\exp\left[-\frac{1}{a}\ln\left(\frac{\Delta\varepsilon(\eta)}{\beta}\right)R\right] \tag{2.5}$$

where N is the number of intermediate groups, R the distance between the electrode and the molecular acceptor *along the particular nearest-neighbor route*, and a the average structural extension of each group. $\beta_{\text{electr},1}$ and β_{NA} are the exchange coupling between the electrode and the nearest intermediate group and between the terminal group and the acceptor molecule, respectively. Equation (2.5) shows the *directional* exponential distance decay associated with tunneling, with faster decay the larger the energy gap and the weaker the coupling.

The following other form is also convenient. By introducing the energy "broadening" of the group linked to the electrode [51, 52, 90]

$$\Delta_1 = \pi(\beta_{\text{electr},1})^2 \rho(\varepsilon_F) \tag{2.6}$$

combination of Eqs. (2.1)–(2.3) with Eq. (2.5) gives

$$j(\eta) = 8e\Gamma_{\text{ox}}^{(1-\alpha)}\Gamma_{\text{red}}^{\alpha}\sqrt{\frac{\pi^3 k_B T}{E_r\hbar^2\omega_{\text{eff}}^2}}\frac{\omega_{\text{eff}}}{2\pi}\left(\frac{\beta_{\text{NA}}}{\beta}\right)^2\Delta_1\exp\left[-\frac{2}{a}\ln\left(\frac{\Delta\varepsilon(\eta)}{\beta}\right)R\right]\exp\left[-\frac{(E_R+e\eta)^2}{4E_R k_B T}\right] \tag{2.7}$$

This form will be used in Section 2.4.

2.3 Theoretical Notions in Bioelectrochemistry towards the Single-Molecule Level

2.3.1 Biomolecules in Nanoscale Electrochemical Environment

Interfacial bioelectrochemistry of monolayers of proteins, DNA-based molecules as well as of amino acids and DNA bases as their building blocks has come to include a range of electrochemical techniques. Linear and cyclic voltammetry remain central [91–93]. Other electrochemical techniques include impedance [94, 95] and electroreflectance spectroscopy [96, 97], ultramicroelectrodes [98], and chronoamperometry [94, 99]. To these can be added spectroscopic techniques such as infrared [100], surface-enhanced Raman and resonance Raman [101, 102], second harmonic generation [103], surface plasmon [104, 105], and X-ray photoelectron spectroscopy [94, 106, 107]. A second line has been to combine state-of-the-art physical electrochemistry with corresponding state-of-the-art microbiology and chemical synthesis, relating to the use of a wide range of designed mutant proteins [108–111], and to *de novo* designed synthetic redox metalloproteins [112–114].

This broad interdisciplinary approach to the new bioelectrochemistry testifies to increasing and multifarious detail in both structural and functional (mechanistic) mapping and control at the monolayer level. Introduction of single-crystal, atomically planar electrode surfaces [94, 115] has opened a basis for the use of the scanning probe microscopies, STM and AFM, also for biological macromolecules. Importantly this extends to the *electrochemical* STM mode where now also biological macromolecules can be mapped directly in their natural aqueous environment (*in situ* STM and AFM) to single-molecule resolution [94, 116–122]. Reactive (bio) molecules combined with *in situ* STM and nanoparticle configurations also show new interfacial electron transfer phenomena, which pose new theoretical challenges. We first give an overview of some of these and then discuss some systems that have been analyzed within these new conceptual frameworks.

2.3.2 Theoretical Frameworks and Interfacial Electron Transfer Phenomena

We focus on the electronic conductivity of molecular monolayers and on single molecules enclosed between a pair of metallic electrodes. We address specifically *in situ* STM of redox (bio)molecules but concepts and formalisms carry over to other metallic nanogap configurations. Importantly, in addition to the substrate and tip, a third electrode serves as reference electrode [42–44] (Figure 2.2). This allows electrochemical potential control of both substrate and tip. The three-electrode configuration is the basis for *two* kinds of tunneling "spectroscopy" unique to electrochemical *in situ* STM. One is the current–bias voltage relation as for STM in air or vacuum, but with the notion that the substrate (over)potential is kept constant. The other is the current–overpotential relation at constant bias

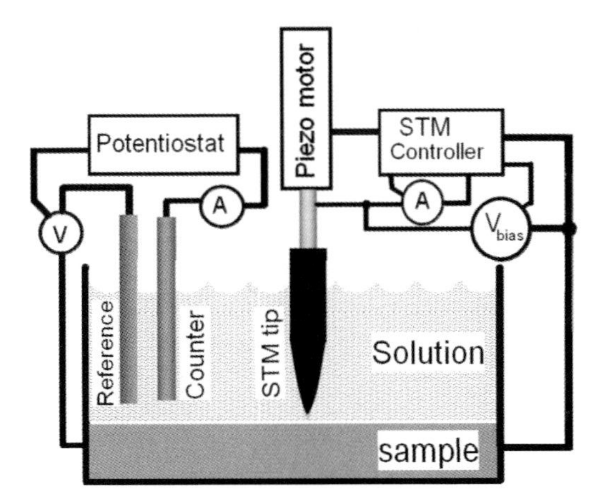

Figure 2.2 Schematic of *in situ* STM. In this four-electrode configuration the working electrode potential is controlled relative to the electrochemical reference electrode ("gate") and the (coated) tip potential relative to the working electrode potential. The counter electrode enables recording of electrochemical processes by the independently controlled tip electrode.

voltage, that is, the electrochemical substrate and tip potentials are varied in parallel relative to the common reference electrode. The two correlations correspond to the current–bias and current–gate voltage relations in solid-state nanoscale transistors. Analogous dual-type correlations in other molecular three-level systems such as molecular Raman spectroscopy have been noted [122].

2.3.2.1 Redox (Bio)molecules in Electrochemical STM and Other Nanogap Configurations

Figure 2.3 shows schematic views of a redox molecule enclosed in a nanoscale or molecular-scale gap between two electrochemically controlled metallic electrodes. In contrast to reported systems that operate in ultrahigh vacuum at cryogenic temperatures [123–127], those represented by Figure 2.3 and addressed below apply to room temperature and condensed matter environments. Figure 2.3 also shows an extension in which a metallic nanoparticle is combined with a redox molecule and the combined supermolecule is inserted in the gap. Configurations corresponding closely to both schemes in Figure 2.3 have been characterized recently [94, 116–119, 121, 126, 128] and prompt the following observations:

1) Figures 2.3a and b extend the scheme for electrochemical ET at a single metallic electrode surface. Two electrode surfaces are now present. Their Fermi levels are separated by the bias voltage, eV_{bias}, at given overpotential η. The overpotential is the substrate potential, E_s, relative to the substrate equilibrium potential, E_s^0, that is, $\eta = E_s - E_s^0$. The bias voltage is viewed as the tip potential relative to the substrate potential, $V_{\text{bias}} = E_t - E_s$.

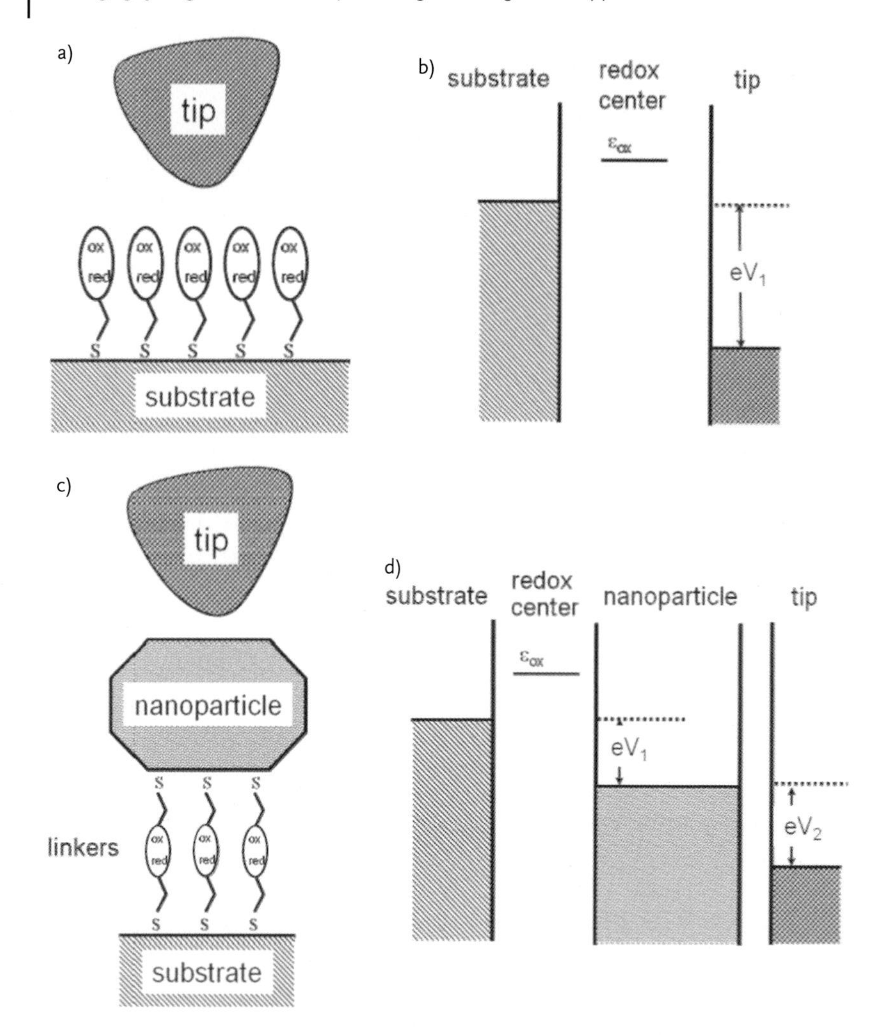

Figure 2.3 Schematic view of a redox molecule in a STM gap or enclosed between two nanogap electrodes. (a) Redox molecule in tunneling gap and (b) electronic energy scheme. (c) Tunneling junction with a redox molecule–metallic nanoparticle hybrid and (d) corresponding electronic energy scheme, with two potential drops.

2) The molecular redox level, say the oxidized, electronically "empty" level at electrochemical equilibrium, is located well above the substrate Fermi energy, that is, by the reorganization free energy E_R. Figure 2.3 shows explicitly the configuration for *positive* bias voltage, so that the tip Fermi level is located *below* the substrate Fermi level.

3) The energies can be controlled by three external factors. The equilibrium locations can be shifted by the electrochemical (over)potential. This raises the Fermi energy of the substrate electrode, $\varepsilon_{Fsubstr}$, relative to the redox level. If

all energies are counted from $\varepsilon_{Fsubstr}$, the empty redox level, ε_{ox}, is lowered relative to $\varepsilon_{Fsubstr}$ with increasing negative η. The bias voltage is a second controlling factor. As the redox site is exposed to part of the bias voltage, the redox level is also shifted relative to $\varepsilon_{Fsubstr}$ on bias voltage variation. Conformational and solvent polarization fluctuations as in molecular ET in homogeneous solution and electrochemical ET are, finally, crucial in the (bio)molecular conduction process. These novel interfacial bioelectrochemical phenomena are now reaching the level of *single-molecule* resolution.

4) The steady-state current through the nanoparticle configuration in Figures 2.3c and d offers similar dual-type current–bias voltage and current–overpotential relations. When the nanoparticle is large enough to behave as bulk metal (diameter ≥ 5 nm) the correlations display "spectroscopic" features resembling those of the redox group alone [128, 129]. The correlations of very small particles, however, display electronic fine structure. This is caused by quantum size effects, that is, discrete electronic structural effects of the particle, which now is more like a molecular structure. These points can be incorporated in a formalism [45, 46, 48–52], a few elements of which we now overview.

2.3.2.2 New Interfacial (Bio)electrochemical Electron Transfer Phenomena

The clearest results in electrochemical *in situ* STM of redox molecules and biomolecules are obtained when the bias voltage and the electronic broadenings are small, that is

$$\gamma|eV_{bias}|,\ \Delta_{substr},\ \Delta_{tip} < E_R - e\eta \tag{2.8}$$

where γ is the fraction of the bias voltage at the site of the redox center. The bias voltage is thus in a sense a "probing energy tip." The "spectral resolution" is better, the narrower the probing tip. As the overpotential is raised, at fixed bias voltage the cathodic current first increases due to more favorable driving force, but decreases as the overpotential is increased further since practically all active species are converted to reduced form with the energy level trapped below the Fermi levels. A different character of the tunneling process arises when the bias voltage is large and the opposite inequality of Eq. (2.8) applies. Either the reduced level or, at even larger bias voltage, both the oxidized and reduced redox level forms are then trapped *between* the Fermi levels. Multi-electron transport (coherent or stepwise) then continues until the overpotential is raised still further and Eq. (2.8) again applies.

An additional distinction between "weak" and "strong" interactions between the redox level and the electrodes is important. In the former limit the steady-state STM process can be viewed as two consecutive, environmentally relaxed interfacial *single-ET steps*, each analogous to electrochemical ET (Section 2.2). This gives the steady-state tunneling current

$$i_{tunn}^{weak} = e\frac{k^{o/r}k^{r/o}}{k^{o/r} + k^{r/o}} \tag{2.9a}$$

$$k^{o/r} \approx 8\kappa_{tip}\rho_{tip}k_B T \frac{\omega_{eff}}{2\pi} \exp\left[-\frac{(E_R - \xi e\eta - e\gamma V_{bias})^2}{4E_R k_B T}\right] \tag{2.9b}$$

$$k^{r/o} \approx 8\kappa_{subst}\rho_{subst}k_B T \frac{\omega_{eff}}{2\pi} \exp\left[-\frac{[E_R + \xi e\eta - (1-\gamma)eV_{bias}]^2}{4E_R k_B T}\right] \tag{2.9c}$$

These forms are equivalent to the interfacial electrochemical ET rate constants (Eqs. (2.2) and (2.3)). Thus $k^{o/r}$ is the rate constant for ET from the reduced molecule to the substrate and $k^{r/o}$ the rate constant for ET from the tip to the oxidized molecule. The parameter ξ represents the potential drop between the electrode and the solution. In the tunneling gap with approximately the same spatial extension as the electrochemical double layer(s), ξ and γ cannot, however, be regarded as independent but are correlated as [48–50, 60]

$$\xi(z) = 1 - \gamma(z) - \gamma(L - z) \tag{2.10}$$

where L is the tunneling gap width and z the distance of the redox center from the substrate electrode surface. This is important in real data analysis.

A different scenario emerges in the opposite limit of strong electronic interactions. Equations (2.9a)–(2.9c) are then replaced by

$$i_{tunn}^{strong} = 2en_{o/r} \frac{k^{o/r}k^{r/o}}{k^{o/r} + k^{r/o}} \tag{2.11a}$$

$$n_{o/r} = \frac{|eV_{bias}|}{\Delta\varepsilon}; \quad \Delta\varepsilon = \frac{1}{\kappa_{substr}\rho_{substr}} + \frac{1}{\kappa_{tip}\rho_{tip}} \ll k_B T \tag{2.11b}$$

$$k^{o/r} \approx \frac{\omega_{eff}}{2\pi} \exp\left[-\frac{(E_R - e\eta - e\gamma V_{bias})^2}{4E_R k_B T}\right] \tag{2.11c}$$

$$k^{r/o} \approx \frac{\omega_{eff}}{2\pi} \exp\left[-\frac{[E_R + e\eta - (1-\gamma)eV_{bias}]^2}{4E_R k_B T}\right] \tag{2.11d}$$

The difference from Eqs. (2.9a)–(2.9c) is that the electronic transmission coefficients now appear in the quantity $n_{o/r}$ in Eq. (2.11a) but the physical meaning is different from the meaning in Eqs. (2.9a)–(2.9c). Due to the strong electronic molecule–electrode interactions, the first single-ET event is followed by a large number (up to or exceeding a hundred or so) of subsequent events, while the occupied/reduced redox level relaxes through the energy window between the two Fermi levels. This is a novel ET phenomenon associated with the three-electrode *in situ* STM process and may be a reason for the frequently observed large tunneling current densities (per molecule), rooted in the large number of electrons transferred.

Equations (2.9) and (2.11) are a useful basis for experimental data analysis. The following simple form derives directly from Eqs. (2.9) and (2.11) for a symmetric contact, $\kappa_{substr}\rho_{substr} = \kappa_{tip}\rho_{tip}$:

$$i_{tunn}^{symm} = \frac{1}{2}en_{o/r} \frac{\omega_{eff}}{2\pi} \exp\left(-\frac{E_R - eV_{bias}}{4k_B T}\right)\left\{\cosh\left[-\frac{(\frac{1}{2}-\gamma)eV_{bias} - \xi e\eta}{2k_B T}\right]\right\}^{-1} \tag{2.12}$$

This form discloses an immediate general implication in Eqs. (2.9)–(2.12), namely a pronounced ("spectroscopic") maximum in the tunneling current–overpotential relation (at given bias voltage). For the symmetric configuration of Eq. (2.12), the maximum appears at the overpotential

$$\eta = \eta_{\text{max}} = \frac{1}{\xi}\left(\frac{1}{2} - \gamma\right)V_{\text{bias}} \tag{2.13}$$

If the redox center in the gap is exposed to half of the bias voltage drop, $\gamma = {}^{1}\!/_{2}$, and the maximum is at the equilibrium redox potential, $\eta_{\text{max}} = 0$. This holds a diagnostic clue regarding the mechanism of single-molecule electronic tunneling. We shall return later to data analysis based on this view. The precise location of the maximum depends, however, rather sensitively on the potential distribution in the tunneling gap, as reflected in the correlation between the parameters ξ and γ (Eq. (2.10)).

Similar "spectroscopic" features are associated with the current–bias voltage relations, that is, new electronic conduction channels open when the redox level is brought to cross into the energy region between the Fermi levels. The "spectroscopic" current–bias voltage peaks are reflected in the *conductivity*, that is, the tunneling current derivative with respect to the bias voltage, rather than in the tunneling current itself. The tunneling current–bias voltage correlations, however, raise other issues as the conduction mechanism changes at large bias voltage where the equilibrated reduced level is trapped between the two Fermi levels [48–50, 60]. We proceed instead to some recent data of single-molecule *in situ* STM imaging and image interpretation based on both computational support and the formalism above.

2.4 *In Situ* Imaging of Bio-related Molecules and Linker Molecules for Protein Voltammetry with Single-Molecule and Sub-molecular Resolution

The two fundamental classes of chemical building blocks of biological macromolecules, proteins and DNA/RNA, have been prominent targets in efforts towards single-biomolecular imaging. The former group includes the amino acids among which thiol-derived cysteine and homocysteine have been particularly important. The latter group is represented by the nucleobases adenine, thymine, guanine, cytosine, and uracil. Extensive recent reviews are available elsewhere [130].

2.4.1 Imaging of Nucleobases and Electronic Conductivity of Short Oligonucleotides

Early *in situ* STM (and AFM) of nucleobases focused on adenine and guanine on molybdenum disulfide and highly oriented pyrolytic graphite [131–133]. Other early studies addressed all the nucleobases and Au(111) electrode surfaces. The bases were concluded to form planar hydrogen-bonded networks on the surfaces and to display a composite pH dependence. *In situ* STM using several low-index Au and Ag surfaces (including also Au(210)) have provided both high-resolution images of thymine and uracil in particular and an interesting surface potential-dependent phase behavior [134–141]. Different chemisorbed and physisorbed phases appear in different potential ranges. The molecules in the former mode are in an upright orientation but lying flat on the surface in the latter. Notably, the phase transitions between highly ordered chemisorbed and physisorbed adlayers could be followed by cyclic voltammetry and directly visualized by *in situ* STM.

High-resolution *in situ* STM as well as phase transition dynamics of nucleobases on Au(111) and other low-index electrode surfaces supported by infrared spectroscopy have been reviewed recently by Nichols and coworkers [142] and Wandlowski and coworkers [143]. We refer to these reviews for details and note instead another aspect of single-molecule dynamics of DNA-based molecules. The observed electronic conductivity of oligonucleotides of variable length and variable base composition has opened almost a "Pandora's box" of novel DNA-based electronic properties. These include particularly photochemical and interfacial electrochemical ET. We refer to other recent reviews [144, 145] for this, still far from settled, issue but note the following STM-based studies that illuminate the conductivity issue at the single-molecule level (Figure 2.4).

Tao and associates [147] and Nichols and associates [146, 148] studied single-molecule conductivity of variable-length single-base short oligonucleotides. The studies of Nichols and associates illuminate both important issues and some puzzles. Short single- and double-strand repetitive oligonucleotides thiolated at both ends were adsorbed on a Au(111) electrode surface. A Au tip was then brought to approach the surface and capture one thiol end of the molecule while the other end kept the molecule linked to the surface. This was followed by conductivity recording of each of the all-A, -T, -G, and -C single-strand bases and of a single GC double-strand.

Hopping of electronic holes has been a strongly advocated DNA conduction mechanism as a rationale for the weak distance dependence of the ET rate along the DNA strands [149]. Hole hopping would, however, imply a strong correlation between the conductivity and the oxidation potential of the bases. The observed single-molecule conductivity displayed in fact a very weak dependence on either distance or nature of the base, say a factor of two or so higher for oligoguanine compared with the other bases, whereas orders of magnitude might have been expected. Local strand conformational fluctuations with accompanying large fluctuations in the hopping electronic energy levels have been proposed as a rationale for the robust thermal conduction [150]. This strongly gated charge transport mechanism seems, however, also to remain a puzzle.

2.4.2
Functionalized Alkanethiols and the Amino Acids Cysteine and Homocysteine

The class of straight or branched, variable-length, and pure or functionalized alkanethiols have been core molecular targets in the development of *in situ* STM

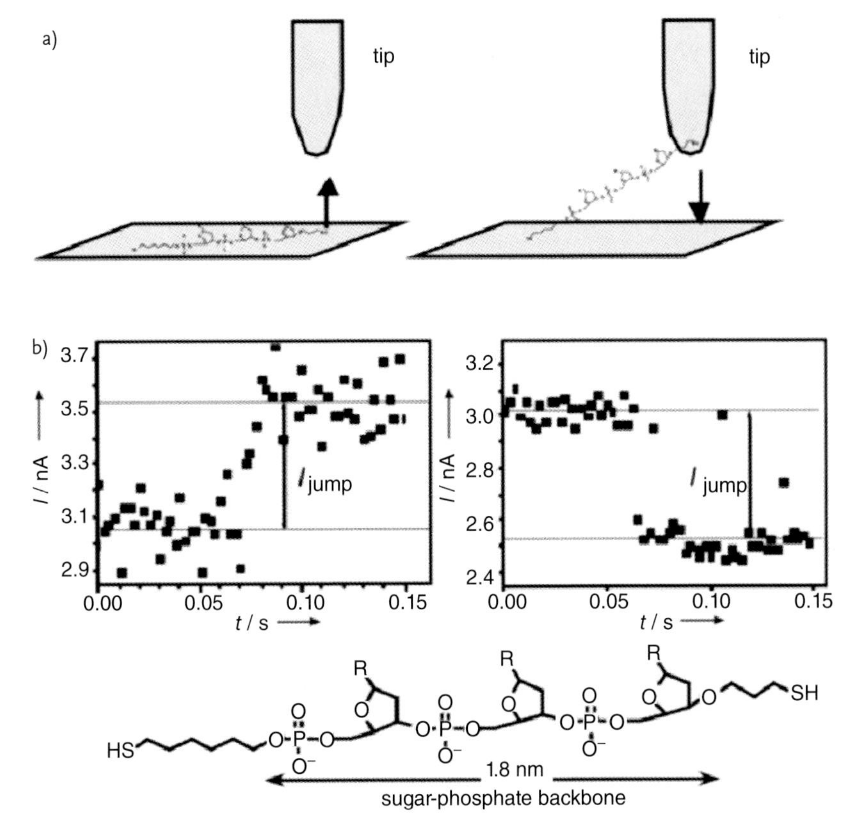

Figure 2.4 (a) Schematic of the measurement technique for single-molecule single- and double-strand oligonucleotide conductivity. The arrows indicate the direction of motion of the molecule. (b) $I(t)$ curves for molecular attachment/detachment to the STM gold tip. Experiments in air. (c) Molecular structure of the oligonucleotides. The bases are denoted by R and the length of the fully extended sugar-phosphate indicated. From [146].

towards single-molecule or even sub-molecular resolution (Figure 2.5) [60, 151]. Imaging has been closely tied to electrochemistry of highly ordered SAMs of alkanethiol-based molecules at single-crystal electrode surfaces where the reductive desorption process

$$\text{Au–S–R} + e^- \rightarrow \text{Au} + {}^-\text{S–R} \tag{2.14}$$

has been crucial. Alkanethiol-based SAMs have been targets in efforts towards imaging optimization regarding, for example, electrolyte composition or anaerobic versus aerobic environments. They have further been subject to state-of-the-art large-scale electronic structure computations that have disentangled the STM contrast to minute detail in both the *ex situ* vacuum environment and in the *in situ* aqueous environment [151].

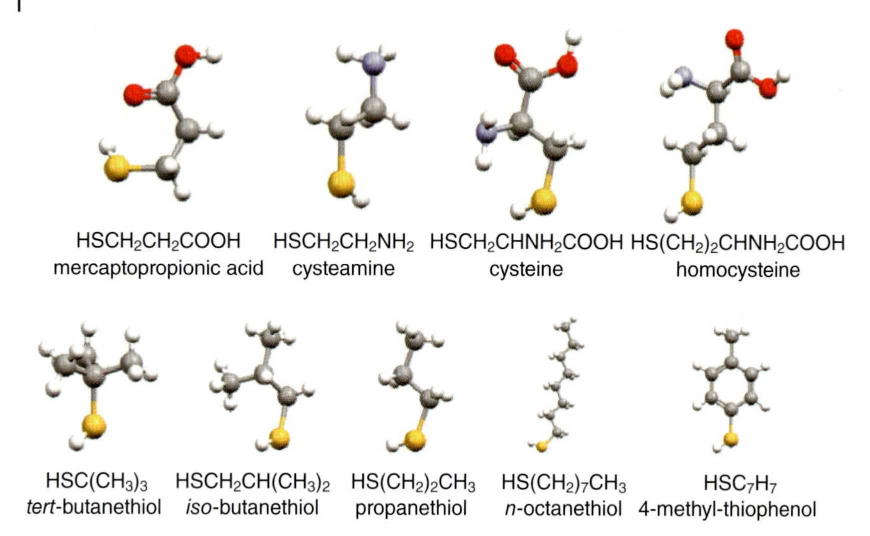

Figure 2.5 Overview of molecular structures of alkanethiol-based molecules imaged by *in situ* STM in various aqueous buffer solutions. For a literature overview, see [151].

Electrochemically controlled SAMs of the alkanethiol class characterized to high voltammetric resolution and to molecular and sub-molecular structural *in situ* STM resolution have been reviewed recently [60, 151]. We note here first some issues of importance to functionalized alkanethiols as linker molecules for gentle immobilization of fully functional redox metalloprotein monolayers on single-crystal Au(111) electrode surfaces. We discuss next specifically the functionalized alkanethiols cysteine (Cys) and homocysteine (Hcy). These two molecules represent a core protein building block and a core metabolite, respectively. The former has been used to display unique sub-molecular *in situ* STM resolution [152]. The latter shows a unique dual surface dynamics pattern that could be followed both by single-molecule *in situ* STM and by high-resolution capacitive voltammetry.

2.4.2.1 Functionalized Alkanethiols as Linkers in Metalloprotein Film Voltammetry

With a few exceptions, redox metalloprotein voltammetry is unstable or not possible unless the electrode is modified by chemisorbed monolayers of linker molecules, or the protein modified by insertion of non-native amino acid residues. The linker molecules are mostly thiol-containing molecules which adsorb strongly on the Au surface. The opposite end holds a functional group that interacts "gently" with the protein, ensuring that the latter is immobilized and retains full functional integrity. Strategies for linker–protein interactions are broadly available but the interactions are often subtle. Closely related linker molecules can induce widely different voltammetric responses, and linker groups with no immediate expectable protein compatibility can cause strong voltammetric signals.

The use of single-crystal electrodes both offers significantly improved voltammetric resolution compared to polycrystalline electrodes and enables surface struc-

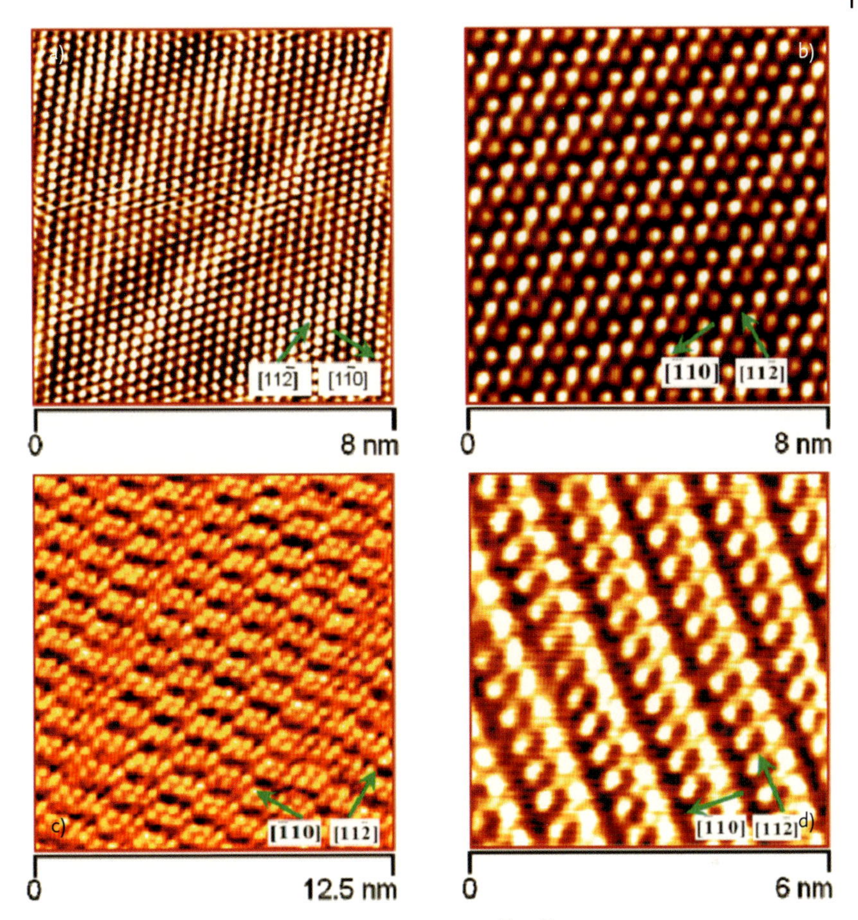

Figure 2.6 High-resolution *in situ* STM images of some pure and functionalized alkanethiols and protein film voltammetry linkers on Au(111) electrode surfaces in various aqueous buffers. (a) Bare reconstructed Au(111) surface; (b) $(2\sqrt{3}\times\sqrt{3})$R30°-4 butanethiol monolayer (hydrophobic) [153]; (c) $(2\sqrt{3}\times6)$R30°-6 monolayer of mercaptopropionic acid (negatively charged) [154]; (d) $(\sqrt{3}\times4)$R30°-2 monolayer of cysteamine (positively charged) [38].

tural characterization at the atomic level for pure electrodes and at the molecular level for modified electrodes. Figure 2.6 shows *in situ* STM images of a reconstructed Au(111) surface and Au(111) surfaces modified by various thiol-based linker molecules. All of them form highly ordered monolayers in aqueous buffer. By virtue of their different terminal groups, the linker molecules are efficient promoters in protein and enzyme voltammetry. Figure 2.6b shows a $(2\sqrt{3}\times\sqrt{3})$R30°-4 monolayer of butanethiol with a hydrophobic surface [153]. Figure 2.6c shows a $(2\sqrt{3}\times5)$R30°-6 monolayer of mercaptopropionic acid [154] which is a highly efficient voltammetric promoter of the strongly negatively charged iron–sulfur protein

Pyrococcus furiosus ferredoxin (see Section 2.6.2.1). Mercaptopropionic acid gives a negatively charged, hydrophilic surface with a lattice structure of clusters of six mercaptopropionic acid molecules. Cysteamine with a terminal ammonium group and an efficient promoter for copper nitrite reductase electrocatalytic voltammetry gives a highly ordered positively charged $(\sqrt{3} \times 4)R30°$-2 monolayer (Figure 2.6d) [38]. The unit cell contains two molecules giving two different *in situ* STM contrasts. These have been addressed comprehensively using molecular dynamics and density functional computations [38]. The two contrasts are assigned to two differently tilted molecular orientations where the strongest contrast arises from the most upright orientation. L-Cysteine, also an efficient promoter molecule for protein voltammetry, is a natural amino acid and protein building block. The zwitterionic nature of the molecule gives highly ordered domains of a more specific $(3\sqrt{3} \times 6)R30°$-6 structure, controlled by subtle electrostatic and hydrogen bond networks on the surface [153]. Each of the clusters contains *six* Cys molecules but the surface structure is specific to the buffer medium used (50 mM ammonium acetate, pH = 4.6), and quite different in other media [151].

Most of the linker molecular monolayers are stable over broad potential ranges, limited by reductive and oxidative desorption and cleavage of the Au–S bond. The adsorption process can be followed in real time through several intermediate phases, such as reported for cysteamine [38] (and 1-propanethiol [155]). The *in situ* STM images shown in Figure 2.6 thus offer an overall impression of the microenvironment for immobilized redox proteins "in voltammetric action."

2.4.2.2 *In Situ* STM of Cysteine and Homocysteine

As the only natural amino acid building block in proteins containing a thiol group, Cys in the form of SAMs on various substrates in different environments [152, 153, 156–161] (Figure 2.7) has attracted great attention, and a number of *in situ* STM studies of Cys adsorption on Au(111) electrode surfaces have been reported [60, 151]. Notably Cys packing appears to depend sensitively on the buffer and electrolyte environment with widely different packing modes in different electrolyte solutions and at different pH values. Cystine is the dimer of Cys, with a disulfide bridge, –S–S–, rather than the –SH group. The cystine disulfide bridge appears to break and single S–Au bonds to form on SAM formation, giving virtually identical *in situ* STM images and reductive desorption peaks for Cys and cystine [153]. Figure 2.7 shows *in situ* STM images of SAMs of Cys/cystine on a Au(111) electrode surface in ammonium acetate (pH = 4.6), showing highly ordered clusters that include six Cys molecules with the same $(3\sqrt{3} \times 6)R30°$ cell.

The dependence of the molecular SAM organization on the interaction with substrate is strikingly illustrated by the organization of the same or similar alkanethiol-based molecules on different low-index Au electrode surfaces. Cys SAMs on Au(111) and Au(110) in both aqueous solution and ultrahigh vacuum (UHV) have been studied in particular detail and illustrate the strong effects of the atomic Au substrate structures on the SAM structures.

Figures 2.7e and f compare *in situ* STM images of Cys SAMs on Au(111) and Au(110) surfaces in ammonium acetate (pH = 4.6). Highly ordered lattices are

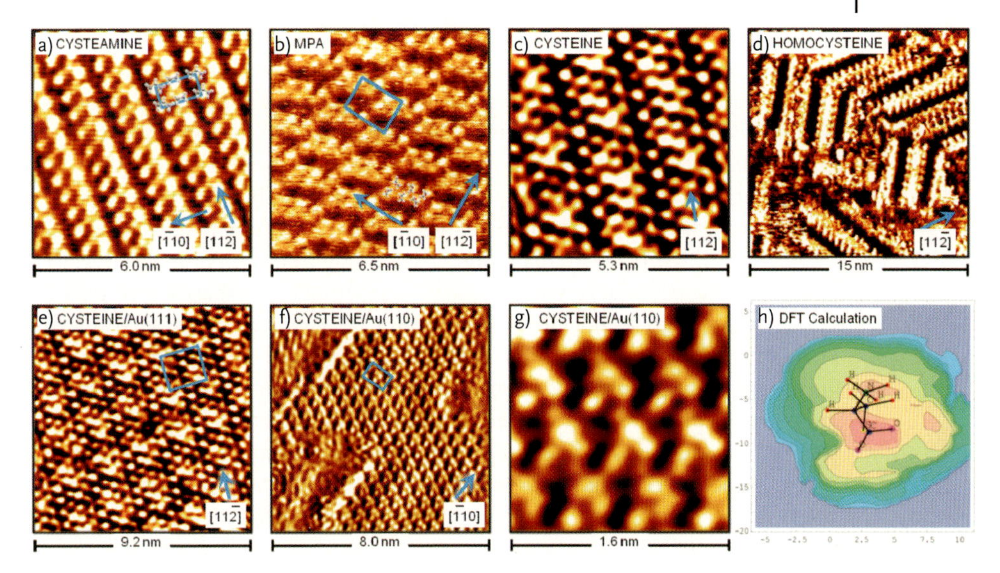

Figure 2.7 High-resolution *in situ* STM images of the amino acid cysteine in different buffers compared with other alkanethiol-based molecules on different low-index Au electrode surfaces. Overview images of (a) cysteamine [38], (b) mercaptopropionic acid (MPA) [154], (c) cysteine [153, 158], and (d) homocysteine [162], all on a Au(111) electrode surface. Cysteine on (e) a Au(111) electrode surface [153, 158] and (f) a Au(110) electrode surface [152]. (g) Zoom-in on cysteine on the Au(110) electrode surface with sub-molecular resolution [152]. The three lobes show the three functional cysteine groups, the AuS unit, the carboxyl group, and the ammonium group. (h) Density functional theory (DFT) calculation of optimized solute cysteine structure of the Au(110) electrode surface [152].

present on both substrate surfaces. The $(\sqrt{3}\times22)R30°$ reconstruction on Au(111) and the (1×3) reconstruction on Au(110) are lifted in the presence of Cys, in contrast to observations in UHV, where both Cys adlayer and Au(111) herringbone reconstruction lines are visible [151, 159, 160]. Cluster structures are found on Au(111) in both UHV [160] and liquid environment [153] but the unit cell and the cluster size are different, with six and four Cys molecules assigned to each cluster in liquid and UHV, respectively. No Cys cluster structure is found in Cys SAMs in liquid environment on Au(110) [152], while clusters with eight Cys molecules are observed in UHV [161]. Both solvation and crystal orientation of the substrate are therefore undoubtedly important in controlling the molecular arrangement in the SAMs. Figure 2.7 also shows a $c(2\times2)$ lattice of L-Cys monolayers on Au(110). Combined with voltammetric surface coverage analysis, each unit cell is found to contain two molecules. Notably, each molecule gives three spots in the *in situ* STM image [152] (Figure 2.7g). Sub-molecular *in situ* STM resolution has thus been reached in this case. First-principles computational support has further led to the assignment of each spot to a particular chemical group (–COOH, $-NH_2$, and –SH), that is, to detailed image interpretation of the origin of the STM contrasts.

The higher Cys homolog Hcy with an additional $-CH_2$ link (Figures 2.5 and 2.7) is also a central metabolite associated with the metabolism of the amino acid methionine and other metabolic processes. In spite of the structural similarity with Cys, Hcy SAMs are packed in a quite different $(\sqrt{3}\times 5)R30°$ structure, even in the same buffer medium, with three Hcy molecules per unit cell (Figure 2.7d). Very notably, in spite of the absence of a redox group, the voltammetry of Hcy monolayers gives a pair of well-defined sharp (24 mV) pH-dependent peaks (−0.06 V vs. SCE, pH = 7.7 [162]). The origin of the peaks, as strongly supported by *in situ* STM, is capacitive and caused by structural reorganization of the Hcy molecules in a narrow potential range around the potential of zero charge [162]. The voltammetric scans could in fact be followed by *in situ* STM all the way across the capacitive peak potential. Highly ordered domains are observed only around the peak potential, disorder appearing reversibly on either side of this potential. A molecular mechanism where the $-COO^-$ and $-NH_3^+$ groups at neutral pH can approach the Au(111) surface around the fixed anchor Au–S during the potential sweep, at potentials positive or negative of the peak potential, respectively, has been proposed [151, 162].

2.4.2.3 Theoretical Computations and STM Image Simulations

Multifarious patterns of differently functionalized alkanethiol SAMs have been mapped to single-molecule and sub-molecular resolution by *in situ* STM in aqueous electrolyte, strongly supported by electrochemical studies of reductive desorption in particular. *In situ* STM is, however, rooted in electronic *conductivity* and quantum mechanical tunneling. Theoretical support is therefore needed in detailed image interpretation of all the many facets of alkanethiol-based SAM packing and *in situ* STM contrasts [163].

The variety of straight versus branched, otherwise nonfunctionalized alkanethiols constitutes one system class where computational support has been decisive. This support has clarified the subtle interplay between Au–S binding sites (hollow, bridge, and a-top sites, or intermediates in between), composite lateral interactions, and Au atom "mining" out of planar Au electrode surfaces [36, 37, 164, 165]. A second class of functionalized alkanethiol SAMs where theoretical and computational support has provided new insight is the strongly solvated L-Cys [158] and cysteamine [38] SAMs. Solvation has been included in different ways. Both continuum models and large-scale molecular dynamics have been combined with quantum chemical computations at the density functional theory (DFT) level, adding immensely to the understanding of the molecular packing and of the *in situ* STM contrasts. The DFT computations of *in situ* STM L-Cys on Au(111) surface models included, for example, a dielectric solvent and representation of the STM contrasts directly in the form of the commonly applied constant current mode with electronic coupling to a model tungsten tip [158]. Image contrasts could be reproduced but not to the same resolution as the data. Maximum electrostatic stability of clusters of exactly *six* zwitterionic Cys molecules as observed in ammonium acetate solution (pH = 4.6) was also found. This kind of computational support applies better in the sub-molecular image interpretation of *in situ* STM of

L-Cys on a Au(110) electrode surface [152]. As observed also for cysteamine [38], the computed lobe positions do not, however, directly accord with the atomic surface structure of L-Cys but rather reflect the dominating electronic densities, and the molecular orbital contributions from the three groups closest to the Fermi levels.

2.4.3 Single-Molecule Imaging of Bio-related Small Redox Molecules

Tao reported the first case of *in situ* STM spectroscopic features using Fe-protoporphyrin IX on highly oriented pyrolytic graphite as target system [166]. Although sophisticated and far from straightforward, single-molecule *in situ* STM imaging and electrochemical scanning tunneling spectroscopy (STS) of redox molecules is now an expanding area of single-molecule science and a wider range of target molecules have been addressed and characterized. Figure 2.8 shows selected examples of redox molecules both imaged to single-molecule resolution and displaying single-molecule tunneling spectroscopy features. These and other reported cases include (i) organic redox molecules (viologens, perylene tetracarboxylic diimide, oligoanilines, tetrathiafulvalenes, and quinones/hydroquinones); (ii) transition metal complexes (metalloporphyrins and metallophthalocyanines, bipyridine and terpyridine complexes of osmium and cobalt); and (iii) molecular-scale metallic nanoparticles in the size range of single-electron charging [60].

In situ STS (Figure 2.8) follows broadly the pattern of sequential two-step interfacial ET. In addition to the expected spectroscopic STS feature, other observations have included (i) conspicuous *in situ* STS resonance features ("molecular transistor" function, "on–off" ratio > 50) and single-molecule rectification ("molecular diode" function, rectification ratio > 20) [167]; (ii) systematic variation of the peak potential with the bias voltage; (iii) comparative *in situ* STS involving different metals and ligands with widely different interfacial electrochemical ET rate constants; (iv) stochastic features in which distributions of single, double, triple, etc., molecular conductivity have been observed; and (v) current–distance correlations from which the notion of coherent multi-ET in a single-molecule *in situ* STS event has received substance. Phenomenological theoretical approaches have been used for successful framing of the observations including the dynamic solvent aspect of the process. New theoretical efforts towards a description of the solvent fluctuational dynamics on a molecular basis are now warranted. We shall readdress some of these aspects below.

The bio-related redox molecular entities shown in Figure 2.8 thus display a pattern which follows consistently the concepts of two-step electrochemical tunneling and the formalism discussed above. Working principles of redox switching, rectification, and amplification at the single-molecule level of interfacial electrochemical ET have thus been achieved. This can be compared with biological redox macromolecules addressed below. We consider first briefly two cases of biomolecules or bio-related molecules intermediate in size between

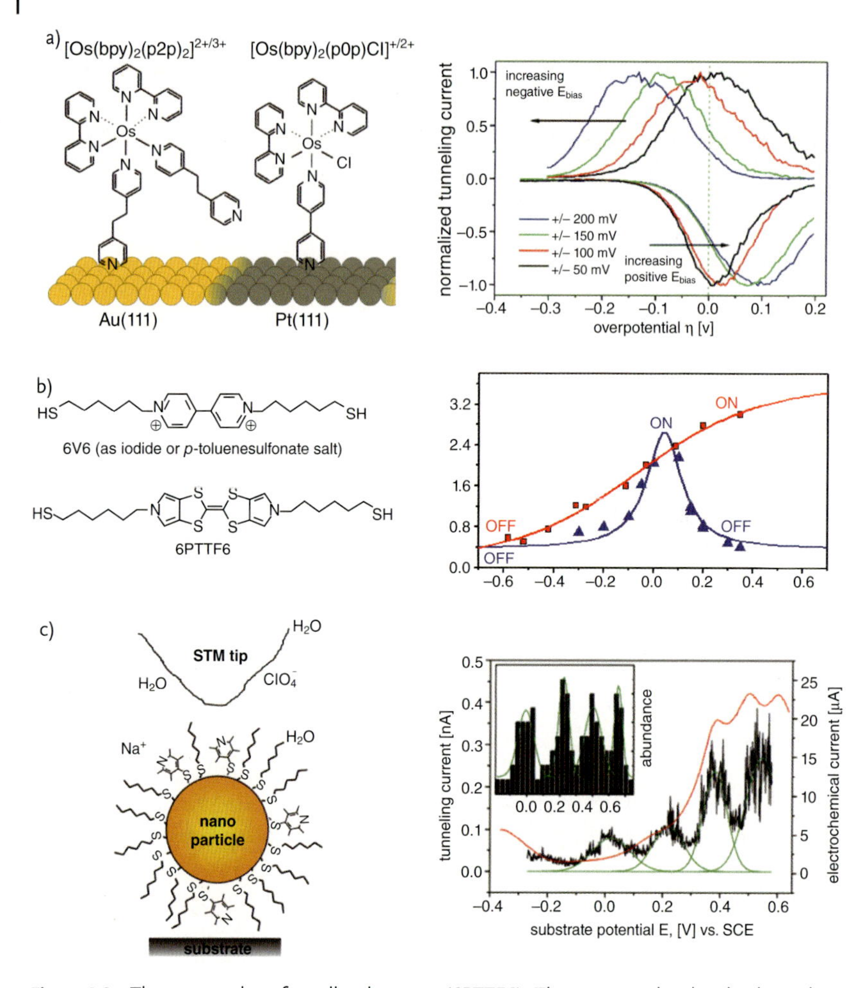

Figure 2.8 Three examples of small redox molecules for which *in situ* STM and single-molecule *in situ* STS have been recorded. Molecular structures and tunneling current–overpotential correlations are shown. (a) Two osmium polypyridine complexes on a Pt(111) electrode surface and tunneling current–overpotential correlations at different bias voltages [55]. (b) Hexanethiol 4,4′-substituted viologen (6V6) and 6-*p*-tetrathiafulvalene (6PTTF6). These two molecules display rather different looking current–overpotential correlations assigned to a softer molecular structure for 6V6 leading to "gated" tunneling, whereas 6PTTF6 has a much more rigid structure [171, 172]. (c) Coated Au_{145} nanoparticle displaying single-electron charging in both differential pulse voltammetry and single-particle *in situ* STS [173]. Inset: histogram showing the peak abundance.

the biological building blocks and the larger "working" metalloproteins and metalloenzymes. The first are membranes immobilized on single-crystal electrode surfaces; the second is human insulin monomer or dimer. *In situ* STM imaging of both, to single-molecule structural resolution, has been accomplished recently [168–170].

2.5
Imaging of Intermediate-Size Biological Structures: Lipid Membranes and Insulin

We discuss here two types of intermediate-size biomolecules imaged in considerable detail by *in situ* STM. The first type consists of mono- and bilayer lipid molecules assembled on Au(111) electrode surfaces. These molecular assemblies resemble biological membranes and offer insight into biomimetic membrane structure and activity. The other system is the protein hormone insulin immobilized on low-index Au(111), Au(100), and Au(110) electrode surfaces. Although a true (but small) protein (molecular mass of 5800 Da), the molecule undergoes drastic deformation and perhaps even decomposition on Au electrode surfaces classifying in a certain sense insulin as an "intermediate-size" biological molecular target.

2.5.1
Biomimetic Mono- and Bilayer Membranes on Au(111) Electrode Surfaces

Phospholipid bilayers assembled on atomically planar Au surfaces rather than on mica or glass surfaces were introduced by Lipkowski and associates [168, 169] as a novel target class of biological molecules for which *in situ* STM has offered single-molecule insight. These molecules, represented by 1,2-dimyristoyl-*sn*-glycero-3-phosphocholine (DMPC) (Figure 2.9) are intermediate in size between amino acids and nucleobases on the one hand and oligonucleotides and proteins on the other hand. Monolayers and bilayers can be assembled by fusion of unilamellar vesicles or using the Langmuir–Blodgett–Schaefer technique but specially prepared ultrasmooth Au surfaces are essential for defect-free adlayers. However, once this is achieved the assembled monolayers or bilayers have offered notable insight regarding structure, assembly dynamics, and biomimetic membrane composition. The latter includes the presence of cholesterol, or insertion of protein-based transport channels and antibiotics such as the short helical protein gramicidin, all at the molecule scale in aqueous biological environment. To this is added the powerful electrochemical aspect by which structural transitions, transmembrane transport, and other biomimetic membrane activity can be addressed.

Figure 2.9a shows the lipid molecule DMPC. Two layers contacted via the hydrophobic tails lead to spontaneous formation of a double-layer biomimetic membrane that can be transferred to a single-crystal ultraplanar electrochemical Au(111) surface. The hydrophilic head groups contact the electrode surface via an intermediate water film. Due to the structurally very well-defined assembly, not only AFM and *in situ* STM but also neutron reflectivity, X-ray diffraction, and infrared reflection absorption spectroscopy (IRRAS) have been employed to support the direct visual *in situ* STM. Electrochemically controlled structural changes, phase transitions, and the effects of the common membrane component cholesterol (Figure 2.9b) and peptide drugs have been investigated in this way.

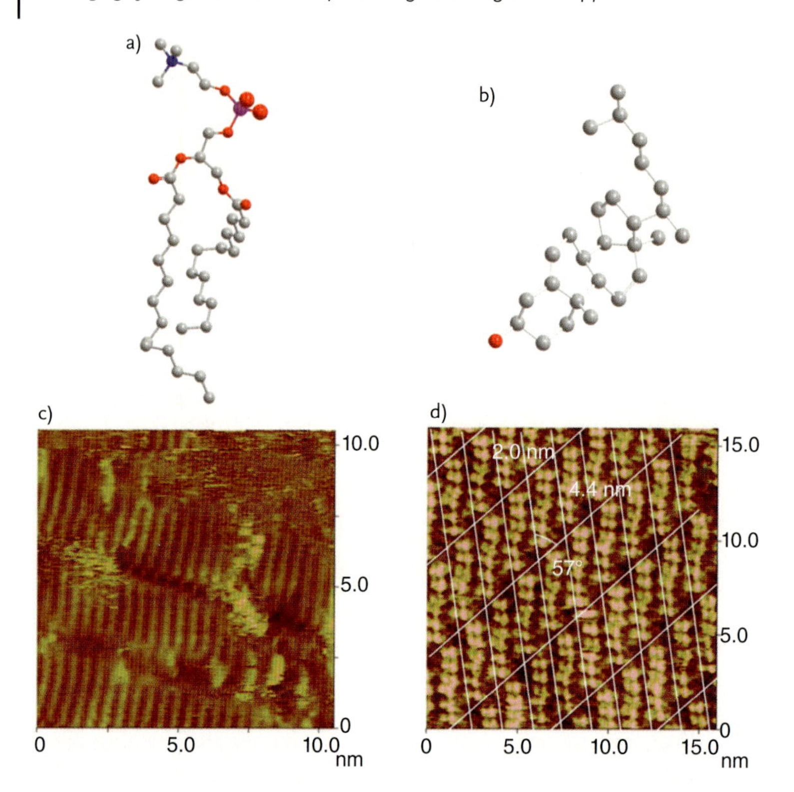

Figure 2.9 *In situ* STM imaging of 1,2-dimyristoyl-*sn*-glycero-3-phosphocholine (DMPC) lipid membranes. Molecular structures of the membrane constituents: (a) DMPC; (b) cholesterol. (c) DMPC monolayer in early phase of immobilization and (d) segregated cholesterol monolayer component in 7 : 3 mixed DMPC/cholesterol monolayer. STM images adapted from [168, 169].

As illustrations of the DMPC/cholesterol-based biomimetic membranes on electrochemical Au(111) electrode surfaces, Figures 2.9c and d show *in situ* STM images just after exposure to a 7 : 3 DMPC/cholesterol solution. Segregated ordered monolayer domains with flat-lying DMPC and cholesterol molecules are clearly seen. After prolonged exposure, bilayers with the molecules in upright orientation form with featureless *in situ* STM images. In comparison, Figure 2.10 shows *in situ* STM images of a differently prepared 1 : 9 molar mixture of (helical) gramicidin and DMPC. Gramicidin appears as dark spots or cavities randomly dispersed inside the highly ordered DMPC monolayer. Together with support from AFM, neutron reflectivity, and IRRAS, images such as those in Figure 2.10 therefore offer powerful information regarding single-molecule membrane dynamic events such as phase transitions, transmembrane transport, and the action of antibiotic peptides.

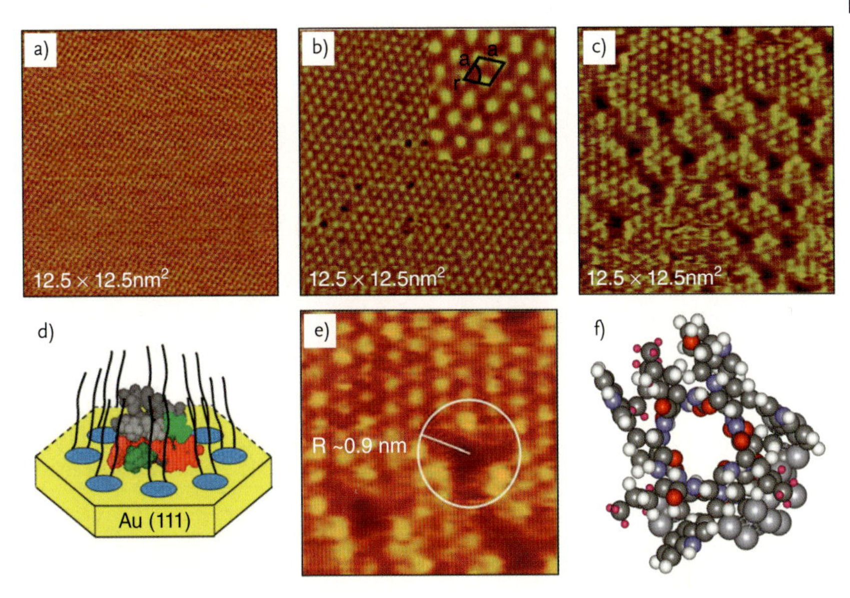

Figure 2.10 *In situ* STM images of mixed gramicidine/DMPC film: (a) pure Au(111) electrode surface; (b) DMPC monolayer on the surface (prepared differently from the layer shown in Figure 2.9); (c) mixed gramicidine/DMPC monolayer. The dark cavities are gramicidine molecules embedded in the membrane matrix. (d) Schematic of gramicidine embedded in the DMPC matrix on the Au(111) surface; (e) zoomed *in situ* STM image of the membrane with gramicidine (dark holes) inserted; (f) structural representation of the gramicidine molecule (PDB 1JNO). Images adapted from [168, 169].

2.5.2 Monolayers of Human Insulin on Different Low-Index Au Electrode Surfaces Mapped to Single-Molecule Resolution by *In Situ* STM

Insulin is one of the most important hormones in the cellular uptake of glucose followed by conversion of glucose to glycogen in the liver [174–176]. The active form is the insulin monomer, a small protein (molecular mass of 5800 Da) with 51 amino acids (human insulin) in two strands, the A- and the B-strands, folded into a globular tertiary structure assisted by three disulfide groups [177–182]. Stimulated by glucose uptake, insulin is released to the blood as the physiologically active monomeric form (Figure 2.11) [177, 178]. Dimer insulin is formed by hydrophobic monomer–monomer interactions at concentrations above 1 ng ml^{-1} [183].

The insulin molecule is exposed to multiple interactions with interfaces *in vitro* and *in vivo*. *In vivo* surface interactions are with protein and lipid serum components, membrane-bound receptors, or membrane surfaces [176, 184–186]. Surface interactions strongly affect the folding of insulin and the interconversion between

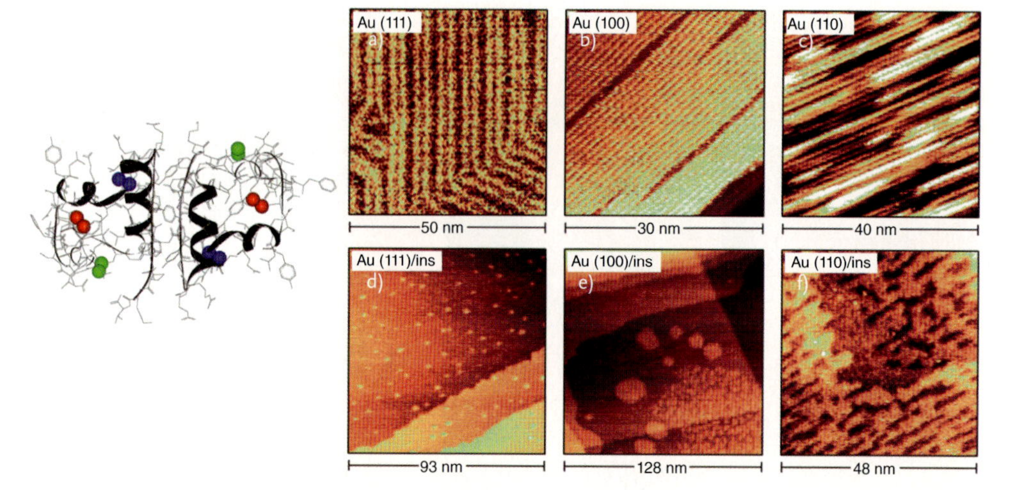

Figure 2.11 *In situ* STM of human insulin on single-crystal Au electrode surfaces [170]. At left is a structural representation of the expected dominating insulin dimer form (PDB 1B9E). The A- and B-chains and the three disulfide groups in each monomer (blue, red, and green) are indicated. *In situ* STM images of the three low-index reconstructed Au surfaces: (a) Au(111); (b) Au(100); (c) Au(110). *In situ* STM images of molecular-scale insulin structures on the three surfaces: (d) Au(111); (e) Au(100); (f) Au(110).

insulin monomer and dimer. Surface structure and dynamics are therefore crucial in physiological action and in insulin therapy.

Insulin monolayers on surfaces (mica, silica, mercury, gold, platinum) have been investigated using spectroscopy (ellipsometry [187, 188], reflectometry [189], mass spectrometry [190], surface plasmon resonance spectroscopy [191]), using electrochemical techniques such as cyclic voltammetry [192–197], electrochemical impedance spectroscopy [196], and electrochemical quartz crystal microbalance techniques [195], and using AFM (insulin fibrils) [198–200]. Following recent progress in redox metalloprotein film voltammetry and single-molecule *in situ* STM [60, 106, 130, 154, 201, 202], Welinder *et al.* [170] studied voltammetry and *in situ* STM of dimeric human insulin using single-crystal Au(111), Au(100), and Au(110) electrode surfaces. The data indicated that structural single-molecule mapping and highly surface-specific packing of insulin dimer had been achieved (Figures 2.11d–f). These observations hold prospects for following insulin self-association such as the interconversion between dimeric and hexameric forms in real time at the single-molecule level. This could be compared with cysteamine and 1-propanethiol monolayer formation and the Hcy orientational interconversion noted above [38, 155, 162].

Figures 2.11a–c show high-resolution *in situ* STM images of the three bare (reconstructed) electrode surfaces. The reconstruction lines and herringbone organization of the Au(111)-$(22 \times \sqrt{3})$ surface structure are clearly visible (Figure 2.11a). Figure 2.11b shows the Au(100)-hex reconstructed electrode surface, dominated by the corrugation lines caused by the incommensurability of the quasi-hexagonal surface layer with the square arrangement of the underlying Au atoms. Figure 2.11c shows the (1 × 3) missing row reconstructed Au(110) electrode surface. These images represent the microscopic environment for insulin adsorption [152, 203, 204].

The surface reconstructions are lifted when insulin is adsorbed leading to patterns that show highly specific organization on the different surfaces. Figures 2.11d–f show insulin adsorption on the Au(111), Au(100), and Au(110) electrode surfaces at low insulin concentration ($0.01\,mg\,ml^{-1}$). *In situ* STM has clearly reached the level of resolution of the single molecule. The data show distinct surface-specific adsorption on the three electrode surfaces and point to the anticipated role of disulfides in both adsorption and *in situ* STM contrast. A common feature is a high surface density of molecular-scale (single-molecule) structures. These represent different insulin adsorbate features on the three surfaces.

Insulin monomer/dimer structures on Au(111) are evenly scattered over the wide terraces. The different sizes reflect different binding modes and Au–S binding units as well as concentration-dependent monomer/dimer distribution in the adsorption of presumably heavily unfolded protein. Different molecular-scale structures appear on Au(100) electrode surfaces. A dense adlayer covers most of the surface. Molecular-scale structures appear both at the Au(100) terraces and close to the terrace edges, where the layers extend across the steps. Larger structures are also scattered over the surface. Their rectangular shape, monatomic height, and alignment along the crystallographic axes of the substrate suggest that these structures are Au(100) islands from the lift of the surface reconstruction. The more open Au(110) surface structure induces still another adsorption mode. Like small thiol-based molecules such as Cys [94, 152, 153, 205], insulin displays much higher adsorption reactivity on Au(110) than on Au(111) and Au(100). Even the smallest insulin concentration used, which leads to low or intermediate coverage on Au(111) and Au(100) electrode surfaces, gives almost complete coverage on Au(110) electrode surfaces. The adsorbate organization follows the Au(110) electrode surface topology with insulin molecules aligned in the Au(110) surface grooves, at some places spilling over and merging into larger structures.

Insulin adsorption on the three low-index Au electrode surfaces thus displays a diversity of patterns resolved to the single-molecule level by *in situ* STM. Together with the voltammetric diversity of insulin monomer/dimer monolayers on the same surfaces, multifarious molecular-scale scenarios emerge but with some common notions. Au–S bond formation and disulfide bond splitting are first key determinants in the adsorption of S-rich, structurally "soft" protein molecules. The adsorption of such soft molecules is therefore invariably accompanied by extensive protein unfolding. More rigid molecules such as the blue copper protein azurin retain functional ET integrity in spite of the structural perturbation caused by disulfide binding to Au(111) electrode surfaces [60, 94, 115, 130]. Solvation in the bulk and adsorbed states can, further, vary widely. As noted, the amino acid Cys is, for example, organized in quite different patterns in different electrolyte solutions [152, 153, 158]. Adsorption of insulin would be accompanied both by strong solvation changes associated with hydrophilic amino acids in folded, unfolded, and adsorbed states, and with competing hydrophobic forces. Hydrolytic or other protein degradation caused by the strong chemical interaction between insulin and kink or other local metallic structures may accompany the adsorption process. Other, smaller molecular fragments can therefore contribute to the adsorption patterns.

In addition to the chemical diversity of insulin on the different Au electrode surfaces, the *in situ* STM contrasts reflect different electronic conductivity of the different insulin molecular fragments. The electronic density of the Au–S unit at the site of the tip thus exceeds significantly the contributions from the rest of the protein [37, 38, 158, 164]. *In situ* STM contrasts can therefore be dominated by the Au–S units while the molecular-scale contrast *distribution* is of course also determined by the rest of the protein.

2.6
Interfacial Electrochemistry and *In Situ* Imaging of Redox Metalloproteins and Metalloenzymes at the Single-Molecule Level

2.6.1
Metalloprotein Voltammetry at Bare and Modified Electrodes

PFV, reviewed extensively elsewhere [60, 92, 93, 206–208], is established as a powerful tool in protein science, including mapping molecular mechanisms of intramolecular and interfacial ET processes in surface-bound protein systems. PFV methodologies include linear, cyclic, fast-scan, square wave and differential pulse voltammetry, electrochemical impedance, X-ray photoelectron spectroscopy, and microcantilever sensor technology. Other approaches are represented by artificial interfacial biological ET chains [109], the use of ultramicroelectrodes [98], and the use of gold nanoparticles [209]. A range of redox metalloproteins including representatives of small ET proteins, the blue copper, heme group, and iron–sulfur proteins, and a range of redox metalloenzymes at the level of protein monolayers have been addressed [210]. The enzymes include glucose oxidase [211, 212], fumarate oxidase [213], succinate dehydrogenase [214], nitrate [215] and nitrite reductases [216, 217], peroxidases [206, 218], DMSO reductase [219], hydrogenases [220], and cytochrome c oxidase [221]. These represent complex interfacial electrocatalytic functions. Novel detail can be expected from nanoscale and single-molecule imaging such as for the smaller ET metalloproteins.

2.6.2
Single-Molecule Imaging of Functional Electron Transfer Metalloproteins by *In Situ* STM

Imaging of single protein molecules on conducting surfaces by STM and AFM was reported early on, but mostly in ambient air environment and on bare surfaces [45]. *In situ* STM of proteins in their natural aqueous biological media under

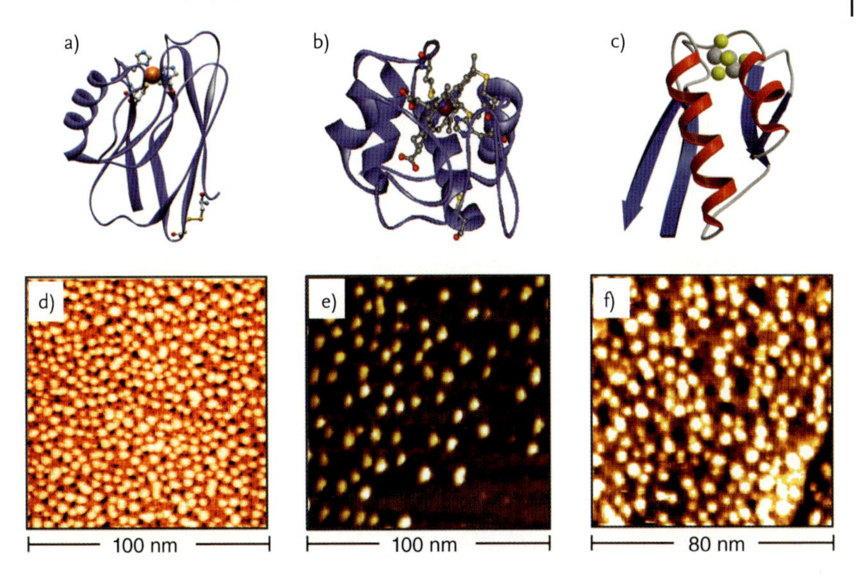

Figure 2.12 Overview of three-dimensional structures and *in situ* STM images of metalloproteins representative of the three ET protein classes characterized by single-crystal PFV and *in situ* STM to single-molecule resolution. (a) Blue copper protein *P. aeruginosa* azurin (PDB 4AZU) [94]; (b) heme protein *S. cerevisiae* cytochrome c (PDB 1YCC) [106]; (c) iron–sulfur protein *P. furiosus* ferredoxin (PDB 1SJ1) [154]. (d) *P. aeruginosa* azurin and (e) *S. cerevisiae* cytochrome c on bare Au(111); (f) *P. furiosus* ferredoxin on Au(111) modified by a mercaptopropionic acid SAM (cf. Figure 2.7).

electrochemical control and in combination with SAM-modified single-crystal electrode surfaces, on which the proteins retain their ET or enzyme function, is much more recent. Figures 2.12–2.14 show overviews of redox metalloproteins on surfaces such as those shown in Figure 2.6 imaged to single-molecule resolution by *in situ* STM. The molecules include representatives of the three major classes of ET metalloproteins, the blue copper proteins [54, 94, 118, 119, 202] (azurin in particular), heme group proteins [106], and iron–sulfur proteins [154] (Figure 2.12). Other single-molecule studies have addressed the multicenter redox metalloenzymes copper and decaheme nitrite reductases [54, 217, 222] (Figure 2.13). To these can be added *de novo* designed 4α-helix proteins without [223, 224] and with a heme group inserted [223, 224]. Single-crystal, bare, and modified electrodes were used, and the proteins in their functional states were mapped to single-molecule resolution. Other studies of azurin [114, 126, 127, 225], plastocyanine [226], and yeast cytochrome c [121] have also been reported. We discuss first briefly *in situ* STM studies of these different system classes [60] which illuminate both the powerful potential and some limitations of *in situ* STM imaging of single functional biological macromolecules. We then proceed to three new cases not reviewed before in the context of single-molecule imaging. These are the wild-type and mutant 4α-helix bundle heme protein cytochrome b_{562}, the bacterial diheme protein cytochrome c_4, and surface-immobilized metalloprotein–nanoparticle hybrid entities.

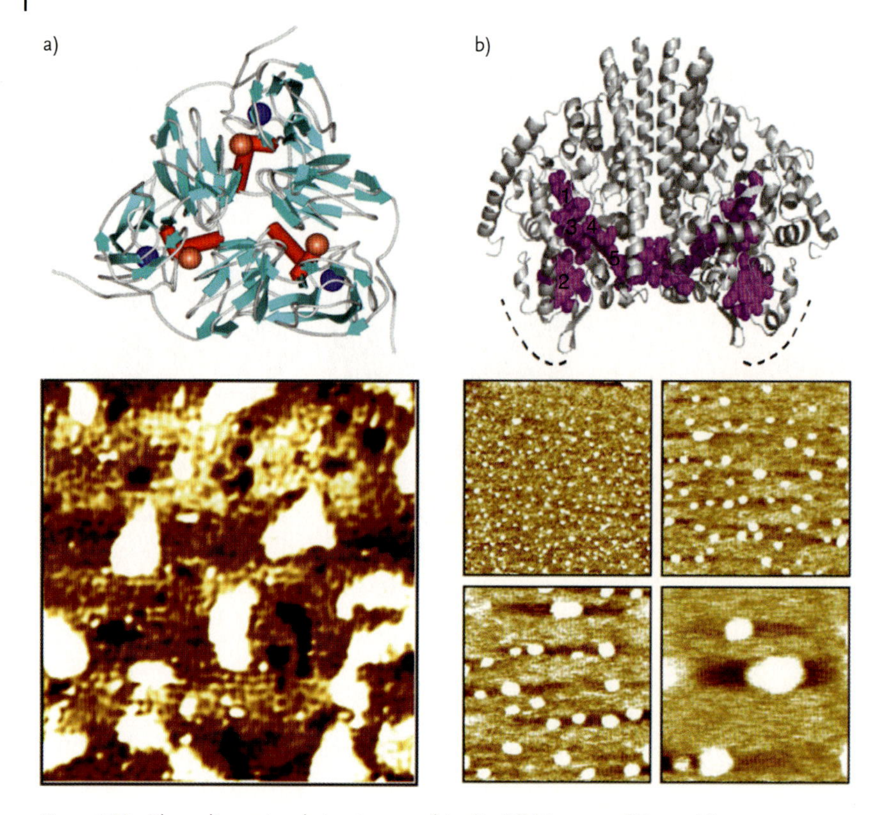

Figure 2.13 Three-dimensional structures and *in situ* STM images of the multicenter metalloenzymes (a) *A. xylosoxidans* copper nitrite reductase (PDB 1HAU) [217] and (b) *E. coli* decaheme nitrite reductase (PDB 1GU6) [227].

2.6.2.1 **Small Redox Metalloproteins: Blue Copper, Heme, and Iron–Sulfur Proteins**

Pseudomonas aeruginosa azurin (Az), horse heart (HHC) and yeast (*Saccharomyces cerevisiae*) cytochrome c (YCC), and *Pyrococcus furiosus* ferredoxin (*Pf*Fd), as representatives of the three classes of small ET metalloproteins (Figure 2.12), have been single-molecule *in situ* STM targets supported by single-crystal voltammetry and other methods. Focus was on conditions where the molecules retain their ET function in the immobilized state on bare (YCC, Az) Au(111) electrode surfaces or on Au(111) electrode surfaces modified by thiol-based SAMs. In separate ways HHC and *P. aeruginosa* Az have emerged as electrochemical paradigms for protein interfacial electrochemical ET [60].

2.6.2.2 **Single-Molecule Tunneling Spectroscopy of Wild-Type and Cys Mutant Cytochrome b_{562}**

4α-Helix heme proteins of approximately the same size as HHC and YCC constitute a second class of redox metalloproteins recently introduced as single-molecule

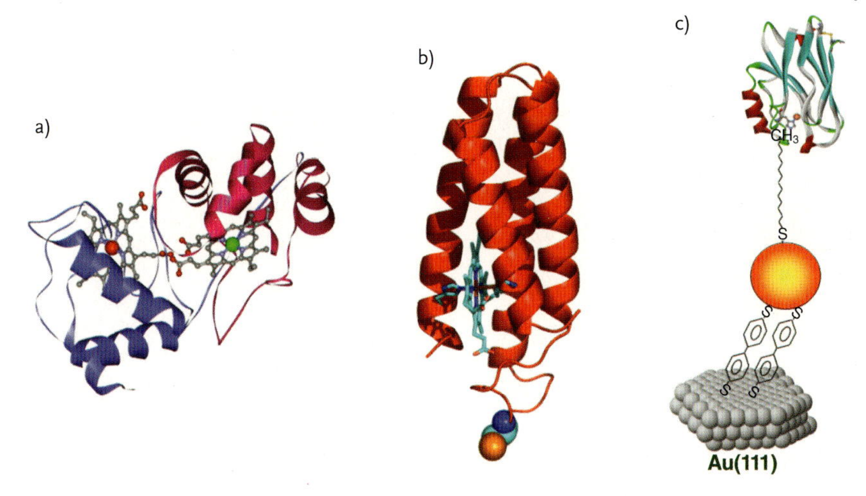

Figure 2.14 Three new target metalloproteins in single-molecule combined voltammetric and *in situ* STM studies recently reported: (a) *P. stutzeri* cytochrome c_4 (PDB 1EPT) [228]; (b) D50C mutant cytochrome b_{562} [229]; (c) *P. aeruginosa* azurin hydrophobically linked to 3 nm coated Au nanoparticle in a metalloprotein–AuNP hybrid immobilized in a 4,4′-biphenyldithiol matrix on a Au(111) electrode surface [230].

in situ STM and STS targets [229]. A totally synthetic 4α-helix heme-free carboprotein with a thiol-based linker group [223] and another totally synthetic MOP-C (modular organized protein; "C" represents Cys) linked via maleimidopropionic acid to Cys, in turn linked to a Au(111) electrode surface [224], are previously reported *in situ* STM cases of 4α-helix proteins. Although feasible, there were limitations in both cases. Molecular resolution could only be obtained for the molecular linker group in the former case probably due to high internal conformational flexibility of the heme-free ("apo-") protein. Notably, single-molecule structures did appear close to the potential of reductive desorption of the S-linked molecules. Single-molecule resolution of the MOP-C protein was achieved but it was inconclusive as to the role of the heme group in the single-molecule conduction process.

The recent study of Della Pia *et al.* [229] offered another, more successful strategy addressing the "natural" wild-type and mutant 4α-helix heme protein class cytochrome b_{562} (Figure 2.15). Cytochrome b_{562} is a class of bacterial (*Escherichia coli*) periplasmic respiratory ET proteins and contains a single heme group attached parallel to the α-helices by axial coordination via Met7 and His102. The reduction potential is slightly lower than those of the c-type cytochromes that also link the heme groups to the protein by thioether groups. The mutant protein obtained by replacing aspartic acid (D) with Cys at position 50, that is, the D50C mutant protein (Figure 2.15), has proved highly suitable for direct linking of the protein to Au(111) electrode surfaces in sub-monolayers diluted in a 1,4-dithiothreitol coadsorbed matrix. Robust voltammetry (in contrast to the wild-type protein), stable

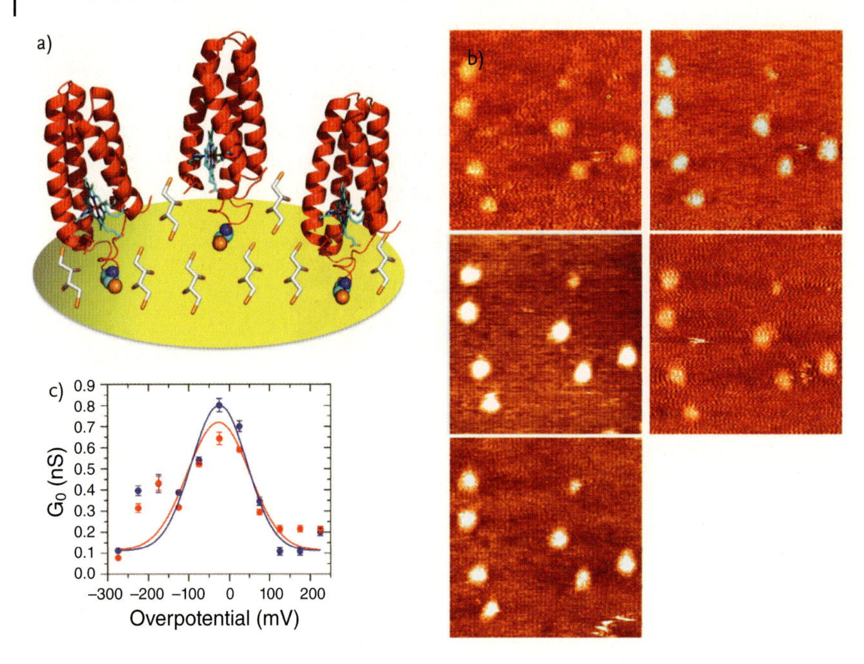

Figure 2.15 (a) D50C cytochrome b_{562} mutant linked directly to a Au(111) electrode surface via the inserted Cys residue [229]. (b) *In situ* STM images at different overpotentials. (c) Conductivity–overpotential correlation of the D50C cytochrome b_{562} mutant linked to the Au(111) electrode surface [229].

single-molecule *in situ* STM imaging, and single-molecule *in situ* STS with a conspicuous spectroscopic feature were obtained (Figure 2.15).

Figure 2.15a shows a schematic of the adsorbed D50C protein and Figure 2.15b the apparent height variation of a small assembly of adsorbed molecules with the electrochemical overpotential at constant bias voltage. The data are recast as a single-molecule normalized *in situ* STS plot in Figure 2.15c. A spectroscopic feature around the equilibrium redox potential with an "on–off" ratio of about five is observed (cf. the azurin case and a range of small molecules). By its gene technological diversity, this redox protein class thus offers a new type of single-molecule *in situ* STM targets.

2.6.2.3 Cytochrome c_4: A Prototype for Microscopic Electronic Mapping of Multicenter Redox Metalloproteins

Core functions in biological ET and redox enzyme function are commonly controlled by large metalloproteins with several transition metal centers. The photosynthetic reaction centers, redox protein complexes such as cytochrome c oxidase, the nitrogen-fixing enzyme nitrogenase, and other large redox enzyme complexes are examples. Intramolecular ET between the metal centers is a key operational function but mutual "cooperativity" among the centers is another key molecular function. This notion refers to the fact that ET to or from a given center affects the microscopic, as opposed to the macroscopic redox potentials and ET rate constants of all the other centers in overall cooperative charge transport. The number of electronic interactions is mostly prohibitive for microscopic mapping, but *two-center* metalloproteins offer the merits of representing prototype multicenter redox metalloproteins, at the same time with a simple enough electronic communication network such that complete microscopic thermodynamic and kinetic ET mapping is within reach. The bacterial respiratory two-heme protein *Pseudomonas stutzeri* cytochrome c_4 has been investigated most comprehensively in the context of cooperative intramolecular, and "gated" interfacial electrochemical ET. *Pseudomonas stutzeri* cytochrome c_4 has also acquired the recent status of a single-molecule *in situ* STM target [228].

Pseudomonas stutzeri cytochrome c_4 is organized in two globular domains, each with a single heme group, and connected with a 12-residue peptide chain (Figure 2.16). The protein is strongly dipolar, with excess positive and negative charge

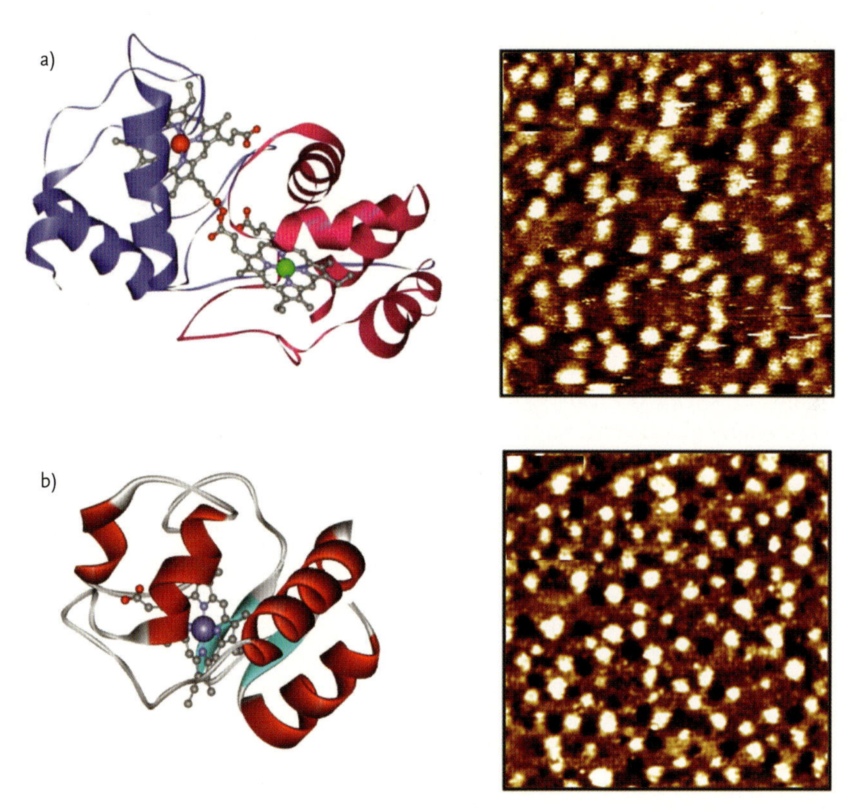

Figure 2.16 Three-dimensional structures with corresponding *in situ* STM images of molecules immobilized on Au(111) electrode surfaces modified by a mercaptodecanoic acid SAM (negatively charged): (a) *P. stutzeri* cytochrome c_4 (PDB 1EPT); (b) HHC (PDB 1HRC) [228].

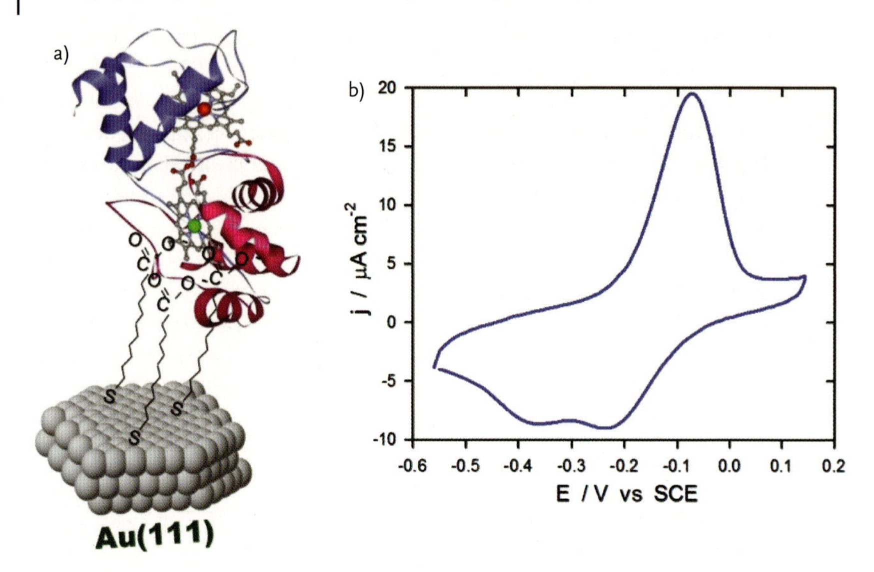

Figure 2.17 (a) *P. stutzeri* cytochrome c_4 immobilized on a Au(111) electrode modified by a mercaptodecanoic acid SAM. The positively charged high-potential domain is marked in red, the negatively charged low-potential domain in blue. (b) Cyclic voltammogram of *P. stutzeri* cytochrome c_4. The asymmetric appearance reflects the orientation of the molecule and intramolecular ET between the heme groups as a key feature [228].

(pH = 7) on the C- and N-terminal domains, respectively. The excess charges are reflected in a redox potential difference of about 100 mV, with the higher potential domain associated with the C-terminal domain and the lower potential with the N-terminal domain. These properties are crucial for electrostatic protein immobilization in specific orientations on SAM-modified Au(111) electrode surfaces and corresponding voltammetric patterns. Spectroscopic and other data are also available [231–233].

The heme groups are strongly hydrogen bonded via two propionates. The 19 Fe–Fe equilibrium distance implies that electronic contact is established but bimolecular ET with inorganic reaction partners discloses no evidence of intramolecular ET in time ranges up to 10–100 s. In contrast, the notably asymmetric cyclic voltammograms (Figure 2.17) indicate a pattern only compatible with fast intramolecular ET (10–100 μs) in the electrochemical two-ET process. Cytochrome c_4 is here oriented with the positively charged, high-potential C-terminal domain adjacent to the negatively charged SAM-modified Au(111) electrode surface used to record the data. This domain is reduced first in a cathodic scan followed by reduction of the remote N-terminal domain at the lower potential of this group, via intramolecular ET through the adjacent high-potential C-terminal domain. Reoxidation of the remote low-potential heme group in the anodic scan, however, only begins when the higher potential of the adjacent high-potential heme group is reached, again via fast intramolecular ET through the latter. Gated intramolecular ET triggered by protein binding to the electrode surface thus appears as a new *P. stutzeri* cytochrome c_4 feature.

The mechanistic view involving upright protein orientation and fast intramolecular ET is strikingly supported by *in situ* STM imaging of the molecule directly on the electrochemical surface (Figure 2.16). The individual molecular structures correspond in size closely to those of a single cytochrome c_4 domain or of the single-heme HHC. This accords with upright rather than horizontal or recumbent cytochrome c_4 orientation. *In situ* STM has therefore provided both single-molecule structural and ET mechanistic support for the interfacial behavior of this protein. The strikingly different intramolecular ET rates of *P. stutzeri* cytochrome c_4 in free solute and surface-bound states suggest further that surface binding triggers an efficient intramolecular ET channel. The latter could involve breaking of the hydrogen bond between the two propionates and translational relocation of the heme groups towards more efficient direct electronic overlap between the heme groups.

2.6.2.4 **Redox Metalloenzymes in Electrocatalytic Action Imaged at the Single-Molecule Level: Multicopper and Multiheme Nitrite Reductases**

Imaging of redox metalloenzymes directly in enzyme action offers fascinating novel *in situ* STM perspectives. Such target molecules pose greater challenges than the smaller ET proteins due to their larger size, fragile nature, and more sophisticated conductivity mechanisms. Recent efforts have illuminated prospects and limitations of imaging single-molecule operational enzymes by *in situ* STM. *In situ* STM of single-molecule electrochemical enzyme dynamics follows more established, optically based stochastic single-molecule enzyme dynamics [234–236]. The – so far very few – *in situ* STM studies of single-molecule electrochemical surface enzyme dynamics illustrate the challenges of this area of single-molecule biological surface science.

The redox nitrite reductase enzymes are central in the bacterially controlled global biological nitrogen cycle where they catalyze the reduction of nitrite to lower oxidation states of nitrogen [237–242]. There are three classes of nitrite reductases: the copper nitrite reductases (CuNiRs), the multiheme-based class, and the two-heme cytochrome cd_1-type nitrite reductases [241, 242]. There are several rationales for the trimeric CuNiRs (each monomer molecular mass about 36 kDa, here represented by *Achromobacter xylosoxidans* CuNiR) as single-molecule *in situ* STM targets. Each monomer contains a type I blue copper center for electron inlet, here from the substrate electrode, and a type II center for catalytic NO_2^- reduction. The two centers are directly linked, offering facile intramolecular ET (cf. patterns of cytochrome c_4 in Section 2.6.2.3). The substrate nitrite is a small molecule structurally not detectable by *in situ* STM on the enzyme background. As frequently observed in enzyme voltammetry, binding of substrate induces, however, significant electronic changes in the enzyme, notably in the contact between the electron acceptance center and the electrode surface [217, 243]. This holds prospects for electronic mapping of the enzyme in action at the single-molecule level.

Achromobacter xylosoxidans CuNiR as a representative of the CuNiR class is electrocatalytically active on modified Au(111) electrode surfaces [217]. The voltammetric patterns are controlled by subtle combinations of hydrophilic and hydrophobic surface properties of many surface linker molecules tested [243]. A notable outcome of these studies [217, 243] is that the enzyme on cysteamine- and benzylthiol-modified Au(111) electrode surfaces can be directly imaged in action at the level of the single molecule (Figure 2.13). Enzyme molecules even with the triangular crystallographic CuNiR substructure are observed. These structures *only* appear when nitrite is present. Single-molecule CuNiR thus offers a case for following electrochemical redox metalloenzyme activity at the single-molecule level.

Butt and coworkers studied the voltammetry of the (*Escherichia coli*) decaheme class of nitrite reductases [244] (Figure 2.13). Well-defined although unstable catalytic multi-electron voltammetric reduction of nitrite by the enzyme immobilized on bare Au(111) electrode surfaces is notable. The decaheme nitrite reductase is a second case for single-molecule *in situ* STM of a redox metalloenzyme, but image interpretation is presently not at the level for CuNiR [227]. Molecular-scale structures can be observed on the Au(111) electrode surface under conditions where the enzyme is electrocatalytically active, with both the natural dimer and surface-dissociated monomer enzyme structures identified. Molecular conductivities (*in situ* STM contrasts) of the enzyme and the active enzyme–substrate states are, however, not very distinctive.

2.6.2.5 Au–Nanoparticle Hybrids of Horse Heart Cytochrome c and *P. aeruginosa* Azurin

Inorganic particle, tube, and other structures of controlled size and shape have reached the size range of biomolecules such as proteins. This is also the size range of nanotechnology where electronic structures of the objects gradually transform from macroscopic to single-molecule behavior. The combination of inorganic metallic or semiconductor structures with comparable-size (bio)molecules into biological–inorganic hybrid structures is a novel core notion [245–256]. Nanoparticle- and nanowire-based electroanalytical chemistry and biological diagnostics have presently gone beyond proof of principle levels [245, 246]. We first address some issues of variable-size gold nanoparticles (AuNPs) at electrochemical surfaces. This is followed by a discussion of electrochemical properties and single-molecule *in situ* STM of AuNP–redox metalloprotein hybrids [230, 257].

AuNPs in Liquid-State Environment Solute pure and monolayer-coated ("capped") AuNPs are central targets in colloid and surface science also with a historical dimension [258–262]. Facile chemical syntheses introduced by Schmid *et al.* [260] and by Brust *et al.* [263] have boosted AuNP and other metal nanoparticle science towards characterization of the physical properties and use of these nanoscale metallic entities by multifarious techniques and in a variety of environments. Physical properties in focus have been the surface plasmon optical extinction band [264–269], scanning and transmission electron microscopy properties, and electrochemical properties of surface-immobilized coated AuNPs [173, 268–276]. To this can be added a variety of AuNP crosslinked molecular and biomolecular structures with applications in (electro)analytical chemistry, biological diagnostics, and others [245, 246, 258, 259]. The smallest (≤1 nm) particles behave like a similar-sized molecule with a discrete electronic spectrum or a wide HOMO/LUMO gap [277–279]. Intermediate-sized AuNPs, say 1.6 nm (Au_{145}) to 2.5 nm (already many hundreds of Au atoms), display electrochemically detectable Coulomb charging effects at room temperature [173, 272–275, 280–283] while the Coulomb energy spacings in larger AuNPs (≥3–5 nm) are too close for discreteness-of-charge effects. Successive Coulomb charging is associated with the electrostatic charging energy increments [280]

$$E_{\text{el.stat.}} = \left(z - \frac{1}{2}\right)\frac{e^2}{C_{\text{NP}}}; \quad C_{\text{NP}} = \varepsilon_s R_{\text{NP}} \tag{2.14}$$

where z is the number of electronic charges (e) already on the particle, C_{NP} the capacitance of a spherical particle of radius R_{NP}, and ε_s the dielectric constant of the surrounding medium. The energy separation between successive charge states exceeds the thermal energy, $k_B T$, if

$$\frac{e^2}{C_{\text{NP}}} > k_B T; \quad R_{\text{NP}} < \frac{e^2}{\varepsilon_s k_B T} \tag{2.15}$$

At room temperature this condition accords with $R_{NP} < 33$ nm in vacuum ($\varepsilon_s = 1$) and $R_{NP} < 6$–7 nm or $R_{NP} < 0.5$ nm if the short-range ($\varepsilon_s = 5$) and bulk static dielectric constant of water ($\varepsilon_s = 80$) applies, respectively. These radii should be corrected by the coating monolayer. AuNPs of 1.6 nm, Au_{145}, are observed to give electrochemical energy spacings of about 0.17 eV [173, 272–274, 283], corresponding to an "effective" dielectric constant of $\varepsilon_{eff} \approx 13$–20.

Quantized electrochemical capacitive charging of variable-size AuNP monolayers with protective monolayer coatings in aqueous and organic solvents has been reported [272, 277–279, 283]. A recent study reported *in situ* STM and Coulomb charging of water-soluble protected 1.6 nm AuNPs by both differential pulse voltammetry (DPV) and electrochemical *in situ* STM under *the same* aqueous electrolyte and room temperature conditions (Figure 2.18) [173]. Multiple 0.15 V spaced peaks were observed on the anodic side of the potential of zero charge both in DPV and *in situ* STM current–overpotential correlation (at fixed bias voltage) (cf. Section 2.3). The peak heights corresponded to 1.1 and 0.9×10^{-18} F per particle from DPV and *in situ* STM, respectively. This discrepancy was addressed in [272, 273, 283]. A recent report by Wandlowski and associates [284] has followed up these observations and reported what appears to be single-ET charging in Au(111) electrode–ferrocene interfacial electrochemical ET. This could reflect ET via molecular-scale roughened Au surface structures caused by chemical interaction of the thiol-modified ferrocene [284]. Observations such as these raise the issue of whether a molecular-sized AuNP in these contexts is appropriately regarded as a ("quasi-")molecule or as a small metallic entity. The environmental dynamics would strongly affect the ET processes in the former case but be of minor importance in the latter. This issue is presently unsettled.

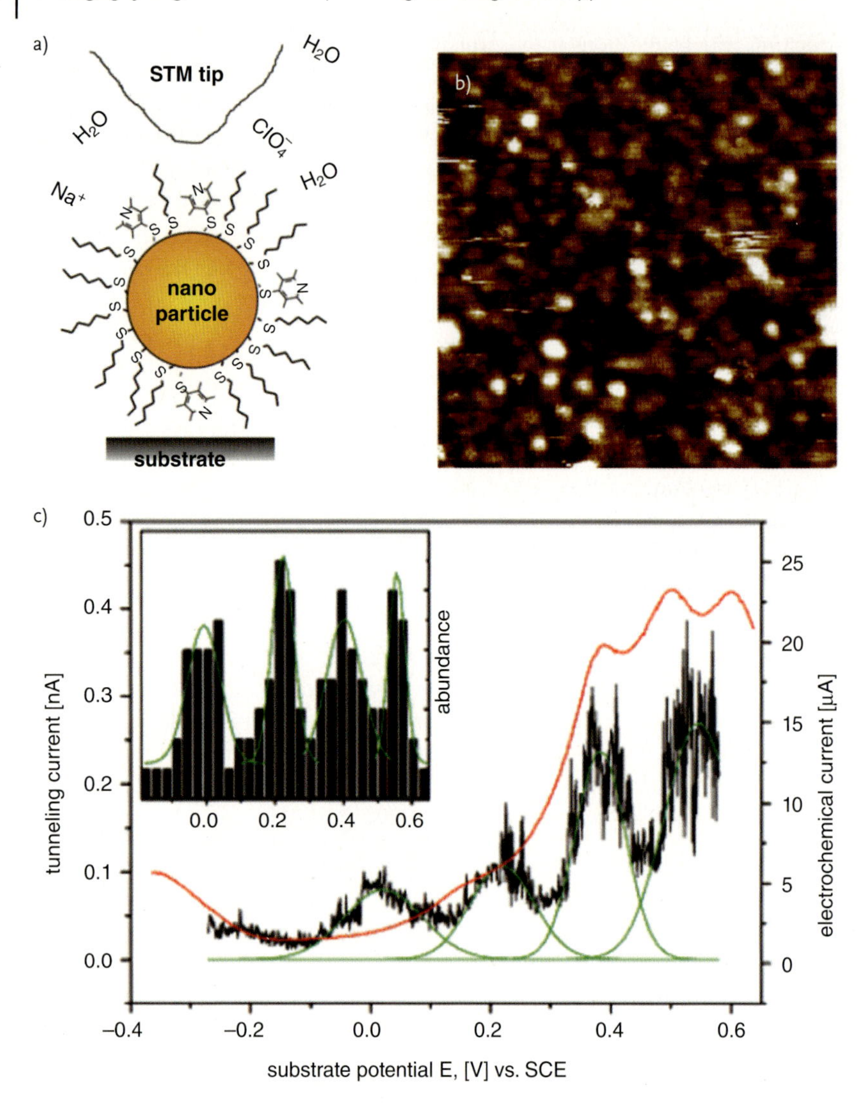

Figure 2.18 Electrochemical single-electron charging of coated 1.5 nm Au_{145} nanoparticle on a Au(111) electrode surface. (a) Schematic view of the AuNP in the *in situ* STM environment. (b) *In situ* STM image. (c) Differential pulse voltammogram of AuNP monolayer (red line) and *in situ* STM tunneling current–overpotential correlation (dark line). The green lines show Gaussian resolution but this is only to guide the eye. Inset: abundance histogram of the observed peak currents. For details, see [173].

Electrocatalysis by AuNPs and *In Situ* STM of AuNP–Metalloprotein Hybrids Protected AuNPs have been combined with redox metalloproteins [230, 257] into covalently, chemisorptively, or electrostatically linked hybrids of artificial supramolecular biostructures. AuNP–protein hybrids are important in electrocatalysis and bioelectrochemistry extending to large protein complexes such as cytochrome c oxidase. Biological recognition and diagnostics [245, 246] with links even to protein-based "nanocircuitry" are other areas [285]. We note here a few issues of the fundamental protein–nanoparticle interactions as reflected in interfacial electrochemical ET and *in situ* STM of protein–nanoparticle hybrids. These issues are illuminated by a recent study of a hybrid between a 3 nm coated AuNP linked to *P. aeruginosa* azurin [230]. An analogous study of a AuNP–cytochrome c hybrid was also reported [257].

Pseudomonas aeruginosa azurin (cf. Section 2.6.2.1) can be linked by strong hydrophobic forces to alkanethiol-protected 3 nm AuNPs in turn immobilized on a Au(111) electrode surface via an aromatic 4,4′-biphenyldithiol linker (Figure 2.19) [230]. Three observations of importance for AuNP-induced electrocatalysis emerge. The first is that the AuNP hydrophobically linked to azurin evokes an electrochemical ET rate enhancement of at least an order of magnitude compared to azurin alone on similar alkanethiol-modified surfaces. Notably, both AuNP-enhanced and direct interfacial ET voltammetric peaks are apparent at high scan rates (Figure 2.19). The second observation is that a two-step, azurin–AuNP and AuNP–electrode ET mechanism accords with the data. Interfacial AuNP–electrode ET appears as virtually activationless compared to protein–AuNP ET. The third observation is that the dual voltammetric pattern accords with dual *in situ* STM contrasts of sub-monolayers of the AuNP-linked azurin molecules (Figure 2.19). The weaker contrast which also displayed contrast fluctuations in time was assigned to the AuNP–azurin hybrid, the stronger robust contrast to individual AuNPs or azurin molecules.

AuNPs inserted between the electrode surface and redox metalloproteins therefore both work as effective molecular linkers and exert efficient electrocatalysis. Recent considerations based on resonance tunneling between the electrode and the molecule via the AuNP as a mechanism for enhanced interfacial ET rates suggest that electronic spillover rather than energetic resonance is a likely origin of the effects (J. Kleis *et al.*, work in progress). Even slightly enhanced spillover compared with a planar Au(111) surface is enough to enhance the ET rate by the observed amount over a 10–15 Å ET distance.

2.7 Some Concluding Observations and Outlooks

Single-crystal, atomically planar electrode surfaces have paved the way for introducing the scanning probe microscopies of STM and AFM in the bioelectrochemical sciences. The powerful *in situ* STM technology has increased the structural resolution of (bio)electrochemical electrode surfaces to the molecular and

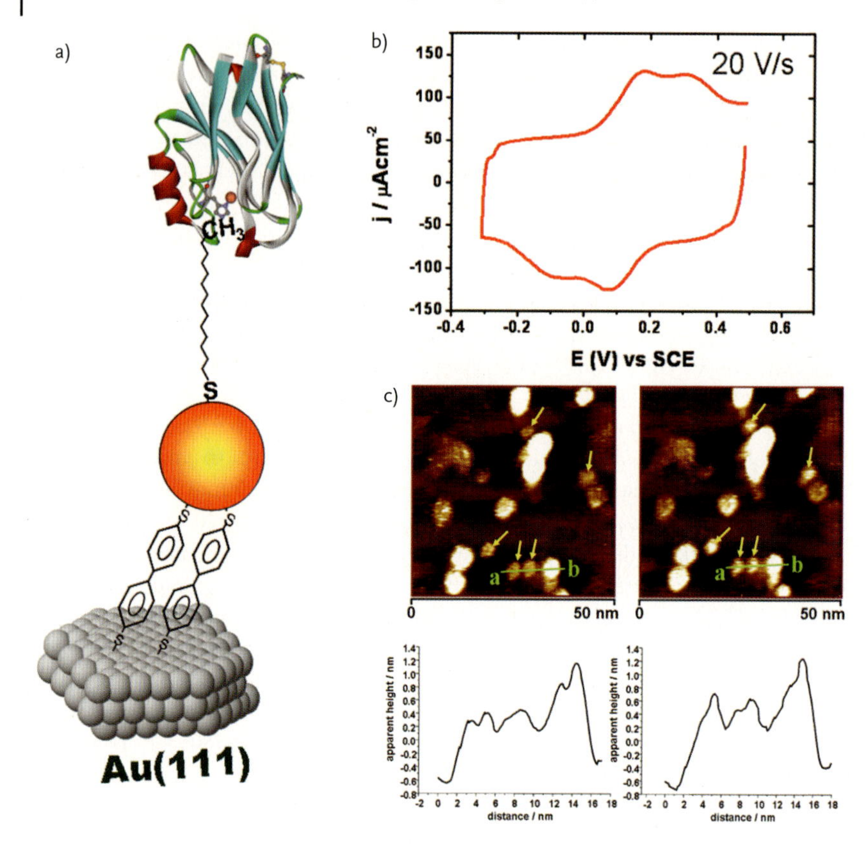

Figure 2.19 Cyclic voltammogram and *in situ* STM images of *P. aeruginosa* azurin–AuNP hybrid. (a) Schematic of the hybrid and of *P. aeruginosa azurin* alone on 4,4′-biphenyldithiol-modified Au(111) electrode. (b) Cyclic voltammogram showing the duality caused by the presence of both hybrid and azurin alone on the surface. The outer pair of voltammetric peaks represents azurin, the inner pair the hybrid with electrochemical ET rate constants higher by up to two orders of magnitude. (c) *In situ* STM images of azurin–AuNP hybrid in the 4,4′-biphenyldithiol matrix with height profiles indicated. The weak and temporally fluctuating contrasts were assigned to the hybrid, the strong and robust contrasts either to azurin or to AuNPs [230].

sometimes sub-molecular levels. High-resolution images have been achieved for the building blocks of biological macromolecules, that is, amino acids and nucleobases, and for lipid monolayers as building blocks of membrane structures. All of these form highly ordered, two-dimensional monolayers on Au(111) or Pt(111) electrode surfaces. Dynamic surface phenomena such as phase transitions in the adlayers and the monolayer formation process can also be followed. The degree of image detail in both the individual adsorbate molecules and in their lateral organization holds clear perspectives for the understanding of the interaction of "biological liquids" broadly with solid surfaces. It is in fact notable that these large and fragile metalloproteins can now be controlled towards the level of resolution of the single molecule and retain close to full enzyme functionality. This offers other technological perspectives for metalloenzyme-based biosensor function where functional units that respond to optical or magnetic signals can be inserted between the electrode and the reactive protein/enzyme.

We have overviewed cases of imaging of redox metalloproteins and metalloenzymes, of their amino acid building blocks, and of lipid membrane structures by *in situ* STM. As *in situ* STM is essentially based on molecular electronic conductivity, we have included some discussion of single-molecule electronic properties and interfacial ET processes. Focus has been on "natural" biomolecular aqueous solution environment and on the target molecules in monolayers on metallic electrode surfaces, where (bio)molecular function is controlled by the electrochemical potential. Redox metalloproteins have been identified as biomolecular target systems, but novel hybrid configurations where proteins are combined with metallic nanoparticles have also been addressed. Single-molecule resolution has been achieved under conditions where the molecules (i.e., particularly proteins) are fully active in ET or enzyme function. This opens the possibility that structural mapping of immobilized redox metalloproteins, DNA-based molecules, and membrane structures can be achieved. Given adequate theoretical support, ET and redox enzyme function as well as cooperative phenomena such as molecular monolayer and membrane formation can also be addressed, at the level of the single molecule. *Pseudomonas aeruginosa* azurin and cytochrome b_{562} have illuminated these perspectives. These could be extended to exploit surface mapping of biological sensors and other bioelectrochemical systems with potential device function.

STM and in *situ* STM are theoretically demanding because the properties recorded, that is, electrical currents through molecules, do not translate directly into molecular shape or topography. Long-range off-resonance molecular electronic conductivity is broadly understood in terms of electron exchange and energy gaps of the atomic or molecular orbitals involved and energy broadening of the orbitals closest to the substrate and tip. The "percolation" of electron density through the protein structure in interfacial protein ET processes, however, still constitutes a challenge. Exponential distance dependence of the tunneling current or conductivity is broadly expected (Section 2.3). This is sometimes observed but unexpected very weak current attenuation emerges in other cases such as for single- and double-strand oligonucleotides linked to enclosing Au electrodes [146, 147]. Hopping of excess electrons or holes by temporary particle accommodation in intermediate molecular orbitals does not immediately account for this due to unfavorable energetics. Configurational and energetic fluctuational effects [145, 286], formation of, so far elusive, electronic surface states [287], and trapping in extended polaron states have been forwarded as possible physical origins [288, 289].

Electronic conductivity of molecules including redox metalloproteins with accessible low-lying redox states in nanogap electrode configurations or *in situ* STM displays quite different patterns. These are dominated by sequential two-step (or multiple-step) hopping through the redox center, induced both by potential variation and environmental configurational fluctuations. Both redox molecules and metalloproteins have been brought to display interfacial electrochemistry at the single-molecule level along these lines. Theoretical notions rest on theory of interfacial electrochemical ET, but the nanogap environments or association with metallic nanoparticles have revealed new ET phenomena. "Switching" or "negative differential resistance," quite different from electrochemical ET at single, semi-infinite electrode surfaces, is an immediate example. "Coherent" multi-ET in a single *in situ* STM event when the redox level is strongly coupled to the electrodes is another novel ET phenomenon. We have discussed these phenomena, and new nanoscale electrochemical and bioelectrochemical systems have been shown to accord with these views.

(Bio)molecular electronics, enzyme electrochemistry, oligonucleotide monolayer organization, and DNA-based biological screening towards the single-molecule level are attractive applied perspectives of the new bioelectrochemistry and *in situ* STM. Networks of hybrid biomolecular structures of biomolecules with nanoparticles and nanowires in electrochemical nanogaps and *in situ* STM could here be novel targets. From a biotechnological perspective, *fundamental* bioelectrochemical innovation including new interfacial ET phenomena and theoretical support remains, however, a prerequisite.

Acknowledgments

This work was supported financially by the Danish Research Council for Technology and Production Sciences, the Lundbeck Foundation, and the Villum Kann Rasmussen Foundation.

References

1 Furtak, T.E., Kliewer, K.L., Lynch, D.W. (eds) (1980) Non-traditional approaches to the study of the solid-electrolyte interface. *Surface Science*, **101**.
2 Wieckowski, A. (ed.) (1999) *Interfacial Electrochemistry. Theory, Experiment and Applications*, Marcel Dekker, New York.
3 Hubbard, A.T. (1980) Electrochemistry of well-defined surfaces. *Accounts of Chemical Research*, **13**, 177–184.
4 Clavilier, J., Faure, R., Guinet, G., and Durand, R. (1980) Preparation of mono-crystalline Pt microelectrodes and electrochemical study of the plane surfaces cut in the direction of the (111) and (110) planes. *Journal of Electroanalytical Chemistry*, **107**, 205–209.
5 Hamelin, A. (1996) Cyclic voltammetry at gold single-crystal surfaces. 1. Behaviour at low-index faces. *Journal of Electroanalytical Chemistry*, **407**, 1–11.
6 Hamelin, A. and Martins, A.M. (1996) Cyclic voltammetry at gold single-crystal surfaces. 2. Behaviour of high-index faces. *Journal of Electroanalytical Chemistry*, **407**, 13–21.
7 Furtak, T.E. and Lynch, D.W. (1977) Bias potential effects on anisotropic electroreflectance of single-crystal silver. *Journal of Electroanalytical Chemistry*, **79**, 1–17.
8 Kolb, D.M. and Kotz, R. (1977) Electroreflectance spectra of Ag(111) electrodes. *Surface Science*, **64**, 96–108.
9 Bewick, A. (1983) In situ infrared-spectroscopy of the electrode electrolyte solution interphase. *Journal of Electroanalytical Chemistry*, **150**, 481–493.

10 Hansen, W.N., Osteryoung, R.A., and Kuwana, T. (1966) Internal reflection spectroscopic observation of electrode-solution interface. *Journal of the American Chemical Society*, **88**, 1062–1063.

11 Furtak, T.E. (1983) Current understanding of the mechanism of surface enhanced Raman scattering. *Journal of Electroanalytical Chemistry*, **150**, 375–388.

12 Jeanmaire, D.L. and Vanduyne, R.P. (1977) Surface Raman spectroelectrochemistry. 1. Heterocyclic, aromatic, and aliphatic amines adsorbed on anodized silver electrode. *Journal of Electroanalytical Chemistry*, **84**, 1–20.

13 Nuzzo, R.G. and Allara, D.L. (1983) Adsorption of bifunctional organic disulfides on gold surfaces. *Journal of the American Chemical Society*, **105**, 4481–4483.

14 Schumacher, R. (1990) The quartz microbalance–a novel approach to the in situ investigation of interfacial phenomena at the solid liquid junction. *Angewandte Chemie (International Edition in English)*, **29**, 329–343.

15 Carnie, S.L. and Torrie, G.M. (1984) The statistical-mechanics of the electrical double-layer. *Advances in Chemical Physics*, **56**, 141–253.

16 Henderson, D. and Blum, L. (1982) A simple theory of the electric double-layer including solvent effects. *Journal of Electroanalytical Chemistry*, **132**, 1–13.

17 Badiali, J.P., Rosinberg, M.L., and Goodisman, J. (1981) Effect of solvent on properties of the liquid–metal surface. *Journal of Electroanalytical Chemistry*, **130**, 31–45.

18 Badiali, J.P. (1986) Contribution of the metal to the differential capacitance of the ideally polarizable electrode. *Electrochimica Acta*, **31**, 149–154.

19 Kornyshev, A.A. (1988) Solvation of a metal surface, in *The Chemical Physics of Solvation. Part C. Solvation Phenomena in Specific Physical, Chemical and Biological Systems* (eds R.R. Dogonadze, E. Kálmán, A.A. Kornyshev, and J. Ulstrup), Elsevier, Amsterdam, pp. 355–400.

20 Lustenberger, P., Rohrer, H., Christoph, R., and Siegenthaler, H. (1988) Scanning tunneling microscopy at potential controlled electrode surfaces in electrolytic environment. *Journal of Electroanalytical Chemistry*, **243**, 225–235.

21 Wiechers, J., Twomey, T., Kolb, D.M., and Behm, R.J. (1988) An in situ scanning tunneling microscopy study of Au(111) with atomic scale resolution. *Journal of Electroanalytical Chemistry*, **248**, 451–460.

22 Engel, A., Lyubchenko, Y., and Muller, D. (1999) Atomic force microscopy: a powerful tool to observe biomolecules at work. *Trends in Cell Biology*, **9**, 77–80.

23 Gewirth, A.A. and Niece, B.K. (1997) Electrochemical applications of in situ scanning probe microscopy. *Chemical Reviews*, **97**, 1129–1162.

24 Kolb, D.M. (2001) Electrochemical surface science. *Angewandte Chemie International Edition*, **40**, 1162–1181.

25 Bunge, E., Port, S.N., Roelfs, B., Meyer, H., Baumgartel, H., Schiffrin, D.J., and Nichols, R.J. (1997) Adsorbate-induced etching of Au(111) surfaces: a combined in-situ infrared spectroscopy and scanning tunneling microscopy study. *Langmuir*, **13**, 85–90.

26 Bunge, E., Nichols, R.J., Roelfs, B., Meyer, H., and Baumgartel, H. (1996) Structure and dynamics of tetramethylthiourea adsorption on Au(111) studied by in situ scanning tunneling microscopy. *Langmuir*, **12**, 3060–3066.

27 Talonen, P., Sundholm, G., Floate, S., and Nichols, R.J. (1999) A combined in situ infrared spectroscopy and scanning tunnelling microscopy study of ethyl xanthate adsorption on Au(111). *Physical Chemistry Chemical Physics*, **1**, 3661–3666.

28 Wandlowski, T. (1995) Phase-transitions in uracil adlayers on Ag, Au and Hg electrodes: substrate effects. *Journal of Electroanalytical Chemistry*, **395**, 83–89.

29 Cunha, F., Jin, Q., Tao, N.J., and Li, C.Z. (1997) Structural phase transition in self-assembled 1,10′-phenanthroline monolayer on Au(111). *Surface Science*, **389**, 19–28.

30 Yoshimoto, S., Tsutsumi, E., Suto, K., Honda, Y., and Itaya, K. (2005) Molecular assemblies and redox reactions of zinc(II) tetraphenylporphyrin and zinc(II) phthalocyanine on Au(111) single crystal surface at electrochemical interface. *Chemical Physics*, **319**, 147–158.

31 Yoshimoto, S. and Itaya, K. (2007) Advances in supramolecularly assembled nanostructures of fullerenes and porphyrins at surfaces. *Journal of Porphyrins and Phthalocyanines*, **11**, 313–333.

32 Yeh, P. and Kuwana, T. (1977) Reversible electrode reaction of cytochrome c. *Chemistry Letters*, 1145–1148.

33 Niki, K., Yagi, T., Inokuchi, H., and Kimura, K. (1979) Electrochemical behavior of cytochrome c3 of *Desulfovibrio vulgaris*, strain Miyazaki, on the mercury electrode. *Journal of the American Chemical Society*, **101**, 3335–3340.

34 Eddowes, M.J. and Hill, H.A.O. (1977) Novel method for investigation of electrochemistry of metalloproteins–cytochrome c. *Journal of the Chemical Society. Chemical Communications*, 771–772.

35 Eddowes, M.J. and Hill, H.A.O. (1979) Electrochemistry of horse heart cytochrome c. *Journal of the American Chemical Society*, **101**, 4461–4464.

36 Wang, Y., Hush, N.S., and Reimers, J.R. (2007) Understanding the chemisorption of 2-methyl-2-propanethiol on Au(111). *Journal of Physical Chemistry C*, **111**, 10878–10885.

37 Wang, Y., Chi, Q.J., Hush, N.S., Reimers, J.R., Zhang, J.D., and Ulstrup, J. (2009) Scanning tunneling microscopic observation of adatom-mediated motifs on gold-thiol self-assembled monolayers at high coverage. *Journal of Physical Chemistry C*, **113**, 19601–19608.

38 Zhang, J.D., Bilič, A., Reimers, J.R., Hush, N.S., and Ulstrup, J. (2005) Coexistence of multiple conformations in cysteamine monolayers on Au(111). *Journal of Physical Chemistry B*, **109**, 15355–15367.

39 Kuznetsov, A.M. (1995) *Charge Transfer in Physics, Chemistry and Biology*, Gordon & Breach, Reading.

40 Kuznetsov, A.M. and Ulstrup, J. (1999) *Electron Transfer in Chemistry and Biology: An Introduction to the Theory*, John Wiley & Sons, Ltd, Chichester.

41 Schmickler, W. (1996) *Interfacial Electrochemistry*, Oxford University Press, New York.

42 Kuznetsov, A.M. and Ulstrup, J. (1991) Dissipative relaxation of a low-energy intermediate electronic state in 3-level electron-transfer. *Chemical Physics*, **157**, 25–33.

43 Kuznetsov, A.M., Sommer Larsen, P., and Ulstrup, J. (1992) Resonance and environmental fluctuation effects in STM currents through large adsorbed molecules. *Surface Science*, **275**, 52–64.

44 Schmickler, W. and Widrig, C. (1992) The investigation of redox reactions with a scanning tunneling microscope–experimental and theoretical aspects. *Journal of Electroanalytical Chemistry*, **336**, 213–221.

45 Zhang, J., Chi, Q., Kuznetsov, A.M., Hansen, A.G., Wackerbarth, H., Christensen, H.E.M., Andersen, J.E.T., and Ulstrup, J. (2002) Electronic properties of functional biomolecules at metal/aqueous solution interfaces. *Journal of Physical Chemistry B*, **106**, 1131–1152.

46 Zhang, J.D., Kuznetsov, A.M., and Ulstrup, J. (2003) In situ scanning tunnelling microscopy of redox molecules. Coherent electron transfer at large bias voltages. *Journal of Electroanalytical Chemistry*, **541**, 133–146.

47 Sumi, H. (1998) V-I characteristics of STM processes as a probe detecting vibronic interactions at a redox state in large molecular adsorbates such as electron-transfer metalloproteins. *Journal of Physical Chemistry B*, **102**, 1833–1844.

48 Medvedev, I.G. (2006) The effect of the electron–electron interaction on the pre-exponential factor of the rate constant of the adiabatic electrochemical electron transfer reaction. *Journal of Electroanalytical Chemistry*, **598**, 1–14.

49 Medvedev, I.G. (2007) The theory of in situ scanning tunneling microscopy of

redox molecules in the case of the fully adiabatic electron transitions. *Journal of Electroanalytical Chemistry*, **600**, 151–170.
50 Kuznetsov, A.M. and Medvedev, I.G. (2007) Does the electron spin affect the rates of electron tunneling in electrochemical systems? *Electrochemistry Communications*, **9**, 1343–1347.
51 Kuznetsov, A.M. and Ulstrup, J. (2000) Mechanisms of in situ scanning tunnelling microscopy of organized redox molecular assemblies. *Journal of Physical Chemistry A*, **104**, 11531–11540.
52 Kuznetsov, A.M. and Ulstrup, J. (2001) Mechanisms of in situ scanning tunnelling microscopy of organized redox molecular assemblies. *Journal of Physical Chemistry A*, **105**, 7494–7494.
53 Albrecht, T., Guckian, A., Kuznetsov, A.M., Vos, J.G., and Ulstrup, J. (2006) Mechanism of electrochemical charge transport in individual transition metal complexes. *Journal of the American Chemical Society*, **128**, 17132–17138.
54 Chi, Q.J., Zhang, J.D., Jensen, P.S., Christensen, H.E.M., and Ulstrup, J. (2006) Long-range interfacial electron transfer of metalloproteins based on molecular wiring assemblies. *Faraday Discussions*, **131**, 181–195.
55 Albrecht, T., Moth-Poulsen, K., Christensen, J.B., Guckian, A., Bjornholm, T., Vos, J.G., and Ulstrup, J. (2006) In situ scanning tunnelling spectroscopy of inorganic transition metal complexes. *Faraday Discussions*, **131**, 265–279.
56 Li, Z., Han, B., Meszaros, G., Pobelov, I., Wandlowski, T., Blaszczyk, A., and Mayor, M. (2006) Two-dimensional assembly and local redox activity of molecular hybrid structures in an electrochemical environment. *Faraday Discussions*, **131**, 121–143.
57 Marcus, R.A. and Sutin, N. (1985) Electron transfers in chemistry and biology. *Biochimica et Biophysica Acta*, **811**, 265–322.
58 Kohen, A. and Limbach, H.H. (eds) (2005) *Isotope Effects in Chemistry and Biology*, CRC Press, Boca Raton.
59 Dogonadze, R.R. and Chizmadzhev, Y.A. (1962) Computation of probability of an elementary act for certain heterogeneous redoxy reactions. *Doklady Akademii Nauk SSSR*, **144**, 1077–1080.
60 Zhang, J.D., Kuznetsov, A.M., Medvedev, I.G., Chi, Q.J., Albrecht, T., Jensen, P.S., and Ulstrup, J. (2008) Single-molecule electron transfer in electrochemical environments. *Chemical Reviews*, **108**, 2737–2791.
61 Kuznetsov, A.M. and Ulstrup, J. (1999) Simple schemes in chemical electron transfer formalism beyond single-mode quadratic forms: environmental vibrational dispersion and anharmonic nuclear motion. *Physical Chemistry Chemical Physics*, **1**, 5587–5592.
62 Marcus, R.A. (1956) On the theory of oxidation–reduction reactions involving electron transfer 1. *Journal of Chemical Physics*, **24**, 966–978.
63 German, E.D. and Kuznetsov, A.M. (1981) Outer sphere energy of reorganization in charge-transfer processes. *Electrochimica Acta*, **26**, 1595–1608.
64 Kharkats, Y.I. and Krishtalik, L.I. (1985) Medium reorganization energy and enzymatic-reaction activation energy. *Journal of Theoretical Biology*, **112**, 221–249.
65 Krishtalik, L.I. and Topolev, V.V. (2000) Effects of medium polarization and pre-existing field on activation energy of enzymatic charge-transfer reactions. *Biochimica et Biophysica Acta*, **1459**, 88–105.
66 Newton, M.D. (1999) Control of electron transfer kinetics: models for medium reorganization and donor–acceptor coupling. *Advances in Chemical Physics*, **106**, 303–375.
67 Kornyshev, A.A. (1985) Nonlocal electrostatics of solvation, in *The Chemical Physics of Solvation. Part A. Theory of Solvation* (eds R.R. Dogonadze, E. Kálmán, A.A. Kornyshev, and J. Ulstrup), Elsevier, Amsterdam, pp. 77–118.
68 Dzhavakhidze, P.G., Kornyshev, A.A., and Krishtalik, L.I. (1987) Activation energy of electrode reactions: the nonlocal effects. *Journal of Electroanalytical Chemistry*, **228**, 329–346.
69 Iversen, G., Kharkats, Y.I., and Ulstrup, J. (1998) Simple dielectric image charge

models for electrostatic interactions in metalloproteins. *Molecular Physics*, **94**, 297–306.

70 Wolynes, P.G. (1987) Linearized microscopic theories of nonequilibrium solvation. *Journal of Chemical Physics*, **86**, 5133–5136.

71 Dogonadze, R.R. and Kuznetsov, A.M. (1975) Dynamic effect of ionic atmosphere on kinetics of elementary act of charge-transfer reactions. *Journal of Electroanalytical Chemistry*, **65**, 545–554.

72 Jensen, T.J., Gray, H.B., Winkler, J.R., Kuznetsov, A.M., and Ulstrup, J. (2000) Dynamic ionic strength effects in fast bimolecular electron transfer between a redox metalloprotein of high electrostatic charge and an inorganic reaction partner. *Journal of Physical Chemistry B*, **104**, 11556–11562.

73 Kuznetsov, A.M. and Ulstrup, J. (1981) Long-range intramolecular electron transfer in aromatic radical-anions and binuclear transition-metal complexes. *Journal of Chemical Physics*, **75**, 2047–2055.

74 Larsson, S. (1981) Electron transfer in chemical and biological systems: orbital rules for non-adiabatic transfer. *Journal of the American Chemical Society*, **103**, 4034–4040.

75 Christensen, H.E.M., Conrad, L.S., Mikkelsen, K.V., Nielsen, M.K., and Ulstrup, J. (1990) Direct and superexchange electron tunneling at the adjacent and remote sites of higher-plant plastocyanins. *Inorganic Chemistry*, **29**, 2808–2816.

76 Gray, H.B. and Winkler, J.R. (2003) Electron tunneling through proteins. *Quarterly Reviews of Biophysics*, **36**, 341–372.

77 Skourtis, S.S. and Beratan, D.N. (1999) Theories of structure–function relationships for bridge-mediated electron transfer reactions. *Advances in Chemical Physics*, **106**, 377–452.

78 Datta, S. (1995) *Electronic Transport in Mesoscopic Systems*, Cambridge University Press, Cambridge.

79 Solomon, G.C., Reimers, J.R., and Hush, N.S. (2005) Overcoming computational uncertainties to reveal chemical sensitivity in single molecule conduction calculations. *Journal of Chemical Physics*, **122**, 224502.

80 Reimers, J.R., Solomon, G.C., Gagliardi, A., Bilič, A., Hush, N.S., Frauenheim, T., Di Carlo, A., and Pecchia, A. (2007) The Green's function density functional tight-binding (gDFTB) method for molecular electronic conduction. *Journal of Physical Chemistry A*, **111**, 5692–5702.

81 Troisi, A. and Ratner, M.A. (2006) Molecular signatures in the transport properties of molecular wire junctions: what makes a junction "molecular"? *Small*, **2**, 172–181.

82 Jones, D.R. and Troisi, A. (2007) Single molecule conductance of linear dithioalkanes in the liquid phase: apparently activated transport due to conformational flexibility. *Journal of Physical Chemistry C*, **111**, 14567–14573.

83 Mcconnell, H. (1961) Intramolecular charge transfer in aromatic free radicals. *Journal of Chemical Physics*, **35**, 508–515.

84 Dogonadze, R.R., Ulstrup, J., and Kharkats, Y.I. (1972) Theory of electrode reactions through bridge transition states – bridges with a discrete electronic spectrum. *Journal of Electroanalytical Chemistry*, **39**, 47–61.

85 Dogonadze, R.R., Kharkats, Y.I., and Ulstrup, J. (1972) Theory of concert reactions of proton transfer in a polar medium. *Doklady Akademii Nauk SSSR*, **207**, 640–644.

86 Dogonadze, R.R., Ulstrup, J., and Kharkats, Y.I. (1973) Theory of polar medium electron-transfer reactions through bridge groups with a quasicontinuous energy spectrum. *Journal of Theoretical Biology*, **40**, 259–277.

87 Kharkats, Y.I., Madumarov, A.K., and Vorotyntsev, M.A. (1974) Application of density matrix method in quantum-mechanical calculation of bridge-assisted electron-transfer probability in polar media. *Journal of the Chemical Society. Faraday Transactions II*, **70**, 1578–1590.

88 Kuznetsov, A.M. and Kharkats, Y.I. (1976) Semiclassical theory of adiabatic and nonadiabatic electron-transfer bridging reactions. *Soviet Electrochemistry*, **12**, 1170–1176.

89 Kuznetsov, A.M. and Kharkats, Y.I. (1977) Method of classical trajectories in theory of bridge-type electron-transfer reactions. *Soviet Electrochemistry*, **13**, 1283–1288.

90 Muscat, J.P. and Newns, D.M. (1978) Chemisorption on metals. *Progress in Surface Science*, **9**, 1–43.

91 Hill, H.A.O. (1996) The development of bioelectrochemistry. *Coordination Chemistry Reviews*, **151**, 115–123.

92 Heering, H.A., Hirst, J., and Armstrong, F.A. (1998) Interpreting the catalytic voltammetry of electroactive enzymes adsorbed on electrodes. *Journal of Physical Chemistry B*, **102**, 6889–6902.

93 Armstrong, F.A. (2002) Insights from protein film voltammetry into mechanisms of complex biological electron-transfer reactions. *Journal of the Chemical Society. Dalton Transactions*, 661–671.

94 Chi, Q.J., Zhang, J.D., Nielsen, J.U., Friis, E.P., Chorkendorff, I., Canters, G.W., Andersen, J.E.T., and Ulstrup, J. (2000) Molecular monolayers and interfacial electron transfer of *Pseudomonas aeruginosa* azurin on Au(111). *Journal of the American Chemical Society*, **122**, 4047–4055.

95 Bard, A.J. and Faulkner, L.R. (2000) *Electrochemical Methods: Fundamentals and Applications*, John Wiley & Sons, Inc., New York.

96 Avila, A., Gregory, B.W., Niki, K., and Cotton, T.M. (2000) An electrochemical approach to investigate gated electron transfer using a physiological model system: cytochrome c immobilized on carboxylic acid-terminated alkanethiol self-assembled monolayers on gold electrodes. *Journal of Physical Chemistry B*, **104**, 2759–2766.

97 Feng, Z.Q., Imabayashi, S., Kakiuchi, T., and Niki, K. (1995) Electroreflectance spectroscopic study of the electron-transfer rate of cytochrome-C electrostatically immobilized on the omega-carboxyl alkanethiol monolayer modified gold electrode. *Journal of Electroanalytical Chemistry*, **394**, 149–154.

98 Heering, H.A., Wiertz, F.G.M., Dekker, C., and de Vries, S. (2004) Direct immobilization of native yeast Iso-1 cytochrome c on bare gold: fast electron relay to redox enzymes and zeptomole protein-film voltammetry. *Journal of the American Chemical Society*, **126**, 11103–11112.

99 Lamle, S.E., Albracht, S.P.J., and Armstrong, F.A. (2004) Electrochemical potential-step investigations of the aerobic interconversions of [NiFe]-hydrogenase from *Allochromatium vinosum*: insights into the puzzling difference between unready and ready oxidized inactive states. *Journal of the American Chemical Society*, **126**, 14899–14909.

100 Ataka, K., and Heberle, J. (2003) Electrochemically induced surface-enhanced infrared difference absorption (SEIDA) spectroscopy of a protein monolayer. *Journal of the American Chemical Society*, **125**, 4986–4987.

101 Murgida, D.H. and Hildebrandt, P. (2004) Electron-transfer processes of cytochrome c at interfaces. New insights by surface-enhanced resonance Raman spectroscopy. *Accounts of Chemical Research*, **37**, 854–861.

102 Seibert, M., Picorel, R., Kim, J.H., and Cotton, T.M. (1992) Surface-enhanced Raman scattering spectroscopy of photosynthetic membranes and complexes. *Methods in Enzymology*, **213**, 31–42.

103 Corn, R.M. and Higgins, D.A. (1994) Optical 2nd-harmonic generation as S probe of surface chemistry. *Chemical Reviews*, **94**, 107–125.

104 Knoll, W. (1998) Interfaces and thin films as seen by bound electromagnetic waves. *Annual Review of Physical Chemistry*, **49**, 569–638.

105 Brockman, J.M., Nelson, B.P., and Corn, R.M. (2000) Surface plasmon resonance imaging measurements of ultrathin organic films. *Annual Review of Physical Chemistry*, **51**, 41–63.

106 Hansen, A.G., Boisen, A., Nielsen, J.U., Wackerbarth, H., Chorkendorff, I., Andersen, J.E.T., Zhang, J.D., and Ulstrup, J. (2003) Adsorption and interfacial electron transfer of *Saccharomyces cerevisiae* yeast cytochrome c monolayers on Au(111) electrodes. *Langmuir*, **19**, 3419–3427.

107 Alessandrini, A., Gerunda, M., Facci, P., Schnyder, B., and Kotz, R. (2003) Tuning molecular orientation in protein films. *Surface Science*, **542**, 64–71.

108 Royal Society of Chemistry (2000) *Faraday Discussions*, **116**, 1–353.

109 Gilardi, G., Fantuzzi, A., and Sadeghi, S.J. (2001) Engineering and design in the bioelectrochemistry of metalloproteins. *Current Opinion in Structural Biology*, **11**, 491–499.

110 Chen, K.S., Hirst, J., Camba, R., Bonagura, C.A., Stout, C.D., Burgess, B.K., and Armstrong, F.A. (2000) Atomically defined mechanism for proton transfer to a buried redox centre in a protein. *Nature*, **405**, 814–817.

111 Davis, J.J., Bruce, D., Canters, G.W., Crozier, J., and Hill, H.A.O. (2003) Genetic modulation of metalloprotein electron transfer at bare gold. *Chemical Communications*, 576–577.

112 Chen, X.X., Discher, B.M., Pilloud, D.L., Gibney, B.R., Moser, C.C., and Dutton, P.L. (2002) De novo design of a cytochrome b maquette for electron transfer and coupled reactions on electrodes. *Journal of Physical Chemistry B*, **106**, 617–624.

113 Willner, I., Heleg-Shabtai, V., Katz, E., Rau, H.K., and Haehnel, W. (1999) Integration of a reconstituted de novo synthesized hemoprotein and native metalloproteins with electrode supports for bioelectronic and bioelectrocatalytic applications. *Journal of the American Chemical Society*, **121**, 6455–6468.

114 Albrecht, T., Li, W.W., Ulstrup, J., Haehnel, W., and Hildebrandt, P. (2005) Electrochemical and spectroscopic investigations of immobilized de novo designed heme proteins on metal electrodes. *ChemPhysChem*, **6**, 961–970.

115 Chi, Q.J., Zhang, J.D., Friis, E.P., Andersen, J.E.T., and Ulstrup, J. (1999) Electrochemistry of self-assembled monolayers of the blue copper protein *Pseudomonas aeruginosa* azurin on Au(111). *Electrochemistry Communications*, **1**, 91–96.

116 Andersen, J.E.T., Moller, P., Pedersen, M.V., and Ulstrup, J. (1995) Cytochrome c dynamics at gold and glassy-carbon surfaces monitored by in-situ scanning tunnel microscopy. *Surface Science*, **325**, 193–205.

117 Zhang, J.D., Chi, Q.J., Dong, S.J., and Wang, E.K. (1996) In situ electrochemical scanning tunnelling microscopy investigation of structure for horseradish peroxidase and its electrocatalytic property. *Bioelectrochemistry and Bioenergetics*, **39**, 267–274.

118 Friis, E.P., Andersen, J.E.T., Madsen, L.L., Moller, P., and Ulstrup, J. (1997) In situ STM and AFM of the copper protein *Pseudomonas aeruginosa* azurin. *Journal of Electroanalytical Chemistry*, **431**, 35–38.

119 Friis, E.P., Andersen, J.E.T., Kharkats, Y.I., Kuznetsov, A.M., Nichols, R.J., Zhang, J.D., and Ulstrup, J. (1999) An approach to long-range electron transfer mechanisms in metalloproteins: in situ scanning tunneling microscopy with submolecular resolution. *Proceedings of the National Academy of Sciences of the United States of America*, **96**, 1379–1384.

120 Facci, P., Alliata, D., and Cannistraro, S. (2001) Potential-induced resonant tunneling through a redox metalloprotein investigated by electrochemical scanning probe microscopy. *Ultramicroscopy*, **89**, 291–298.

121 Bonanni, B., Alliata, D., Bizzarri, A.R., and Cannistraro, S. (2003) Topological and electron-transfer properties of yeast cytochrome c adsorbed on bare gold electrodes. *ChemPhysChem*, **4**, 1183–1188.

122 Friis, E.P., Andersen, J.E.T., Madsen, L.L., Moller, P., Nichols, R.J., Olesen, K.G., and Ulstrup, J. (1998) Metalloprotein adsorption on Au(111) and polycrystalline platinum investigated by in situ scanning tunneling microscopy with molecular and submolecular resolution. *Electrochimica Acta*, **43**, 2889–2897.

123 Liang, W.J., Shores, M.P., Bockrath, M., Long, J.R., and Park, H. (2002) Kondo resonance in a single-molecule transistor. *Nature*, **417**, 725–729.

124 Park, J., Pasupathy, A.N., Goldsmith, J.I., Chang, C., Yaish, Y., Petta, J.R., Rinkoski, M., Sethna, J.P., Abruna,

H.D., McEuen, P.L., and Ralph, D.C. (2002) Coulomb blockade and the Kondo effect in single-atom transistors. *Nature*, **417**, 722–725.

125 Kubatkin, S., Danilov, A., Hjort, M., Cornil, J., Bredas, J.L., Stuhr-Hansen, N., Hedegard, P., and Bjornholm, T. (2003) Single-electron transistor of a single organic molecule with access to several redox states. *Nature*, **425**, 698–701.

126 Davis, J.J., Hill, H.A.O., and Bond, A.M. (2000) The application of electrochemical scanning probe microscopy to the interpretation of metalloprotein voltammetry. *Coordination Chemistry Reviews*, **200**, 411–442.

127 Davis, J.J. and Hill, H.A.O. (2002) The scanning probe microscopy of metalloproteins and metalloenzymes. *Chemical Communications*, 393–401.

128 Gittins, D.I., Bethell, D., Schiffrin, D.J., and Nichols, R.J. (2000) A nanometre-scale electronic switch consisting of a metal cluster and redox-addressable groups. *Nature*, **408**, 67–69.

129 Kornyshev, A.A., Kumetsov, A.M., and Ulstrup, J. (2005) Double-tunnel nanoscale switch with a redox mediator: operational principles and tunneling spectroscopy. *ChemPhysChem*, **6**, 583–586.

130 Hammerich, O. and Ulstrup, J. (eds) (2008) *Bioinorganic Electrochemistry*, Springer, Dordrecht.

131 Allen, M.J., Balooch, M., Subbiah, S., Tench, R.J., Siekhaus, W., and Balhorn, R. (1991) Scanning tunneling microscope images of adenine and thymine at atomic resolution. *Scanning Microscopy*, **5**, 625–630.

132 Allen, M.J., Balooch, M., Subbiah, S., Tench, R.J., Balhorn, R., and Siekhaus, W. (1992) Analysis of adenine and thymine adsorbed on graphite by scanning tunneling and atomic force microscopy. *Ultramicroscopy*, **42**, 1049–1053.

133 Heckl, W.M., Smith, D.P.E., Binnig, G., Klagges, H., Hansch, T.W., and Maddocks, J. (1991) Two-dimensional ordering of the DNA base guanine observed by scanning tunneling microscopy. *Proceedings of the National Academy of Sciences of the United States of America*, **88**, 8003–8005.

134 Sowerby, S.J., and Petersen, G.B. (1997) Scanning tunneling microscopy of uracil monolayers self-assembled at the solid/liquid interface. *Journal of Electroanalytical Chemistry*, **433**, 85–90.

135 Tao, N.J., Derose, J.A., and Lindsay, S.M. (1993) Self-assembly of molecular superstructures studied by in situ scanning tunneling microscopy: DNA bases on Au(111). *Journal of Physical Chemistry*, **97**, 910–919.

136 Cavallini, M., Aloisi, G., Bracali, M., and Guidelli, R. (1998) An in situ STM investigation of uracil on Ag(111). *Journal of Electroanalytical Chemistry*, **444**, 75–81.

137 Holzle, M.H., Wandlowski, T., and Kolb, D.M. (1995) Structural transitions in uracil adlayers on gold single-crystal electrodes. *Surface Science*, **335**, 281–290.

138 Roelfs, B., Bunge, E., Schroter, C., Solomun, T., Meyer, H., Nichols, R.J., and Baumgartel, H. (1997) Adsorption of thymine on gold single-crystal electrodes. *Journal of Physical Chemistry B*, **101**, 754–765.

139 Dretschkow, T., Dakkouri, A.S., and Wandlowski, T. (1997) In-situ scanning tunneling microscopy study of uracil on Au(111) and Au(100). *Langmuir*, **13**, 2843–2856.

140 Dretschkow, T. and Wandlowski, T. (1998) In-situ scanning tunneling microscopy study of uracil on Au(100). *Electrochimica Acta*, **43**, 2991–3006.

141 Wandlowski, T., Lampner, D., and Lindsay, S.M. (1996) Structure and stability of cytosine adlayers on Au(111): an in-situ STM study. *Journal of Electroanalytical Chemistry*, **404**, 215–226.

142 Nichols, R.J., Haiss, W., Fernig, D.G., van Zalinge, H., Schiffrin, D.J., and Ulstrup, J. (2008) *In situ* STM studies of immobilized biomolecules at the electrode–electrolyte interface, in *Bioinorganic Electrochemistry* (eds O. Hammerich and J. Ulstrup), Springer, Dordrecht, pp. 207–247.

143 Han, B., Li, Z.H., Li, C., Pobelov, I., Su, G.J., Aguilar-Sanchez, R., and Wandlowski, T. (2009) From self-assembly to charge transport with single molecules – an electrochemical approach. *Templates in Chemistry III*, **287**, 181–255.

144 Kuznetsov, A.M. and Ulstrup, J. (2008) Charge transport of solute oligonucleotides in metallic nanogaps – observations and some puzzles, in *Bioinorganic Electrochemistry* (eds O. Hammerich and J. Ulstrup), Springer, Dordrecht, pp. 161–205.

145 Genereux, J.C. and Barton, J.K. (2010) Mechanisms for DNA charge transport. *Chemical Reviews*, **110**, 1642–1662.

146 van Zalinge, H., Schiffrin, D.J., Bates, A.D., Haiss, W., Ulstrup, J., and Nichols, R.J. (2006) Single-molecule conductance measurements of single- and double-stranded DNA oligonucleoticles. *ChemPhysChem*, **7**, 94–98.

147 Xu, B.Q., Zhang, P.M., Li, X.L., and Tao, N.J. (2004) Direct conductance measurement of single DNA molecules in aqueous solution. *Nano Letters*, **4**, 1105–1108.

148 van Zalinge, H., Schiffrin, D.J., Bates, A.D., Starikov, E.B., Wenzel, W., and Nichols, R.J. (2006) Variable-temperature measurements of the single-molecule conductance of double-stranded DNA. *Angewandte Chemie International Edition*, **45**, 5499–5502.

149 Wagenknecht, H.A. (ed.) (2005) *Charge Transfer in DNA*, Wiley-VCH Verlag GmbH, Weinheim.

150 Voityuk, A.A., Siriwong, K., and Rosch, N. (2004) Environmental fluctuations facilitate electron–hole transfer from guanine to adenine in DNA pi stacks. *Angewandte Chemie International Edition*, **43**, 624–627.

151 Zhang, J.D., Welinder, A.C., Chi, Q., and Ulstrup, J. (2011) Electrochemically controlled self-assembled monolayers characterized with molecular and sub-molecular resolution. *Physical Chemistry Chemical Physics*, **13**, 5526–5545.

152 Zhang, J.D., Chi, Q.J., Nazmutdinov, R.R., Zinkicheva, T.T., and Bronshtein, M.D. (2009) Submolecular electronic mapping of single cysteine molecules by in situ scanning tunneling imaging. *Langmuir*, **25**, 2232–2240.

153 Zhang, J.D., Chi, Q.J., Nielsen, J.U., Friis, E.P., Andersen, J.E.T., and Ulstrup, J. (2000) Two-dimensional cysteine and cystine cluster networks on Au(111) disclosed by voltammetry and in situ scanning tunneling microscopy. *Langmuir*, **16**, 7229–7237.

154 Zhang, J.D., Christensen, H.E.M., Ooi, B.L., and Ulstrup, J. (2004) In situ STM imaging and direct electrochemistry of *Pyrococcus furiosus* ferredoxin assembled on thiolate-modified Au(111) surfaces. *Langmuir*, **20**, 10200–10207.

155 Zhang, J.D., Chi, Q.J., and Ulstrup, J. (2006) Assembly dynamics and detailed structure of 1-propanethiol monolayers on Au(111) surfaces observed real time by in situ STM. *Langmuir*, **22**, 6203–6213.

156 Dakkouri, A.S., Kolb, D.M., EdelsteinShima, R., and Mandler, D. (1996) Scanning tunneling microscopy study of L-cysteine on Au(111). *Langmuir*, **12**, 2849–2852.

157 Xu, Q.M., Wan, L.J., Wang, C., Bai, C.L., Wang, Z.Y., and Nozawa, T. (2001) New structure of L-cysteine self-assembled monolayer on Au(111): studies by in situ scanning tunneling microscopy. *Langmuir*, **17**, 6203–6206.

158 Nazmutdinov, R.R., Zhang, J.D., Zinkicheva, T.T., Manyurov, I.R., and Ulstrup, J. (2006) Adsorption and in situ scanning tunneling microscopy of cysteine on Au(111): structure, energy, and tunneling contrasts. *Langmuir*, **22**, 7556–7567.

159 Kuhnle, A., Linderoth, T.R., and Besenbacher, F. (2003) Self-assembly of monodispersed, chiral nanoclusters of cysteine on the Au(110)-(1 × 2) surface. *Journal of the American Chemical Society*, **125**, 14680–14681.

160 Kuhnle, A., Linderoth, T.R., Schunack, M., and Besenbacher, F. (2006) L-cysteine adsorption structures on Au(111) investigated by scanning tunneling microscopy under ultrahigh vacuum conditions. *Langmuir*, **22**, 2156–2160.

161 Kuhnle, A., Linderoth, T.R., Hammer, B., and Besenbacher, F. (2002) Chiral

recognition in dimerization of adsorbed cysteine observed by scanning tunnelling microscopy. *Nature*, **415**, 891–893.

162 Zhang, J.D., Demetriou, A., Welinder, A.C., Albrecht, T., Nichols, R.J., and Ulstrup, J. (2005) Potential-induced structural transitions of DL-homocysteine monolayers on Au(111) electrode surfaces. *Chemical Physics*, **319**, 210–221.

163 Sprik, M., Delamarche, E., Michel, B., Rothlisberger, U., Klein, M.L., Wolf, H., and Ringsdorf, H. (1994) Structure of hydrophilic self-assembled monolayers – a combined scanning-tunneling-microscopy and computer-simulation study. *Langmuir*, **10**, 4116–4130.

164 Wang, Y., Hush, N.S., and Reimers, J.R. (2007) Formation of gold–methanethiyl self-assembled monolayers. *Journal of the American Chemical Society*, **129**, 14532–14533.

165 Bilič, A., Reimers, J.R., and Hush, N.S. (2005) The structure, energetics, and nature of the chemical bonding of phenylthiol adsorbed on the Au(111) surface: implications for density-functional calculations of molecular-electronic conduction. *Journal of Chemical Physics*, **122**, 094708.

166 Tao, N.J. (1996) Probing potential-tuned resonant tunneling through redox molecules with scanning tunneling microscopy. *Physical Review Letters*, **76**, 4066–4069.

167 Albrecht, T., Moth-Poulsen, K., Christensen, J.B., Hjelm, J., Bjornholm, T., and Ulstrup, J. (2006) Scanning tunneling spectroscopy in an ionic liquid. *Journal of the American Chemical Society*, **128**, 6574–6575.

168 Lipkowski, J. (2010) Building biomimetic membrane at a gold electrode surface. *Physical Chemistry Chemical Physics*, **12**, 13874–13887.

169 Sek, S., Laredo, T., Dutcher, J.R., and Lipkowski, J. (2009) Molecular resolution imaging of an antibiotic peptide in a lipid matrix. *Journal of the American Chemical Society*, **131**, 6439–6444.

170 Welinder, A.C., Zhang, J.D., Steensgaard, D.B., and Ulstrup, J. (2010) Adsorption of human insulin on single-crystal gold surfaces investigated by in situ scanning tunnelling microscopy and electrochemistry. *Physical Chemistry Chemical Physics*, **12**, 9999–10011.

171 Haiss, W., Albrecht, T., van Zalinge, H., Higgins, S.J., Bethell, D., Hobenreich, H., Schiffrin, D.J., Nichols, R.J., Kuznetsov, A.M., Zhang, J., Chi, Q., and Ulstrup, J. (2007) Single-molecule conductance of redox molecules in electrochemical scanning tunneling microscopy. *Journal of Physical Chemistry B*, **111**, 6703–6712.

172 Leary, E., Higgins, S.J., van Zalinge, H., Haiss, W., Nichols, R.J., Nygaard, S., Jeppesen, J.O., and Ulstrup, J. (2008) Structure–property relationships in redox-gated single molecule junctions – a comparison of pyrrolo-tetrathiafulvalene and viologen redox groups. *Journal of the American Chemical Society*, **130**, 12204–12205.

173 Albrecht, T., Mertens, S.F.L., and Ulstrup, J. (2007) Intrinsic multistate switching of gold clusters through electrochemical gating. *Journal of the American Chemical Society*, **129**, 9162–9167.

174 Aschenbrenner, D.S. and Venable, S.J. (2008) *Drug Therapy in Nursing* (ed. D.S. Aschenbrenner), Lippincott Williams & Wilkins, Philadelphia, PA.

175 Saltiel, A.R. and Kahn, C.R. (2001) Insulin signalling and the regulation of glucose and lipid metabolism. *Nature*, **414**, 799–806.

176 Duckworth, W.C., Bennett, R.G., and Hamel, F.G. (1998) Insulin degradation: progress and potential. *Endocrine Reviews*, **19**, 608–624.

177 Bocian, W., Sitkowski, J., Bednarek, E., Tarnowska, A., Kawecki, R., and Kozerski, L. (2008) Structure of human insulin monomer in water/acetonitrile solution. *Journal of Biomolecular NMR*, **40**, 55–64.

178 Yao, Z.P., Zeng, Z.H., Li, H.M., Zhang, Y., Feng, Y.M., and Wang, D.C. (1999) Structure of an insulin dimer in an orthorhombic crystal: the structure analysis of a human insulin mutant (B9 Ser→Glu). *Acta Crystallographica D*, **55**, 1524–1532.

179 Smith, G.D. and Blessing, R.H. (2003) Lessons from an aged, dried crystal of T-6 human insulin. *Acta Crystallographica D*, **59**, 1384–1394.

180 Smith, G.D. (2003) *Handbook of Metalloproetins* (eds A. Messersmidt, W. Bode, and M. Cygler), John Wiley & Sons, Ltd, Chichester, pp. 367–377.

181 Chang, X.Q., Jorgensen, A.M.M., Bardrum, P., and Led, J.J. (1997) Solution structures of the R-6 human insulin hexamer. *Biochemistry*, **36**, 9409–9422.

182 Schlichtkrull, J. (1956) Insulin crystals 1. The minimum mole-fraction of metal in insulin crystals prepared with Zn^{++}, Cd^{++}, Co^{++}, Ni^{++}, Cu^{++}, Mn^{++}, or Fe^{++}. *Acta Chemica Scandinavica*, **10**, 1455–1458.

183 Pekar, A.H. and Frank, B.H. (1972) Conformation of proinsulin – comparison of insulin and proinsulin self-association at neutral pH. *Biochemistry*, **11**, 4013–4016.

184 Nilsson, P., Nylander, T., and Havelund, S. (1991) Adsorption of insulin on solid surfaces in relation to the surface properties of the monomeric and oligomeric forms. *Journal of Colloid and Interface Science*, **144**, 145–152.

185 Hovorka, R., Powrie, J.K., Smith, G.D., Sonksen, P.H., Carson, E.R., and Jones, R.H. (1993) 5-Compartment model of insulin kinetics and its use to investigate action of chloroquine in niddm. *American Journal of Physiology*, **265**, E162–E175.

186 Brange, J., Ribel, U., Hansen, J.F., Dodson, G., Hansen, M.T., Havelund, S., Melberg, S.G., Norris, F., Norris, K., Snel, L., Sorensen, A.R., and Voigt, H.O. (1988) Monomeric insulins obtained by protein engineering and their medical implications. *Nature*, **333**, 679–682.

187 Arnebrant, T. and Nylander, T. (1988) Adsorption of insulin on metal surfaces in relation to association behavior. *Journal of Colloid and Interface Science*, **122**, 557–566.

188 Elwing, H. (1998) Protein absorption and ellipsometry in biomaterial research. *Biomaterials*, **19**, 397–406.

189 Mollmann, S.H., Jorgensen, L., Bukrinsky, J.T., Elofsson, U., Norde, W., and Frokjaer, S. (2006) Interfacial adsorption of insulin – conformational changes and reversibility of adsorption. *European Journal of Pharmaceutical Sciences*, **27**, 194–204.

190 Buijs, J., Vera, C.C., Ayala, E., Steensma, E., Hakansson, P., and Oscarsson, S. (1999) Conformational stability of adsorbed insulin studied with mass spectrometry and hydrogen exchange. *Analytical Chemistry*, **71**, 3219–3225.

191 Gobi, K.V., Iwasaka, H., and Miura, N. (2007) Self-assembled PEG monolayer based SPR immunosensor for label-free detection of insulin. *Biosensors & Bioelectronics*, **22**, 1382–1389.

192 Markus, G. (1964) Electrolytic reduction of disulfide bonds of insulin. *Journal of Biological Chemistry*, **239**, 4163–4170.

193 Stankovich, M.T. and Bard, A.J. (1977) Electrochemistry of proteins and related substances 2. Insulin. *Journal of Electroanalytical Chemistry*, **85**, 173–183.

194 Zahn, H., and Gattner, H.G. (1968) Partial reduction of insulin. *Hoppe-Seylers Zeitschrift fur Physiologische Chemie*, **349**, 373–384.

195 MacDonald, S.M. and Roscoe, S.G. (1996) Electrochemical studies of the interfacial behavior of insulin. *Journal of Colloid and Interface Science*, **184**, 449–455.

196 Wright, J.E.I., Cosman, N.P., Fatih, K., Omanovic, S., and Roscoe, S.G. (2004) Electrochemical impedance spectroscopy and quartz crystal nanobalance (EQCN) studies of insulin adsorption on Pt. *Journal of Electroanalytical Chemistry*, **564**, 185–197.

197 Zong, W.S., Liu, R.T., Sun, F., Wang, M.J., Zhang, P.J., Liu, Y.H., and Tian, Y.M. (2010) Cyclic voltammetry: a new strategy for the evaluation of oxidative damage to bovine insulin. *Protein Science*, **19**, 263–268.

198 Zhu, M., Souillac, P.O., Ionescu-Zanetti, C., Carter, S.A., and Fink, A.L. (2002) Surface-catalyzed amyloid fibril formation. *Journal of Biological Chemistry*, **277**, 50914–50922.

199 Khurana, R., Ionescu-Zanetti, C., Pope, M., Li, J., Nielson, L., Ramirez-Alvarado, M., Regan, L., Fink, A.L., and Carter, S.A. (2003) A general model for amyloid fibril assembly based on morphological studies using atomic force microscopy. *Biophysical Journal*, **85**, 1135–1144.

200 Jansen, R., Dzwolak, W., and Winter, R. (2005) Amyloidogenic self-assembly of insulin aggregates probed by high resolution atomic force microscopy. *Biophysical Journal*, **88**, 1344–1353.

201 Bonanni, B., Andolfi, L., Bizzarri, A.R., and Cannistraro, S. (2007) Functional metalloproteins integrated with conductive substrates: detecting single molecules and sensing individual recognition events. *Journal of Physical Chemistry B*, **111**, 5062–5075.

202 Chi, Q.J., Zhang, J.D., Andersen, J.E.T., and Ulstrup, J. (2001) Ordered assembly and controlled electron transfer of the blue copper protein azurin at gold (111) single-crystal substrates. *Journal of Physical Chemistry B*, **105**, 4669–4679.

203 Berman, H.M., Westbrook, J., Feng, Z., Gilliland, G., Bhat, T.N., Weissig, H., Shindyalov, I.N., and Bourne, P.E. (2000) The protein data bank. *Nucleic Acids Research*, **28**, 235–242.

204 Kolb, D.M. (1993) *Structure of Electrified Interfaces* (eds J. Lipkowski and P.N. Ross), John Wiley & Sons, Inc., New York, pp. 65–102.

205 Kakiuchi, T., Usui, H., Hobara, D., and Yamamoto, M. (2002) Voltammetric properties of the reductive desorption of alkanethiol self-assembled monolayers from a metal surface. *Langmuir*, **18**, 5231–5238.

206 Ferapontova, E. and Gorton, L. (2003) Bioelectrocatalytical detection of H_2O_2 with different forms of horseradish peroxidase directly adsorbed at polycrystalline silver and gold. *Electroanalysis*, **15**, 484–491.

207 Vincent, K.A., Parkin, A., and Armstrong, F.A. (2007) Investigating and exploiting the electrocatalytic properties of hydrogenases. *Chemical Reviews*, **107**, 4366–4413.

208 Leger, C. and Bertrand, P. (2008) Direct electrochemistry of redox enzymes as a tool for mechanistic studies. *Chemical Reviews*, **108**, 2379–2438.

209 Willner, I. and Katz, E. (2000) Integration of layered redox proteins and conductive supports for bioelectronic applications. *Angewandte Chemie International Edition*, **39**, 1180–1218.

210 Armstrong, F.A. (2005) Recent developments in dynamic electrochemical studies of adsorbed enzymes and their active sites. *Current Opinion in Chemical Biology*, **9**, 110–117.

211 Heller, A. (2004) Miniature biofuel cells. *Physical Chemistry Chemical Physics*, **6**, 209–216, and references therein.

212 Gorton, L., Bremle, G., Csöregi, E., Jönsson-Pettersson, G., and Persson, B. (1991) Amperometric glucose sensors based on immobilized glucose-oxidizing enzymes and chemically modified electrodes. *Analytica Chimica Acta*, **249**, 43–54.

213 Heering, H.A., Weiner, J.H., and Armstrong, F.A. (1997) Direct detection and measurement of electron relays in a multicentered enzyme: voltammetry of electrode-surface films of *E. coli* fumarate reductase, an iron-sulfur flavoprotein. *Journal of the American Chemical Society*, **119**, 11628–11638.

214 Leger, C., Heffron, K., Pershad, H.R., Maklashina, E., Luna-Chavez, C., Cecchini, G., Ackrell, B.A.C., and Armstrong, F.A. (2001) Enzyme electrokinetics: energetics of succinate oxidation by fumarate reductase and succinate dehydrogenase. *Biochemistry*, **40**, 11234–11245.

215 Anderson, L.J., Richardson, D.J., and Butt, J.N. (2001) Catalytic protein film voltammetry from a respiratory nitrate reductase provides evidence for complex electrochemical modulation of enzyme activity. *Biochemistry*, **40**, 11294–11307.

216 Angove, H.C., Cole, J.A., Richardson, D.J., and Butt, J.N. (2002) Protein film voltammetry reveals distinctive fingerprints of nitrite and hydroxylamine reduction by a cytochrome c nitrite reductase. *Journal of Biological Chemistry*, **277**, 23374–23381.

217 Zhang, J.D., Welinder, A.C., Hansen, A.G., Christensen, H.E.M., and Ulstrup, J. (2003) Catalytic monolayer voltammetry and in situ scanning tunneling microscopy of copper nitrite reductase on cysteamine-modified Au(111) electrodes. *Journal of Physical Chemistry B*, **107**, 12480–12484.

218 Mondal, M.S., Fuller, H.A., and Armstrong, F.A. (1996) Direct measurement of the reduction potential of catalytically active cytochrome c peroxidase compound: I. Voltammetric detection of a reversible, cooperative two-electron transfer reaction. *Journal of the American Chemical Society*, **118**, 263–264.

219 Heffron, K., Leger, C., Rothery, R.A., Weiner, J.H., and Armstrong, F.A. (2001) Determination of an optimal potential window for catalysis by *E. coli* dimethyl sulfoxide reductase and hypothesis on the role of Mo(V) in the reaction pathway. *Biochemistry*, **40**, 3117–3126.

220 Armstrong, F.A. and Albracht, P.J. (2005) [NiFe]-hydrogenases: spectroscopic and electrochemical definition of reactions and intermediates. *Philosophical Transactions of the Royal Society A: Mathematical Physical and Engineering Sciences*, **363**, 937–954.

221 Haas, A.S., Pilloud, D.L., Reddy, K.S., Babcock, G.T., Moser, C.C., Blasie, J.K., and Dutton, P.L. (2001) Cytochrome c and cytochrome c oxidase: monolayer assemblies and catalysis. *Journal of Physical Chemistry B*, **105**, 11351–11362.

222 Hansen, A.G., Zhang, J.D., Christensen, H.E.M., Welinder, A.C., Wackerbarth, H., and Ulstrup, J. (2004) Electron transfer and redox metalloenzyme catalysis at the single-molecule level. *Israel Journal of Chemistry*, **44**, 89–100.

223 Brask, J., Wackerbarth, H., Jensen, K.J., Zhang, J.D., Chorkendorff, I., and Ulstrup, J. (2003) Monolayer assemblies of a de novo designed 4-alpha-helix bundle carboprotein and its sulfur anchor fragment on Au(111) surfaces addressed by voltammetry and in situ scanning tunneling microscopy. *Journal of the American Chemical Society*, **125**, 94–104.

224 Albrecht, T., Li, W., Haehnel, W., Hildebrandt, P., and Ulstrup, J. (2006) Voltammetry and in situ scanning tunnelling microscopy of de novo designed heme protein monolayers on Au(111) electrode surfaces. *Bioelectrochemistry*, **69**, 193–200.

225 Alessandrini, A., Gerunda, M., Canters, G.W., Verbeet, M.P., and Facci, P. (2003) Electron tunnelling through azurin is mediated by the active site Cu ion. *Chemical Physics Letters*, **376**, 625–630.

226 Andolfi, L., Bonanni, B., Canters, G.W., Verbeet, M.P., and Cannistraro, S. (2003) Scanning probe microscopy characterization of gold-chemisorbed poplar plastocyanin mutants. *Surface Science*, **530**, 181–194.

227 Gwyer, J.D., Zhang, J.D., Butt, J.N., and Ulstrup, J. (2006) Voltammetry and in situ scanning tunneling microscopy of cytochrome c nitrite reductase on Au(111) electrodes. *Biophysical Journal*, **91**, 3897–3906.

228 Chi, Q.J., Zhang, J.D., Arslan, T., Borg, L., Pedersen, G.W., Christensen, H.E.M., Nazmudtinov, R.R., and Ulstrup, J. (2010) Approach to interfacial and intramolecular electron transfer of the diheme protein cytochrome c(4) assembled on Au(111) surfaces. *Journal of Physical Chemistry B*, **114**, 5617–5624.

229 Della Pia, E., Chi, Q., Jones, D.D., Macdonald, J.E., Ulstrup, J., and Elliott, M. (2011) Single-molecule mapping of long-range electron transport for a cytochrome b_{562} variant. *Nano Letters*, **11**, 176–182.

230 Jensen, P.S., Chi, Q., Zhang, J., and Ulstrup, J. (2009) Long-range interfacial electrochemical electron transfer of Pseudomonas aeruginosa azurin–gold nanoparticle hybrid systems. *Journal of Physical Chemistry C*, **113**, 13993–14000.

231 Raffalt, A.C., Schmidt, L., Christensen, H.E.M., Chi, Q., and Ulstrup, J. (2009) Electron transfer patterns of the di-heme protein cytochrome c(4) from *Pseudomonas stutzeri*. *Journal of Inorganic Biochemistry*, **103**, 717–722.

232 Andersen, N.H., Christensen, H.E.M., Iversen, G., Nørgaard, A., Scharnagle, C., Thuesen, M.H., *et al.* (2001) Cytochrome c_4, in *Handbook of Metalloproteins* (eds A. Messerschmidt, T.L. Poulos, and K. Wieghardt), John Wiley & Sons, Ltd, Chichester, pp. 100–109.

233 Chi, Q.J., Zhang, J.D., Jensen, P.S., Nazmudtinov, R.R., and Ulstrup, J. (2008) Surface-induced intramolecular electron transfer in multi-centre redox metalloproteins: the di-haem protein cytochrome c_4 in homogeneous solution and at electrochemical surfaces. *Journal of Physics: Condensed Matter*, **20**, 374124.

234 Min, W., English, B.P., Luo, G.B., Cherayil, B.J., Kou, S.C., and Xie, X.S. (2005) Fluctuating enzymes: lessons from single-molecule studies. *Accounts of Chemical Research*, **38**, 923–931.

235 Comellas-Aragones, M., Engelkamp, H., Claessen, V.I., Sommerdijk, N.A.J., Rowan, A.E., Christianen, P.C.M., Maan, J.C., Verduin, B.J.M., Cornelissen, J.J.L.M., and Nolte, R.J.M. (2007) A virus-based single-enzyme nanoreactor. *Nature Nanotechnology*, **2**, 635–639.

236 Hulsken, B., Van Hameren, R., Gerritsen, J.W., Khoury, T., Thordarson, P., Crossley, M.J., Rowan, A.E., Nolte, R.J.M., Elemans, J.A.A.W., and Speller, S. (2007) Real-time single-molecule imaging of oxidation catalysis at a liquid–solid interface. *Nature Nanotechnology*, **2**, 285–289.

237 Richardson, D.J. and Watmough, N.J. (1999) Inorganic nitrogen metabolism in bacteria. *Current Opinion in Chemical Biology*, **3**, 207–219.

238 Adman, E.T. and Murphy, M.E.P. (2001) *Handbook of Metalloproteins* (eds A. Messerschmidt, R. Huber, T.L. Poulos, and K. Wieghardt), John Wiley & Sons, Ltd, Chichester, p. 1381.

239 Simon, J. (2002) Enzymology and bioenergetics of respiratory nitrite ammonification. *FEMS Microbiology Reviews*, **26**, 285–309.

240 Bamford, V.A., Angove, H.C., Seward, H.E., Thomson, A.J., Cole, J.A., Butt, J.N., Hemmings, A.M., and Richardson, D.J. (2002) Structure and spectroscopy of the periplasmic cytochrome c nitrite reductase from *Escherichia coli. Biochemistry*, **41**, 2921–2931.

241 Williams, P.A., Fulop, V., Garman, E.F., Saunders, N.F.W., Ferguson, S.J., and Hajdu, J. (1997) Haem-ligand switching during catalysis in crystals of a nitrogen-cycle enzyme. *Nature*, **389**, 406–412.

242 Gordon, E.H.J., Sjogren, T., Lofqvist, M., Richter, C.D., Allen, J.W.A., Higham, C.W., Hajdu, J., Fulop, V., and Ferguson, S.J. (2003) Structure and kinetic properties of *Paracoccus pantotrophus* cytochrome cd(1) nitrite reductase with the d(1) heme active site ligand tyrosine 25 replaced by serine. *Journal of Biological Chemistry*, **278**, 11773–11781.

243 Welinder, A.C., Zhang, J., Hansen, A.G., Moth-Poulsen, K., Christensen, H.E.M., Kuznetsov, A.M., Bjornholm, T., and Ulstrup, J. (2007) Voltammetry and electrocatalysis of achrornobacter xylosoxidans copper nitrite reductase on functionalized Au(111) electrode surfaces. *Zeitschrift fur Physikalische Chemie – International Journal of Research in Physical Chemistry & Chemical Physics*, **221**, 1343–1378.

244 Gwyer, J.D., Richardson, D.J., and Butt, J.N. (2006) Inhibiting *Escherichia coli* cytochrome c nitrite reductase: voltammetry reveals an enzyme equipped for action despite the chemical challenges it may face *in vivo*. *Biochemical Society Transactions*, **34**, 133–135.

245 Palacek, E., Schneller, F., and Wang, J. (2005) *Electrochemistry of Nucleic Acids and Proteins: Towards Electrochemical Sensors for Genomics and Proteomics*, Elsevier, Amsterdam.

246 Katz, E., and Willner, I. (2004) Integrated nanoparticle–biomolecule hybrid systems: synthesis, properties, and applications. *Angewandte Chemie International Edition*, **43**, 6042–6108.

247 Cui, Y., Zhong, Z.H., Wang, D.L., Wang, W.U., and Lieber, C.M. (2003) High performance silicon nanowire field effect transistors. *Nano Letters*, **3**, 149–152.

248 Zheng, G.F., Lu, W., Jin, S., and Lieber, C.M. (2004) Synthesis and fabrication of

high-performance n-type silicon nanowire transistors. *Advanced Materials*, **16**, 1890–1893.

249 Patolsky, F., Zheng, G.F., and Lieber, C.M. (2006) Nanowire-based biosensors. *Analytical Chemistry*, **78**, 4260–4269.

250 Chen, D., Wang, G., and Li, J.H. (2007) Interfacial bioelectrochemistry: fabrication, properties and applications of functional nanostructured biointerfaces. *Journal of Physical Chemistry C*, **111**, 2351–2367.

251 Besteman, K., Lee, J.O., Wiertz, F.G.M., Heering, H.A., and Dekker, C. (2003) Enzyme-coated carbon nanotubes as single-molecule biosensors. *Nano Letters*, **3**, 727–730.

252 Day, T.M., Wilson, N.R., and Macpherson, J.V. (2004) Electrochemical and conductivity measurements of single-wall carbon nanotube network electrodes. *Journal of the American Chemical Society*, **126**, 16724–16725.

253 Love, J.C., Estroff, L.A., Kriebel, J.K., Nuzzo, R.G., and Whitesides, G.M. (2005) Self-assembled monolayers of thiolates on metals as a form of nanotechnology. *Chemical Reviews*, **105**, 1103–1169.

254 Gooding, J.J., Chou, A., Liu, J., Losic, D., Shapter, J.G., and Hibbert, D.B. (2007) The effects of the lengths and orientations of single-walled carbon nanotubes on the electrochemistry of nanotube-modified electrodes. *Electrochemistry Communications*, **9**, 1677–1683.

255 Yang, W., Thordarson, P., Gooding, J.J., Ringer, S.P., and Braet, F. (2007) Carbon nanotubes for biological and biomedical applications. *Nanotechnology*, **18**, 412001.

256 Mattoussi, H., Mauro, J.M., Goldman, E.R., Anderson, G.P., Sundar, V.C., Mikulec, F.V., and Bawendi, M.G. (2000) Self-assembly of CdSe–ZnS quantum dot bioconjugates using an engineered recombinant protein. *Journal of the American Chemical Society*, **122**, 12142–12150.

257 Jensen, P.S., Chi, Q., Grumsen, F.B., Abad, J.M., Horsewell, A., Schiffrin, D.J., and Ulstrup, J. (2007) Gold nanoparticle assisted assembly of a heme protein for enhancement of long-range interfacial electron transfer. *Journal of Physical Chemistry C*, **111**, 6124–6132.

258 Daniel, M.C. and Astruc, D. (2004) Gold nanoparticles: assembly, supramolecular chemistry, quantum-size-related properties, and applications toward biology, catalysis, and nanotechnology. *Chemical Reviews*, **104**, 293–346.

259 Templeton, A.C., Wuelfing, M.P., and Murray, R.W. (2000) Monolayer protected cluster molecules. *Accounts of Chemical Research*, **33**, 27–36.

260 Schmid, G., Pfeil, R., Boese, R., Bandermann, F., Meyer, S., Calis, G.H.M., and Vandervelden, W.A. (1981) Au55[p(c6h5)3]12cl6 – a gold cluster of an exceptional size. *Chemische Berichte – Recueil*, **114**, 3634–3642.

261 Schmid, G. (ed.) (2004) *Nanoparticles: From Theory to Applications*, Wiley-VCH Verlag GmbH, Weinheim.

262 Rotello, V. (ed.) (2004) *Nanoparticles. Building Blocks for Nanotechnology*, Springer, New York.

263 Brust, M., Walker, M., Bethell, D., Schiffrin, D.J., and Whyman, R. (1994) Synthesis of thiol-derivatized gold nanoparticles in a 2-phase liquid–liquid system. *Journal of the Chemical Society. Chemical Communications*, 801–802.

264 Mulvaney, P. (1996) Surface plasmon spectroscopy of nanosized metal particles. *Langmuir*, **12**, 788–800.

265 Yguerabide, J. and Yguerabide, E.E. (1998) Light-scattering submicroscopic particles as highly fluorescent analogs and their use as tracer labels in clinical and biological applications: I. Theory. *Analytical Biochemistry*, **262**, 137–156.

266 Bohren, C.F. and Huffman, D.R. (1998) *Absorption and Scattering of Light by Small Particles*, John Wiley & Sons, Ltd, Chichester.

267 Alvarez, M.M., Khoury, J.T., Schaaff, T.G., Shafigullin, M.N., Vezmar, I., and Whetten, R.L. (1997) Optical absorption spectra of nanocrystal gold molecules. *Journal of Physical Chemistry B*, **101**, 3706–3712.

268 Hicks, J.F., Zamborini, F.P., and Murray, R.W. (2002) Dynamics of electron transfers between electrodes and monolayers of nanoparticles. *Journal of Physical Chemistry B*, **106**, 7751–7757.

269 Horswell, S.L., O'Neil, I.A., and Schiffrin, D.J. (2003) Kinetics of electron transfer at Pt nanostructured film electrodes. *Journal of Physical Chemistry B*, **107**, 4844–4854.

270 Su, B. and Girault, H.H. (2005) Redox properties of self-assembled gold nanoclusters. *Journal of Physical Chemistry B*, **109**, 23925–23929.

271 Ranganathan, S., Guo, R., and Murray, R.W. (2007) Nanoparticle films as electrodes: voltammetric sensitivity to the nanoparticle energy gap. *Langmuir*, **23**, 7372–7377.

272 Chen, S.W. and Murray, R.W. (1999) Electrochemical quantized capacitance charging of surface ensembles of gold nanoparticles. *Journal of Physical Chemistry B*, **103**, 9996–10000.

273 Chen, S.W. (2000) Self-assembling of monolayer-protected gold nanoparticles. *Journal of Physical Chemistry B*, **104**, 663–667.

274 Quinn, B.M., Liljeroth, P., Ruiz, V., Laaksonen, T., and Kontturi, K. (2003) Electrochemical resolution of 15 oxidation states for monolayer protected gold nanoparticles. *Journal of the American Chemical Society*, **125**, 6644–6645.

275 Su, B., Eugster, N., and Girault, H.H. (2005) Reactivity of monolayer-protected gold nanoclusters at dye-sensitized liquid/liquid interfaces. *Journal of the American Chemical Society*, **127**, 10760–10766.

276 Galletto, P., Girault, H.H., Gomis-Bas, C., Schiffrin, D.J., Antoine, R., Broyer, M., and Brevet, P.F. (2007) Second harmonic generation response by gold nanoparticles at the polarized water/2-octanone interface: from dispersed to aggregated particles. *Journal of Physics. Condensed Matter*, **19**, 375108.

277 Chen, S.W., Murray, R.W., and Feldberg, S.W. (1998) Quantized capacitance charging of monolayer-protected Au clusters. *Journal of Physical Chemistry B*, **102**, 9898–9907.

278 Chen, S.W., Ingram, R.S., Hostetler, M.J., Pietron, J.J., Murray, R.W., Schaaff, T.G., Khoury, J.T., Alvarez, M.M., and Whetten, R.L. (1998) Gold nanoelectrodes of varied size: transition to molecule-like charging. *Science*, **280**, 2098–2101.

279 Ingram, R.S., Hostetler, M.J., Murray, R.W., Schaaff, T.G., Khoury, J.T., Whetten, R.L., Bigioni, T.P., Guthrie, D.K., and First, P.N. (1997) 28 kDa alkanethiolate-protected Au clusters give analogous solution electrochemistry and STM Coulomb staircases. *Journal of the American Chemical Society*, **119**, 9279–9280.

280 Devoret, M.H., Esteve, D., and Urbina, C. (1992) Single-electron transfer in metallic nanostructures. *Nature*, **360**, 547–553.

281 Likharev, K.K. and Claeson, T. (1992) Single electronics. *Scientific American*, **266**, 80–85.

282 Grabert, H. and Devoret, M.H. (eds) (1992) *Single Charge Tunneling*, Plenum Press, New York.

283 Chen, S.W. and Pei, R.J. (2001) Ion-induced rectification of nanoparticle quantized capacitance charging in aqueous solutions. *Journal of the American Chemical Society*, **123**, 10607–10615.

284 Li, Z., Liu, Y., Mertens, S.F.L., Pobelov, I.V., and Wandlowski, T. (2010) From redox gating to quantized charging. *Journal of the American Chemical Society*, **132**, 8187–8193.

285 Xiao, Y., Patolsky, F., Katz, E., Hainfeld, J.F., and Willner, I. (2003) "Plugging into enzymes": nanowiring of redox enzymes by a gold nanoparticle. *Science*, **299**, 1877–1881.

286 O'Neill, M.A. and Barton, J.K. (2004) Sequence dependant DNA dynamics: the regulator of DNA-mediated charge transport, in *Charge Transfer in DNA: From Mechanisms to Application* (ed. H.A. Wagenknecht), Wiley-VCH Verlag GmbH, Weinheim, pp. 27–75.

287 Shapir, E., Yi, J.Y., Cohen, H., Kotlyar, A.B., Cuniberti, G., and Porath, D.

(2005) The puzzle of contrast inversion in DNA STM imaging. *Journal of Physical Chemistry B*, **109**, 14270–14274.
288 Conwell, E.M. and Rakhmanova, S.V. (2000) Polarons in DNA. *Proceedings of the National Academy of Sciences of the United States of America*, **97**, 4556–4560.
289 Conwell, E. (2004) Polarons and transport in DNA. *Topics in Current Chemistry*, **237**, 73–101.

3
Applications of Neutron Reflectivity in Bioelectrochemistry

Ian J. Burgess

3.1
Introduction

The assembly of biomimetic phospholipid layers on solid substrates is a means to create sensors and platforms to study the biophysics of membrane processes. If the solid substrate upon which the film is assembled is an electronic conductor then careful manipulation of the surface charge density allows the physical and chemical properties of the adsorbed organic film to be altered in a controlled fashion. Interest in biomimetic-based research has led to the need to extract more detailed information pertaining to adsorbed organic films and their electrical variable-dependent molecular structure. Electrochemical measurements alone are often insufficient to extract detailed information with molecular-scale resolution of the components of the adsorbed films. To achieve this level of understanding, electrochemists have long recognized that traditional experimental tools need to be coupled with *in situ* experimental methods. For example, infrared and Raman spectroscopy [1–3], scanning probe microscopy [4], surface X-ray scattering [5], and surface plasmon resonance [6] have all been adopted by electrochemists in the pursuit of finding new surface-sensitive means to probe the structure and composition of adsorbed organic films *in situ*.

The specular reflection of neutrons is a surface technique that, unlike X-ray methods, is sensitive to low atomic number elements and can distinguish between isotopes of the same element. This makes neutron reflectivity (NR) especially effective for the study of biologically relevant material due to the ubiquitous nature of carbon and hydrogen in these systems. However, until relatively recently the technique had evaded the attention of the electrochemical community. The reasons for this stem largely from the limited number of facilities where neutron scattering measurements can be made and the unfamiliarity of the technique to scientists outside the field of physics. The creation of dedicated research facilities and the encouragement of external, user-oriented beam time allocation have led to increasing interest in the field of slow and thermal neutron scattering. In the last decade, NR has begun to attract increasing attention from electrochemical researchers. The intent of this contribution is to provide an introduction to the field of NR

Advances in Electrochemical Science and Engineering. Edited by Richard C. Alkire, Dieter M. Kolb, and Jacek Lipkowski

ISBN: 978-3-527-32885-7

studies of biomimetic thin organic films adsorbed on electrode surfaces both in theoretical and practical terms. The scope and context are addressed to future possible users of NR in the bioelectrochemistry community particularly those interested in studying biomimetic lipid layers supported on electrode surfaces. There exist several excellent reviews in the literature concerning the concepts of NR [7–11]; however, the starting level and the complexity of these papers makes them difficult reading for non-physicists. It is hoped that while being slightly less rigorous, this contribution will provide electrochemists with an accessible introduction to the theoretical and experimental applications of NR.

3.2
Theoretical Aspects of Neutron Scattering

3.2.1
Why Use Neutrons?

The purpose of a scattering-based experiment is to determine the spatial and/or temporal correlation of the atomic or molecular components of the object under study. If one wishes to extract spatial information pertaining to collections of atoms or molecules it is expected that the probe chosen for a scattering experiment must have wavelengths comparable in magnitude to the distance between the constituent units. Similarly, in order to study temporal correlations, the incident radiation should have comparable energies to those associated with atomic or molecular motions. Beyond these criteria, there are several other requirements of the probe which are particularly relevant to the study of thin organic films such as biomimetic membranes.

1) The probe must be capable of penetrating condensed matter if the sample of interest is not at an accessible interface.
2) The probe should respond equally to light atoms as well as heavier ones.
3) The probe should not deposit destructive levels of energy in the sample when absorbed.

Two types of scattering probes that have wavelengths small enough to provide information on the atomic or molecular length scale are electromagnetic radiation in the form of X-rays and de Broglie waves in the form of neutrons. It should be stated that although these two probes are often considered complementary, only neutrons satisfy all the above listed desired criteria. This is shown in Figure 3.1, where the energies of photons and neutrons of the same wavelength are displayed. In the range of wavelengths useful for studying interatomic or intramolecular spacings in condensed matter, photons have energy roughly five orders of magnitude higher than that of neutrons and are usually far too energetic to probe temporal correlation functions in condensed matter. In addition, upon absorption, the keV range of energy carried by a photon is sufficient to damage the material under study, particularly when soft materials are involved. Another distinction between

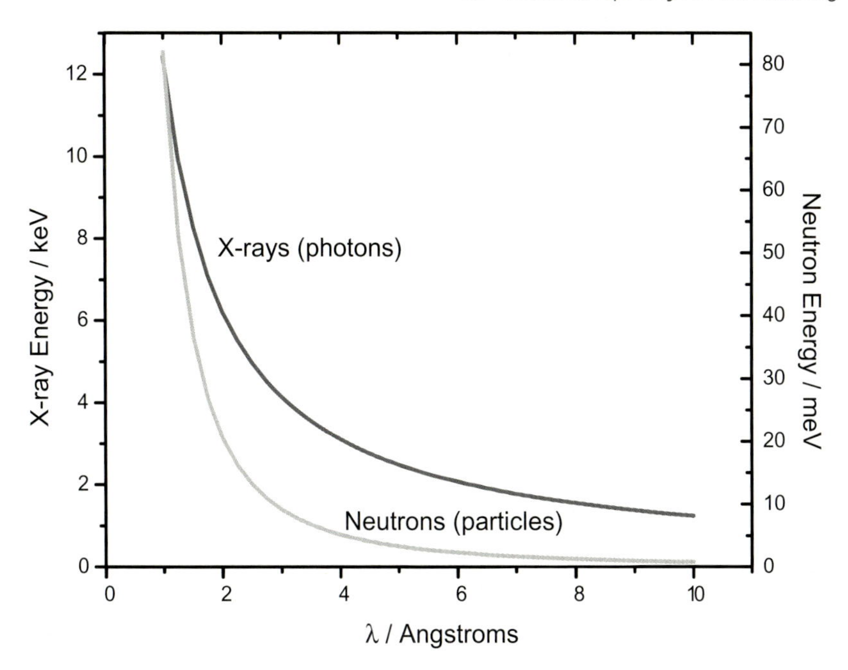

Figure 3.1 Kinetic energies of X-rays (photons) and neutrons as a function of wavelength.

neutrons and X-rays involves the fundamental difference between the interactions of these two forms of radiation with matter. X-rays are scattered by electron density, and thus the probability for a scattering event to occur is directly proportional to the atomic number of the matter causing the scattering. This means that light atoms such as hydrogen are barely visible to X-ray radiation. Whereas X-rays probe matter via an electromagnetic interaction, the next section describes how the primary interaction in a neutron–matter scattering event is via the strong nuclear force. Two important consequences follow from this fact. First, unlike X-rays, neutrons are strongly scattered from low atomic number atoms; second, a neutron has a different scattering potential with isotopes of the same element.

3.2.2
Scattering from a Single Nucleus

The essence of NR is the scattering of incident neutrons from the nuclei of the material being studied due to the strong nuclear force. As the wavelength of the incident neutron (about 1–10 Å) is much greater than the range of the strong nuclear force (about $1\text{–}10 \times 10^{-5}$ Å) the result of a single scattering event can be envisioned as a spherical wave analogous to the scattering of water waves from a pole projecting from the surface of water. An analytical form of the scattered wave requires an adequate description of the strong nuclear potential energy function. For the moment, assume that the strong nuclear interaction potential for a given

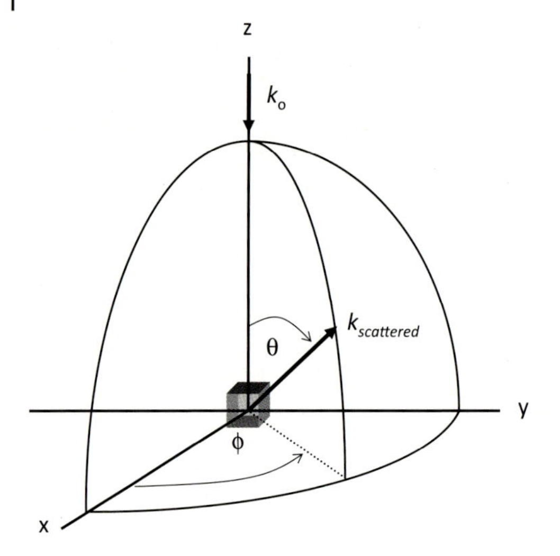

Figure 3.2 Coordinate system used to describe a single scattering event between a neutron of initial wave vector (k_0) and a bound nucleus positioned at the origin.

nucleus can be defined by a square well potential of known barrier height and interaction radius, *R*. The coordinate system, as shown in Figure 3.2, is defined such that the nucleus lies at the origin and the *z*-axis is along the direction of k_0, the wave vector of the incident neutron. It is assumed that the nucleus is fixed at the origin and has no freedom to recoil from the collision with the impinging neutron. In the absence of any potential energy, the incident free neutron can be described by a traveling plane wave propagating along the *z*-coordinate:

$$\psi_{\text{in}} = \exp(ik_0 z) \tag{3.1}$$

The wave scattered from the nucleus is spherically symmetric and the wavefunction of the scattered neutrons at any point *r* can be written in the form

$$\psi_{\text{sc}} = -\frac{b}{r}\exp(ikr) \tag{3.2}$$

where *b* is a constant, independent of the angles θ and ϕ. The minus sign in Eq. (3.2) was originally chosen by Fermi and Marshall [12] so that a repulsive potential would correspond to a positive value of *b*. The quantity *b* is known as the scattering length and along with the closely related scattering length density parameter, ρ, is found ubiquitously in all forms of neutron scattering (see below). Strictly speaking the scattering length is a complex parameter. In physical terms, the imaginary contribution to the scattering length represents a measure of neutron capture via a resonance phenomenon and is highly dependent on the incident neutron's energy, whereas the real component is energy independent. The majority of nuclei have relatively small capture cross sections, and for these nuclei the scattering length can be treated as a real-valued quantity. There are some notable exceptions such as ^{103}Rh, ^{113}Cd, and ^{157}Gd which are strong neutron absorbers but the remaining discussion will be limited to nuclei of the former description. As there is no proper theory to describe the strong nuclear interaction, it is impossible to predict values of b from other properties of the nucleus and consequently the scattering length parameters need to be determined experimentally. Empirically it is seen that the value of b depends on the spin state and the mass of the combined neutron–nucleus system. This opens up the field of isotopic substitution as a method of improving a great number of neutron scattering experiments.

3.2.2.1 The Fermi Pseudo Potential

It was Fermi who realized that it was possible to invoke an equivalent potential, which can be used to calculate the changes in the wavefunction outside the interaction by perturbation theory [13]. The unknown form of the strong nuclear interaction can be replaced by a new potential, which gives the same scattered wavefunction as the square well potential. In the derivation of Fermi's equivalent or pseudo potential [14] it is seen that the magnitude of the scattering potential depends on the scattering length of the nucleus and the mass of the neutron, m:

$$V_{\mathrm{f}}(r) = \frac{2\pi\hbar^2}{m} b\delta(r) \tag{3.3}$$

where the delta function is defined to be $\delta(r) = 1$ for $r = b$ and $\delta(r) = 0$ for all other values of r. It should be clearly stated that the pseudo potential does not describe the true strong nuclear–nucleus interaction but rather is a mathematical construct that produces the correct form of the scattered wavefunction at large distances from the nucleus and depends entirely on the scattering length parameter of the nucleus from which the neutron scatters.

3.2.3 Scattering from a Collection of Nuclei

3.2.3.1 Neutron Scattering Cross Sections

If a collimated beam of neutrons (having initial energy E) is directed along the z-axis onto a target composed of a collection of nuclei then it is possible to measure the number of scattered neutrons in a given direction using a detector that is also capable of discriminating the energies, E', of the scattered neutrons. If Ω represents a solid angle along the direction defined by θ and ϕ and σ represents the total number of scattered neutrons in all directions then the *partial differential scattering cross section* is

$$\frac{\mathrm{d}^2\sigma}{\mathrm{d}\Omega\mathrm{d}E'} = \frac{\text{No. of neutrons per second scattered into } \mathrm{d}\Omega \text{ with energy between } E' \text{ and } E'+\mathrm{d}E'}{\Phi\mathrm{d}\Omega\mathrm{d}E'} \tag{3.4}$$

where Φ is the flux of incident neutrons.

3.2.3.2 Coherent and Incoherent Scattering

In the multinuclear target system described above, the relative occurrence of scattering length b_i is defined to be f_i such that

$$\sum_i f_i = 1 \tag{3.5}$$

and the average value for the scattering length of the system is

$$\bar{b} = \sum_i f_i b_i \tag{3.6}$$

Note that the scattering length has been treated as a purely real-valued number in Eqs. (3.5) and (3.6). If the number of scattering centers in the target system is sufficiently large, then the partial differential cross section is the cross section averaged over all the scattering centers:

$$\frac{d^2\sigma}{d\Omega dE'} = \frac{k'}{k}\frac{1}{2\pi\hbar}\sum_{jj'} \overline{b_{j'}b_j}\int \langle j', j\rangle \exp(-i\omega t)dt \tag{3.7}$$

where $\omega = (E - E')/h$ and the matrix element $\langle j', j\rangle$ represents a complex function relating the positional vectors of the nuclei (j and j') before and after the scattering event. The derivation of Eq. (3.7) is given in [14].

Noting that

$$\begin{aligned} \overline{b_{j'}b_j} &= \left(\bar{b}\right)^2, \quad j \neq j' \\ \overline{b_{j'}b_j} &= \left(\overline{b^2}\right), \quad j = j' \end{aligned} \tag{3.8}$$

it is possible to rewrite Eq. (3.7) as follows:

$$\frac{d^2\sigma}{d\Omega dE'} = \frac{k'}{k}\frac{1}{2\pi\hbar}\left(\bar{b}\right)^2 \sum_{jj', j\neq j'} \int \langle j', j\rangle \exp(-i\omega t)dt + \frac{k'}{k}\frac{1}{2\pi\hbar}\left(\overline{b^2}\right) \sum_{jj', j=j'} \int \langle j', j\rangle \exp(-i\omega t)dt \tag{3.9}$$

The first term in Eq. (3.9) defines what is known as the *coherent scattering cross section* whereas the second term defines the *incoherent scattering cross section*. It is seen that coherent scattering arises from the positional correlation of different nuclei at any given time. Physically, this means coherent scattering results in interference effects. In contrast, incoherent scattering depends only on the correlation between the positions of the same nucleus at different times. It does not give interference effects and its contribution to the scattered neutron intensity should be removed when attempting to perform structural studies.

3.2.3.3 Effective Potential and Scattering Length Density

According to Eq. (3.3), a neutron incident on a solid or liquid would see an array of δ-function potentials

$$V_f(r) = \frac{2\pi\hbar^2}{m}\sum_i b_i \delta(r - r_i) \tag{3.10}$$

where r_i is the position of the ith nucleus and the sum over i includes all nuclei in the system. Thus the total wavefunction of the neutron consists of the incident wavefunction plus the sum of all the spherical waves scattered by each nucleus. It is important to realize that the wave incident on each nucleus is not the same as the incident wave on the entire sample. A good discussion on treating the problem of multiple scattering has been given by Sears [15] but falls beyond the scope of this introduction to neutron scattering theory. It can be shown [16] that the effective potential for a neutron scattered by a collection of nuclei is simply the volume average of the Fermi potentials

$$V = \frac{2\pi\hbar^2 bN}{m} = \frac{2\pi\hbar^2\rho}{m} \tag{3.11}$$

where b is the effective scattering length of the medium and N is the effective number density of scattering nuclei. The product bN is known as the scattering length density (SLD), ρ. Knowledge of the SLD provides a direct measure of the nuclear composition of the scattering medium.

3.2.4 Theoretical Expressions for Specular Reflectivity

3.2.4.1 The Continuum Limit

Consider the reflection of neutrons elastically scattered from the interface between vacuum (the incident medium) and a layer of condensed matter. The interface is chosen to lie in the xy plane and is assumed to be perfectly flat such that the only variation in the SLD occurs along the z-direction. The problem to address is the formulation of a description of the reflected wave as a function of its initial wave vector. The scattering is considered to be specular in the sense that the angle of the incident wave vector, k_{in}, is equal to the angle of the reflected wave vector, k_{f}. The momentum transfer wave vector Q is perpendicular to the interface, that is, the z-direction, and is given as the difference between the wave vectors of the outgoing and incoming plane waves:

$$Q_z = \mathbf{k}_{\text{f},z} - \mathbf{k}_{\text{in},z} \tag{3.12}$$

If Q_z is sufficiently small such that $2\pi/Q_z$ is much larger than interatomic distances in the condensed phase, then the medium can be treated as a continuum, though its nuclear density need be neither constant nor homogeneous. In vacuum, the total energy of the neutron outside the medium is simply its kinetic energy

$$E = \frac{\hbar^2 k_0^2}{2m} \tag{3.13}$$

In the continuum limit, the effective potential energy in the condensed phase is given by Eq. (3.11) and the total neutron energy is

$$E = \frac{\hbar^2 k^2}{2m} + \frac{2\pi\hbar^2\rho}{m} \tag{3.14}$$

where k is the wave vector of the neutron inside the non-vacuum medium. By rearranging Eqs. (3.13) and (3.14) for the wave vectors k_0 and k, respectively, one can determine the refractive index in the direction perpendicular to the interfacial plane:

$$n(z) = \frac{k_z}{k_{0,z}} = \sqrt{1 - \frac{4\pi\rho(z)}{k_{0,z}^2}} \tag{3.15}$$

The one-dimensional wave equation describing the propagation of the neutron wave inside the medium arises from the time-independent Schrödinger equation

$$\frac{d^2\psi}{dz^2} + \frac{2m}{h^2}[E - V(z)]\psi = 0 \tag{3.16}$$

Substituting in Eqs. (3.11) and (3.15) yields the following differential equation:

$$\psi'' + k_z^2\psi = 0 \tag{3.17}$$

with the general solution

$$\psi(z) = a_1 \exp(ik_z z) + a_2 \exp(-ik_z z) \tag{3.18}$$

where the coefficients a_1 and a_2 describe the amplitudes of the forward going and scattered wave, respectively. The reflection amplitude (r) and transmission amplitude (t) are defined in terms of the limiting forms of Eq. (3.18):

$$\psi(z) = \exp(ik_{0,z}z) + r\exp(-ik_{0,z}z) \tag{3.19}$$

$$\psi(z) = t\exp(ik_z z) \tag{3.20}$$

The procedure can be expanded in a piecewise fashion [17, 18] to obtain the reflection and transmission amplitudes arising from the reflection of neutrons from an arbitrary potential or SLD profile if the potential is divided into a discrete number (j) of rectangular lamellae. The reflection and transmission amplitudes are obtained from a pair of simultaneous equations which, when written in matrix notation, define the transfer matrix:

$$\begin{pmatrix} t \\ in_b t \end{pmatrix} = \left[\prod_{j=N}^{1} \begin{pmatrix} \cos\theta_j & \sin\theta_j / n_j \\ -n_j \sin\theta_j & \cos\theta_j \end{pmatrix}\right] \begin{pmatrix} 1+r \\ in_f(1-r) \end{pmatrix} \tag{3.21}$$

where $\theta_j(Q) = (Q/2)n_j d_j$ and $n(Q) = \sqrt{1 - 16\pi\rho / Q^2}$ for the fronting medium (n_f), the backing medium (n_b), and the interstitial layers (n_j). The transmission and reflection amplitudes are seen to be complex numbers and consequently the *transmissivity* and *reflectivity* are defined as $t * t = |t|^2 = T$ and $r * r = |r|^2 = R$, respectively. In a NR experiment, it is the complex conjugate term that is measurable and all information concerning the imaginary component (or phase) of the reflectivity is lost. It is apparent from Eq. (3.21) that the reflectivity for any known SLD profile can be calculated directly. However, as will be discussed later, the converse of this statement is not necessarily true.

3.2.4.2 The Kinematic Approach

The previous derivation of the reflection and transmission coefficients correctly describes the intensity of reflected neutrons at any value of momentum transfer vector. However, there is a useful alternative derivation, which gives a highly analytical function describing the reflectivity. This derivation is based on the Born approximation and is often referred to as reflectivity in the kinematic limit. Suppose there are two arbitrary but different SLD profiles $\rho_1(z)$ and $\rho_2(z)$ and one wishes to determine the separate reflectivities $R_1(Q_z)$ and $R_2(Q_z)$ for the two scattering potentials. The solution to the problem is described by combining Eqs. (3.15) and (3.17)

$$\psi_j'' + (k_0^2 - 4\pi\rho_j)\psi_j = 0, \quad j = 1, 2 \tag{3.22}$$

from which the Wronskian function

$$W(z) \equiv W[\psi_1(z), \psi_2(z)] = \psi_1(z)\psi_2'(z) - \psi_2(z)\psi_1'(z) \tag{3.23}$$

can be constructed. Differentiating both sides of Eq. (3.23) with respect to z and applying Eq. (3.22) yields

$$W'(z) = -\psi_1(z)\psi_2(z)4\pi[\rho_1(z) - \rho_2(z)] \tag{3.24}$$

Inspection of Eq. (3.24) reveals that the original Wronskian function, $W(z)$, must be a constant whenever $\rho_1 - \rho_2 = 0$ (i.e., whenever the two SLD profiles are equal). If the argument is extended to a real situation, where the two different SLD profiles share a common fronting medium and a common backing medium, then it is evident that $\rho_1 - \rho_2 = 0$ for $z < \tau_1$ and $\rho_1 - \rho_2 = 0$ for $z > \tau_2$, where τ_1 and τ_2 mark the end of the fronting medium and the beginning of the backing medium, respectively. For simplicity the derivation continues for a vacuum fronting ($\rho_1 = \rho_2 = 0$) although this restriction is not necessary. The wavefunctions for $z < \tau_1$ can be written

$$\psi_j(z) = \exp(ik_0 z) + r_j \exp(-ik_0 z), \quad j = 1, 2 \tag{3.25}$$

where r_j is the reflection coefficient for each SLD profile. As the two profiles share a common backing such that $\rho_1 = \rho_2 = \rho_{\text{backing}}$ for $z > \tau_2$, then the wavefunctions in this region are

$$\psi_j(z) = t_j \exp(iKz), \quad j = 1, 2 \tag{3.26}$$

where t_j is the transmission coefficient for each problem and K is the wave vector after the transmission through the scattering system:

$$K^2 = k_0^2 - 4\pi\rho_{\text{backing}} \tag{3.27}$$

By substituting Eq. (3.26) into the Wronskian function (Eq. (3.23)) one obtains

$$W(z) = 0; \quad z \geq \tau_2 \tag{3.28}$$

which arises from the fact that the wavefunctions are linearly dependent for $z \geq \tau_2$. However, substitution of Eq. (3.25) into Eq. (3.23) leads to

$$W(z) = 2ik_0[r_1 - r_2]; \quad z \leq \tau_1 \tag{3.29}$$

which is a complex constant and applies for all $z < \tau_1$. Finally, for $\tau_1 < z < \tau_2$ integration of Eq. (3.24) yields

$$\int_{\tau_1}^{\tau_2} W'\mathrm{d}z = W(\tau_2) - W(\tau_1) = -\int_{\tau_1}^{\tau_2} \psi_1(z)\psi_2(z)4\pi[\rho_1(z) - \rho_2(z)]\mathrm{d}z \tag{3.30}$$

and as $W(\tau_2) = 0$ (via Eq. (3.28)) and $W(\tau_1) = 2ik_0[r_1 - r_2]$ (via Eq. (3.29)), Eq. (3.30) can be rewritten as

$$r_1 = r_2 + \frac{1}{\mathrm{i}Q_z}\int_{\tau_1}^{\tau_2} \psi_1(z)\psi_2(z)4\pi[\rho_1(z) - \rho_2(z)]\mathrm{d}z \tag{3.31}$$

where Q_z is equal to $2k_0$, as k_0 has been previously defined to be the z-component of the incident wave vector. By choosing to set $\rho_2(z) = 0$ for all values of z, then $\psi_2 = \exp(ik_0z)$ and $r_2 = 0$. Accordingly,

$$r_1 = \frac{4\pi}{\mathrm{i}Q_z}\int_{-\infty}^{\infty} \psi(z)\exp(\mathrm{i}k_0z)\rho(z)\mathrm{d}z \tag{3.32}$$

where the subscript has been omitted and the integration has been formally extended over all values of z even though it is evident that the only contribution to the integral exists for the interval $\tau_1 < z < \tau_2$. Note that Eq. (3.32) is written in terms of the wavefunction $\psi(z)$ that describes the neutron *inside* the scattering medium. From Eq. (3.32) the expression for the reflectivity of the neutrons is

$$R = |r|^2 = \frac{16\pi^2}{Q_z^2}\left|\int_{-\infty}^{\infty} \psi(z)\exp(\mathrm{i}k_0z)\rho(z)\mathrm{d}z\right|^2 \tag{3.33}$$

The difficulty in applying Eq. (3.33) lies in finding an appropriate expression for $\psi(z)$. An important simplification which has been widely applied in X-ray and neutron reflectivity is obtained if $\psi(z)$ is replaced with the free space wavefunction $\exp(ik_0z)$. This is known as the Born approximation and provides

$$R = \frac{16\pi^2}{Q_z^2}\left|\int_{-\infty}^{\infty} \exp(\mathrm{i}Q_zz)\rho(z)\mathrm{d}z\right|^2 \tag{3.34}$$

This equation is commonly referred to as the "kinematic" expression and is valid whenever the perturbation of the incident free space wavefunction by the scattering medium is sufficiently small. A highly reflective scattering event greatly perturbs the incident wavefunction and consequently the kinematic approximation is inadequate for such interactions.

To evaluate the adequacy of the kinematic approximation one can compare the reflectivity as calculated via Eq. (3.34) and compare it to the output from Eq. (3.21) for a simple SLD profile. The SLD profile for a smooth interface between air and quartz is shown in Figure 3.3a where the function $\rho(z)$ is seen to be a Heaviside function

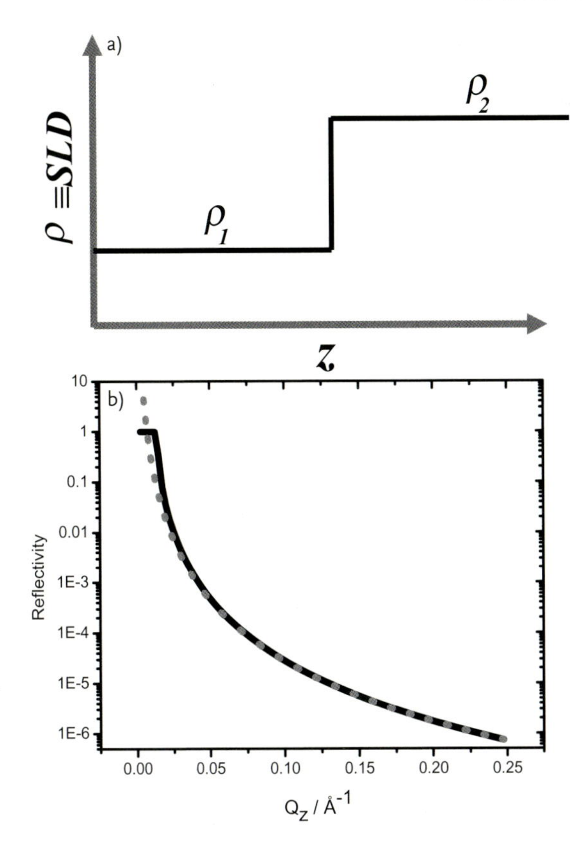

Figure 3.3 (a) SLD profile for a smooth interface between air and a single-crystal substrate. As the interface is stated to be infinitely sharp, the function $\rho(z)$ is a Heaviside function. (b) Comparison of reflectivity curves for the SLD profile using the matrix method (solid curve) and the kinematic approximation (dotted curve).

$$\rho(z) = \begin{Bmatrix} \rho_{\text{air}} & z < 0 \\ \rho_{\text{quartz}} & z > 0 \end{Bmatrix} \tag{3.35}$$

Substitution of Eq. (3.35) into Eq. (3.34) yields the Fourier integral for the Heaviside function $H(z)$

$$\int_{-\infty}^{\infty} \exp(\mathrm{i}Q_z z) H(z) \mathrm{d}z = \left(\pi\delta(Q_z) + \frac{1}{Q_z} \right) \Delta\rho \tag{3.36}$$

where $\delta(Q_z)$ is the Dirac delta function, which is equal to zero for all values of Q_z except for $Q_z = 0$ where $\delta(Q_z) = 1$. From Eq. (3.36), the reflectivity for any Q_z greater than zero is

$$R = \frac{16\pi^2}{Q_z^4} (\rho_{\text{quartz}} - \rho_{\text{air}})^2 \tag{3.37}$$

Figure 3.3b plots $R(Q_z)$ calculated from Eq. (3.37) and $R(Q_z)$ calculated from Eq. (3.21). At values of Q_z greater than about 0.03 Å^{-1} the two curves almost perfectly coincide. For smaller momentum transfer vectors the kinematic approximation significantly diverges from the correct reflectivity for reasons explained above. To reduce this discrepancy a distorted wave Born approximation (DWBA) has been applied to high-reflection domains of Q space. As the DWBA only modifies the low-Q range of the reflectivity curve it does not change the analytical usefulness of the kinematic formalism. Therefore, the DWBA will not be discussed herein but the interested reader is directed elsewhere for details [19, 20].

3.3 Experimental Aspects

Four main aspects are addressed in turn: (i) neutron sources and reflectometer operation; (ii) choice of substrate, substrate preparation, and characterization; (iii) cell design and assembly; and (iv) data acquisition and analysis.

3.3.1 Experimental Aspects of Reflectometer Operation

Neutron intensities sufficiently large to perform scattering experiments are only available at large-scale research facilities. These facilities are either reactor- or spallation-based neutron sources. A recent review provides a comprehensive description of new neutron sources with particular emphasis on instruments dedicated for biological studies [21]. Reactor-sourced reflectometers such as the NG-7 and NG-1 instruments at the NIST Centre for Neutron Research (NCNR) at the National Institute of Standards and Technology operate at a fixed neutron wavelength and usually rely on variation of the incident angle to access momentum transfer space. Time-of-flight instruments are usually found at spallation sources such as the SPEAR reflectometer at the Los Alamos Neutron Science Centre (LANSCE) and make use of pulsed polychromatic neutron radiation. The simultaneous acquisition of a wide range of Q_z has been shown to be very advantageous for studying kinetic processes in thin films [22–25]. The horizontal NG-7 reflectometer at NCNR is shown schematically in Figure 3.4. The intensity of neutrons is determined using two ^{3}He scintillators, one placed before the sample (denoted the monitor) and one placed after the sample (the detector). In some instances a multichannel detector can be used to collect off-specular as well as specular reflected neutrons. In the simplest applications the off-specular signal can be used for background correction but this signal also carries information about the SLD profile in the plane of the substrate and has been used to measure in-plane order [26–28] but will not be discussed further. The ratio of the specular detector count to the monitor count gives the measured specular reflectivity. As the wavelength is fixed for reactor-source neutrons, the momentum transfer vector is systematically changed by varying the angle of the incident (and, consequently,

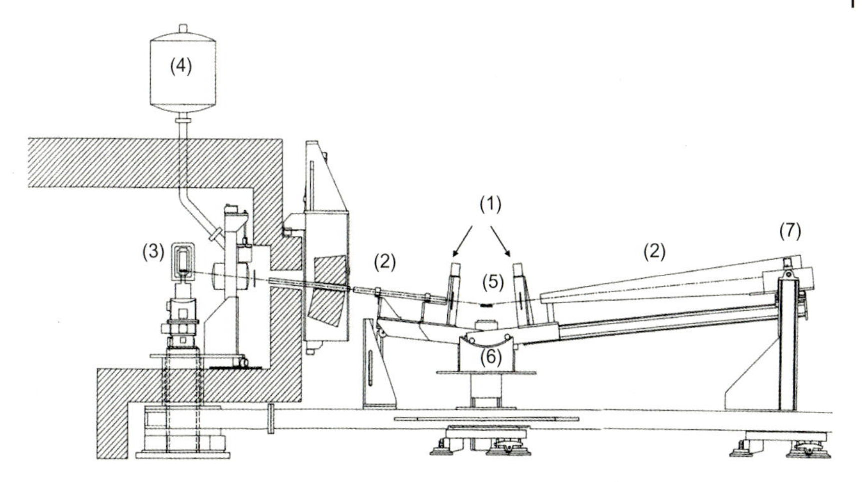

Figure 3.4 Schematic of the NG-7, horizontal, reflectometer used for NR studies at NCNR. 1. *Collimating slits*–these are located on either side of the sample. A beam monitor on the precollimating slit allows the primary detector to be scaled to absolute reflectivites. 2. *Pre- and post-sample flights*–both have LiF slits for collimation of the incident or reflected beam. Flight tubes have $2\theta_c$ (^{58}Ni) supermirrors and are evacuated to minimize losses. 3. *Tilting pyrolitic graphite monochromator*–provides four possible wavelengths: 2.35, 4.10, 4.75, and 5.50 Å. 4. *Thermal neutron filter*–a liquid nitrogen-cooled beryllium filter is placed after the monochromator to remove thermal neutrons. 5. *Sample stage*–available space is approximately 30 cm between the collimating slits. 6. *Translation stage*–computer-driven stepper motors provide precise alignment of the sample with the beam. 7. *Detector*–a 2.5 cm diameter, cylindrical ^{3}He proportional counter is used as the detector.

the specularly reflected) neutrons ($Q_z = 4\pi \sin\theta/\lambda$). This requires concerted movement of the main sample elevator, the pre- and post-sample flights, and the collimating slits. Slits are required to minimize the angular divergence in the scattering plane of the incident neutrons. Angular distribution leads to multiple angles of incidence, and consequently a distribution of momentum transfer vectors rather than a single Q_z value. The angular divergence in Q_z space defines the instrumental resolution dQ_z/Q_z of the reflectometer. When the neutron source is monochromatic (as opposed to a "white" source used with a time-of-flight instrument) the slit apertures are increased with increasing incident angle in order to maintain a constant instrumental resolution. If the substrate supporting the film of interest is curved rather than perfectly flat then the effective angular divergence of the incident beam will be increased leading to reduced instrumental resolution. Thus low curvature, also referred to as bow or warp, is an additional mandatory feature for a suitable substrate (see Section 3.3.2). In a typical experiment the angle of incidence can be varied such that the Q_z range can extend beyond about 0.5 Å^{-1}. However, for most experiments the loss of measurable signal above background at $R \approx 10^{-8}$–10^{-7} limits $Q_{z,\max}$ to the range of about 0.20 to about

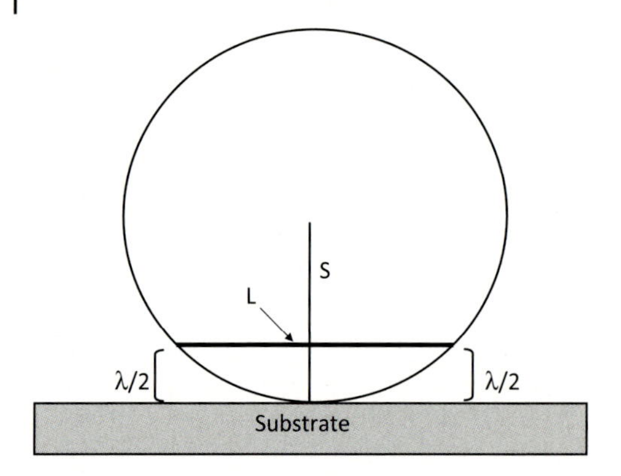

Figure 3.5 A spherical wave of radius *S*, impinging on the interface of interest. The line of length *L* lies parallel to the substrate and defines the length scale where the distance between the wave front and the substrate remains less than or equal to half the wavelength of the incident neutron. This defines the lateral coherence length of the incident neutron.

0.25 Å^{-1}. To improve the statistical quality of data, typical counting times can exceed 8–10 h which requires that systems being studied are sufficiently stable.

The neutron beam incident upon the sample is an incoherent collection of individual neutrons whose lateral coherence length is determined by the method of preparation (how it was emitted by the fissile material, its collision with atoms in the cold source, diffraction by the collimating slits, etc.). The lateral coherence length is defined as the spatial extent over which an incident neutron wave front has a constant phase in its interaction with the sample. Geometrically, this is only achieved over the dimension of the wave front which can be considered sufficiently flat. A simplified explanation follows. Imagine that a thermal neutron is generated in a reactor core and its final interaction before proceeding along the guide tube was a collision with a single hydrogen nucleus in the cold source moderator. The scattered wave from this interaction is spherical with a wavelength λ. Subsequently, after traveling along the guide tube, it reaches the sample at a distance S away from the hydrogen atom of origin. Referring to Figure 3.5, a line segment can be inscribed in the sphere, which is parallel to the interface, but whose terminal points are still a distance $\lambda/2$ from the interface. The length L of such a line is given by

$$\left(\frac{L}{2}\right)^2 = \lambda\left(S - \frac{\lambda^2}{4}\right) \tag{3.38}$$

The end points define the lateral distance in the sample plane across which a given spherical wave front becomes out of phase by π degrees. For a spherical wavelength of 5 Å emanating from a source 5 m away, the length L is of the order of 100 μm. In practice, the lateral coherence length depends on the energy spread and the wave vector spread of the incident neutron wave front. If the instrumental resolution is kept constant during the scan this means that the lateral coherence length varies with Q_z but in all instances it is a macroscopic distance of the order of micrometers to hundreds of micrometers. The finite coherence length can influence the interpretation of NR data. There are two limiting cases to be addressed. If the lateral inhomogeneities, or deviations from perfect flatness, are on length scales significantly smaller than the coherence length of the incident neutron, this scale of roughness gives rise to diffuse reflection, but as long as the contribution to the total reflection by this off-specular component is significantly smaller than the specular signal then a one-dimensional approach as described in Section 3.2 remains valid. The in-plane roughness will be manifest in the SLD profile by slabs of thickness dz which represent average values over the area that the incident neutron wave is effectively coherent or in-phase. In the other limiting scenario, the in-plane inhomogeneities are on a length scale much greater than the incident neutron coherence length scale. In such a case the observed reflectivity will be proportional to a sum of independent, area-weighted reflected intensities for each different region. The analysis of data of this type is very complex and usually cannot be successfully treated with simple one-dimensional approaches.

3.3.2
Substrate Preparation and Characterization

There are several parameters that dictate which materials are viable substrates for NR studies of the solid–liquid interface. If the momentum transfer vector is being varied by adjustment of angle, neutron beam footprints are of the order of several square centimeters at lowest grazing angles. At larger angles (higher Q_z), the dramatic decrease in reflectivity requires wider slits (maintaining comparable sample footprint areas) to maintain a measurable intensity of reflected neutrons. Generally speaking, a large-dimension surface is required for NR experiments due to the comparatively low incident neutron flux. For example, the neutron flux on the NG-1 beamline at the NCNR is 1–5 × 10^4 neutrons $cm^{-2} s^{-1}$ [29] compared to the X-ray flux of roughly 10^{17} photons $cm^{-2} s^{-1}$ at the Brookhaven National Light Source Synchrotron [30]. New and proposed next-generation neutron sources have increased time-averaged fluxes to more than 10^{13} neutrons $cm^{-2} s^{-1}$ [21].

Apart from the special case of large losses due to incoherent scattering from proton-containing materials such as water, the main loss of neutron intensity in bulk media is due to scattering from diffraction. It is impossible to avoid large diffraction losses if the neutron beam passes through an appreciable thickness of liquid. For supported films in aqueous electrolytes, the incoherent scattering of neutrons by water also greatly attenuates intensity such that experiments are almost always run in an inverted configuration. In this mode, neutrons are incident on the interface through the supporting medium rather than through the electrolyte. As the neutrons must travel through a sizeable path length in the supporting material (on the centimeter scale), one may also expect high losses due to multiple diffraction processes. This can be avoided if the substrate is an appropriately oriented single crystal and/or suitable neutron wavelengths are chosen that do not lead to Bragg diffraction. These arguments lead to the criteria that the substrate material should be single-crystalline and transparent to neutrons. In principle a wide range of solid crystals would be suitable, but in practice there are few materials that are available as large single crystals (typical dimensions of suitable quartz blocks are of the order of $10\,cm \times 5\,cm \times 1\,cm$ thickness) for an accessible price. Three readily available substrates suitable for neutron reflectivity are silicon, quartz (SiO_2), and sapphire (Al_2O_3). Fortunately, it is possible to chemically modify these surfaces to change their surface properties or one can coat them with thin metallic films in order to use them as electrodes.

Proper characterization of the substrate is imperative when data analysis is based upon fitting analyses (see Section 3.3.4). As most neutron facilities have ready access to off-line X-ray reflectivity infrastructure, it is possible to characterize the roughness of the surface prior to any chemical and/or physical modifications. The typical roughness of highly polished substrates such as silicon and quartz should be less than 5 Å (root-mean-square). A very smooth substrate is critical for reflectivity experiments because roughness greatly attenuates specular reflection causing the measured reflectivity curves to drop to background levels at prematurely low momentum transfer vectors. For a detailed description of the effect of roughness on specular reflectivity, see [8]. To convert single-crystal substrates into electrodes, thin layers of chromium and gold can be sequentially sputtered on the substrates. A chromium layer has to be present in order to ensure adhesion of gold to quartz. Other means to achieve the adherence of smooth gold films on inorganic substrates include chemically modifying quartz with mercaptopropylsiloxane [31] and this has been successfully used to adhere gold onto quartz substrates for electrochemical NR studies [32]. Before using chemically or physically modified substrates it is imperative to once again characterize them in air with neutron and/or X-ray reflectivity. These pre-measurements provide valuable information on the roughness and SLDs of chromium and gold layers. Cyclic voltammetry (CV) curves recorded on these thin-film gold electrodes often resemble CV profiles recorded at the Au(111) surface suggesting that the sputtered gold layer is usually preferentially (111) oriented. Substrates can be reused by stripping the chromium and gold layers in a heated mixture of H_2SO_4 and HNO_3 (1 : 1 ratio); however, it is desirable to measure a final X-ray reflectivity curve after electrochemical NR experiments and before stripping. Experimentally, it has been observed that the gold and chromium layers can sometimes be affected structurally when a negative potential is applied to the substrate. In general, this potential-induced alteration of modified quartz substrates has a much greater rate of occurrence when the quartz block is being reused (i.e., it had been previously coated with chromium and gold). It is hypothesized that the stripping procedure may not lead to complete removal of the old metal films and the reused quartz substrate may leave patches of chromium oxide on its surface when it is recoated. The chromium oxide is then reduced when the electrode is biased negatively and consequently the observed parameters for the substrate differ for the before and after measurements. Figure 3.6 shows a comparison of the before and after reflectivity measurements for a virgin coated quartz substrate and a reused substrate. Figure

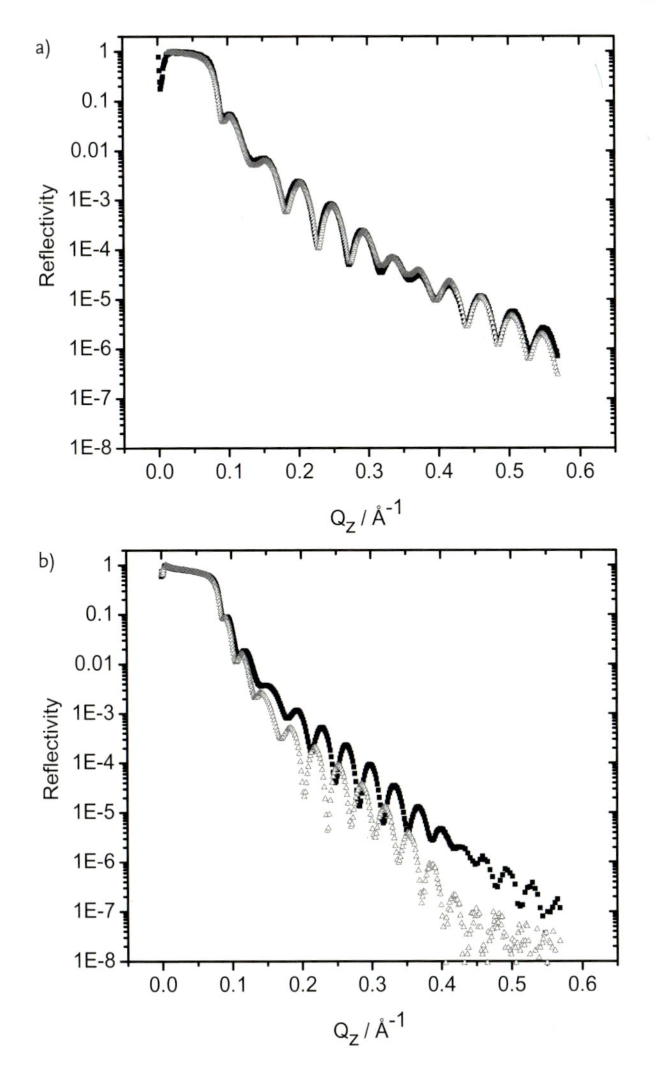

Figure 3.6 Experimental X-ray reflectivity curves in air for gold- and chromium-modified quartz substrates measured before (filled squares) and after (open triangles) electrochemical experiments. (a) Nearly identical curves for a virgin-coated substrate indicative of a metal layer that is not perturbed by the potential bias. (b) Recycled substrates lead to metal layers that are significantly altered by cathodic biases.

3.6a shows that the X-ray reflectivity curves for a pristine quartz crystal coated with chromium and gold measured before and after electrochemical experiments are almost superimposable. For comparison, Figure 3.6b plots the X-ray reflectivity curves, measured before and after electrochemical experiments, for a quartz block recoated with chromium and gold. Note that the “before” curve in Figure 3.6b is marked by a pronounced loss in the oscillation amplitude at momentum transfer vectors larger than 0.45 $Å^{-1}$ which is an indication of a rough interface. The curve for the same crystal after electrochemical experiments shows marked changes and indicates that the structure of the gold and chromium films has been altered. In such a case, the *in situ* NR data cannot be properly analyzed and experiments need to be repeated on a stable substrate.

Thin silicon wafers appear to be an attractive alternative to thicker quartz blocks for several reasons. Due to their prominence in the semiconductor industry, large-diameter, thin (500 μm) silicon wafers are commercially available from a large number of sources. Wafers, 10 cm in diameter, easily meeting the demanding flatness criteria required for NR experiments, can be purchased for as little as US$15–20 per wafer, which is roughly 20 times lower in price compared to polished quartz blocks. Such silicon wafers could be treated as disposable samples whereby each wafer is coated only once and not stripped and recoated. An additional advantage of circular silicon wafers over rectangular quartz blocks is an increased neutron footprint area. The larger the area illuminated by the incident neutron beam the better the signal-to-noise ratio for a given count time duration. As a proviso it should be stated that this area advantage only occurs if the illuminated surface is uniformly coated. Quartz blocks have a maximum working surface of approximately 32 cm^2 compared to 79 cm^2 in the case of a 10 cm diameter silicon wafer. In theory, switching from quartz blocks to silicon wafers could result in a nearly 2.5 times increase in the surface area and, as the signal-to-noise ratio is proportional to the square root of the area, a more than 50% increase in signal-to-noise ratio could be achieved. Disadvantages to using silicon wafers include the presence of the native oxide film inherently found on all untreated silicon surfaces. This oxide film is usually between 10 and 50 Å thick depending on the precise method used to grow the single-crystal wafer [33]. For X-rays, the difference in the SLD of silicon and its oxide is very small, but for neutrons the SLD difference is more appreciable (3.5×10^{-6} $Å^{-2}$ for the oxide versus 2.1×10^{-6} $Å^{-2}$ for silicon). As a result, the native oxide film on silicon contributes to the NR response in an electrochemical experiment and the measured reflectivity would have contributions from three layers (SiO_2, chromium, gold) from the substrate alone. Fortunately, the oxide layer is very stable to electrochemical perturbation and should not be influenced when a potential is applied to the modified substrate. Most commercially available silicon is doped with group III or group V elements. As long as the level of doping is not extremely high, then the scattering from elements such as antimony and phosphorus is insignificant. However, silicon doped with boron should be avoided due to its strong scattering cross section.

3.3.3 Cell Design and Assembly

Like many other *in situ* coupled techniques, the ideal experimental conditions for the collection of specularly reflected neutrons are not ideal for maintaining electrochemical control. This mandates the improvisation of new cell designs. As detailed above, measurements performed in the presence of electrolyte should use an inverted cell geometry. The cell needs to be constructed so that the neutrons are incident on a transparent substrate, which has been modified in such a way to fashion a working electrode upon which the films of interest can be formed. The major difficulty in cell design for *in situ* electrochemistry–NR studies is that the two techniques have conflicting requirements. An optimal electrochemical cell for studying thin-film adsorption would have the following characteristics:

1) A small, ideally polarized working electrode.
2) The working electrode should be constructed out of a pure material, ideally a single crystal.
3) The reference electrode should be electrically connected to the cell via high ionic conductivity.
4) A large-surface-area counter electrode.
5) An inert atmosphere to remove faradaic current arising from dissolved oxygen in the electrolyte.

On the other hand, an ideal cell for studying NR would have the following features:

1) A large-area working electrode to increase the number of reflected neutrons.
2) The working electrode should be derived from a material optically transparent to neutrons (e.g., quartz, silicon, sapphire).
3) The amount of solvent in the cell should be minimized to reduce incoherent scattering and lower the background level.
4) A quiescent solution to minimize incoherent scattering with no void volume or trapped air bubbles.

Comparing the two lists, the greatest challenge in cell design is the electrolyte volume. From a NR perspective, a great enhancement in the signal above background is achieved by reducing the thickness of the electrolyte to less than 1 mm. However, too thin a layer of electrolyte provides large ohmic impedance and may have disastrous electrochemical consequences if it completely prevents conductivity between the working and reference electrodes. Often a thin film of electrolyte leads to the formation of bubbles in the cell. Achieving an inert atmosphere for long-duration NR experiments can be problematic. Electrolyte introduced into the cell can be pre-purged with argon but once the cell is filled there is no headspace, preventing an inert atmosphere from being maintained over the electrolyte. CV curves obtained immediately after filling a cell typically show very little oxygen contamination but over the course of 30 min or so a progressively larger oxygen reduction wave is often observed in the CV curves as oxygen permeates the Teflon components of the cell. Depending on the electrochemical application and techniques employed, the presence of oxygen in the cell may not greatly affect the experiment.

Two types of cells designed and used for inverted-geometry electrochemistry–NR experiments are shown in Figure 3.7. Figure 3.7a shows a large-cavity cell used in early applications of NR in bioelectrochemistry. The advantages of such a cell are its ease of assembly, reliable electrical contact to the working and counter electrodes, and its overall electrochemical stability due to the large electrolyte reservoir. The disadvantages of this cell are the large amounts of Teflon and

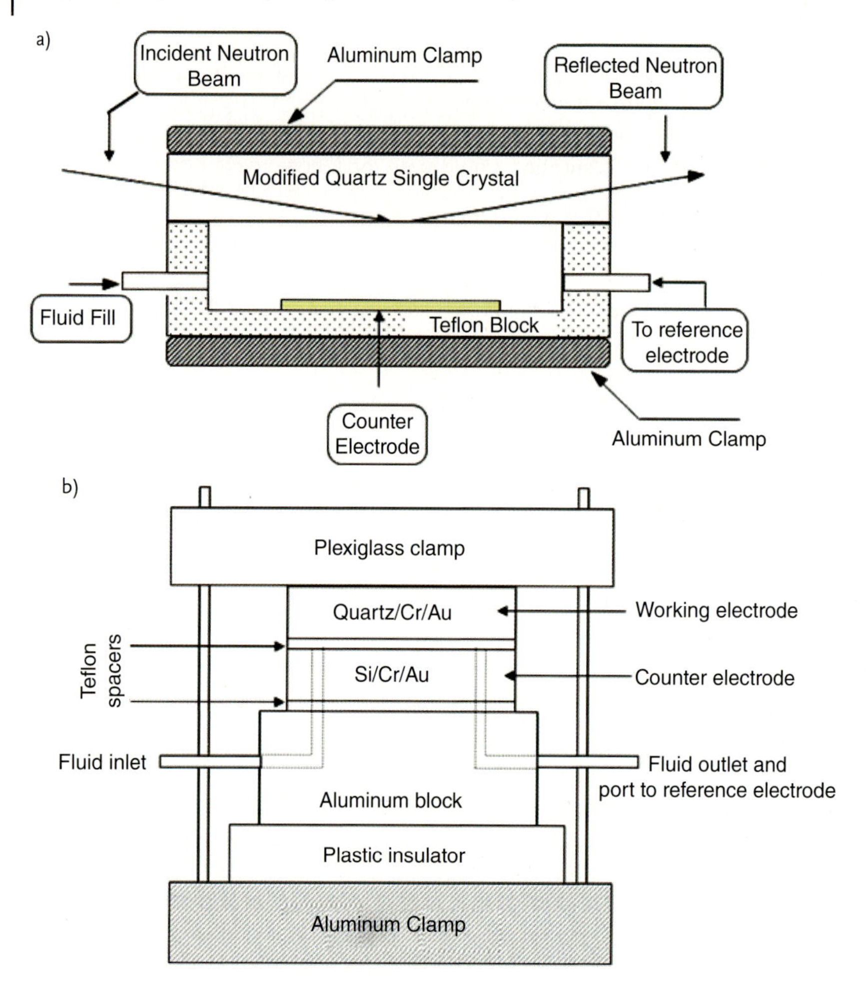

Figure 3.7 Schematics of two electrochemical cell configurations used in NR experiments: (a) larger electrolyte cavity design; (b) thin electrolyte layer configuration. Figure adapted from [79, 84].

aqueous solvent which yield large levels of incoherent scattering. This cell configuration is limited to only being capable of measuring reflectivities greater than 10^{-6}. Figure 3.7b shows a thin-layer cell which extends the lower limit of measurable reflectivities by almost an order of magnitude due to the removal of the Teflon and the reduction of incoherent scattering from the aqueous solvent.

3.3.4 Data Acquisition and Analysis

In measured neutron reflectivity curves, the error points in the experimentally measured data represent the statistical errors in the measurements (standard

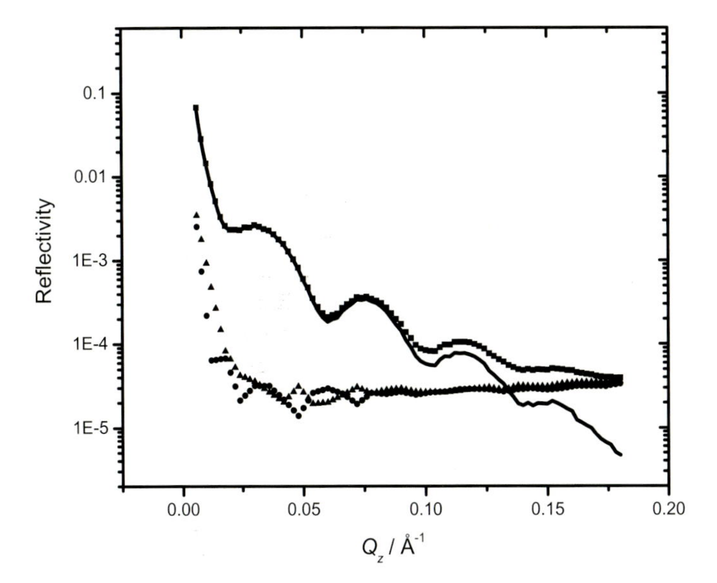

Figure 3.8 Influence of the background correction on the specular reflectivity data. The uncorrected specular data are shown as the square data points, while the triangles and circles show the background reflectivity determined with two off-specular measurements. The solid line is the result of subtracting the averaged, off-specular data from the specular measurement.

deviation, σ). The standard deviation scales inversely with the square root of the number of neutron counts and consequently lower reflectivities carry a larger degree of uncertainty than high reflectivities. Data reduction requires accounting for neutron beam transmission losses through the substrate and correction for the background (off-specular) signal. In the absence of a multichannel detector, the background measurement can be made by deliberately misaligning the crystal so that the off-specular reflection is measured on both sides of the specular angle. The averaged off-specular signal is then subtracted from the measured specular signal. To illustrate the background correction procedure, Figure 3.8 shows a representative reflectivity curve, two background curves, and the background-corrected curve. The SLD profile is then usually determined by fitting a model to the measured reflectivity data.

A more detailed description of the analysis of neutron reflectivity data is provided with the aid of hypothetical examples. The reflection of neutrons from a multilaminar interface can be calculated in an analogous fashion to the optical matrix method for electromagnetic radiation [34, 35], provided the SLD, roughness, and thickness of each layer are known. For neutrons, the refractive index of a given phase may be calculated from the knowledge of the SLD, ρ, using Eq. (3.15). The dependence of the SLD on the composition of a given phase is described by the formula

$$\rho = \sum_j b_j n_j \tag{3.39}$$

where b_j is the scattering length and n_j is the number density of atomic species j. Neutron scattering lengths are empirically determined and the values for elements and their isotopes are tabulated.

When the SLD profile of an interface is known, a matrix method or a recursion method can easily be used to calculate reflectivity curves. However, for pedagogical purposes the relationship between the reflectivity and the structure of the interface is better revealed by the analytical expressions derived with the help of the kinematic approximation. The kinematic approximation has been shown to describe the reflection of neutrons from stratified media very well when the reflectivity is significantly less than unity. When a film of SLD ρ_1 and thickness τ_1 is sandwiched between two phases of identical SLD ρ, the expression for the reflectivity derived from a kinematic approach is [36]

$$R = \frac{16\pi^2}{Q_z^4}\left[4(\rho_1 - \rho)^2 \sin^2\left(\frac{Q_z\tau}{2}\right)\right] \tag{3.40}$$

and the reflectivity is seen to be proportional to the square of the difference between the SLD of the film and the backing and fronting phases and inversely proportional to the fourth power of the momentum transfer vector.

In an electrochemistry–NR experiment, the reflection of neutrons takes place at an interface consisting of five parallel phases as schematically shown in Figure 3.9. Table 3.1 lists the numerical values of the theoretical SLDs of the materials used in typical electrochemical studies. Each phase contributes to the overall measured NR, and to understand the shape of the experimental reflectivity curves it is instructive to examine the contribution of each individual lamina. To do this, a recursion scheme for stratified media described by Parratt [18] can be used to calculate the reflectivity of a simulated interface. These calculated reflectivities are then compared to the reflectivities predicted by the kinematic approximation. Consider a 20 Å thick film of a hydrocarbon-based surfactant deposited on a gold/chromium-modified quartz sample. To simplify the analysis, a mixed D_2O/H_2O

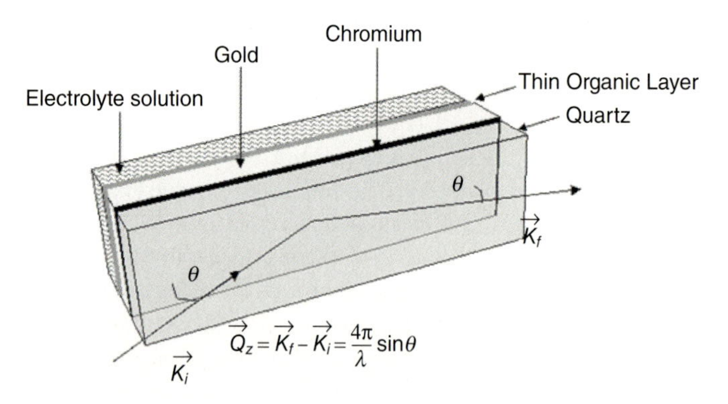

Figure 3.9 Schematic of an interface consisting of five parallel phases employed in electrochemistry–NR studies. Figure taken from [80].

Table 3.1 Calculated SLD values for the materials used in electrochemistry–NR studies of thin organic films.

Material/substance	$\rho \times 10^6$ (Å^{-2})
Quartz	4.18
Chromium	3.03
Gold	4.50
$-CH_2-$	−0.49 (determined for $C_{10}H_{22}$)
$-CD_2-$	6.60 (determined for $C_{10}D_{22}$)
D_2O	6.33
H_2O	−0.56

solvent is simulated such that the SLD for the solvent is contrast-matched to the gold (gold contrast-matched water). The presence of the quartz (SiO_2) and the thin film of chromium is temporarily neglected. Curve 1 of Figure 3.10a plots the SLD profile in the direction normal to the interface. The reflectivity curve calculated for this profile is presented in Figure 3.10b, which shows that the reflectivity decreases quickly with Q_z. This decay is modulated by the interference between neutrons reflected from the front and the back sides of the organic film lamina. In Figure 3.10c the same set of data are plotted as $Q_z^2\sqrt{R}/8\pi$ against Q_z. This presentation eliminates the Q_z^{-4} dependence of the reflectivity and allows a better analysis of the interference fringes. Consistent with Eq. (3.40), the period of the oscillation is $\Delta Q_z = 2\pi/\tau_1$ which allows the thickness of the film to be readily determined. The amplitude of the oscillation gives the absolute value of the difference between SLDs, $|\rho_1 - \rho|$, from which information on the composition of the film can be determined, with the help of Eq. (3.40). The presence of the chromium layer can also be examined. Curve 2 in Figure 3.10a plots the SLD profile for an interface that consists of quartz, 20 Å chromium, gold, and gold-matched solvent in the absence of the organic film. The SLD contrast between chromium and either adjacent layer is very weak (gold and quartz have very similar SLDs) and, as Figure 3.10b shows, the reflectivity is much lower in this case. The effect of the SLD contrast on the amplitude can be conveniently seen when the data are plotted as $Q_z^2\sqrt{R}/8\pi$ in Figure 3.10c. Curves 1 and 2 in Figure 3.10c have the same periodicity; however, curve 2 has much lower amplitude consistent with the lower contrast shown in the SLD profile in Figure 3.10a. To simulate an actual biomimetic membrane, a film of organic molecules is added to a 100 Å thick gold film placed on top of the chromium layer. The SLD profile at the interface is now represented by curve 3 in Figure 3.10a. The reflectivity curve corresponding to this profile shows a new interference fringe pattern with higher frequency due to the thick gold layer. When the gold is sandwiched between the different SLDs of the chromium and organic film layer its presence becomes visible in the reflectivity profile. Curve 3 in Figure 3.10c demonstrates that this pattern results from a sum of two periodic functions: a low-frequency component whose periodicity is determined by the

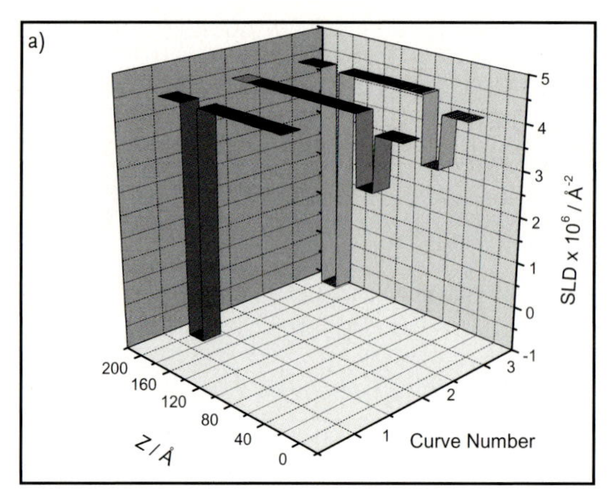
a)
SLD x 10^6 / Å^{-2}
Z / Å
Curve Number

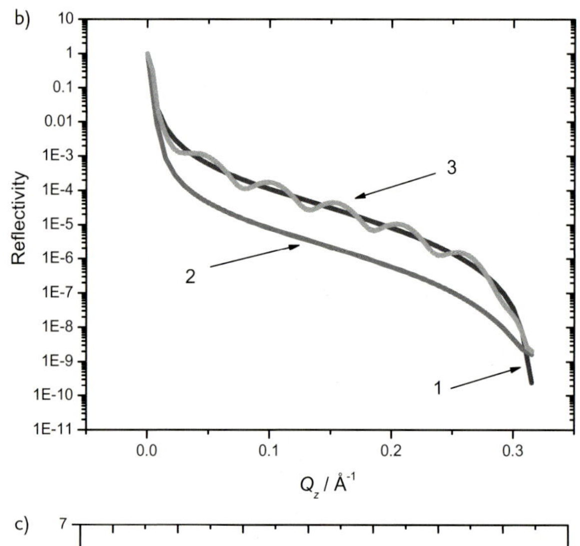
b)
Reflectivity
Q_z / Å^{-1}
1
2
3

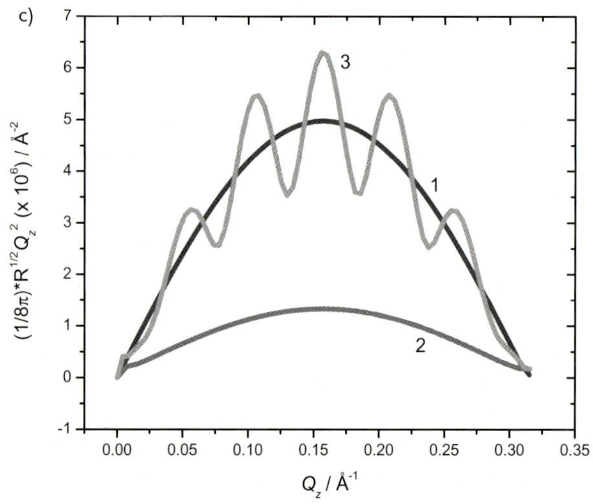
c)
$(1/8\pi)*R^{1/2}Q_z^2$ (x 10^6) / Å^{-2}
Q_z / Å^{-1}
1
2
3

Figure 3.10 (a) SLD profiles for components of the interface shown in Figure 3.9 assuming that the SLD for the solution phase is contrast-matched to gold. Curve 1 represents a hydrocarbon film adsorbed on a gold film and purposely omits the chromium layer and the quartz substrate. Curve 2 includes the chromium and quartz but omits the organic film. Curve 3 shows the SLD profile for the entire interface. (b) Reflectivity curves calculated from the SLD profiles in (a). (c) Plot of $Q_z^2\sqrt{R}/8\pi$ versus Q_z calculated from the reflectivity curves shown in (b). Figure taken from [80].

thickness of the organic film and to a lesser extent by the chromium (which has the same thickness but much lower amplitude); and a high-frequency component determined by the thickness of the gold film.

The above discussion reveals how NR can be used to provide direct information concerning both the thickness and the composition via the SLD of the various layers in the interface. This constitutes an advantage over ellipsometry or surface plasmon resonance methods where the value of the refractive index (composition) of the film is usually assumed in order to determine its thickness. However, to use the kinematic approach to directly determine these quantities, the reflectivity has to be measured in a sufficiently large range of Q_z so that at least one-half of the longest period interference fringe, corresponding to the thinnest layer in the interface, is observed on the reflectivity curve. The spatial resolution for a film of thickness τ is defined as $\tau = \pi/Q_{z,\max}$, where $Q_{z,\max}$ is the maximum momentum transfer vector accessible to the experiment [37]. Thus, access to reliable reflectivities at increasingly larger momentum transfer vectors is required to probe the finer details of thin films. The NG-7 line at the NCNR allows reflectivity to be measured up to $Q_{z,\max} = 0.25\,\text{Å}^{-1}$. At the NG-1 reflectometer at the NCNR, experiments can be run routinely up to $0.3\,\text{Å}^{-1}$ [38, 39] and in an exceptionally well-designed experiment even up to $0.7\,\text{Å}^{-1}$ [40]. For $Q_{z,\max} = 0.25\,\text{Å}^{-1}$, the spatial resolution is about 12 Å and for $Q_{z,\max} = 0.15\,\text{Å}^{-1}$ it is about 20 Å. This resolution is sufficient to study polymer films or corrosion layers with thickness of the order of 100 Å. However, it imposes a limitation on studies of monolayers and bilayers of surfactants and phospholipids where the thickness of the adsorbed film is comparable to the spatial resolution. This does not prevent one from studying aspects of the organic film less than the spatial resolution as, although a pure kinematic-based analysis of the data is insufficient to resolve a very thin component of the film, each discrete layer will contribute to the overall measured reflectivity. If the contrast is high enough, even a very thin layer (about 5 Å) can have a pronounced effect on the entire reflectivity curve which then can be accounted for in a fitting procedure.

The common procedure used to calculate the SLD profile from the reflectivity curve is to assume a model profile, calculate the theoretical reflectivity curve using the optical matrix or recursion method, and compare calculated and experimental curves. A least-squares iterative procedure is then used to vary the parameters of the SLD profile until a good fit between the calculated curve and the experimental data is achieved. Although the inversion of the reflectivity data is not unique and the SLD profile calculated from the reflectivity curve is not the only mathematical function that fits the reflectivity curve, it usually gives a good physical model of the interface. In order to reduce the ambiguity and the potential non-uniqueness of the fitting procedure, one can measure a system of interest with several different contrasts by varying the nuclear composition (isotopic variation) of either the organic or solvent phase. This is usually done through the use of H_2O/D_2O but many perdeuterated phospholipids and surfactants are commercially available. To further reduce the number of adjustable parameters it is critical to determine a portion of the structural information from independent experiment. The thickness, the roughness, and the SLD of chromium and gold layers sputtered onto quartz crystals can be independently determined by measuring the NR and/or X-ray reflectivity of the crystals in air. These parameters can then be fixed when the substrates are used for *in situ* electrochemistry–NR experiments.

3.4 Selected Examples

Although Thomas and coworkers proposed NR as a sensitive technique for investigating surface structure in 1981 [41], the first papers describing successful application of the methodology did not appear until the late 1980s [42–44]. Since these early papers describing applications of NR to surface and interfacial problems, the technique has slowly started to blossom as more dedicated instruments are brought on-line throughout the world. The increased popularity of the technique is manifest in a recent special issue of *Langmuir* (the American Chemical Society's journal for interfacial science) dedicated to applications of NR [45]. In recent years there have been many new NR studies of soft biological samples supported on solid supports but only a limited number of these were performed with potential control. Some instances of the former are reviewed in Section 3.4.1 as these systems remain of interest to the bioelectrochemcial community. A case study of electric field-driven transformations in a model biomembrane is discussed in Section 3.4.2.

3.4.1 Supported Proteins, Peptides, and Membranes without Potential Control

3.4.1.1 Quartz- and Silicon-Supported Bilayers

In 1991, Johnson *et al.* reported one of the first NR studies of phospholipid bilayers at the solid–solution interface [46]. Although these measurements were not the first to employ NR to study molecules adsorbed at the solid–liquid interface, they did constitute the first measurements of a supported bilayer using NR. A bilayer of dimyristoylphosphatidylcholine (DMPC) was spread on a quartz surface by the fusion and rupturing of small unilamellar vesicles. The very smooth, single-component substrate allowed a complex model of the interface to be constructed from layers corresponding to (i) the quartz, (ii) a thin film of water on the quartz surface, (iii) the lipid head group directed toward the quartz, (iv) the hydrocarbon core of the bilayer, (v) the lipid head group directed toward the solvent, and (vi) the pure solvent itself. With access to perdeuterated DMPC and by varying the ratio of D_2O and H_2O in the subphase several different neutron-contrast measurements could be performed to increase the reliability of their fitting analyses. Information concerning the extent of head group hydration and the presence of solvent cushion layers was extracted from the NR data but a major limitation of this early effort was the limited accessible Q space ($Q_{max} \approx 0.11\,\text{Å}^{-1}$). Koenig *et al.* [38] also studied the spontaneous assembly of lipid bilayers on a solid support using NR. They studied dipalmitoyl glycero-3-phosphocholine (DPPC) and distearoyl glycero-3-phosphocholine on silicon single crystals. Their novel cell design allowed access to much higher momentum vector space and consequently their model-fitting parameters carry much less uncertainty. Koenig *et al.* found a much thinner water layer separating the native surface silicon oxide from the head group region of the lipid molecules and reported the thickness of the water layer to be 6–12 Å compared to the 30 Å found in [46]. The inner core of the DPPC lipid bilayer was altered by 4–6 Å by crossing the phase transition temperature which is consistent with increased gauche conformations in the melted hydrocarbon tails. Analysis of the lipid layer's SLD indicated a patchy bilayer structure was formed on the silicon–silicon oxide surface. An assessment of the kinetics of bilayer film formation showed that the patchy bilayer is rapidly formed on the large substrate (about 3–5 min) and that a small apparent increase in the surface coverage was the result of vesicles loosely associating with the bilayer film. Washing the substrate with vesicle-free buffer returned the surface coverage to the values observed at short times (3–5 min). After removing adsorbed vesicles, the surface coverage was found to be invariant with time in the aqueous environment.

For supported membranes to maintain the structural and dynamic properties of free biomembranes, the interaction between the membrane and the substrate should be minimized. One strategy for decoupling the biomembrane from the underlying surface is to support the biomembrane upon a soft hydrated polymer or polyelectrolyte film. Various approaches to achieve this have been reviewed by Sackmann [47]. Majewski and coworkers [48–50] have employed NR to perform structural studies of polyethyleneimine (PEI)-cushioned lipid bilayers. As characterized by NR, DMPC vesicles added to PEI-coated quartz substrates led to an inhomogeneous structure consisting of a mixture of vesicles and multilayers of lipids. The lipid multilayer was dramatically illustrated by high-frequency oscillations in the reflectivity curve at low momentum transfer vectors. It was postulated that, as opposed to quartz, glass, or mica, the fusion of the DMPC lipids on a polymer surface is hindered by the absence of a strong van der Waals interaction between the phospholipid molecules and the substrate. As a result, the interface becomes a complicated structure consisting of multilayer complexes of PEI with bilayers as well as intact bilayers adsorbed to the surface. In contrast, Majewski and coworkers were able to successfully spread DMPC bilayer films on the unmodified quartz substrate from vesicles and observed qualitatively similar results to those of Johnson *et al.* [46]. Remarkably, when PEI was added to the cell it was noted that the PEI polymer appeared underneath the previously deposited DMPC bilayer. The authors explained their observation thermodynamically by noting that the positively charged PEI is drawn to the negatively charged quartz substrate.

Recently, Brzozowska *et al.* used NR and *ex situ* electrochemical techniques to characterize an innovative type of monolayer system intended to serve as a support for a bilayer lipid membrane on a gold electrode surface [51]. Zr^{4+} ions were used to noncovalently couple a phosphate-terminated self-assembled monolayer (SAM) formed on a gold surface to the carboxylate groups of negatively charged phosphatidylserine (PS). This tethered surface was then used for the formation of a PS lipid bilayer structure formed by vesicle fusion and spreading. NR studies revealed the presence of an aqueous environment associated with the tether layer which arises from nonstoichiometric water associated with the zirconium phosphate moieties [52].

3.4.1.2 **Hybrid Bilayers on Solid Supports**

Another method used to create biomimetic membranes on solid supports involves the use of so-called hybrid bilayers. This method was first reported by Tamm and McConnell [53] and involves a monolayer-coated substrate being immersed through a film of lipid spread at the air–solution interface. Kuhl *et al.* employed such a grafting procedure to study poly(ethylene glycol) lipids on octadecyltrichlorosilane (OTS)-modified substrates [54] with NR. The procedure involved silanizing the quartz substrates and then, using Langmuir–Blodgett deposition, adding a single layer of distearoylphosphatidylcholine with poly(ethylene glycol) covalently bound to the head group. After passing the silane-modified substrate through the air–water interface the cell was assembled under water. From NR measurements it was determined that this procedure led to continuous and tightly packed lipid monolayers on the substrates. Meuse *et al.* slightly modified the above procedure in their infrared spectroscopy and NR studies of hybrid bilayer membranes (HBMs) [55]. In this experimental approach, the lipid layer was added to the monolayer-covered substrate with a single horizontal touch and was subsequently withdrawn without passing the substrate entirely through the interface. This eliminated the need to assemble the cell under water. Meuse *et al.* were able to access momentum transfer vectors as high as $0.25\,Å^{-1}$ allowing sufficient resolution to observe differences in the thickness of the single phospholipid layer above and below the phase transition temperature. A thiol layer self-assembled on gold-coated silicon was used rather than an OTS layer chemically grafted to silicon. The presence of the 50 Å thick layer of gold boosted the overall reflectivity of the interface helping access higher Q_z values. Meuse *et al.* were able to demonstrate that a hybrid bilayer approach leads to nearly 100% surface coverage by the membrane. In comparison, earlier work by the same group estimated the surface coverage of pure lipids to be between 70 and 95% depending on the sample [38]. From a biophysical point of view, the extended spatial resolution of these measurements led to much greater detail in the description of the lipid head group. The head group of a single bilayer is hydrated to a much greater extent as compared to the average degree of hydration for multilamellar systems. Hybrid bilayers composed of covalently bound thiol-modified lipid inner layers and physisorbed distal leaflets have been used in several NR studies [56–59] with particular interest devoted to designing hydrophilic tethering groups that hydrate the head groups of both sides of the model bilayer.

Krueger and coworkers worked to further improve the resolution of NR measurements and their work culminated in one of the most elegant applications of NR to the study of biological layers [40]. They investigated the interaction of the peptide toxin melittin with the head group of a phospholipid. The main purpose of this work was to determine the effect of melittin insertion into a hybrid bilayer membrane. Melittin is believed to enhance ion transfer across biological membranes which, for example, can lead to lysis of red blood cells. However, the precise structure of the channel-forming pores is the subject of some debate. Krueger and coworkers were able to perform NR experiments up to an unprecedented momentum transfer vector maximum and were therefore able to measure very small changes in the hybrid bilayer (down to 2–3 Å) due to the high resolution of their experiment. This work is impressive from both a technical aspect and the scientific information garnered. Background scattering that arises due to scattering from the substrate, the air environment, and the aqueous backing was greatly reduced with the use of very thin (0.5 mm) silicon wafers and the placement of the entire cell assembly in a helium-filled chamber. Importantly, a very thin (about 15–20 μm) aqueous reservoir of D_2O was used which greatly attenuated the incoherent scattering arising from the much thicker aqueous environments commonly used in NR studies of aqueous systems. Using this experimental set-up, data were obtained out to a maximum Q_z value of 0.7 Å^{-1} and reflectivities as low as 10^{-8}. A schematic of the cell design used in [40] is shown in Figure 3.11.

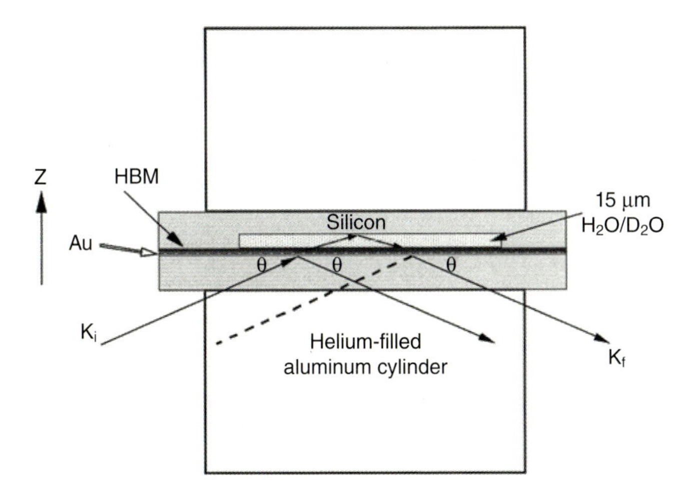

Figure 3.11 Schematic of the novel cell design used to measured the reflectivity of a hybrid bilayer system down to the level of about 10^{-7}. Figure taken from [40].

The hybrid bilayer was built by first derivatizing the silicon substrate with a thin gold film (50 Å) layered on top of a 25 Å thick layer of chromium. A SAM of octadecanethiol or a thiahexa(ethylene oxide) alkane (THEO) was formed on the gold surface followed by the transfer of a monolayer of deuterated DMPC. In the absence of the melittin protein, NR measurements indicated that the roughness of the gold layer sputtered onto the substrate is not transferred across the SAM to the phospholipid leaflet. Additionally NR measurements indicated that the DMPC molecules comprising the outer half of the hybrid bilayer are sufficiently fluid to effectively anneal the roughness of the inner layer which is conformal to the roughness of the deposited gold film. In the presence of melittin, the region of the head group of the HBM's deuterated DMPC leaflet showed the greatest changes. The SLD profiles were determined from a model independent fitting routine and are reproduced in Figure 3.12. Figure 3.12 clearly demonstrates that the addition of melittin displaces a large quantity of solvent (D_2O in this instance) from the lipid head group. Although the SLD profiles do not offer strong evidence of the insertion of the melittin into the deuterated DMPC hydrocarbon core, this occurrence cannot be ruled out because the SLD of the hydrated melittin pore is expected to be similar to the SLD of the $-CD_2-$ tails. Furthermore, it was noted that the overall thickness of the $-CH_2-$ layer of the thiol becomes thicker by about 3 Å. This is consistent with molecular dynamics

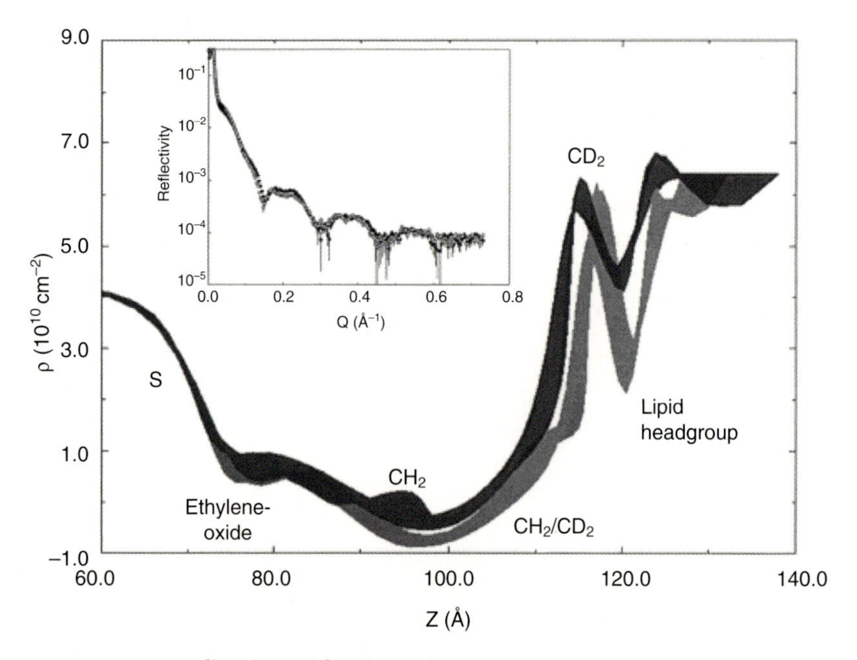

Figure 3.12 SLD profiles obtained for a hybrid bilayer in the absence (black curve) and in the presence (grey curve) of the protein melittin. The experimental NR curves used to generate the SLD profiles are shown in the inset (taken from [53]). Figure taken from [40].

simulations of a single melittin molecule in a DMPC bilayer and it implies that the insertion of melittin not only affects the lipid leaflet in which it is present but also the hydrocarbon tail of the innermost leaflet of the HBM. However, the SLD profiles also clearly demonstrate that the presence of melittin does not lead to hydration of the ethylene oxide region in the THEO-based HBM. This surprising result indicates that melittin does not alter the HBM in such a way as to allow D_2O to fully penetrate the membrane. It was proposed that alternate tethering chemistries need to be considered to lead to more biomimetic hybrid bilayers (see above). In any event, this very impressive paper deserves emphasis as it is one of the most methodologically advanced NR studies of biological systems reported in the literature. The small changes in the inner layer thickness of the HBM upon insertion of melittin would not be measurable using "traditional" NR measurements.

Valincius *et al.* used NR and *ex situ* electrochemical impedance spectroscopy to study the effect of amyloid β-peptide insertion on supported lipid bilayers [60]. It has been speculated that the interaction between these oligomeric proteins and bilayer membranes plays an important role in the aggregation and protein misfolding that leads to various neurodegenerative diseases including Alzehimer's and Parkinson's diseases [61]. Using a variety of solvent and phospholipid contrasts, including the use of perdeuterated phospholipids, NR demonstrated that amyloid β-peptides inserted into lipid bilayers tethered to a gold-modified silicon substrate. The NR data revealed that the protein fully inserts into the hydrophobic core, disrupting both lipid leaflets but does not lead to significant disruption of the head group region as determined by the absence of increased water content in the SLD profiles. This information was vital for explaining the results of *ex situ* impedance spectroscopy and modeling the increased ion transport capabilities of the supported membranes in the presence of amyloid β-peptides.

3.4.1.3 **Protein Adsorption and DNA Monolayers**

Many recent NR studies of protein and peptide assembly on various solid and liquid interfaces have been reviewed by Zhao *et al.* [62]. The interested reader is directed to that paper for a more comprehensive outline of these NR studies. The first NR study of protein–membrane interactions was reported by Schmidt *et al.* in the early 1990s [63]. The interfacial binding reaction between streptavidin and a solid-supported lipid monolayer partly functionalized by biotin moieties was evaluated. NR was sensitive enough to detect the presence of biotin proteins attached to the membrane, but again a poor range of accessible data prevented high-resolution details pertaining to protein adsorption. Fragneto *et al.* were able to provide much greater detail in their analysis of bovine β-casein adsorbed on a silicon surface derivatized with deuterated octadecyltrimethylsiloxane [64]. A hydrophobic surface was chosen as the conformation of proteins is in most cases determined by the hydrophobic interactions in the nonpolar residues of the peptide chains. A deuterated OTS SAM was chosen to enhance the contrast between the solvent and the nondeuterated protein. Measurements were repeated using three different solvent contrasts, pure D_2O, pure H_2O, and silicon contrast-matched water, and a two-layer model of the protein layer was proposed to fit the reflectivity data for all three contrasts. The inner layer of adsorbed β-casein was relatively protein dense (about 60% by volume fraction) and was 23 ± 1 Å thick. The outer layer was thicker (35 ± 1 Å) and was only 12% protein by volume fraction. The results from the reflectivity experiments can be explained in terms of the charge distribution of the protein molecule. β-Casein resembles a surfactant with a polar head and a nonpolar tail. The high concentration of charged amino acids in the head group makes it essentially negatively charged whereas the remainder of the protein is hydrophobic. The inner layer observed by Fragneto *et al.* is consistent with a hydrophobic layer, containing little water, whereas the outer layer corresponds to the hydrophilic head group which is heavily solvated and protrudes into the bulk of the aqueous phase. Fragneto *et al.* continued their studies of protein adsorption on hydrophobic surfaces by investigating the influence of pH on the adsorption of β-casein and β-lactoglobulin on silicon [65]. It was shown that at pH greater than 7, the two-layer structure of β-casein applied but β-lactoglobulin formed a single uniform layer. At lower pH the adsorption of both proteins was greatly enhanced but the β-lactoglobulin could no longer be described as a single uniform layer. As before, the observed behavior was readily explained in terms of the degree of ionization of the proteins.

Levicky *et al.* [66] used NR to investigate how self-assembly can control the structure of DNA monolayers adsorbed onto gold surfaces. This report was an extension of an earlier effort to use single-stranded DNA monolayers as model DNA chips for diagnostic applications. The strategy employed by Levicky *et al.* was to prepare mixed monolayers of 6-mercapto-1-hexanol (MCH) and thiol-terminated single-stranded DNA (HS-ssDNA) on gold. NR clearly demonstrated that the oligomeric DNA is oriented parallel with the gold-coated substrate in the absence of MCH. However, upon introduction of MCH, the HS-ssDNA is displaced from the surface in favor of the MCH. Furthermore, the replacement reaction leaves the HS-ssDNA directed to the aqueous solution and oriented perpendicular to the substrate. This orientation is greatly favored for hybridization of the ssDNA with its complementary sequence. Similar work with thiolated hairpin DNA strands immobilized on gold-modified substrates was performed by Steichen *et al.* [67]. NR confirmed the opening of the stem-loop configuration of the ssDNA upon hybridization associated with helix formation. More recent work [68–70] has demonstrated how NR is ideally suited to follow transfection-related interactions between lipoplexes (vesicles of cationic lipid–DNA) and model membranes. A primarily DMPC bilayer was deposited on a silicon substrate and lipoplexes containing 1 : 1 complexes of dimethyldioctadecylammonium bromide and DNA were used as transfection mimics. With clever use of contrast variation and selective deuteration of lipid components it was shown that the rate of lipid exchange between the cell membrane and the lipid–DNA complex is vital in determining transfection efficiency. All of the studies described above are well suited for electrochemical measurements as protein binding, DNA hybridization, and drug transfection would be expected to be highly dependent on the electrical state of the supporting substrate.

3.4.2
Electric Field-Driven Transformations in Supported Model Membranes

Until about 10 years ago, electrochemical applications of NR had largely been directed toward hard material problems such as corrosion [71, 72] and oxide formation [5, 73–75]. More recently electrochemical NR has been used to study the hydration of Nafion in proton-exchange membranes for fuel cell applications [76]. The characterization of soft materials at electrode surfaces began in the late 1990s when Lipkowski and coworkers began reporting NR studies of electric field-driven transformations in several model surfactant monolayer and bilayer systems including dodecylsulfate [77, 78], pentadecylpyridine [79, 80], and octadecanol [81]. This was concurrent with Hillman and coworkers using NR to profile the composition (particularly solvent permeation) in electropolymerized films including dynamic measurements of polymer swelling upon redox cycling [22, 25, 32, 82]. The technical development of these works has laid the foundation for bioelectrochemical applications, particularly the study of supported phospholipid films under the influence of applied AC [83] and static [84–86] electric fields. In the latter case, both Burgess *et al.* [84, 85] and Hillman *et al.* [86] have characterized changes in biomimetic layers upon variation of the electrical state of the supporting electrode. A case study of the former is provided below.

Lipids and proteins in natural biological membranes are frequently exposed to static electric fields of the order 10^7 to $10^8\,\mathrm{V\,m^{-1}}$ [87] and a model membrane supported on a conductive substrate may be used to study voltage-gated membrane proteins and lipid–lipid and lipid–protein interactions [88–91]. A model lipid system consisting of a mixed DMPC–cholesterol bilayer (70:30 mol% ratio) was formed on a gold surface by fusion and spreading of small unilaminar vesicles and electrochemically characterized as shown in Figure 3.13. The membrane is stable at the electrode when the surface charge density is in the range $-8\,\mu\mathrm{C\,cm^{-2}} < \sigma_m < 8\,\mu\mathrm{C\,cm^{-2}}$. Outside this region, the charge density curve for the membrane-covered electrode abruptly rises or falls to approach the corresponding curve of a lipid-free interface. This behavior indicates that the membrane becomes detached from the gold surface at sufficiently large polarizations. An interesting feature is evident in the charge density measurements at applied potentials of about −0.8 V where the charge of the electrode in the presence of the detached film approaches but does not quite merge with the curve for the film-free surface. It was inferred that this was caused by either incomplete desorption or the possible formation of intact vesicles that remain in very close proximity to the electrode surface. It is impossible from the electrochemical experiments alone to adequately describe the film structure and water distribution at the interface for either the detached or contact adsorbed layer. However, this problem is ideally suited for electrochemistry–NR studies as described below.

The substrates were formed from virgin quartz blocks modified with nominally 3 nm thick chromium and 12 nm thick gold layers. The thickness, roughness, and SLD of chromium and gold layers sputtered onto the quartz crystals were determined independently by measuring the X-ray reflectivity of the crystals in air

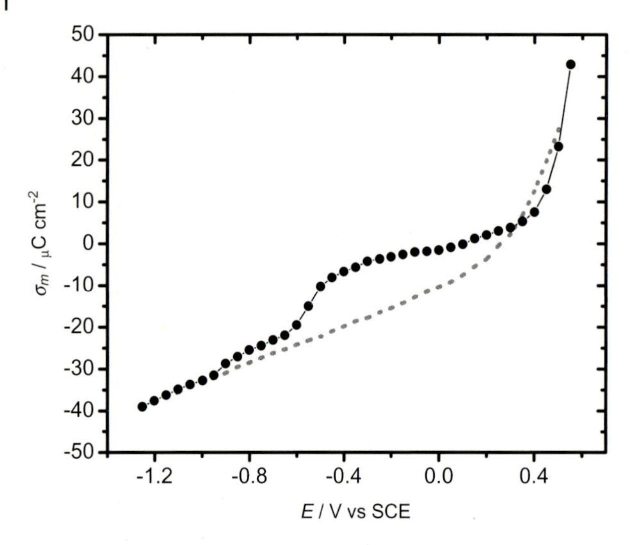

Figure 3.13 Surface charge density on a gold electrode surface plotted versus the electrode potential for (dotted curve) 50 mM NaF supporting electrolyte and (circles) mixed 7:3 DMPC–cholesterol bilayer spread from vesicle solution. Figure taken from [84].

before and after the electrochemical NR experiments. Figure 3.14 shows both the experimental and fitted X-ray reflectivity curves for the experiments before the electrochemical experiments. The curve obtained for the crystal after the *in situ* electrochemical measurements was nearly identical demonstrating that the chromium and gold layers remained essentially unaffected by the electrochemistry. The inset to Figure 3.14 shows the electron density (as determined from the real component of the SLDs) as a function of the distance normal to the electrode surface. From these data the parameters for the chromium and gold layers were determined and, as expected, the differences between the fit parameters for the chromium and gold layers determined before and after the experiment are small.

Curve 2 in Figure 3.15 shows the NR curve for the membrane formed by fusing mixed DMPC–cholesterol vesicles at the gold electrode at $E = 50\,mV$ ($\sigma_m \approx -1\,\mu C\,cm^{-2}$). For comparison, curve 1 shows the reflectivity curve recorded at the gold-coated quartz electrode in D_2O in the absence of DMPC. The reflectivity is significantly higher in the presence of vesicles, indicating that vesicles fuse and form a membrane at the gold surface. The data show that reflectivities could be measured with good statistics down to about 2×10^{-6} and $Q_z = 0.18\,Å^{-1}$. Figure 3.16 shows reflectivity plots for different applied potentials as well as the SLD profiles determined from the best-fit models of the reflectivity data. The reflectivity was best described by a single layer of proton-containing molecules deposited directly on the metal surface. This fit requires a limited number of adjustable parameters but averages the SLDs for the phosphatidyl head groups and the hydrocarbon tail regions. In order to extract more detailed information concerning the membrane, the same reflectivity data were fitted to a three-layer model assum-

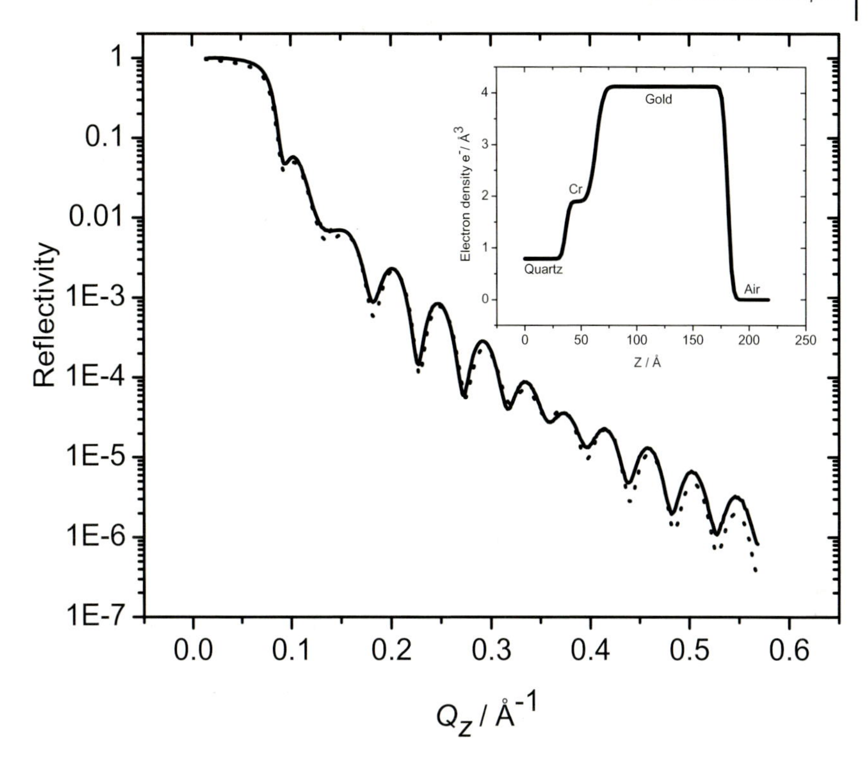

Figure 3.14 X-ray reflectivity as a function of momentum transfer vector for a gold–chromium-covered quartz substrate. The curve corresponds to the measurement made before electrochemistry. The curve for the post-electrochemical measurement is essentially identical and is omitted for clarity. The dotted curve represents the results of modeling. The inset shows the electron density profile for the model that best fits the data. Figure taken from [84].

ing that the membrane consists of two 8 Å thick polar head regions and a middle section composed of hydrocarbon acyl chains. While the three-layer model is physically more realistic and statistically gives a slightly better fit to the experimental data, it involves at least four additional adjustable parameters. The contrast for the polar head region and the backing gold or D_2O phases is weak. In addition, the SLD profiles calculated for a single-layer film allowing for adjustable roughness of the film were very close to the SLD profiles calculated for the three-layer model of the film. For these reasons a single layer was considered to be a sufficient approximation. The thickness and the SLD values for the acyl chain region of the membrane, determined from the fit to the single-layer model, are compiled in Table 3.2. The SLD for the organic layer was then used to determine the occupancy of phospholipid and cholesterol at the electrode surface assuming that the modeled layer contains only the deuterated solvent and the hydrocarbon adsorbate. The total SLD of the organic layer (ρ_{total}) is, therefore, given by the equation $\rho_{total} = x\rho_{org} + (1-x)\rho_{D_2O}$, where ρ_{org} and ρ_{D_2O} are the SLDs of the organic and

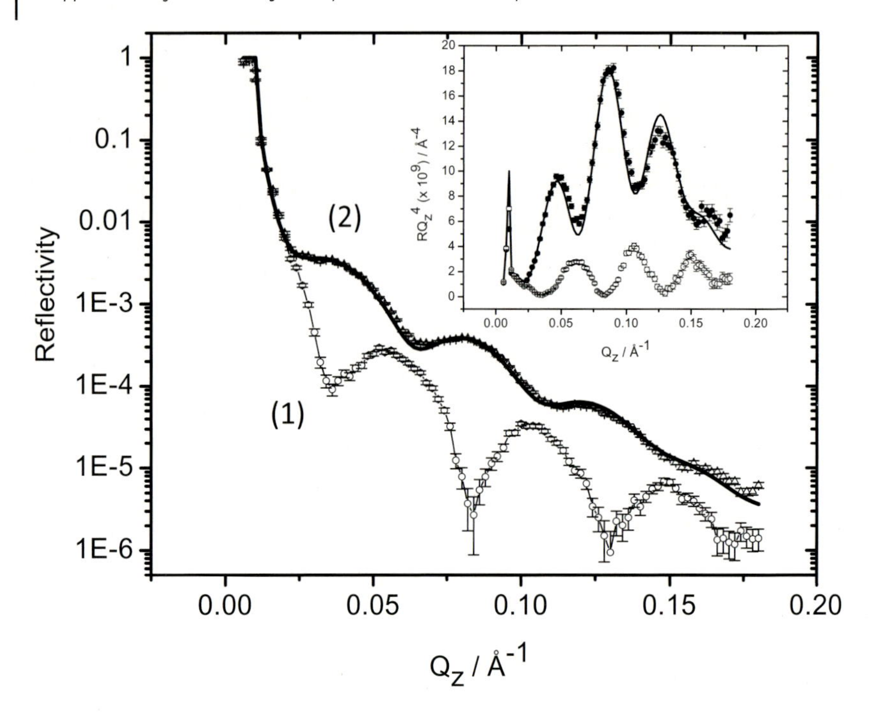

Figure 3.15 Experimentally determined reflectivity curves (points with associated error bars) for $E = 50\,\text{mV}$ in 50 mM NaF in D_2O: curve 1, the film-free electrode surface; curve 2, the electrode covered by a bilayer of a 7 : 3 mixture of h-DMPC and cholesterol. The inset shows plots of RQ^4 versus Q_z calculated from the data presented in the main part of the figure. The solid curve shows the calculated reflectivity curves from the best-fit model whose parameters are given in Table 3.2. Figure taken from [84].

solvent material, respectively, and x is the volume fraction of hydrocarbon in the film. The value for ρ_{D_2O} is known ($6.33 \times 10^{-6}\,\text{Å}^{-2}$) and the value for ρ_{org} was estimated using the densities and molecular formulae of cholesterol and hydrocarbons. The values for the best-fit parameters of the NR data obtained for potentials between +50 mV and −600 mV exhibit an interesting trend. As the potential applied to working electrode becomes increasingly negative the phospholipid–cholesterol layer thickness increases by nearly 28% with a significant concomitant increase in the SLD. This thickening of the phospholipid layer can be attributed to swelling of the organic layer by ingress of progressively larger amounts of solvent molecules with decreasing electrode potential. Figure 3.17 plots the RQ_z^4 versus Q_z curves for E equal to −700, −800, and −950 mV and the SLD profiles determined from the best-fit analysis. The reflectivity of the electrode surface remains very high indicating that the bilayer remains near the electrode surface even though the charge density measurements indicate the film has delaminated. These curves could only be adequately fitted to a two-box model with the box directly adjacent to the gold surface consisting of a film a few angstroms thick having a very high SLD. Moving

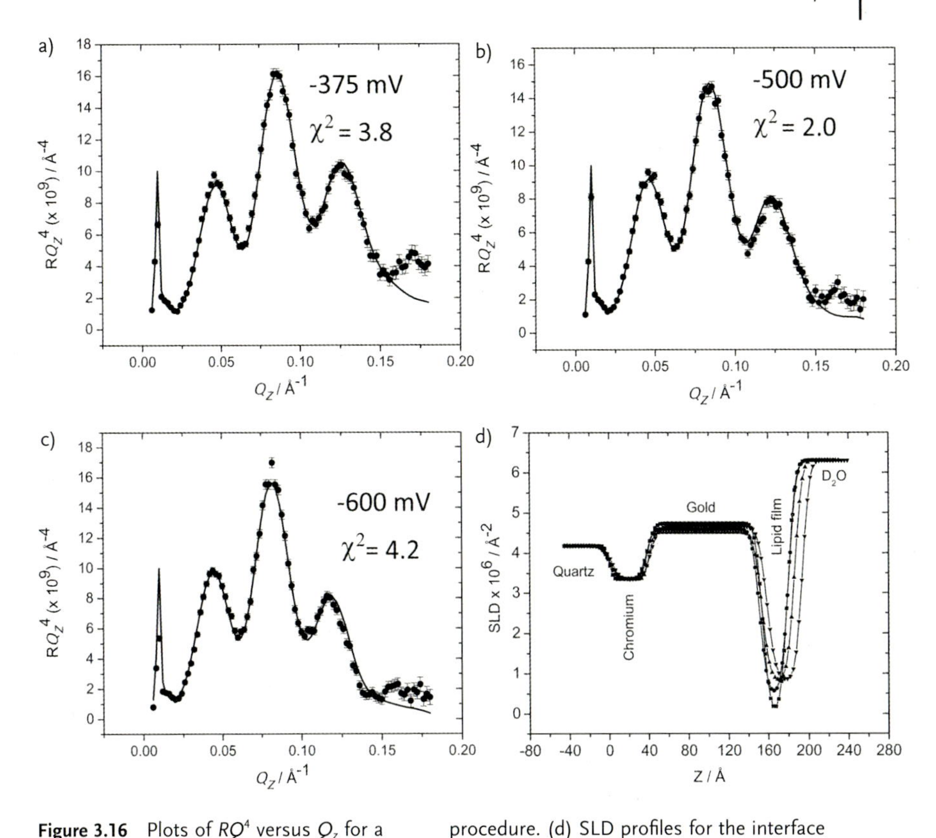

Figure 3.16 Plots of RQ^4 versus Q_z for a bilayer of a 7 : 3 mixture of h-DMPC and cholesterol in 50 mM NaF in D_2O: (a) $E = -375$ mV; (b) $E = -500$ mV; (c) $E = -600$ mV. Points with associated error bars show the experimental data. Solid curves show the reflectivity calculated from the parameters obtained from the fitting procedure. (d) SLD profiles for the interface for $E = 50$ mV (squares), $E = -375$ mV (circles), $E = -500$ mV (up triangles), and $E = -600$ mV (down triangles). The best-fit model parameters corresponding to the SLD profiles are listed in Table 3.2. Figure taken from [84].

the potential in the negative direction, the thickness of this box increases and its SLD reaches a limiting value very close to that of pure D_2O. Apparently, the bilayer becomes separated from the metal surface by a thin layer of solvent at negative potentials. In an effort to further verify the validity of the observation of a solvent cushion layer the NR measurements for $E = -800$ mV were repeated using a different isotopic contrast. The bilayer was formed by fusion of vesicles prepared from perdeuterated DMPC mixed with nondeuterated cholesterol and the solvent was H_2O. Figure 3.18 shows the RQ_z^4 versus Q_z curve and SLD profile for this system. The result confirms that the bilayer is separated from the gold surface by an approximately 10 Å film of water at this negative potential. The bilayer is 38 ± 1 Å thick and the SLD for the bilayer is within the experimental error equal

Table 3.2 Best-fit results describing a mixed DMPC–cholesterol bilayer deposited at the electrode surface as a function of the applied potential. The results are obtained from the best-fit results of the measured neutron reflectivity curves in Figures 3.16–3.18.

System (*E* (mV) vs. SCE)	Organic layers							
	Water layer				Lipid–cholesterol layer			
	τ[a] (Å)	SLD[b] ($\times 10^6$ Å^{-2})	Volume fraction of solvent	σ[c] (Å)	τ[a] (Å)	SLD[b] ($\times 10^6$ Å^{-2})	Volume fraction of solvent	σ[c] (Å)
h-DMPC+ chol (50)	N/A	N/A	N/A	N/A	25.6 ± 0.9	0.03 ± 0.22	0.08	5.6
h-DMPC+ chol (−375)	N/A	N/A	N/A	N/A	29.3 ± 1.0	0.51 ± 0.12	0.15	5.9
h-DMPC+ chol (−500)	N/A	N/A	N/A	N/A	32.8 ± 1.0	0.82 ± 0.18	0.19	6.3
h-DMPC+ chol (−600)	N/A	N/A	N/A	N/A	35.3 ± 0.7	0.85 ± 0.17	0.20	6.8
h-DMPC+ chol (−700)	4.1 ± 0.2	5.38 ± 0.40	0.86	6.1	35.8 ± 0.7	0.05 ± 0.20	0.08	5.4
h-DMPC+ chol (−800)	6.0 ± 1.4	6.33 ± 0.41	1.00	5.5	37.0 ± 0.9	−0.43 ± 0.32	0.01	5.6
h-DMPC+ chol (−950)	10.1 ± 1.0	6.28 ± 0.32	1.00	5.3	36.8 ± 0.8	−0.51 ± 0.19	0.00	5.8
d-DMPC+ chol (−800)	10.4 ± 1.3	−0.56 ± 0.21	1.00	2.1	37.2 ± 1.0	5.22 ± 0.28	0.05	4.1

a) Thickness of layer.
b) Scattering length density.
c) Roughness.
Source: Taken from [40].

to the theoretical value of the SLD calculated for the water-free film estimated to be 5.5×10^{-6} Å^{-2}. The good agreement between the results of NR experiments performed using hydrogenated and deuterated DMPC gives credibility to the SLD profiles calculated from the reflectivity curves. The main results of this work are summarized pictorially in Figure 3.19. It should be emphasized that Figure 3.19 is not a model but rather a schematic representation of the DMPC–cholesterol film consistent with the analysis of the specular NR data. The NR analysis does not provide information about the lateral distribution of individual solvent molecules nor does it provide sufficient resolution to give details concerning the fine structure of the film normal to the gold surface due to a limited maximum Q_z.

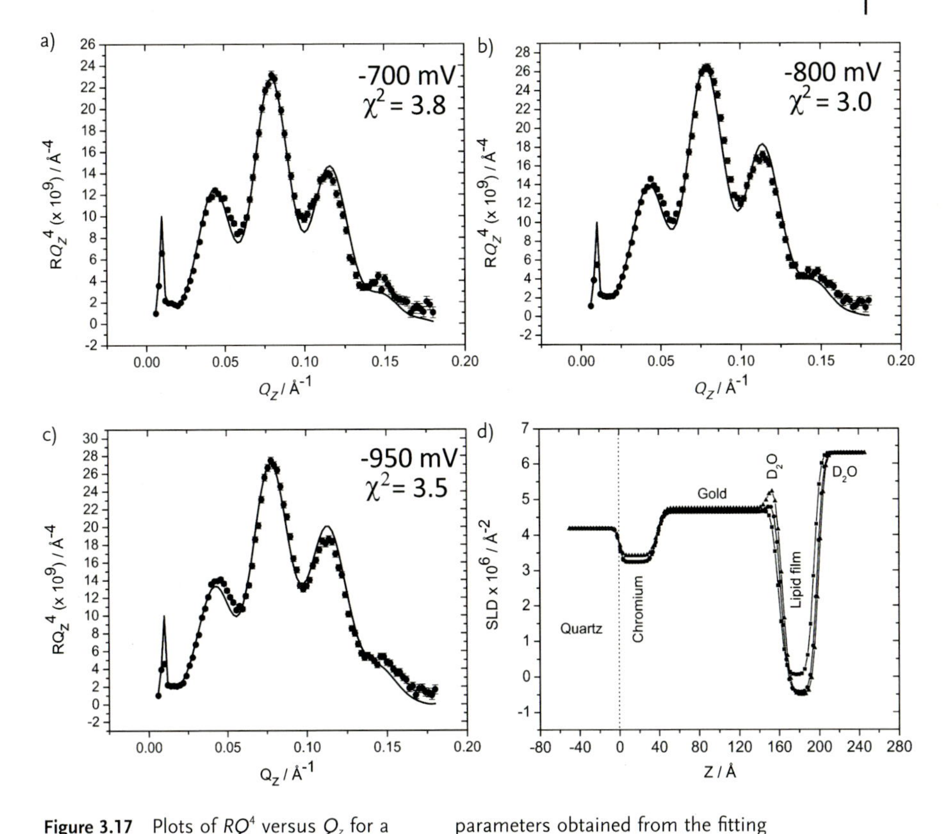

Figure 3.17 Plots of RQ^4 versus Q_z for a bilayer of a 7 : 3 mixture of h-DMPC and cholesterol in 50 mM NaF in D_2O: (a) $E = -700$ mV; (b) $E = -800$ mV; (c) $E = -950$ mV. Points with associated error bars show the experimental data. Solid curves show the reflectivity calculated from the parameters obtained from the fitting procedure. (d) SLD profiles for the interface for $E = -700$ mV (squares), $E = -800$ mV (circles), and $E = -950$ mV (triangles).The best-fit model parameters corresponding to the SLD profiles are listed in Table 3.2. Figure taken from [84].

The analysis of specularly reflected neutrons provides a compositional profile, in the directional normal to the film, averaged over the transverse coherence length of the incident neutrons, which for the range of momentum transfer vectors accessed in these measurements spans roughly between 10 and 100 μm. As long as the dimension of the lateral inhomogeneities is less than the coherence length of the incident neutrons then the measured reflectivity curve accurately probes the area-averaged interface [39]. As such, the water molecules shown incorporated in the biological films in Figure 3.19 qualitatively represent the volume fraction of water inside the lipid film but they do not represent the precise location or distribution of the solvent molecules. On the other hand, when water exists as a discrete, uniform layer, as is the case for negative electrode potentials, NR analysis can accurately place the slab of water within the overall SLD profile of the entire

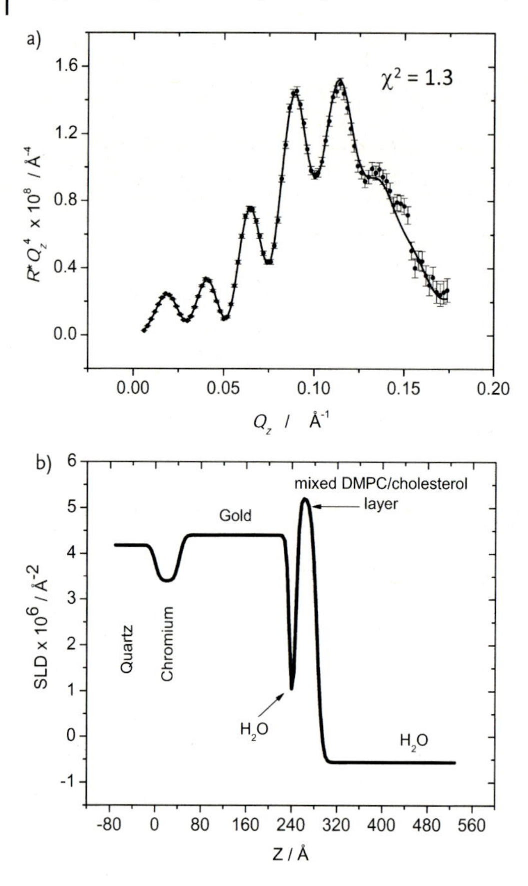

Figure 3.18 (a) Plot of RQ^4 versus Q_z for a bilayer of a 7:3 mixture of deuterated DMPC and cholesterol in 50 mM NaF in H_2O at $E = -800$ mV. Points with associated error bars show the experimental data. Solid curve shows the reflectivity calculated from the SLD profile presented in (b). The best-fit model parameters corresponding to the SLD profile are listed in Table 3.2. Figure taken from [84].

interface. This is a unique advantage of NR as no other technique can probe buried structures in soft materials.

3.5 Summary and Future Aspects

NR is a tool that has yet to capture the imaginations of bioelectrochemists despite several examples of successful implementation of the technique. NR provides a unique means to study soft interfaces under potential control and its remarkable sensitivity to water is ideally suited for structural studies of supported phospholi-

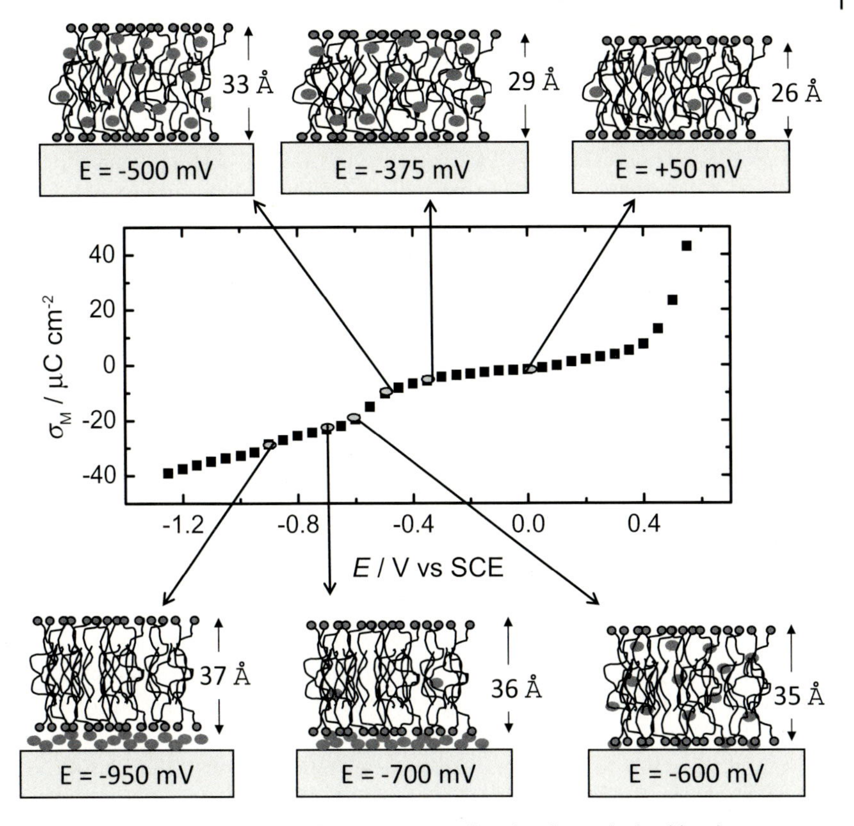

Figure 3.19 Pictorial description of the changes in the structure of a mixed DMPC–cholesterol bilayer deposited at the electrode surface as a function of the applied potential. Small circles depict the lipid head groups, whereas the large circles represent solvent molecules. Figure adapted from [84].

pid layers. Many of the technical difficulties associated with interfacing potential control with a suitable cell design for NR measurements have been outlined and discussed herein. Several seminal papers emphasizing the ability of neutrons to probe electrified interfaces would establish the field as a unique tool to investigate aspects of adsorbed molecular films. The particular area where this breakthrough will most likely occur is the field of biomimetic research such as membrane bilayers supported on electrode surfaces where many opportunities remain unrealized. For example, isotopic variation of the inner and outer layers of a membrane would allow a user to discriminate between field-driven processes occurring in the proximal and distal leaflets of the bilayer. Probing the structure of buried components of biomimetic films such as solvent cushions and interstitial layers as a function of the potential drop across the membrane seems particularly well suited for NR studies. However, even among proponents of NR the reliance on model-dependent fitting is seen as a potential weakness of the technique, especially if there is no pre-existing, physical intuition of the system under study. One can greatly reduce the uncertainties in fitting-based analysis by careful characterization of the substrate prior to NR measurements to limit the number of fitting parameters and by measuring multiple contrast systems. Model-independent means of extracting SLD profiles rely on obtaining phase information which is lost in a single measurement of reflectivity. Several methodologies to obtain phase information have been proposed including direct inversion [92, 93]; however, these are technically demanding and have yet to be implemented for electrochemical NR. A more fundamental problem is the truncation effects imposed by limited accessible momentum transfer space. The consequences of limited Q_z space measurements have been discussed in terms of the kinematic and model-fitting approaches but also lead to oscillations appearing in SLD profiles obtained through direct inversion analysis.

The growth of electrochemical NR will most likely develop along two fronts: (i) methodological advances made by physicists who work at dedicated neutron science facilities and (ii) increased exposure from users who wish to apply the method to their areas of expertise. These two avenues of development are discussed in turn below. Expanding the range of accessible momentum vectors is the critical element in terms of increasing the resolution of the method. In the last 15 years the lower limit of reflectivity has already been reduced from about 10^{-5} to about 10^{-7}. This has been achieved by implementing higher cold neutron fluxes and better experimental sample design. Next-generation neutron reflectometers such as the Liquid Reflectometer at Oak Ridge National Laboratories in Tennessee, USA, and the advanced neutron diffractometer/reflectometer [94] at the NCNR can lower the level of measured reflectivites by a further one to two orders of magnitude. Lower measurable reflectivities lead to larger accessible Q_z domains and ultimately to more unique fits and higher spatial resolution. NR studies of electrified interfaces pose an exciting and challenging direction in the field of interfacial science. The method requires a great deal of patience and high levels of effective collaboration between physicists, physical chemists, synthetic chemists, and materials scientists. Though conducting experiments is often taxing because it requires "suitcase-science" it is hoped that this work will persuade both electrochemists and those interested in interfacial architecture that NR at electrified interfaces should be added to their repertoire of surface-sensitive techniques.

Acknowledgments

The author acknowledges the many helpful conversations with past collaborators in the field of neutron reflectivity, especially Zin Tun (NRC, Chalk River) Sushil Satija (NCNR), Jaroslaw Majewski (LANL), and especially Chuck Majkrzak (NCNR). The author would like to thank Jacek Lipkowski for providing the means and opportunity to learn neutron reflectivity.

References

1 Pettinger, B. (1992) *In situ* Raman spectroscopy at metal electrodes, in *Adsorption of Molecules at Metal Electrodes* (eds J. Lipkowski and P.N. Ross), John Wiley & Sons, Inc., New York, pp. 285–346.

2 Nichols, R.J. (1992) IR spectroscopy of molecules at the solid–solution interface, in *Adsorption of Molecules at Metal Electrodes* (eds J. Lipkowski and P.N. Ross), John Wiley & Sons, Inc., New York, pp. 347–389.

3 Korzeniewski, C. (2006) Recent advances in *in-situ* infrared spectroscopy and applications in single-crystal electrochemistry and electrocatalysis, in *Diffraction and Spectroscopic Methods in Electrochemistry* (eds R.C. Alkire, D.M. Kolb, J. Lipkowski, and P.N. Ross), Wiley-VCH Verlag GmbH, Weinheim, pp. 233–268.

4 Itaya, K. (2006) *Scanning Tunneling Microscopy of Electrode Surfaces*, Taylor & Francis, pp. 5489–5509.

5 Noel, J.J., Jensen, H.L., Tun, Z., and Shoesmith, D.W. (2000) *Electrochemical and Solid-State Letters*, **3** (10), 473–476.

6 Brockman, J.M., Nelson, B.P., and Corn, R.M. (2003) *Annual Review of Physical Chemistry*, **51** (1), 41–63.

7 Majkrzak, C.F. (1995) *Materials Research Society Symposium Proceedings*, **376**, 143.

8 Penfold, J. and Thomas, R.K. (1990) *Journal of Physics: Condensed Matter*, **2** (6), 1369.

9 Fermon, C., Ott, F., and Menelle, A. (2009) Neutron reflectometry, in *X-Ray and Neutron Reflectivity: Principles and Applications*, vol. 770 (eds J. Daillant and A. Gibault), Springer, Heidelberg, pp. 183–234.

10 Penfold, J. (2002) *Current Opinion in Colloid and Interface Science*, **7** (1–2), 139–147.

11 Thomas, R.K. (1999) Neutron reflectivity at liquid–vapor, liquid–liquid and solid–liquid interfaces, in *Modern Characterization Methods of Surfactant Systems* (ed. B.P. Binks), Marcel Dekker, New York, pp. 417–479.

12 Fermi, E. and Marshall, L. (1947) *Physical Review*, **71** (10), 666.

13 Fermi, E. (1936) *Ricerca Scientifica*, **7** (II), 13–52.

14 Squires, G.L. (1978) *Introduction to the Theory of Thermal Neutron Scattering*, Cambridge University Press, Cambridge.

15 Sears, V.F. (1989) *Neutron Optics*, Oxford University Press, Oxford.

16 Golub, R., Richardson, D., and Lamoreaux, S.K. (1991) *Ultra-Cold Neutrons*, Adam Hilger Press, Boston.

17 Croce, P. and Pardo, B. (1970) *Nouvelle Revue d'Optique Applique*, **1** (4), 229–232.

18 Parratt, L.G. (1954) *Physical Review*, **95** (2), 359.

19 Zabel, H. and Robinson, I.K. (eds) (1992) *Surface X-ray and Neutron Scattering*, Springer Proceedings in Physics, vol. 61, Springer-Verlag, Berlin.

20 Majkrzak, C.F., Berk, N.F., Ankner, J.F., Satija, S.K., and Russell, T.P. (1992) *Proceedings of SPIE*, **1738**, 282–304.

21 Teixeira, S.C.M., Zaccai, G., Ankner, J., Bellissent-Funel, M.C., Bewley, R., Blakeley, M.P., Callow, P., Coates, L., Dahint, R., Dalgliesh, R., Dencher, N.A., Forsyth, V.T., Fragneto, G., Frick, B., Gilles, R., Gutberlet, T., Haertlein, M., Hauss, T., Haussler, W., Heller, W.T., Herwig, K., Holderer, O., Juranyi, F., Kampmann, R., Knott, R., Krueger, S., Langan, P., Lechner, R.E., Lynn, G., Majkrzak, C., May, R.P., Meilleur, F., Mo, Y., Mortensen, K., Myles, D.A.A., Natali, F., Neylon, C., Niimura, N., Ollivier, J., Ostermann, A., Peters, J., Pieper, J., Ruhm, A., Schwahn, D., Shibata, K., Soper, A.K., Strassle, T., Suzuki, J., Tanaka, I., Tehei, M., Timmins, P., Torikai, N., Unruh, T., Urban, V., Vavrin, R., and Weiss, K. (2008) *Chemical Physics*, **345** (2–3), 133–151.

22 Cooper, J.M., Cubitt, R., Dalgliesh, R.M., Gadegaard, N., Glidle, A., Hillman, A.R., Mortimer, R.J., Ryder, K.S., and Smith, E.L. (2004) *Journal of the American Chemical Society* **126** (47), 15362–15363.

23 Mueller-Buschbaum, P., Bauer, E., Maurer, E. and Cubitt, R. (2006) *Physica B: Condensed Matter*, **385–386**, 703–705.

24 Ogawa, H., Kanaya, T., Nishida, K., Matsuba, G., Majewski, J.P., and Watkins, E. (2009) *Journal of Chemical Physics*, **131** (10), 104907.

25 Glidle, A., Hillman, A.R., Ryder, K.S., Smith, E.L., Cooper, J., Gadegaard, N., Webster, J.R.P., Dalgliesh, R., and Cubitt, R. (2009) *Langmuir*, **25** (7), 4093–4103.

26 Singh, S. (2009) *Journal of Physics: Condensed Matter*, **21** (5), 055010.

27 Richardson, R. (1997) *Journal of Applied Crystallography*, **30** (6), 943.

28 Lavery, K.A., Prabhu, V.M., Lin, E.K., Wu, W., Satija, S.K., Choi, K.W., and Wormington, M. (2008) *Applied Physics Letters*, **92** (6), 064106.

29 NIST Center for Neutron Research, www.ncnr.nist.gov (accessed 14 June 2011).

30 National Synchrotron Light Source at the Brookhaven National Laboratory,http:// www.nsls.bnl.gov/facility/ (accessed 14 June 2011).

31 Goss, C.A., Charych, D.H., and Majda, M. (1991) *Analytical Chemistry*, **63** (1), 85–88.

32 Glidle, A., Bailey, L., Hadyoon, C.S., Hillman, A.R., Jackson, A., Ryder, K.S., Saville, P.M., Swann, M.J., Webster, J.R.P., Wilson, R.W., and Cooper, J.M. (2001) *Analytical Chemistry*, **73** (22), 5596–5606.

33 Grunthaner, F. (1986) *Materials Science Reports*, **1** (2), 65.

34 Hansen, W.N. (1968) *Journal of the Optical Society of America*, **58** (3), 380–388.

35 Zamlynny, V. and Lipkowski, J. (2006) Quantitative SNIFTIRS and PM IRRAS of organic molecules at electrode surfaces, in *Diffraction and Spectroscopic Methods in Electrochemistry*, vol. 9 (eds R.C. Alkire, D.M. Kolb, J. Lipkowski, and P.N. Ross), Wiley-VCH Verlag GmbH, Weinheim, pp. 315–376.

36 Als-Nielsen, J. (1985) *Zeitschrift fur Physik B*, **61** (4), 411.

37 Lu, J.R., Lee, E.M., and Thomas, R.K. (1996) *Acta Crystallographica A*, **52** (1), 11–41.

38 Koenig, B.W., Krueger, S., Orts, W.J., Majkrzak, C.F., Berk, N.F., Silverton, J.V., and Gawrisch, K. (1996) *Langmuir*, **12** (5), 1343–1350.

39 Majkrzak, C.F., Berk, N.F., Krueger, S., Dura, J.A., Tarek, M., Tobias, D., Silin, V., Meuse, C.W., Woodward, J., and Plant, A.L. (2000) *Biophysical Journal*, **79** (6), 3330–3340.

40 Krueger, S., Meuse, C.W., Majkrzak, C.F., Dura, J.A., Berk, N.F., Tarek, M., and Plant, A.L. (2000) *Langmuir*, **17** (2), 511–521.

41 Hayter, J.B., Highfield, R.R., Pullman, B.J., Thomas, R.K., McMullen, A.I., and Penfold, J. (1981) *Journal of the Chemical Society, Faraday Transactions 1*, **77** (6), 1437–1448.

42 Highfield, R.R., Thomas, R.K., Cummins, P.G., Gregory, D.P., Mingins, J., Hayter, J.B., and Schärpf, O. (1983) *Thin Solid Films*, **99** (1–3), 165–172.

43 Bradley, J.E., Lee, E.M., Thomas, R.K., Willatt, A.J., Penfold, J., Ward, R.C., Gregory, D.P., and Waschkowski, W. (1988) *Langmuir*, **4** (4), 821–826.

44 Lee, E.M., Thomas, R.K., Penfold, J., and Ward, R.C. (1989) *Journal of Physical Chemistry*, **93** (1), 381–388.

45 Whitten, D.G., Rowell, •R.L., and Penfold, •J. (eds) (2009) Special issue on neutron reflectivity. *Langmuir*, **25** (7), 3917–4242.

46 Johnson, S.J., Bayerl, T.M., McDermott, D.C., Adam, G.W., Rennie, A.R., Thomas, R.K., and Sackmann, E. (1991) *Biophysical Journal*, **59** (2), 289–294.

47 Sackmann, E. (1996) *Science*, **271** (5245), 43–48.

48 Majewski, J., Wong, J.Y., Park, C.K., Seitz, M., Israelachvili, J.N., and Smith, G.S. (1998) *Biophysical Journal*, **75** (5), 2363–2367.

49 Wong, J.Y., Majewski, J., Seitz, M., Park, C.K., Israelachvili, J.N., and Smith, G.S. (1999) *Biophysical Journal*, **77** (3), 1445–1457.

50 Smith, H.L., Jablin, M.S., Vidyasagar, A., Saiz, J., Watkins, E., Toomey, R., Hurd, A.J., and Majewski, J. (2009) *Physical Review Letters*, **102** (22), 228102.

51 Brzozowska, M., Oberts, B.P., Blanchard, G.J., Majewski, J., and Krysinski, P. (2009) *Langmuir*, **25** (16), 9337–9345.

52 Major, J.S. and Blanchard, G.J. (2001) *Langmuir*, **17** (4), 1163–1168.

53 Tamm, L.K. and McConnell, H.M. (1985) *Biophysical Journal*, **47** (1), 105–113.

54 Kuhl, T.L., Majewski, J., Wong, J.Y., Steinberg, S., Leckband, D.E., Israelachvili, J.N., and Smith, G.S. (1998) *Biophysical Journal*, **75** (5), 2352–2362.

55 Meuse, C.W., Krueger, S., Majkrzak, C.F., Dura, J.A., Fu, J., Connor, J.T., and Plant, A.L. (1998) *Biophysical Journal*, **74** (3), 1388–1398.

56 Heinrich, F., Ng, T., Vanderah, D.J., Shekhar, P., Mihailescu, M., Nanda, H., and Losche, M. (2009) *Langmuir*, **25** (7), 4219–4229.

57 Junghans, A. and Kolper, I. (2010) *Langmuir*, **26** (13), 11035–11040.

58 McGillivray, D.J., Valincius, G., Heinrich, F., Robertson, J.W.F., Vanderah, D.J., Febo-Ayala, W., Ignatjev, I., Losche, M., and Kasianowicz, J.J. (2009) *Biophysical Journal*, **96** (4), 1547–1553.

59 McGillivray, D.J., Valincius, G., Vanderah, D.J., Febo-Ayala, W., Woodward, J.T., Heinrich, F., Kasianowicz, J.J., and Losche, M. (2007) *Biointerphases*, **2** (1), 21–33.

60 Valincius, G., Heinrich, F., Budvytyte, R., Vanderah, D.J., McGillivray, D.J., Sokolov, Y., Hall, J.E., and Lösche, M. (2008) *Biophysical Journal*, **95** (10), 4845–4861.

61 Hardy, J.A. and Higgins, G.A. (1992) *Science*, **256** (5054), 184–185.

62 Zhao, X., Pan, F., and Lu, J.R. (2009) *Journal of the Royal Society Interface*, **6** (Suppl. 5), S659–S670.

63 Schmidt, A., Spinke, J., Bayerl, T., Sackmann, E., and Knoll, W. (1992) *Biophysical Journal*, **63** (5), 1385–1392.

64 Fragneto, G., Thomas, R.K., Rennie, A.R., and Penfold, J. (1995) *Science*, **267** (5198), 657–660.

65 Fragneto, G., Su, T.J., Lu, J.R., Thomas, R.K., and Rennie, A.R. (2000) *Physical Chemistry Chemical Physics*, **2** (22), 5214–5221.

66 Levicky, R., Herne, T.M., Tarlov, M.J., and Satija, S.K. (1998) *Journal of the American Chemical Society*, **120** (38), 9787–9792.

67 Steichen, M., Brouette, N., Buess-Herman, C., Fragneto, G., and Sferrazza, M. (2009) *Langmuir*, **25** (7), 4162–4167.

68 Callow, P., Fragneto, G., Cubitt, R., Barlow, D.J., Lawrence, M.J., and Timmins, P. (2005) *Langmuir*, **21** (17), 7912–7920.

69 Talbot, J.P., Barlow, D.J., Lawrence, M.J., Timmins, P.A., and Fragneto, G. (2009) *Langmuir*, **25** (7), 4168–4180.

70 Callow, P., Fragneto, G., Cubitt, R., Barlow, D.J., and Lawrence, M.J. (2008) *Langmuir*, **25** (7), 4181–4189.

71 Singh, S. (2009) *Corrosion Science*, **51** (3), 575.

72 John, D., Blom, A., Bailey, S., Nelson, A., Schulz, J., De, M., and Kinsella, B. (2006) *Physica B*, **385–386**, 924–926.

73 Wiesler, D.G. and Majkrzak, C.F. (1994) *Physica B*, **198** (1–3), 181–186.

74 Noel, J.J., Shoesmith, D.W., and Tun, Z. (2008) *Journal of the Electrochemical Society*, **155** (8), C444–C454.

75 Tun, Z., Noel, J.J., and Shoesmith, D.W. (1999) *Journal of the Electrochemical Society*, **146** (3), 988–994.

76 Wood, D.L., Chlistunoff, J., Majewski, J., and Borup, R.L. (2009) *Journal of the American Chemical Society*, **131** (50), 18096–18104.

77 Majewski, J., Smith, G.S., Burgess, I., Zamlynny, V., Szymanski, G., Lipkowski, J., and Satija, S. (2002) *Applied Physics A*, **74** (Suppl., Pt. 1), S364–S367.

78 Burgess, I., Zamlynny, V., Szymanski, G., Lipkowski, J., Majewski, J., Smith, G., Satija, S., and Ivkov, R. (2001) *Langmuir*, **17** (11), 3355–3367.

79 Zamlynny, V., Burgess, I., Szymanski, G., Lipkowski, J., Majewski, J., Smith, G., Satija, S., and Ivkov, R. (2000) *Langmuir*, **16** (25), 9861–9870.

80 Burgess, I., Zamlynny, V., Szymanski, G., Schwan, A.L., Faragher, R.J., Lipkowski, J., Majewski, J., and Satija, S. (2003) *Journal of Electroanalytical Chemistry*, **550–551**, 187–199.

81 Zawisza, I., Burgess, I., Szymanski, G., Lipkowski, J., Majewski, J., and Satija, S. (2004) *Electrochimica Acta*, **49** (22–23), 3651–3664.

82 Glidle, A., Cooper, J., Hillman, A.R., Bailey, L., Jackson, A., and Webster,

J.R.P. (2003) *Langmuir*, **19** (19), 7746–7753.

83 Lecuyer, S., Fragneto, G., and Charitat, T. (2006) *European Physics Journal E*, **21** (2), 153–159.

84 Burgess, I., Li, M., Horswell, S.L., Szymanski, G., Lipkowski, J., Majewski, J., and Satija, S. (2004) *Biophysical Journal*, **86** (3), 1763–1776.

85 Burgess, I., Li, M., Horswell, S.L., Szymanski, G., Lipkowski, J., Satija, S., and Majewski, J. (2005) *Colloids and Surfaces B*, **40** (3–4), 117–122.

86 Hillman, A.R., Ryder, K.S., Madrid, E., Burley, A.W., Wiltshire, R.J., Merotra, J., Grau, M., Horswell, S.L., Glidle, A., Dalgliesh, R.M., Hughes, A., Cubitt, R., and Wildes, A. (2010) *Faraday Discussions*, **145**, 357–379.

87 Tsong, T.Y. and Astumian, R.D. (1988) *Annual Review of Physiology*, **50**, 273–290.

88 Jones, S.W. (1998) *Journal of Bioenergetics and Biomembranes*, **30** (4), 299–312.

89 Olivotto, M., Arcangeli, A., Carla, M., and Wanke, E. (1996) *Bioessays*, **18** (6), 495–504.

90 Terlau, H. and Stuhmer, W. (1998) *Naturwissenschaften*, **85** (9), 437–444.

91 Naumann, R., Schmidt, E.K., Jonczyk, A., Fendler, K., Kadenbach, B., Liebermann, T., Offenhausser, A., and Knoll, W. (1999) *Biosensors & Bioelectronics*, **14** (7), 651–662.

92 Berk, N.F. and Majkrzak, C.F. (2009) *Langmuir*, **25** (7), 4132–4144.

93 Majkrzak, C.F., Berk, N.F., Kienzle, P., and Perez-Salas, U. (2009) *Langmuir*, **25** (7), 4154–4161.

94 Dura, J.A., Pierce, D.J., Majkrzak, C.F., Maliszewskyj, N.C., McGillivray, D.J., Losche, M., O'Donovan, K.V., Mihailescu, M., Perez-Salas, U., Worcester, D.L., and White, S.H. (2006) *Review of Scientific Instruments*, **77** (7), 074301–074311.

4
Model Lipid Bilayers at Electrode Surfaces

Rolando Guidelli and Lucia Becucci

4.1
Introduction

Biological membranes are by far the most important electrified interfaces in living systems. They consist of a bimolecular layer of lipids (the lipid bilayer) incorporating proteins. Lipid molecules are "amphiphilic," that is, they consist of a hydrophobic section (the hydrocarbon tail) and a hydrophilic section (the polar head). In biological membranes the two lipid monolayers are oriented with the hydrocarbon tails directed toward each other and the polar heads turned toward the aqueous solutions that are in contact with the two sides of the membrane. The resulting lipid bilayer is a matrix that incorporates different proteins performing a variety of functions.

Biomembranes form a highly selective barrier between the inside and the outside of living cells. They are highly insulating to inorganic ions, and large electrochemical potential gradients can be maintained across them. The permeability and structural properties of biological membranes are sensitive to the chemical nature of the membrane components and to events that occur at the interface or within the bilayer. For example, biomembranes provide the environmental matrix for proteins that specifically transport certain ions and other molecules, for receptor proteins, and for signal transduction molecules.

4.2
Biomimetic Membranes: Scope and Requirements

In view of the complexity and diversity of the functions performed by the various proteins embedded in a biomembrane (the integral proteins), it has been found convenient to incorporate single integral proteins or smaller lipophilic biomolecules into experimental models of biological membranes, so as to isolate and investigate their functions. This serves to reduce complex membrane processes to well-defined interactions between selected proteins, lipids, and ligands. There is

Advances in Electrochemical Science and Engineering. Edited by Richard C. Alkire, Dieter M. Kolb, and Jacek Lipkowski

ISBN: 978-3-527-32885-7

great potential for application of experimental models of biomembranes (so-called biomimetic membranes) for the elucidation of structure–function relationships of many biologically important membrane proteins. These proteins are the key factors in cell metabolism, for example, in cell–cell interactions, signal transduction, and transport of ions and nutrients. Because of this important function, membrane proteins are a preferred target for pharmaceuticals. Biomimetic membranes are also useful for the investigation of phase stability (e.g., lipid–lipid phase separation, lipid raft formation, lateral diffusion), protein–membrane interactions (e.g., receptor clustering and co-localization), and membrane–membrane processes such as fusion, electroporation, and intercellular recognitions. They are also relevant to the design of membrane-based biosensors and devices, and to analytical platforms for assaying membrane-based processes.

With only a few exceptions, metal-supported biomimetic membranes consist of a more or less complex architecture that includes a lipid bilayer. In order of increasing complexity, they can be classified into: solid-supported bilayer lipid membranes (sBLMs), tethered bilayer lipid membranes (tBLMs), polymer-cushioned bilayer lipid membranes (pBLMs), S-layer stabilized bilayer lipid membranes (ssBLMs), and protein-tethered bilayer lipid membranes (ptBLMs).

To incorporate integral proteins in a functionally active state, biomembrane models consisting of lipid bilayers should meet a number of requirements: (i) they should be robust enough for long-term stability, and be easily and reproducibly prepared; (ii) they should have the lipid bilayer in the liquid crystalline state, and such as to allow lateral mobility; (iii) they should have water (or, at least, a highly hydrated hydrophilic region) on both sides of the lipid bilayer; and (iv) they should be sufficiently free from pinholes and other defects that might provide preferential pathways for electron and ion transport across the lipid bilayer. Requirements (ii) and (iii) are necessary for the incorporation of integral proteins into the lipid bilayer in a functionally active state. In fact, integral proteins have a hydrophobic domain buried inside the biomimetic membrane, which must be sufficiently fluid to accommodate this domain. Often, they also have hydrophilic domains protruding by several nanometers outside the lipid bilayer. To avoid their denaturation and to promote their function, incorporation of integral proteins into biomimetic membranes must ensure that their extramembrane hydrophilic domains are accommodated in a hydrophilic medium on both sides of the lipid bilayer. Moreover, the transport of hydrophilic ions across a solid-supported lipid bilayer via ion channels or ion pumps is only possible if an aqueous or hydrophilic layer is interposed between the bilayer and the support. Requirement (iv) is needed to make the biomembrane model sufficiently blocking so as to enable characterization of ion channel or ion pump activity by electrochemical means without the disturbing presence of stray currents due to defects.

Apart from lipid molecules, the molecules that are most commonly employed for the fabrication of biomimetic membranes are "hydrophilic spacers" and "thiolipids." Hydrophilic spacers consist of a hydrophilic chain (e.g., a polyethyleneoxy or oligopeptide chain) terminated at one end with an anchor group for tethering to a support and, at the other end, with a hydrophilic functional group (e.g., a

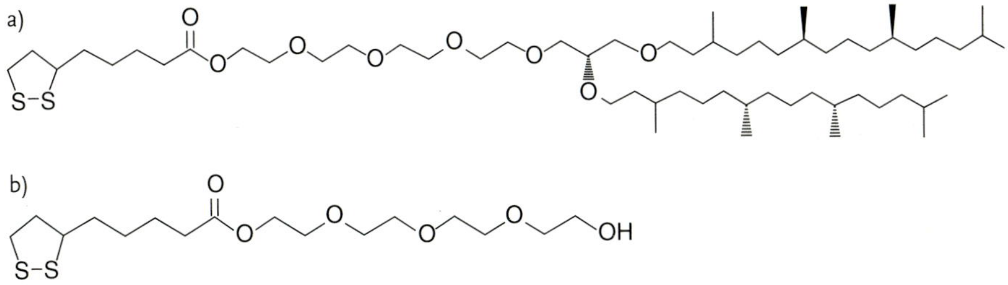

Figure 4.1 (a) Structure of a widely adopted thiolipid, called DPTL [1]. (b) Structure of the corresponding hydrophilic spacer (TEGL), in which the two phytanyl chains are replaced by a hydroxyl group.

hydroxyl group). Sulfhydryl or disulfide groups are employed as anchor groups for tethering to metals such as gold, silver, or mercury (Figure 4.1b); methyl-, methyloxy- or chloride-substituted silane groups are used for tethering to glass, quartz, silica, or mica. The latter supports are nonconducting and cannot be investigated by electrochemical techniques. Hydrophilic spacers serve to separate the lipid bilayer from a solid support, to compensate for surface roughness effects, to prevent any incorporated peptides or proteins from touching the support surface (thus avoiding loss of their functionality due to denaturation), and to provide an ionic reservoir underneath the lipid bilayer.

Thiolipids differ from hydrophilic spacers in that the hydrophilic chain is covalently linked to one or, more frequently, two alkyl chains at the opposite end with respect to the anchor group. A convenient and widely used thiolipid, first employed by Schiller *et al.* [1] and denoted by the abbreviation DPTL, consists of a tetraethyleneoxy hydrophilic chain covalently linked at one end to a lipoic acid residue, for anchoring to a metal via a disulfide group, and bound at the other end via ether linkages to two phytanyl chains (Figure 4.1a). The alkyl chains simulate the hydrocarbon tails of a lipid molecule and provide one half of the lipid bilayer to the biomimetic membrane. When tethered to a support, hydrophilic spacers expose to the bulk aqueous phase a hydrophilic surface, while thiolipids expose a hydrophobic surface. Clearly, lipid bilayers formed on top of hydrophilic spacers are noncovalently linked to them and can be regarded as "freely suspended." Conversely, lipid monolayers self-assembled on top of thiolipid monolayers form lipid bilayers that are tethered to the support.

Before discussing advantages and disadvantages of these systems, we will briefly describe the electrochemical technique that is commonly employed for their investigation, namely electrochemical impedance spectroscopy (EIS), as well as some fabrication methodologies. In fact, attention is focused on those biomimetic membranes that are amenable to investigation by electrochemical methods. Biomimetic membranes that are investigated exclusively by nonelectrochemical surface-sensitive techniques, such as those formed on insulating supports (e.g., glass, mica, quartz, silica, etc.), are outside the scope of this chapter.

4.3 Electrochemical Impedance Spectroscopy

Many membrane proteins are "electrogenic," that is, translocate a net charge across a membrane. Consequently, it is possible to monitor their function directly by measuring the current flowing along an external electrical circuit upon their activation. The techniques of choice for these measurements are EIS and potential-step chronoamperometry or chronocoulometry, because the limited volume of the ionic reservoir created by a hydrophilic spacer in solid-supported biomimetic membranes cannot sustain a steady-state current.

EIS applies an AC voltage of given frequency to the system under study and measures the resulting current that flows with the same frequency. Both the amplitude of the AC current and its phase shift with respect to the AC voltage are measured. The frequency is normally varied gradually from 10^{-3} to 10^5 Hz. To interpret measured impedance spectra, it is necessary to compare them with the electrical response of an "equivalent circuit" assembled from resistors and capacitors, capable of simulating the system under investigation. In general, a metal-supported self-assembled mono- or multilayer can be regarded as consisting of a series of slabs with different dielectric properties. When ions flow across each slab, they give rise to an ionic current $\boldsymbol{J}_{\text{ion}} = \sigma \boldsymbol{E}$, where $\boldsymbol{E}$ is the electric field and σ is the conductivity. Ions may also accumulate at the boundary between contiguous dielectric slabs, causing a discontinuity in the electric displacement vector $\boldsymbol{D} = \boldsymbol{\varepsilon E}$, where ε is the dielectric constant. Under AC conditions, the accumulation of ions at the boundary of the dielectric slabs varies in time, and so does the electric displacement vector, giving rise to a capacitive current $\boldsymbol{J}_c = \mathrm{d}\boldsymbol{D}/\mathrm{d}t$. The total current is, therefore, given by the sum of the ionic current and of the capacitive current. In this respect, each dielectric slab can be simulated by a parallel combination of a resistance, accounting for the ionic current, and of a capacitance, accounting for the capacitive current, namely by an "*RC* mesh." Accordingly, the impedance spectrum of a self-assembled layer can be simulated by a series of *RC* meshes.

Application of an AC voltage of amplitude V and frequency f to a pure resistor of resistance R yields a current of equal frequency f and of amplitude V/R, in phase with the voltage. Conversely, application of the AC voltage to a pure capacitor of capacitance C yields a current of frequency f and amplitude $2\pi fC$, out of phase by $-\pi/2$ with respect to the voltage, that is, in quadrature with it. This state of affairs can be expressed by stating that the admittance Y of a resistance element equals $1/R$, while that of a capacitance element equals $-\mathrm{i}\omega C$, where $\omega = 2\pi f$ is the angular frequency and i is the imaginary unit. More generally, in an equivalent circuit consisting of resistances and capacitances, Y is a complex quantity, and the impedance Z is equal to $1/Y$, by definition. Hence, Z equals R for a resistance element, and $\mathrm{i}/\omega C$ for a capacitance element. Impedance spectra are often displayed on a Bode plot, namely a plot of $\log|Z|$ and phase angle Φ against $\log f$, where $|Z|$ is the magnitude of the impedance. This plot is relatively featureless. Over the frequency range where the equivalent circuit is exclusively controlled by a capacitance, the $\log|Z|$ versus $\log f$ plot is a linear segment of slope -1, and the phase angle equals

$-90°$. Conversely, in the case of pure control by a resistance, the $\log|Z|$ versus $\log f$ plot is horizontal and Φ equals zero. Another commonly used plot, called a Nyquist plot, reports the quadrature component, Z'', of the impedance Z against the corresponding in-phase component, Z'. In the case of a single RC mesh, the Nyquist plot yields a semicircle of diameter R and center of coordinates $(R/2, 0)$. The angular frequency, ω, at the maximum of this semicircle equals the reciprocal of the time constant RC of the mesh. In the presence of a series of RC meshes, their time constants are often close enough to cause the corresponding semicircles to overlap partially. In this case, if the mesh of highest time constant has also the highest resistance, R_1, as is often the case, then the Nyquist plot of the whole impedance spectrum exhibits a single well-formed semicircle, R_1 in diameter. The semicircles of the remaining meshes are compressed in a very narrow area close to the origin of the Z'' versus Z' plot, and can be visualized only by enlarging this area. Therefore, the Nyquist plot of the whole spectrum is conveniently employed if one is interested in pointing out the resistance R_1 of the dielectric slab of highest resistance.

To better visualize all semicircles, we have found it convenient to represent impedance spectra on a $\omega Z'$ versus $\omega Z''$ plot [2]. Henceforth, this plot will be briefly referred to as an "M plot," since $\omega Z'$ and $\omega Z''$ are the components of the modulus function M. It is possible to demonstrate that a single RC mesh yields a semicircle even on a $\omega Z'$ versus $\omega Z''$ plot; its diameter equals C^{-1} and its center has coordinates $(C^{-1}/2, 0)$. Moreover, ω at the maximum of the semicircle is again equal to the reciprocal of the time constant RC of the mesh. While ω decreases along the positive direction of the abscissas on a Nyquist plot, it increases on an M plot. Therefore, for a series of RC meshes, the last semicircle on the M plot is characterized by the lowest time constant. This is, unavoidably, the semicircle simulating the solution that bathes the self-assembled film, due to its very low capacitance. Figure 4.2 shows the M plot for a biomimetic membrane consisting of a DPTL monolayer anchored to a mercury electrode, with a diphytanoylphosphatidylcholine (DPhyPC) monolayer on top of it; the tBLM incorporates the ion carrier valinomycin, a hydrophobic depsipeptide that cages a desolvated potassium ion, shuttling it across the lipid bilayer [2]. The plot shows four partially overlapping semicircles. The solid curve is the best fit of the plot by an equivalent circuit consisting of four RC meshes in series. The semicircles overlap only to a moderate extent, thus allowing their straightforward deconvolution. This is due to an appreciable difference between the time constants of the four RC meshes, which are evenly distributed over a frequency range covering seven orders of magnitude. Proceeding along the positive direction of the abscissas, the four semicircles are ascribable to the lipoid acid residue, the tetraethyleneoxy hydrophilic spacer, the lipid bilayer moiety, and the aqueous solution bathing the lipid bilayer.

A plot that has been frequently adopted in the literature to display an impedance spectrum as rich in features as the M plot is the Y'/ω versus Y''/ω plot, sometimes called a Cole–Cole plot [3–5]. Here Y' and Y'' are the in-phase and quadrature component of the electrode admittance. However, it can be shown that this plot yields a semicircle for a series combination of a resistance and a capacitance, and

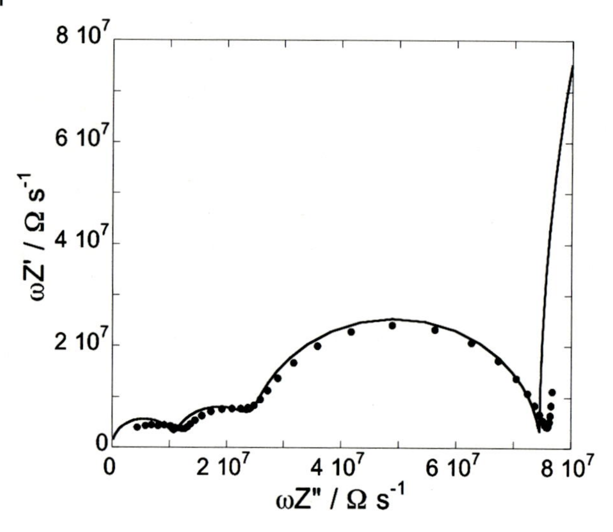

Figure 4.2 Plot of $\omega Z'$ against $\omega Z''$ (M plot) for a mercury-supported DPTL–DPhyPC bilayer incorporating gramicidin from its 1×10^{-7} M solution in aqueous 0.1 M KCl at −0.600 V vs. Ag/AgCl (0.1 M KCl) [2]. The solid curve is the best fit of the impedance spectrum by four *RC* meshes in series.

not for their parallel combination. For such a series combination, the Cole–Cole plot yields a semicircle of diameter C and center of coordinates $(C/2, 0)$. Here too, ω at the maximum of the semicircle equals $1/RC$. Strictly speaking, a Cole–Cole plot is not suitable for verifying the fitting of an impedance spectrum by a series of *RC* meshes. Thus, if we force a Y'/ω versus Y''/ω plot to fit a single *RC* mesh, we obtain "formally" a radius of the semicircle that is given by $(C/2)(1 + 1/\omega^2R^2C^2)$. Consequently, the Cole–Cole plot for a single *RC* mesh yields a semicircle of diameter C only for ω values high enough to make $\omega^2R^2C^2 \gg 1$. Figure 4.3 shows the Cole–Cole plot for the same impedance spectrum displayed on the M plot of Figure 4.2. It is apparent that the fitting of the experimental spectrum by a series of four *RC* meshes shows appreciable deviations at the lower frequencies, which correspond to the higher values of Y''/ω.

4.4
Formation of Lipid Films in Biomimetic Membranes

Methodologies for the fabrication of biomimetic membranes vary somewhat from one biomimetic membrane to another. However, a number of experimental procedures for the formation of lipid monolayers and bilayers on solid supports are common to several biomimetic membranes. The most popular procedures are vesicle fusion, Langmuir–Blodgett and Langmuir–Schaefer transfers, and rapid solvent exchange. The formation of lipid monolayers and bilayers on gold and

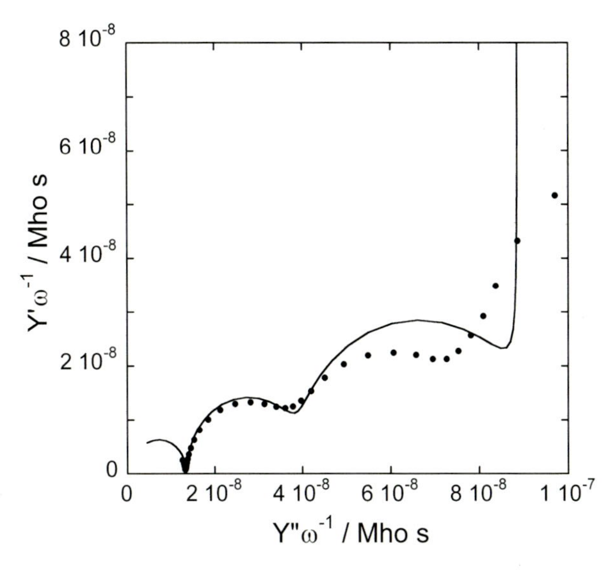

Figure 4.3 Plot of Y'/ω against Y''/ω (Cole–Cole plot) for the same tBLM as in Figure 4.2. The solid curve is the best fit of the impedance spectrum by four *RC* meshes in series.

silver substrates is commonly monitored by surface plasmon resonance (SPR). Surface plasmons are collective electronic oscillations in a metal layer, about 50 nm in thickness, excited by photons from a laser beam [6]. The beam is reflected by the back surface of the metal layer, while its front surface supports a dielectric film (e.g., a lipid bilayer), usually in contact with an aqueous solution. The evanescent electromagnetic field thereby generated in the metal layer can couple with the electronic motions in the dielectric film. The intensity of the electromagnetic field associated with surface plasmons has its maximum at the metal–dielectric film surface and decays exponentially into the space perpendicular to it, extending into the metal and the dielectric. This makes SPR a surface-sensitive technique particularly suitable for the measurement of the optical thickness of ultrathin films adsorbed on metals. As a rule, the incident angle of the laser beam is varied with respect to the back surface of the metal layer, and the reflected light intensity is measured. At a certain angle of incidence there exists a resonance condition for the excitation of surface plasmons, which causes the energy of the incident laser light to be absorbed by the surface plasmon modes and the reflectivity to attain a minimum. The angular position of the minimum of the SPR reflectivity curves (i.e., the curves of reflectivity versus incident angle) is critically dependent on the thickness of the layer adsorbed on the support surface, and is used to estimate such a thickness. Both SPR and EIS allow an evaluation of film thickness, based on a reasonable estimate of the refractive index of the film in the case of SPR, or of its dielectric constant in the case of EIS. However, it must be borne in mind that the two techniques are sensitive to different features of a film.

4.4.1
Vesicle Fusion

Vesicles (or, more precisely, unilamellar vesicles) are spherical lipid bilayers that enclose an aqueous solution. The procedure for vesicle fusion consists of adsorbing and fusing small unilamellar vesicles (SUVs; 20 to 50 nm in diameter) on a suitable substrate from their aqueous dispersion. If the substrate is hydrophilic, vesicle fusion gives rise to a lipid bilayer by rupture of the vesicles and their "unrolling" and spreading onto the substrate, as shown in Figure 4.4b. Conversely, if the substrate is hydrophobic, a lipid monolayer with the hydrocarbon tails directed toward the substrate is formed by rupture of the vesicles, splitting of the vesicular membrane into its two monomolecular leaflets and their spreading [7], as shown in Figure 4.4a. This is confirmed by the different increase in thickness following vesicle fusion on a hydrophobic substrate (2–2.5 nm) with respect to a hydrophilic substrate (4.5–5 nm), as estimated by SPR [8].

The kinetics of vesicle fusion, followed by monitoring the position of the minimum of the SPR reflectivity curves, depends on the composition and molecular shape of the vesicular lipids and on the nature of the substrate. As a rule, bilayer formation by vesicle unrolling onto a hydrophilic surface is faster than monolayer formation by vesicle fusion onto a hydrophobic surface. This is probably due to the fact that the processes involved in forming a planar bilayer starting from a vesicular bilayer are considerably less complex than those involved in forming a planar monolayer [8, 9].

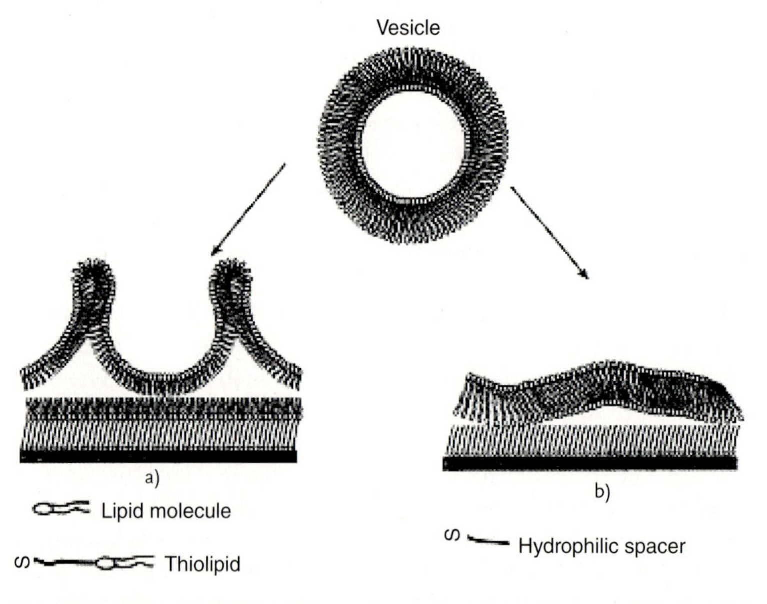

Figure 4.4 Schematics of (a) splitting and spreading of a vesicle on a solid-supported thiolipid monolayer and (b) unrolling and spreading of a vesicle on a solid-supported hydrophilic spacer.

It has been suggested that the initial rapid stage of vesicle adsorption on hydrophilic surfaces is controlled by vesicle adsorption at free sites of the surface, according to a Langmuirian-type behavior. A second, later stage is ascribed to vesicle unrolling and spreading processes. Among hydrophilic substrates, those allowing the formation of lipid bilayers by vesicle fusion more easily are freshly oxidized surfaces of silica, glass, quartz, and mica [10]. However, hydrophilicity is a necessary but not a sufficient condition to promote vesicle fusion. Surfaces of oxidized metals and metal oxides (e.g., TiO_2, platinum, and gold) allow adsorption of intact vesicles but resist the formation of bilayers, presumably due to weak surface interactions [11]. Electrostatic, van der Waals, hydration, and steric forces cause a noncovalently supported lipid bilayer to be separated from a solid surface by a nanometer layer of water [12]. This water layer prevents the support from interfering with the lipid bilayer structure, thus preserving its physical attributes, such as lateral mobility of the lipid molecules. The quartz crystal microbalance with dissipation monitoring (QCM-D) has proved quite valuable for monitoring the macroscopic features of vesicle deposition [13]. Use of the QCM-D permitted confirmation of the formation of lipid bilayers on silica and the adsorption of intact vesicles on oxidized gold [14].

Initial adsorption of vesicles on hydrophobic surfaces is energetically disfavored, due to the presence of the hydrophilic polar heads on the outer surface of the vesicular membrane. Therefore, the vesicular membrane must split to allow its inner hydrophobic tails to get in contact with the hydrophobic surface. A possible pathway for vesicle fusion involves vesicle splitting, unrolling, and spreading on the hydrophobic surface, as shown in Figure 4.4a. The kinetics of vesicle fusion on the hydrophobic surface of gold-supported alkanethiol self-assembled monolayers was followed by SPR [15]. In the initial stage, the adsorbed layer thickness d increases linearly with the square root of time t, denoting control by vesicle diffusion to the surface according to Fick's first law. In a second stage, d increases roughly linearly with $\log t$. Finally, the time dependence of d becomes typical of an adsorption process on an almost fully occupied surface. The curve of the surface coverage by vesicles against time was also monitored by SPR at different vesicle concentrations [7]; it was fitted to an equation practically identical to that derived for an electrode process controlled by diffusion and by a heterogeneous electron transfer step [16]. The resulting kinetic constant was ascribed to some surface reorganization of the vesicles.

When vesicles fuse spontaneously on a support, some of them may remain only partially fused, or even intact in the adsorbed state. When the surface density of vesicles is sufficiently high, their presence is revealed by an anomalously high thickness attained by the lipid film after vesicle fusion, as monitored by SPR. This phenomenon is particularly evident with supports consisting of mixtures, with comparable molar ratios, of two different molecules exposing to the bulk aqueous phase hydrophobic alkyl chains and hydrophilic functional groups, respectively [5, 17]. With very high vesicle concentrations ($5\,mg\,ml^{-1}$), monolayer thicknesses greater than 8 nm were reported on hydrophobic surfaces [9]. Thus, fusion of SUVs onto a binary mixture of hydrophobic cholesteryl-terminated molecules and hydrophilic 6-mercaptohexanol molecules yields atomic force microscopy (AFM) images showing heightened areas; their diameter being close to that of the SUVs denotes the presence of adsorbed vesicles [3]. The presence of a membrane protein, such as cytochrome bo3, in these vesicles increases the number density and the size of the heightened structures ascribable to adsorbed vesicles. Analogously, fusion of large unilamellar vesicles (LUVs) on a hydrophobic support exposing dipalmitoylphosphatidylethanolamine alkyl chains to the aqueous phase yields tapping-mode AFM images with a number of dome-shaped structures [18]. Some of these structures, whose diameter is close to that of the LUVs, tend to disappear after an hour, due to complete vesicle fusion. However, if the vesicles contain the membrane protein Na,K-ATPase, all the domelike structures are stable even after three hours. In general, adsorption of proteoliposomes (namely, vesicles incorporating membrane proteins) on hydrophobic surfaces prevents their complete spreading and fusion, due to the presence of protein molecules with extramembrane domains and to the hydrophilicity of the outer polar heads of the proteoliposomes [19, 20].

The presence of heightened areas in metal-supported lipid films formed by fusion of vesicles labeled with a fluorophore can also be monitored by fluorescence microscopy. In fact, the energy of a photoexcited fluorophore is transferred non-radiatively to the metal support (quenching) if the fluorophore is at a distance from the metal surface shorter than a critical transfer distance (the Foerster radius). This critical distance amounts to about 20–30 nm on gold. Therefore, unquenched bright spots in the fluorescence microscopy images of gold-supported lipid films obtained by fusion of fluorophore-labeled vesicles mark the presence of heightened areas ascribable to adsorbed or hemifused vesicles [21].

4.4.2
Langmuir–Blodgett and Langmuir–Schaefer Transfer

Another procedure for forming a lipid monolayer on a hydrophobic substrate or a lipid bilayer on a hydrophilic substrate makes use of a Langmuir trough equipped with a movable barrier. By spreading a lipid dissolved in an organic solvent on the surface of the aqueous electrolyte contained in the trough and by allowing the solvent to evaporate, a lipid film is formed at the air–water interface. This film is compressed with the movable barrier until it is brought close to the liquid crystalline state by adjusting the surface pressure. To form a lipid monolayer on a hydrophobic slab, the slab is immersed vertically through the lipid monolayer. This brings the hydrocarbon tails of the lipid monolayer, turned toward the air, in direct contact with the hydrophobic surface of the slab, which remains coated by a lipid monolayer (Figure 4.5a). This technique, called Langmuir–Blodgett (L-B) transfer, is also used to form a lipid bilayer on a hydrophilic slab [22, 23]. In this case, the slab is initially immersed vertically through the lipid monolayer into the trough. No significant change in surface pressure is observed at this stage. The slab is then withdrawn at a speed slow enough to permit water to drain from the surface. During the withdrawal, the polar heads of the lipid monolayer are turned

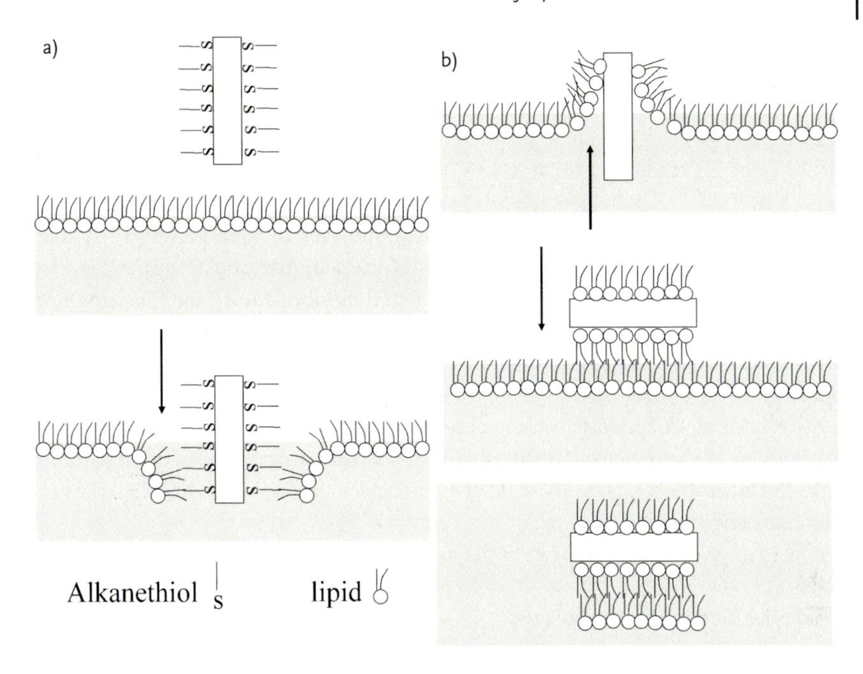

Figure 4.5 (a) Formation of a lipid monolayer on a hydrophobic alkanethiol-coated slab by L-B transfer. (b) Formation of a lipid bilayer on a hydrophilic slab by L-B transfer and subsequent L-S transfer.

toward the surface of the hydrophilic slab, giving rise to noncovalent self-assembly (Figure 4.5b). A resulting decrease in the surface pressure of the lipid monolayer at the air–water interface is prevented by reducing its area with the moving barrier. A further lipid monolayer is then self-assembled on top of the first one by Langmuir–Schaefer (L-S) transfer. In practice, the lipid-coated slab is brought into horizontal contact with the lipid monolayer at the air–water interface, after compressing it at a preset surface pressure. Finally, the slab covered by the lipid bilayer is detached from the aqueous subphase. The combination of L-B and L-S transfers is used principally for lipid bilayer deposition on hydrophilic surfaces of glass, mica, quartz, polymeric materials, graphite, and, to a minor extent, metals such as gold [24].

An advantage of L-B and L-S transfers over vesicle fusion is that the strict composition of a mixed lipid monolayer on the trough surface is maintained after its transfer. Conversely, the composition of a mixed lipid layer on a substrate does not necessarily correspond to that of the vesicles used to form it, and varies strongly with the history of the sample preparation. On the other hand, an advantage of vesicle fusion over L-B and L-S transfers is that vesicles may easily incorporate membrane proteins, forming proteoliposomes. To this end, membrane proteins are usually stabilized in detergent micelles and then incorporated into the lipid vesicles upon removing the detergent molecules by dialysis. Fusion of proteoliposomes on a metal-supported thiolipid monolayer may then cause the insertion of the membrane proteins in the resulting lipid bilayer.

4.4.3 Rapid Solvent Exchange

A further procedure for depositing a lipid monolayer is rapid solvent exchange. This method is usually employed to self-assemble a lipid monolayer onto a hydrophobic surface exposing alkyl chains to the bulk aqueous phase. The method involves placing a small amount of lipid dissolved in a water-miscible solvent, such as ethanol, onto the hydrophobic substrate and incubating it for a few minutes [25, 26]. The ethanol solution is then vigorously displaced in a few seconds by a large excess of an aqueous buffer solution, taking care to avoid the formation of air bubbles at the surface. During the lipid addition and the subsequent rinsing with the aqueous solution, the lipid self-assembles to form a lipid monolayer and the excess lipid is rinsed away.

4.4.4 Fluidity in Biomimetic Membranes

A fundamental property of biological membranes is the long-range lateral mobility of the lipid molecules. The fluidity of the plasma membrane should be preserved in supported biomimetic membranes. Free movement of lipid molecules enables the biomimetic membrane to react to the presence of proteins, charges, and physical forces in a dynamic and responsive manner. A satisfactory fluidity allows biomimetic membranes to reorganize upon interaction with external perturbations, mimicking the functionality of living cell membranes. In particular, lateral mobility enables a biomimetic membrane to incorporate bulky membrane proteins from their detergent solutions by making space for them; it also determines the spontaneous separation of the components of a lipid mixture (demixing), giving rise to the formation of important lipid microdomains, called lipid rafts.

The fluidity and lateral mobility of biomimetic membranes can be characterized quantitatively by AFM [27] and fluorescence microscopy. Fluorescence recovery after photobleaching (FRAP) is one of the most popular ways of measuring molecular diffusion in membranes [28]. It relies on introducing a small amount of fluorescent probe molecules, usually covalently bound to lipids, into the membrane. A short burst of intense excitation light is projected onto the membrane, destroying the fluorescence of the fluorophore molecules in a well-defined spot, a photochemical process called photobleaching. The gradual fluorescence recovery within the given spot is followed as a function of time, thus permitting an estimate of the lipid diffusion coefficient. In supported membranes, this coefficient typically ranges from 1 to 10 $\mu m^2 s^{-1}$. When applied to lipid bilayers self-assembled on smooth supports such as silica, glass, quartz, mica, or indium tin oxide (ITO), FRAP usually confirms a satisfactory lateral mobility of lipid molecules [10, 29]. Conversely, biomimetic membranes consisting of a thiolipid monolayer tethered to a gold electrode, with a self-assembled lipid monolayer on top of it, do not exhibit lateral mobility. This is also true for the distal lipid monolayer noncovalently linked to the thiolipid monolayer, no matter if obtained by vesicle fusion or by L-B transfer [5, 21]. A biomimetic membrane consisting of a hydrophilic spacer tethered to gold, with a lipid bilayer formed on top of it, was reported to exhibit FRAP, if the noncovalently bound lipid bilayer was formed by L-B and L-S transfers [5]. Fluorescence recovery in the gold-supported biomimetic membrane could only be observed for a short period of time, due to the gradual energy transfer from the fluorophore molecules to gold (quenching). No fluorescence recovery could be observed on forming the lipid bilayer on top of the hydrophilic spacer by vesicle fusion. Evidently, the unavoidable presence of adsorbed and hemifused vesicles prevents the lateral mobility of the lipid molecules of the distal monolayer.

As distinct from solid supports such as gold or silver, mercury imparts lateral mobility to lipid monolayers directly self-assembled on its surface, because of its liquid state. This is demonstrated by rapid spontaneous phase separation, with microdomain formation, in a lipid mixture monolayer self-assembled on top of a DPTL thiolipid monolayer tethered to a mercury microelectrode [30]. The presence of microdomains was directly verified from the images of the distal lipid monolayer obtained using two-photon fluorescence lifetime imaging microscopy.

4.5 Various Types of Biomimetic Membranes

Biomimetic membranes comprising a lipid bilayer are discussed here in the order of approximately increasing complexity.

4.5.1 Solid-Supported Bilayer Lipid Membranes

The term sBLM or, simply, "solid-supported membrane," is commonly used to denote a biomimetic membrane consisting of a lipid bilayer in direct contact with a solid support. These biomimatic membranes are typically formed on a hydrophilic solid support by immersing the support in an aqueous dispersion of SUVs, which slowly rupture and spread on the surface of the support. Alternatively, they can be formed by L-B and L-S transfers. When appropriately formed, these sBLMs are separated from the support surface through a hydration layer of water, estimated to be between 6 and 15 Å thick [31, 32]. Several theoretical and experimental investigations of water near polar hydrophilic surfaces suggest that it is more ordered than bulk water, with higher viscosity and lower dielectric constant [33, 34]. In accordance with these predictions, the characteristics of the lipid bilayer have been found to be structurally coupled with the support properties, such as its charge [35], wettability [36], and topography [37]. The lubrification effect of the water layer imparts a significant long-range lateral mobility to the lipid bilayer. However, significant frictional coupling between the bilayer and the underlying substrate slows down lateral diffusion, which may be accompanied by a breakdown of the two-dimensional fluid nature of the membrane.

The above biomembrane models are not suitable for studying the function of integral proteins. In fact, these proteins have hydrophilic domains protruding outside the lipid bilayer. To avoid their denaturation and to promote their function, the incorporation of integral proteins into biomembrane models must ensure that their protruding hydrophilic domains are accommodated in a hydrophilic medium on both sides of the lipid bilayer. Not surprisingly, embedded membrane-spanning proteins usually show no lateral diffusion, because of their interaction with the substrate [38], even though some of them maintain their function if their active site is far from the solid substrate.

The majority of sBLMs are formed on nonconducting supports such as silica, glass, or mica [10, 23, 29, 38, 39]. In what follows, a few examples of sBLMs investigated by electrochemical techniques will be examined. Phospholipid bilayers on Au(111) single-crystal faces have been prepared by Lipkowski and coworkers either by vesicle fusion [40] or by L-B and L-S transfers [24]. The latter procedure yields bilayers with a higher packing density and with smaller tilt angles of the alkyl chains with respect to the surface normal. These bilayers have been characterized by charge density measurements, photon polarization modulation infrared reflection absorption spectroscopy (PM-IRRAS) [24, 40], and neutron reflectivity (NR) [41]. The minimum differential capacitance of these bilayers is about $2\,\mu F\,cm^{-2}$, and is attained at charge densities σ_M on the metal higher than $-8\,\mu C\,cm^{-2}$. The differential capacitance being greater than that of a solvent-free BLM denotes the presence of a number of defects, while NR reveals the presence of water molecules within the lipid bilayer. As σ_M becomes more negative than $-8\,\mu C\,cm^{-2}$, the lipid bilayer starts to detach from the electrode, but remains in close proximity to the electrode surface. In fact, NR indicates that the lipid bilayer is suspended on a thin cushion of the aqueous electrolyte, which screens the metal charge and contains a large fraction of the overall interfacial potential difference; consequently, the electric field across the lipid bilayer becomes weak. This suspended lipid bilayer is essentially defect-free and its structure resembles that of lipid bilayers supported by a quartz surface. The PM-IRRAS data demonstrate that the carbonyl and phosphate groups are more hydrated when the bilayer is adsorbed on the electrode surface at $\sigma_M > -8\,\mu C\,cm^{-2}$ than when it is detached from the electrode at more negative charge densities [40]. NR shows that a significant amount of water enters the polar head region of the lipid bilayer when it is in contact with the metal surface. The detachment of the lipid bilayer from the electrode surface is accompanied by a decrease in the tilt angle of the alkyl chains with respect to the surface normal from about 55° to 35°, with a resulting increase in bilayer thickness. The PM-IRRAS measurements indicate that, as long as the lipid bilayer remains attached to the metal surface, changes in the local electric field by several orders of magnitude have only a small effect on the orientation of the phospholipid molecules [42]. Electrochemical scanning tunneling microscopy images of adsorption of dimyristoylphosphatidylcholine (DMPC) on Au(111) show that the lipid molecules are initially adsorbed flat, with the alkyl chains oriented parallel to the surface

[43]; the resulting ordered monolayer resembles that formed by alkanes. With time, the molecules reorient and the monolayer is transformed into a hemimicellar film. In the presence of a high vesicle concentration in solution, the hemimicellar state is transformed further into a bilayer.

A sBLM formed by fusing vesicles consisting of a mixture of 50 mol% cholesterol and 50 mol% dihexadecyldimethylammonium bromide on boron-doped (p-type) silicon covered with native oxide was investigated by EIS [44]. The positively charged lipid was chosen to favor vesicle spreading on the negatively charged silicon/SiO_2 surface. The impedance spectrum was simulated by an equivalent circuit consisting of two *RC* meshes in series, with the resistance R_Ω of the aqueous solution in series with them. One *RC* mesh simulates the electrode, regarded as a combination of the silicon space-charge region, the SiO_2, and the thin water layer interposed between the semiconductor surface and the lipid bilayer; the other *RC* mesh simulates the lipid bilayer. The resistance and capacitance of the electrode amount to about 19 $M\Omega\,cm^2$ and 2.2 $\mu F\,cm^{-2}$, respectively; those of the lipid bilayer to 0.98 $M\Omega\,cm^2$ and 0.75 $\mu F\,cm^{-2}$, respectively. Incorporation of gramicidin from an aqueous solution decreases the resistance of the lipid bilayer by more than one order of magnitude in the presence of Na^+ ions, and even more in the presence of K^+ ions. This agrees with the cation selectivity scale of gramicidin. A sBLM formed by fusing DMPC–cholesterol vesicles on the optically transparent semiconductor ITO was investigated by EIS [45]. The vesicles also contained 10 mol% of a positively charged lipid to favor their interaction with the negatively charged ITO surface. The impedance spectrum was analyzed by the same equivalent circuit used for the silicon/SiO_2 substrate. Upon incorporating gramicidin, the resistance R_m of the lipid bilayer was strongly reduced in the presence of a Na^+ salt with an organic anion not permeating membranes, whereas its capacitance did not change markedly. Conversely, R_m was practically unaffected in the presence of a Cl^- salt with an organic cation not permeating membranes. This experiment confirms the functional activity and ion selectivity of gramicidin, which is known to be highly selective toward monovalent inorganic cations. The outer membrane proteins OmpF and OmpA from *Escherichia coli* were reconstituted in vesicles that were fused on ITO [45]. In an aqueous solution of NaCl, the resistance of the ITO-supported lipid bilayer was strongly reduced by incorporating OmpF, whereas it was only slightly decreased by incorporating OmpA. Such a behavior was explained by the fact that only OmpF is a pore-forming protein, whereas OmpF does not form pores. Note that the functional activity of OmpF was preserved because, on the cytosolic side of the bacterial membrane, it has no extramembrane domain that might be endangered by direct contact with the ITO surface.

4.5.2 Tethered Bilayer Lipid Membranes

As a rule, tBLMs refer to architectures in which the lipid bilayer is separated from the support through a monomolecular layer tethered to the support via a sulfhydryl or disulfide group (for gold, silver, or mercury supports) or via a silane group (for silica and glass supports). The monolayer interposed between the support surface and the lipid bilayer should have a well-defined composition and geometrical arrangement, as distinct from pBLMs. In general, tBLMs can be classified on the basis of the nature of the molecules composing the tethered monolayer, as follows: (i) spacer-based tBLMs; (ii) thiolipid-based tBLMs; and (iii) thiolipid–spacer-based tBLMs. We will examine these three types separately.

4.5.2.1 **Spacer-Based tBLMs**

The spacer may consist of an alkanethiol, functionalized with a hydrophilic group (e.g., hydroxyl, carboxyl, or amino group) at the opposite end of the alkyl chain with respect to the sulfhydryl group. More frequently, it consists of a thiolated or disulfidated hydrophilic chain. The spacer is tethered to a support via its sulfhydryl or disulfide group and a lipid bilayer is self-assembled on top of it, often by vesicle fusion. The interior of ω-functionalized alkanethiol monolayers is hydrophobic, as opposed to that of thiolated hydrophilic chains. However, even the latter spacers do not necessarily ensure a satisfactory hydration if they are too closed packed. In both cases, a thin water layer, about 1 nm thick, is normally interposed between the spacer monolayer and the lipid bilayer, as in the case of sBLMs; this provides a small ionic reservoir for channel-forming peptides incorporated in the lipid bilayer moiety. Since the lipid bilayer is not covalently linked to the support, it is expected to be sufficiently fluid to accommodate relatively bulky membrane proteins, unless they have large extramembrane domains on both the cytosolic and the extracellular side. The lateral mobility of lipid molecules on a hydrophilic spacer was verified by FRAP in about 10 min on smooth gold surfaces [5], but a sufficient fluidity may be verified at longer times even on rougher surfaces (Becucci *et al.*, unpublished data). The capacitance of the lipid bilayer in these tBLMs is close to $1\,\mu F\,cm^{-2}$, but its resistance assumes relatively low values of the order of $0.5\,M\Omega\,cm^2$.

A tBLM consisting of 3-mercaptopropionic acid (MPA) tethered to gold, with a bilayer of the positively charged lipid dimethyldioctadecylammonium bromide on top, was used to incorporate gramicidin [46]. The lipid bilayer was stabilized by electrostatic interactions with the spacer. To this end, a pH of 8.6 was used to completely deprotonate the MPA monolayer, and the ionic strength of the solution was kept sufficiently low by using 1,1-valent electrolytes of concentration less than 50 mM. The ion-channel activity of gramicidin was verified by EIS. The same tBLM was used to incorporate the 25 kDa *Clavibacter* ion channel, which exhibits anion selectivity [47].

Another procedure for preparing spacer–bilayer assemblies on electrodes consists in anchoring a polyethyleneoxy hydrophilic spacer to a hanging mercury drop electrode via a terminal sulfhydryl group by immersing the mercury drop in an ethanol solution of the spacer for about 20 min [48]. After extracting the spacer-coated mercury drop from the ethanol solution, it is slowly brought into contact with a lipid film previously spread on the surface of an aqueous electrolyte, by taking care to keep the drop neck in contact with the lipid reservoir. This disposition allows a free exchange of lipid material between the lipid reservoir on the surface of the aqueous electrolyte and the spacer-coated drop. This procedure gives rise to the formation of a lipid bilayer in contact with the spacer-coated drop, by exploiting the spontaneous tendency of a lipid film to form a bilayer when interposed between two hydrophilic phases. The fluidity of the tBLM was tested by recording the cyclic voltammogram for the electroreduction of ubiquinone-10 incorporated in the lipid bilayer. The voltage-dependent ion channel activity of melittin incorporated in the tBLM was verified by EIS. A similar mercury-supported tBLM was fabricated by using a thiolated hexapeptide molecule with a high tendency to form a 3_{10}-helical structure as a spacer [49].

4.5.2.2 Thiolipid-Based tBLMs

A thiolipid molecule consists of a hydrophilic polyethyleneoxy or oligopeptide hydrophilic chain terminated at one end with a sulfhydryl or disulfide group for anchoring to the support and covalently linked at the other end to two alkyl chains simulating the hydrocarbon tails of a lipid (Figure 4.1a).

Gold-Supported Thiolipid-Based tBLMs Polyethyleneoxy-based thiolipid monolayers and (polyethyleneoxy-based thiolipid)–phospholipid bilayers tethered to gold have been characterized by SPR, EIS, and cyclic voltammetry, and the synthesis of the thiolipids has been described [50–52]. The cross-sectional area of a hydrophilic polyethyleneoxy chain is smaller than that of the two alkyl chains, if it is in its fully extended conformation, but not if it is coiled. In the former case it is sufficiently hydrated to provide a satisfactory ionic reservoir; conversely, in the latter case, it may accommodate only a limited number of water molecules. Whether the conformation is extended or coiled depends both on the interfacial electric field and on the nature of the metal support. tBLMs fabricated with polyethyleneoxy-based thiolipids exhibit capacitances in the range from 0.5 to 0.7 $\mu F\,cm^{-2}$ and resistances in the range from 5 to 10 $M\Omega\,cm^2$, which are comparable with the values found for conventional BLMs interposed between two aqueous phases; however, somewhat higher capacitance values (1 $\mu F\,cm^{-2}$) have been reported for long spacers [53]. A particularly convenient tBLM of this type is DPTL (see Figure 4.1a). Its cross-sectional area is of about 55 $Å^2$.

"Thiolipopeptides" consist of an oligopeptide chain terminated at one end with a sulfhydryl group and covalently linked at the other end to the polar head of a phospholipid. They are considered to assume a helical structure [54]. In this case, their cross-sectional area amounts to about 75 $Å^2$. They are usually obtained by tethering to a gold electrode a "thiopeptide" consisting of an oligopeptide chain terminated with a sulfhydryl group at one end and with a carboxyl group at the other end. This thiopeptide monolayer is then coupled *in situ* with dimyristoylphosphatidylethanolamine [55]. tBLMs fabricated with thiolipopeptides have been employed for the incorporation of a number of integral proteins; they exhibit high capacitances (from 2 to 10 $\mu F\,cm^{-2}$) and low resistances, of the order of $10^4\,\Omega\,cm^2$, which are three orders of magnitude lower than those of conventional BLMs [21, 56–58]. The low resistance is ascribed to a not perfectly homogeneous coverage of the thiolipid monolayer by the distal lipid monolayer, which is estimated at about 70% of full coverage and makes the tBLMs insufficiently insulating [55]. This is indicative of the presence of an appreciable number of pinholes and other defects in the bilayer, as well as of a certain disorder of the hydrocarbon chains. It was suggested that the thiolipopeptide monolayer actually consists of a mixture of the thiolipopeptide and of the corresponding uncoupled thiopeptide. This conclusion was based on the observation that gold-supported tBLMs formed from a pre-synthesized thiolipopeptide did not permit the functional activity of incorporated proteins [21]. Such a functionality was recovered only by mixing the thiolipopeptide with a hydrophilic thiol, such as mercaptohexanol.

A structural and functional characterization of a DPTL monolayer tethered to gold was recently reported by Vockenroth *et al.* [59] using NR and EIS. The tetraethyleneoxy moiety was found to be only partly hydrated at the more positive potentials. However, at −0.600 V vs. Ag/AgCl (0.1 M KCl) a pronounced increase in the neutron scattering length density of the spacer was observed, denoting an increased amount of water transferred into this region. Leitch *et al.* [60] drew similar conclusions using PM-IRRAS. Thus, the fraction of nonhydrated C=O of the lipoic acid ester group was found to be about 50% at the more positive potentials and to reach a value of about 30% at −0.600 V, which denotes an increasing hydration of the spacer at these negative potentials. Analogous conclusions where also drawn by McGillivray *et al.* [26] by using NR, EIS, and Fourier-transform IRRAS (FT-IRRAS) to investigate a gold-supported thiolipid monolayer similar to DPTL, with a hydrophilic spacer moiety consisting of a hexaethyleneoxy chain directly bound to a sulfhydryl group. FT-IRRAS revealed a significant disorder in the spacer region and a substantial order in the hydrocarbon tail region. Moreover, NR showed that the spacer region had a thickness smaller than its fully extended length and only 5 vol% exchangeable water, despite its significant disorder. Since the incorporation of proteins with extramembrane domains requires a significant hydration of the spacer, the thiolipid monolayer was then diluted with short β-mercaptoethanol molecules. This permitted water molecules to be accommodated in the more spacious thiolipid–β-mercaptoethanol mixture. By self-assembling a lipid monolayer on top of this mixed monolayer, McGillivray *et al.* [26] obtained a tBLM with a differential capacitance comparable to that of conventional BLMs. Moreover, NR data revealed the presence of an appreciable amount of exchangeable water in the spacer moiety of this tBLM.

Thiolipid-based tBLMs, when anchored to solid supports such as gold or silver, do not meet the requirement of fluidity and lateral mobility. The thiolipid molecules are rigidly bound to the metal surface atoms. In principle, the lipid molecules on top of the thiolipid monolayer might be free to move laterally. In practice, however, their lateral mobility is hindered by the presence of adsorbed or hemifused vesicles and by the roughness of the metal support (see Section 4.4.4). Moreover, the hydration of the polyethyleneoxy moiety of thiolipids anchored to gold is low, while the incorporation of proteins with extramembrane domains requires a significant hydration of the spacer. Only small channel-forming peptides and ion carriers can be accommodated in the lipid bilayer moiety of polyethyleneoxy-based tBLMs, via incorporation from their aqueous solutions.

Typical examples are the depsipeptide valinomicin, an ion carrier that complexes a potassium ion with its carbonyl groups and shuttles it across the bilayer, and gramicidin, which spans the bilayer by forming a channel-forming dimer. The resulting impedance spectra are usually fitted by an equivalent circuit consisting of three circuit elements in series: an R_mC_m mesh simulating the lipid bilayer, a capacitance C_s simulating the spacer [53, 61] or (more generally) an *RC* mesh simulating the remaining part of the tBLM [62], and a resistance R_Ω simulating the aqueous solution. In a KCl solution, the incorporation of these ionophores decreases the resistance, R_m, of the lipid bilayer moiety, while, as a rule, the corresponding conductance, C_m, remains practically constant [53, 61]. A gold-supported DPTL–DPhyPC tBLM was also used to incorporate the M2 peptide, a segment of the nicotinic acetylcholine receptor, which spans the lipid bilayer by forming a hydrophilic pore consisting of a bundle of M2 α-helixes [63]. In spite of the lack of lateral mobility and low hydration of these tBLMs, attempts have been made to incorporate the exotoxin α-hemolysin from *Staphylococcus aureus*, a water-soluble, monomeric, 293-residue polypeptide that forms heptameric pores in lipid bilayers [64, 65]. Upon incorporating α-hemolysin, the resistance of the lipid bilayer moiety of a DPTL–DPhyPC tBLM decreases by one order of magnitude [65]. In view of the well-known lysing effect of α-hemolysin on biomembranes, we cannot exclude a similar effect on these tBLMs, with membrane breakdown.

By replacing the lipoic acid residue of DPTL with a trichloropropylsilane group, a supramolecule (DPTTC) was obtained, which was self-assembled on a SiO_2 surface via the silane tether [62]. The SiO_2 layer was very thin (about 0.2 nm), being the native oxide layer of a highly *p*-doped silicon wafer. Consequently, a tBLM obtained by forming a lipid monolayer on top of the DPTTC monolayer tethered to the silicon wafer could be investigated by EIS. A tBLM was also formed on AlO_x sputtered on template-stripped gold [66]. An "anchor lipid" was immobilized on the AlO_x film, about 2.7 nm thick, via L-B transfer. This anchor lipid differs from DPTL by the replacement of the lipoic acid residue with a $-P{=}O(OEt)_2$ group for anchoring to the AlO_x surface. A DPhyPC monolayer was deposited on top of the anchor-lipid monolayer by vesicle fusion. The thin AlO_x film on gold allowed EIS and SPR characterizations of the system. The high capacitance, $10\,\mu F\,cm^{-2}$, of the lipid bilayer suggests that the tBLM adheres to some extent to the pores of the AlO_x layer.

Gold-supported thiolipopeptide-based tBLMs have been reported to incorporate a few bulky proton pumps. Thus, Naumann *et al.* incorporated the proton pumps ATPase CF_0F_1 from chloroplasts and ATPase EF_0F_1 from *Escherichia coli* in a tBLM fabricated with a hydrophilic oligopeptide spacer [21, 56, 57]. Activation of these ATPases by ATP causes an increase in the reduction peak for hydrogen evolution, which takes place at −0.7 V vs. Ag/AgCl (sat. KCl) on gold. This increase was ascribed to proton pumping into the hydrophilic oligopeptide spacer; it is suppressed by tentoxin, a specific inhibitor of the ATPase from chloroplasts. Another proton pump, cytochrome c oxidase (COX), was incorporated in an analogous gold-supported, thiolipopeptide-based tBLM [58]. COX is the terminal component of the respiratory electron transport chain and spans the inner mitochondrial membrane. It catalyzes the redox reaction between the peripheral protein ferrocytochrome c and oxygen, with formation of ferricytochrome c and water; it also pumps protons from the mitochondrion matrix to the intermembrane space. Increasing additions of ferrocytochrome c cause a progressive decrease of the reduction peak for hydrogen evolution; this decrease was ascribed to a decrease in the proton concentration within the thiopeptide spacer, due to proton pumping from the spacer to the aqueous solution. Addition of the inhibitor cyanide eliminates the effect of ferrocytochrome c. The above conclusions as to proton pump incorporation in thiolipopeptide-based tBLMs should be examined in the light of the observation that vesicles have a low propensity to fuse on the hydrophobic surface exposed to the aqueous solution by a thiolipid monolayer, especially if they incorporate an integral protein; rather, they are adsorbed or partially fused (see Section 4.4.1). Incorporation of the above proton pumps from their solutions in detergent may easily take place in the membrane of adsorbed or partially fused vesicles, since the vesicular membrane is clearly interposed between two aqueous phases. In this respect, the functional activity of the above proton pumps might well be due to vesicles or proteoliposomes adsorbed or partially fused on a thiolipid monolayer, without the need of protein incorporation in the lipid bilayer moiety of the tBLM. In fact, their activation would cause an increase in the proton concentration on top of the thiolipid monolayer (in the case of F_0F_1 ATPase activated by ATP) or its decrease (in the case of COX activated by ferrocytochrome c). In view of the relative permeability of the leaky thiolipopeptide monolayers to protons, this would determine an increase or a decrease in the proton electroreduction current on gold, as actually observed.

A tBLM consisting of a gold-supported thiolipopeptide, with a soybean-PC monolayer on top, was used to incorporate the odorant receptor OR5 from *Rattus norvegicus* during its *in vitro* synthesis [67, 68]. The vectorial insertion of the protein into the tBLM in a functional and oriented form was verified. A gold-supported thiolipopeptide-based tBLM was also employed to incorporate the acetylcholine receptor (AChR), a ligand-gated channel protein present in the postsynaptic membrane of muscle cells [69]. Addition of the neurotoxin α-bungarotoxin peptide causes a slow but appreciable increase in the thickness of the tBLM incorporating AChR, thus denoting the binding of this toxin to the binding sites of AChR. The inhibitory effect of α-bungarotoxin on the ion-channel activity of AchR was not verified.

Gold-supported DPTL–DPhyPC tBLMs on microchips have been employed to measure single-channel currents of peptides and proteins. To this end, a microelectrode array device consisting of many ($100 \times 100\,\mu m^2$) "sensor" pads was employed (Figure 4.6) [70–72]. A DPTL monolayer was tethered to the gold-coated sensor pads from a DPTL solution in ethanol; then, a lipid monolayer was formed on top of it by vesicle fusion. In view of the very small surface area of the pad, the resistance of the resulting tBLM ranged from 1.5 to $17\,G\Omega$. This resistance was high enough to reduce the background electrical noise to the low level required for the use of the patch-clamp technique. This device allowed the recording of single-channel currents of gramicidin A [70], the high-conducting Ca^{2+}-

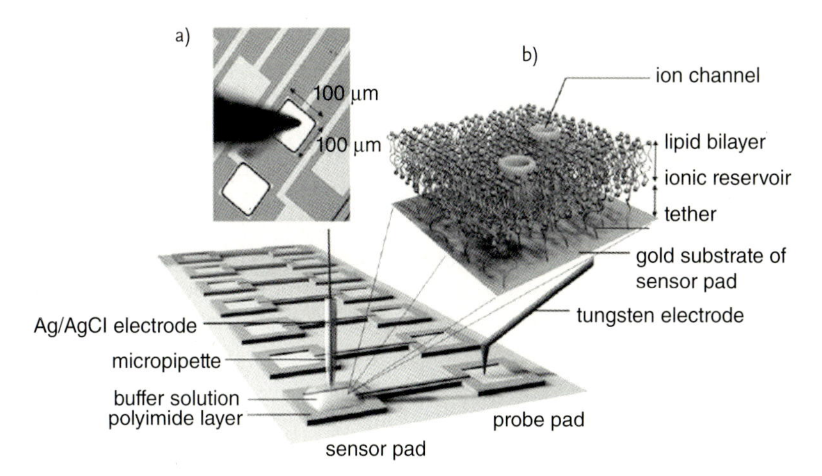

Figure 4.6 Tethered bilayer membrane array. (a) An optical microscope image of the probe pad and the tungsten electrode tip. (b) Schematic of the tethered bilayer membrane array. The lower left corner shows the gold sensor pad covered with a tBLM that incorporates ion channels. An Ag/AgCl electrode situated in a patch micropipette was used as an electrode and was inserted into the buffer drop. The gold substrate of the sensor pad was connected by a thin gold line to the probe pad onto which a tungsten tip was lowered as a counter electrode. The inset shows the tBLM formed at the gold surface of the sensor pad. Reprinted from [72] with kind permission from Elsevier.

activated K^+ (BK or Maxi-K) channel, the synthetic M2δ ion channel [71], and the mechanosensitive channel of large conductance from *Escherichia coli* [72]. All these peptides and proteins were incorporated into lipid vesicles, before fusing them onto the DPTL-coated sensor pad. With the exclusion of the gramicidin channel, the unitary conductance of the remaining ion channels was found to be from one-third to one-tenth that obtained with conventional BLMs. In this connection one cannot exclude, even in this case, the possibility that the channels responsible for single-channel currents be located in the membrane of adsorbed or partially fused vesicles. In fact, in this case, the capacitive coupling between the vesicular membrane and the tBLM is expected to decrease the unitary conductance of the channels. The capacitance and resistance of DPTL–DPhyPC tBLMs on gold microelectrodes of circular shape, with diameters ranging from 4000 to 8 μm, were examined as a function of the electrode size [73]. For the larger electrodes, the capacitance is directly proportional to the electrode area, while the resistance is inversely proportional to it. For the smaller electrodes, the capacitance decreases linearly and the resistance increases linearly with a decrease in the electrode diameter. This indicates that the capacitance and resistance of the larger electrodes are dominated by the electrode area, while the circumference seems to have the dominant role at smaller electrodes. Consequently, for small electrode sizes, a slight disorder of the bilayer structure at the edge of the electrode has a major influence on the electrical properties of the membrane. This conclusion is supported by the consideration that the micro-tBLMs used for the recording of single-channel currents have resistances ranging from 1.5 to 15 GΩ. These resistances are high enough to reduce the level of the background noise down to the range of a few picoamps. However, the corresponding specific resistances are much less than those attained with identical tBLMs anchored to macroscopic gold electrodes of areas of the order of $10^{-2}\,cm^2$. In fact, the specific resistance of the latter tBLMs amounts to about $10\,M\Omega\,cm^2$. If a $100 \times 100\,\mu m^2$ micro-tBLM had such a specific resistance, it should have a resistance of 100 GΩ. The loose packing of micro-tBLMs may explain why they may incorporate relatively bulky proteins much more easily than macro-tBLMs, without having to dilute the thiolipid with a short spacer.

Mercury-Supported Thiolipid-Based tBLMs As distinct from gold-supported thiolipid-based tBLMs, mercury-supported thiolipid-based tBLMs do not require the use of thiolipid–spacer mixtures to incorporate channel-forming proteins. Because of the fluidity imparted to the thiolipid monolayer by the liquid mercury surface, these tBLMs may incorporate bulky proteins, such as OmpF porin from *Escherichia coli* [74] and the HERG potassium channel [75], in a functionally active state. Upon incorporating gramicidin [76] or valinomicin [2], the tetraethyleneoxy (TEO) moiety of DPTL in aqueous KCl solution undergoes a conformational change ascribable to its elongation, as the applied potential is stepped from a fixed initial value of −0.200 V vs. SCE to a final value of −0.500 V vs. SCE [2]. As the final value of this potential step becomes progressively more negative, the charge of K^+ ions accommodated in the TEO spacer increases rapidly, attaining a maximum limiting value of about $45\,\mu C\,cm^{-2}$ at −0.8 V vs. SCE [77]. This corresponds to three potassium ions per DPTL molecule, denoting an appreciable hydration of the spacer. Moreover, EIS measurements of the surface dipole potential of the TEO spacer tethered to mercury yield values that compare favorably with the dipole moment of TEO molecules measured in organic solvents [78]; this suggests a substantial ordering of the TEO chains in mercury-supported tBLMs. Incidentally, tBLMs supported by gold [21, 57] or silver [79] do not allow ionic charge measurements carried out by stepping the applied potential to final values negative of about −0.650 V. In fact, the resulting charge versus time curves show a linear section with a relatively high and constant slope that is maintained for an indefinitely long time [21, 57, 79]. The constant current responsible for this linear increase in charge is ascribed to a slight water electroreduction with hydrogen formation. The high hydrogen overpotential of mercury avoids this inconvenience. When comparing interfacial phenomena on different metals, rational potentials should be used, namely potentials referred to the potential of zero charge (PZC) of the given metal in contact with a nonspecifically adsorbed 1,1-valent electrolyte. The PZC equals −0.435 V vs. SCE for mercury [80] and −0.040 V vs. SCE for polycrystalline gold [81]. Therefore, an appreciable hydration of the TEO moiety, possibly accompanied by its elongation [2], takes place in the proximity of a rational potential of about zero on mercury, but at a much more negative rational potential of about −0.600 V on gold [59, 60], close to the DPTL desorption from this metal. A drawback of the use of mercury-supported tBLMs is represented by the notable difficulty in using surface-sensitive techniques for their structural characterization.

With respect to solid metal supports, mercury has the advantage of providing a defect-free, fluid, and readily renewable surface to the self-assembling thiolipid–lipid bilayer. Moreover, it imparts lateral mobility to the whole mixed bilayer. In addition, the self-assembly of a lipid monolayer on top of a thiolipid monolayer is readily carried out by simply immersing a thiolipid-coated mercury drop in an aqueous electrolyte on whose surface a lipid film has been previously spread [2]. Because of the hydrophobic interactions between the alkyl chains of the thiolipid and those of the lipid, this simple procedure gives rise to a lipid bilayer anchored to the mercury surface via the hydrophilic spacer moiety of the thiolipid. By avoiding the use of vesicles, this procedure excludes any artifacts due to partially fused vesicles. These advantageous features make the incorporation of membrane proteins in mercury-supported thiolipid-based tBLMs easier and safer than in solid-supported tBLMs.

The impedance spectra of mercury-supported DPTL–DPhyPC tBLMs incorporating the ion carrier valinomycin [2], the channel-forming peptides melittin [82] and gramicidin [76], and the channel protein OmpF porin [74] were fitted by four *RC* meshes in series. Fitting was particularly straightforward for the spectrum of the tBLM incorporating valinomicin, where four partially overlapping semicircles are clearly distinguishable in the M plot (see Figure 4.2). The *R* and *C* values relative to the four *RC* meshes vary appreciably with the applied potential. To this end, a generic approximate approach was developed, which applies the concepts of impedance spectroscopy to a model of the electrified interface and to the kinetics of potassium ion transport assisted by valinomycin across the tBLM [2]. This permits the four *RC* meshes to be ascribed to four different slabs composing the tBLM. The first semicircle has the highest resistance and was ascribed to the lipoic acid residue, in direct contact with the electrode surface. The second semicircle has a capacitance of about $7\,\mu F\,cm^{-2}$, close to that of a monolayer of tetraethyleneoxythiol self-assembled on mercury, and was ascribed to the TEO moiety. The third semicircle in the absence of valinomycin has a capacitance close to $1\,\mu F\,cm^{-2}$, a value typical of a conventional solvent-free BLM, and was reasonably ascribed to the lipid bilayer moiety. Finally, the last semicircle has the same resistance as the electrolyte solution and a very low capacitance of the order of $1\,nF\,cm^{-2}$, and was ascribed to the electrolyte solution adjacent to the tBLM.

The sigmoidal charge versus time curves following a potential step from a value at which melittin channels are not formed to one at which they are formed were interpreted on the basis of a generic kinetic model [83]. This model accounts for the potential-independent disruption of melittin clusters adsorbed flat on the lipid bilayer, induced by the potential-dependent penetration of the resulting monomers into the lipid bilayer; the potential-independent aggregation of the monomers inside the bilayer with channel formation is then treated on the basis of a mechanism of nucleation and growth. This explains the initial induction period responsible for the sigmoidal shape of charge versus time curves. Gramicidin is a peculiar channel-forming peptide whose helical structure differs from the α-helix of common peptides and membrane proteins by the fact that its lumen is large enough to allow the passage of simple desolvated monovalent cations. Since its length is about one-half that of a biomembrane, it spans it by forming a dimeric channel. The N-terminuses of the two monomeric units are directed toward each other, in the middle of the lipid bilayer, just as their dipole moments. A transmembrane potential different from zero is, therefore, expected to favor electrostatically one monomeric orientation at the expense of the other, thus destabilizing the dimer. Nonetheless, the stationary current due to the flow of potassium ions along gramicidin channels incorporated in a mercury-supported tBLM increases with an increase in the transmembrane potential, exhibiting an almost quadratic dependence [76]. To explain this behavior, it was assumed that the rate constant for dimer formation increases in parallel with an increase in the ionic flux. In fact, when the time elapsed between the passage of two consecutive cations through the junction between the two monomers forming the conducting dimer starts to become comparable with, and ultimately shorter than, the time required for the dissociation of the two monomers, such a dissociation becomes increasingly less probable.

The HERG K^+ channel is present in the plasma membrane and consists of four identical subunits, each spanning the membrane with six α-helixes [84]. Upon reconstituting the HERG K^+ channel in the tBLM, the potential difference across the lipid bilayer moiety (i.e., the transmembrane potential) was caused to pass from a value more positive than the resting potential of a plasma membrane to a value more negative, by choosing the appropriate potential step [75]. This step causes the channel to pass from an inactive to an open state, which lasts for about 100 ms, followed by a closed state. To separate the "inward" current, which flows during the open state, from the very high capacitive current, the current recorded after blocking the HERG channel with a specific inhibitor was subtracted from that previously recorded in the absence of the inhibitor. The inward current decays monoexponentially in time, with a time constant in good agreement with that obtained by the patch-clamp technique with conventional BLMs.

Sarcolipin (SLN) and phospholamban (PLN) are two small membrane proteins that modulate the function of Ca-ATPase of the sarcoplasmic reticulum. They were incorporated in a mercury-supported DPTL–DPhyPC tBLM to verify whether they may form ion-selective pores in biomembranes [85, 86]. SLN consists of a single α-helix that spans the whole membrane. By measuring the conductance of a tBLM incorporating SLN by EIS, it was shown that SLN forms channels highly selective toward small inorganic anions, such as chloride ions [85]. Incorporation of SLN in a tBLM immersed in an aqueous solution (pH = 5.3) of 0.05 M NaH_2PO_4 causes only a slight decrease in the resistance of the lipid bilayer, as measured by the diameter of a Z'' versus Z' plot (Nyquist plot; Section 4.3); however, submicromolar additions of ATP decrease the resistance of the tBLM to an appreciable extent, as shown in Figure 4.7 [86]. Plotting the conductance against the ATP concentration yields a curve that tends asymptotically to a limiting value and can be satisfactorily fitted by the Michaelis–Menten equation, with an association constant for the SLN–ATP complex of about 0.1 μM (inset of Figure 4.7). This behavior was

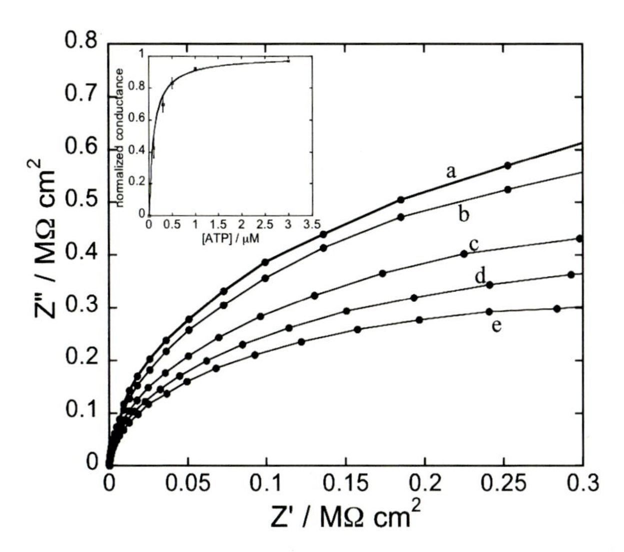

Figure 4.7 Filled circles are experimental points on a Z'' versus Z' plot for a tBLM in an aqueous solution of 0.05 M NaH_2PO_4 (pH = 5.3) at −0.500 V vs. Ag/AgCl (0.1 M KCl) (a) in the absence of SLN, (b) after incorporation of SLN from its 0.7 μM solution, and after subsequent additions of (c) 0.1, (d) 0.3, and (e) 3 μM ATP. The solid curves are fits of the solid circles by a series of four *RC* meshes. The values of the resistance, R_m, of the lipid bilayer moiety resulting from the fits are (a) 0.90, (b) 0.70, (c) 0.53, (d) 0.42, and (e) 0.35 MΩ cm². The corresponding C_m values are all close to 1 μF cm^{-2}. The inset shows conductance, $1/R_m$, normalized to its maximum value equated to unity, as a function of the ATP concentration. The error bars denote standard deviations. The solid curve is a fit of the experimental points by the Michaelis–Menten equation. Reprinted from [86] with kind permission from Elsevier.

explained by assuming that SLN forms a hydrophilic pore consisting of a bundle of four or five SLN α-helixes that turn their hydrophilic side, containing two hydrophilic threonine residues, toward the interior of the bundle. On the basis of this and other features of SLN, it was proposed that SLN is just the "P_i transporter" described by Lee and coworkers [87] in 1991 and whose nature was not known up to now. PLN exists in equilibrium between the pentameric and monomeric forms. The pentamer releases to Ca-APTase a monomer, which forms a 1 : 1 complex inhibiting the affinity of Ca-ATPase for Ca^{2+} [88]. Phosphorylation of the monomer by ATP removes the inhibition. Both SLN and the PLN monomer have a single α-helix that spans the membrane. The main difference between the SLN and PLN helixes is represented by the absence of two hydrophilic threonines in the PLN helix. Incorporation of PLN in a mercury-supported DPTL–DPhyPC tBLM does not affect its resistance [89]. It was concluded that the absence of the two threonine residues prevents the PLN pentamer from forming a hydrophilic pore.

A mercury-supported tBLM was formed at the tip of a microelectrode for measuring single-channel activity [90]. To this end, use was made of a platinum wire embedded in a thin glass capillary and terminated with a platinum microdisc,

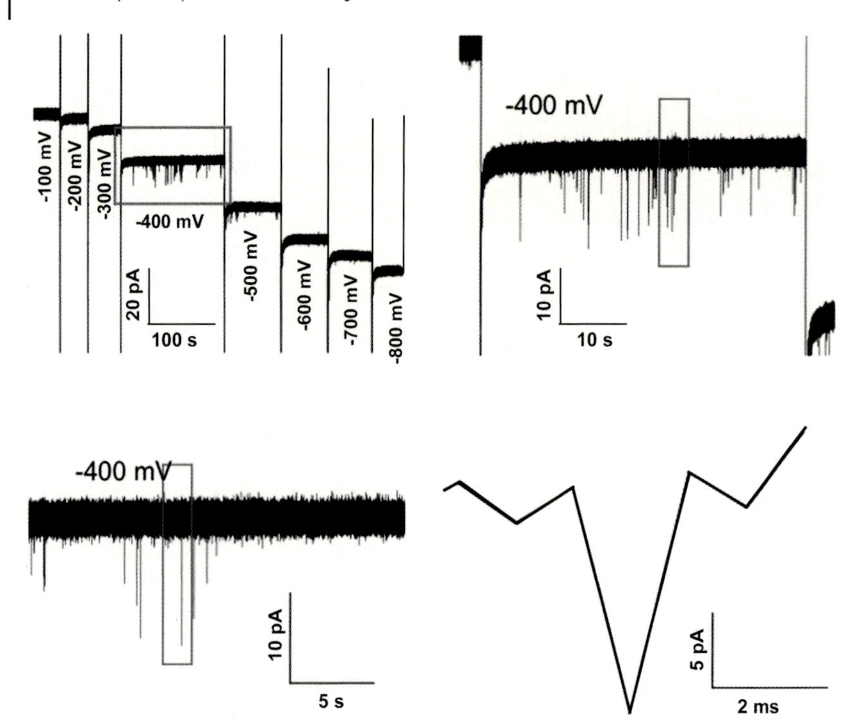

Figure 4.8 Single-channel traces of OmpF porin from *Escherichia coli* incorporated in a mercury-supported tethered bilayer lipid micromembrane immersed in aqueous 0.1 M KCl, at different applied potentials measured vs. a Ag/AgCl (3 M KCl) reference electrode. Potentials must be decreased by 50 mV to refer them to the SCE. Reprinted from [90] with kind permission from Elsevier.

about 20 μm in diameter. Mercury was electrodeposited on this microdisc from a mercurous nitrate aqueous solution (pH = 1), giving rise to a mercury spherical cap that was coated with a DPTL–DPhyPC bilayer. The high resistance of this "tethered bilayer lipid micromembrane," about 5 GΩ, allows the recording of single-channel currents by the patch-clamp technique. Figure 4.8 shows the single-channel currents due to the opening of the OmpF porin channel from *Escherichia coli*, reconstituted in this micromembrane, as a function of time at different applied potentials. A burst of single-channel currents is observed at −400 mV vs. SCE, and to a minor extent at −500 mV, which corresponds to a zero transmembrane potential. This behavior is consistent with a well-known property of OmpF porin, according to which this channel protein undergoes inactivation on both sides of the zero transmembrane potential.

This micromembrane was also used to investigate the spontaneous formation of microdomains, when the distal lipid monolayer is made up of a lipid mixture. Microdomains are in the gel state when they consist primarily of glycolipids and sphingolipids, in a "liquid-ordered" state (so-called "lipid rafts") when they also contain cholesterol, and in a "liquid-disordered" state when they consist primarily

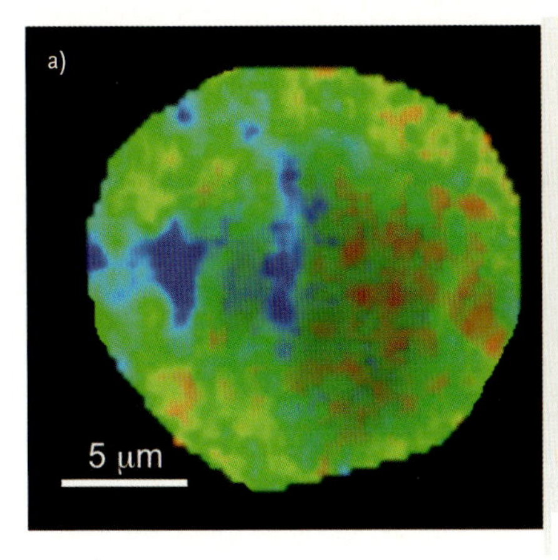

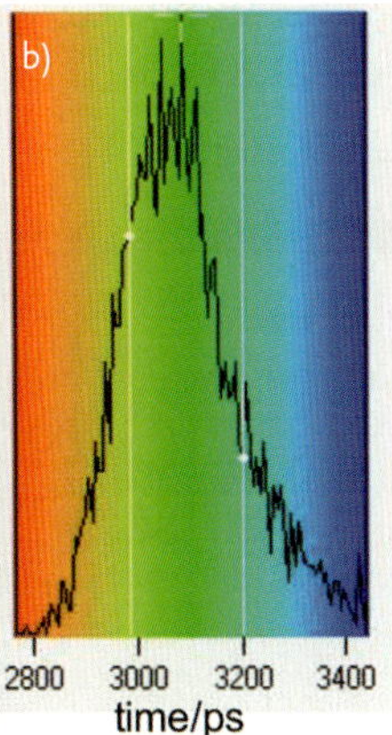

Figure 4.9 (a) Two-photon (excitation wavelength = 760 nm) fluorescence lifetime 21 × 21 μm image and (b) fluorescence lifetime distribution histogram of a distal monolayer of dioleoylphosphatidylcholine–palmitoylsphingomyelin–cholesterol (47 : 47 : 6) mixture, labeled with 1 mol% laurdan, at room temperature. The mixture is the distal monolayer of the micromembrane. The color code in the fluorescence lifetime imaging microscopy image is that indicated in the histogram. Reprinted from [30] with kind permission from the Royal Chemical Society.

of unsaturated phosphatidylcholines and a small percentage of cholesterol [91]. Rafts are receiving increasing attention since they are considered to regulate the membrane function in eukaryotic cells. Figure 4.9 shows a two-photon fluorescence lifetime image of a distal monolayer consisting of a palmitoylsphingomyelin–dioleoylphosphatidylcholine–cholesterol (47 : 47 : 6) mixture, with the addition of 1 mol% of the fluorophore laurdan [30]. The bluish microdomains with an irregular percolative-like shape and lifetimes of 3450 ± 50 ps are ascribed to the gel phase. The small roundish orange microdomains immersed in the surrounding green matrix have lifetimes of 2830 ± 50 ps and are attributed to the liquid-ordered phase. Finally, the matrix, with intermediate fluorescence lifetimes, is ascribed to the coexistence of the liquid-disordered phase and of liquid-ordered microdomains of size below the resolution of the microscope.

4.5.2.3 **Thiolipid–Spacer-Based tBLMs**

A mixture of a short hydrophilic spacer anchored to gold via a sulfhydryl or disulfide group (e.g., mercaptoethanol or dithiodiglycolic acid) and of a thiolipid with its terminal hydrophobic group (e.g., a cholesteryl [8, 9] or a phytanyl group [17]) capable of pinning a lipid bilayer gives rise to a practically aqueous region sandwiched between the short spacer and the lipid bilayer (Figure 4.10); this may favor the accommodation of the extramembrane domains of relatively bulky membrane proteins. Deposition of a lipid bilayer onto such a mixture of a short spacer and of a thiolipid is usually carried out by vesicle fusion or by rapid solution exchange. The resulting lipid bilayer has a capacitance of about 0.5–0.6 $\mu F\,cm^{-2}$

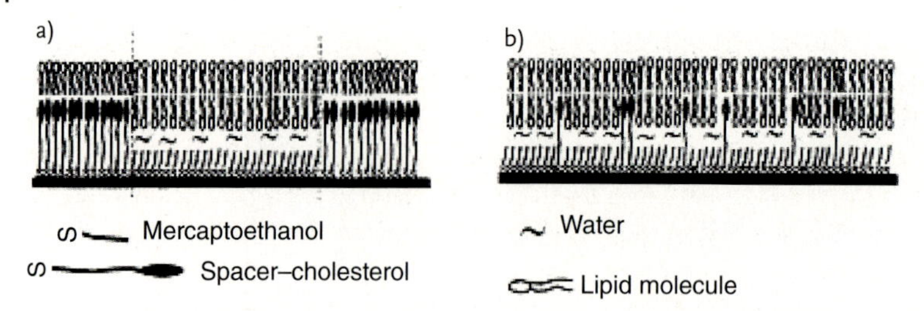

Figure 4.10 (a) Schematic of a micropatterned film consisting of distinct areas covered by a mercaptoethanol monolayer with a lipid bilayer on top, and by a monolayer of a hydrophilic spacer covalently bound to a cholesteryl group with a lipid monolayer on top. (b) Schematic of a mixed monolayer of mercaptoethanol and spacer–cholesterol molecules with a lipid bilayer on top.

and a resistance higher than 5 MΩ cm². These values compare favorably with those of conventional BLMs. Ion-channel activity of gramicidin and melittin incorporated in such a gold-supported tBLM and ion-selectivity of gramicidin were verified by EIS [17]. However, here too, the lateral mobility of the lipid bilayer is hindered by the hydrophobic group of the thiolipid molecules, which provides a fraction of the proximal leaflet of the lipid bilayer.

The ability of the volume enclosed between the gold electrode and the lipid bilayer moiety of thiolipid–spacer-based tBLMs to accommodate water molecules and inorganic ions was estimated from the level of conductance induced by the incorporation of a given amount of the ionophore valinomycin. This ionic reservoir was simulated by a capacitance C_s and the lipid bilayer by an $R_m C_m$ mesh. By using a number of thiolipids and spacers of variable length, Cornell and coworkers [25, 92] found that the conductance, $1/R_m$, of the lipid bilayer increases with an increase in the length of the thiolipid and with a decrease in the length of the spacer.

Relatively bulky membrane proteins with extramembrane domains can be accommodated in the bilayer if the areas covered by the short thiolated spacer are sufficiently large. Differences in chain length [93] and chemistry [94] between thiolipids and spacers may hopefully favor the formation of phase-demixed domains on the nanoscale, facilitating protein incorporation. This seems to be the case with mixed monolayers of cholesteryl-based thiolipids and short thioalcohol spacers on gold [4, 8, 95]. A micropatterned layer, consisting of regions of thiolipid molecules with a lipid monolayer on top alternated with regions of mercaptoundecanoic acid with a lipid bilayer on top, was used to incorporate rhodopsin [96, 97], an integral protein that contributes to the closing of the Na^+ channels of the plasma membrane of the outer segment of vertebrate rod cells.

4.5.3
Polymer-Cushioned Bilayer Lipid Membranes

The spaciousness of the ionic reservoir of tBLMs may not be sufficient to accommodate bulky extramembrane domains of membrane proteins. The problem is particularly serious with cell-adhesion receptors, whose functional extramembrane domains can extend to several tens of nanometers. This problem can be circumvented by separating the lipid bilayer from the solid substrate using soft polymeric materials of typically less than 100 nm thickness, which rest on the substrate and support the bilayer. These stratified films are often referred to as polymer-cushioned or polymer-supported bilayer lipid membranes (for a review, see [98]). This approach reduces the nonspecific binding of proteins to the solid support and the frictional coupling between proteins and the support, preventing the risk of protein denaturation due to direct contact between protein subunits and the bare support surface. In some cases, the cushion may assist self-healing of local defects in lipid bilayers deposited on macroscopically large supports.

To form a thermodynamically stable polymer–lipid composite film on a solid support in aqueous solution and to avoid the formation of polymer blisters, several requirements should be fulfilled. The interaction between the lipid bilayer and the substrate surface must be repulsive. In fact, if the net force acting per unit area (disjoining pressure) is negative, continuous thinning of the interlayer results in film collapse, that is, dewetting, giving rise to regions of tight local contact between the lipid bilayer and the substrate surface (so-called pinning centers). Polymer-cushioned membranes are fairly unstable if the attractive interfacial forces between the polymer and the lipid bilayer are relatively weak. In this case, the bilayer can easily detach from the polymer cushion. On the other hand, attractive forces that are too strong may decrease the lateral mobility of the bilayer. A compromise should therefore be found between a sufficient stability of the polymer–lipid interface and lateral mobility. A possible strategy consists of enhancing the stability of polymer-cushioned membranes via attractive electrostatic interactions [99]. A different approach to stabilization of polymer-cushioned membranes was adopted by Naumann *et al.* [100] and by Wagner and Tamm [101] by tethering the polymer both to the substrate and to the membrane.

Usually, polymer cushions are anchored to supports such as glass, silica, and mica using polymers derivatized with alkyl silanes [12] or triethoxysilane for covalent linkage to silanols at the surface of silicate substrates [102]. Polymer-cushioned lipid bilayers on conducting supports have been investigated only rarely. Spinke *et al.* [103] and Erdelen *et al.* [104] described a polymer-supported lipid bilayer anchored to gold. These authors used a methacrylic terpolymer consisting of a hydrophilic main chain that acts as a spacer, a disulfide unit that anchors the polymer to the gold surface, and a hydrophobic lipid-like part that forms a first lipid monolayer upon self-assembly. A second lipid monolayer was formed on top of the first by fusion with DMPC vesicles. Polymer-cushioned lipid bilayers have frequently been investigated on the ITO semiconductor. In the case of a lipid bilayer deposited on a regenerated (and thus hydrophilic) cellulose cushion, the swollen polymer film behaves like an aqueous electrolyte. Selective ion transport via ion channels and carrier proteins incorporated in the membrane was quantitatively evaluated by determining the electric resistance of the membrane [105].

4.5.4
S-Layer Stabilized Bilayer Lipid Membranes

One of the common surface structures of archea and bacteria are monomolecular crystalline arrays of protein subunits, called S-layers [106–109]. They constitute the outermost component of the cell envelope of these procaryotic organisms. S-layer subunits can be aligned in lattices with oblique, square, or hexagonal symmetry. Since S-layers are monomolecular assemblies of identical protein subunits, they exhibit pores of identical size and morphology. A group of nonclassical cell wall polymers, called "secondary cell wall polymers" (SCWPs), are attached noncovalently, presumably by a lectin-type interaction, to the S-layer proteins.

Since S-layer subunits of most bacteria interact with each other through non-covalent forces, they can be set free with high concentrations of agents that break hydrogen bonds, such as guanidine hydrochloride or urea. Once the S-layer lattice of a bacterial cell is completely disintegrated and the disintegrating agent is removed by dialysis, the S-layer subunits have the unique ability to reassemble spontaneously in suspension, at the liquid–air interface, on solid surfaces, on spread lipid monolayers, and on liposomes. Recrystallization starts at several distant nucleation points on the surface and proceeds until neighboring crystalline areas meet. In this way, a closed mosaic of differently oriented monocrystalline domains is formed. Recrystallization of isolated S-layer proteins on differently charged solid supports reveals an electroneutral outer S-layer surface and a net negative or net positive inner S-layer surface.

The natural tendency of S-layers to interact with membranes has been exploited to insert them as an intermediate layer between a lipid bilayer and a substrate, giving rise to the so-called ssBLMs. Thus, in the case of bacterial S-layer proteins, it has been demonstrated that protein domains or functional groups of the S-layer lattice interact via electrostatic forces with some head groups of lipid molecules. In addition, the affinity of S-layer proteins for the corresponding SCWPs, which are recognized as specific binding sites, can be used to fabricate complex architectures (Figure 4.11). Thus, a monolayer of SCWPs, suitably thiolated at one end, can be anchored to a gold support, and an S-layer can be spontaneously recrystallized on top of it; finally, a lipid bilayer can be self-assembled on top of the S-layer. Optionally, the lipid bilayer so formed can be further stabilized by recrystallizing an additional S-layer on top of it. Alternatively, a loose monolayer of lipidated SCWP molecules can be bound to an S-layer previously recrystallized on a substrate; a lipid bilayer self-assembled on top of the SCWP monolayer is then firmly anchored by the lipid moieties of the SCWPs, which penetrate the inner lipid leaflet. Optionally, the lipidated SCWP molecules present on the outer lipid leaflet can be bound to a further S-layer. All these architectures have a stabilizing effect on the associated lipid bilayer, leading to an improvement in its lifetime and robustness. The assembly of S-layer structures from solution to a solid substrate, such as a gold-coated glass slide, can be followed using SPR or a QCM-D [106]. The self-assembling process is completed after approximately 45 min. The mass

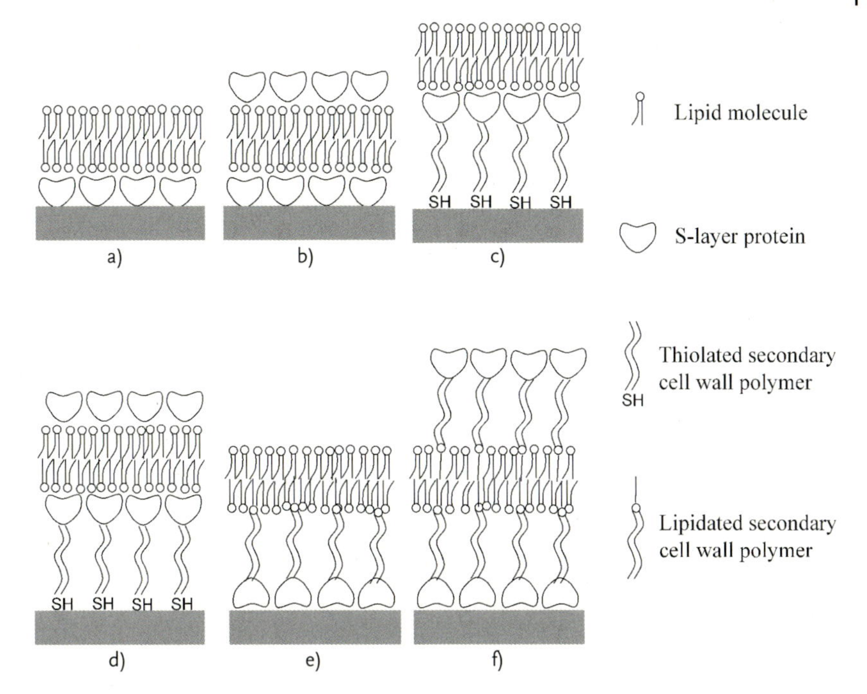

Figure 4.11 Schematic of S-layer stabilized solid supported lipid membranes. (a) S-layer directly recrystallized on gold, with a lipid bilayer on top. (b) Same as (a), with an additional S-layer recrystallized on top of the lipid bilayer. (c) Thiolated SCWPs directly bound to gold and interacting with an S-layer, with a lipid bilayer on top. (d) Same as (c), with an additional S-layer recrystallized on top of the lipid bilayer. (e) S-layer directly recrystallized on gold and interacting with lipidated SCWPs, which anchor a lipid bilayer. (f) Same as (e), with lipidated SCWPs inserted into the distal lipid monolayer and interacting with an additional S-layer.

increase followed using a QCM-D corresponds to a thickness of about 8–9 nm, in agreement with the value estimated by SPR.

A well-characterized S-layer protein, SbqA from *Bacillus sphaericus* CCM 2177, was used as an ultrathin crystalline water-containing hydrophilic layer between a gold electrode and a lipid bilayer [107]. The SbqA protein recrystallizes in monomolecular square lattices with the neutral outer surface exposed to the aqueous phase and the negatively charged inner surface attached to the gold electrode. Membrane resistances of up to 80 MΩ cm^2 were observed for DPhyPC bilayers on SbpA. In addition, lipid bilayers supported by SbpA exhibited a long-term robustness of up to two days. Upon incorporating the voltage-gated ion channel alamethicin at open circuit, the membrane resistance dropped from about 80 MΩ cm^2 to about 950 Ω cm^2, whereas the capacitance did not change.

4.5.5
Protein-Tethered Bilayer Lipid Membranes

In all the biomimetic membranes previously described and allowing the incorporation of proteins, the protein orientation in the membrane is purely casual. At most, if one of the two extremembrane domains of the protein is much bulkier than the other, incorporation in a tBLM occurs preferentially with the bulkier domain turned toward the aqueous phase, in view of the limited spaciousness of the hydrophilic moiety of the tBLM. Moreover, the packing density of the reconstituted proteins in the lipid bilayer is not well controlled. The need for a well-defined protein orientation with respect to the electrode surface is particularly felt with redox membrane proteins, in which the electrons involved in a chain of redox couples are conveyed across the membrane in a well-defined direction.

To overcome this problem, Naumann and coworkers have developed a novel methodology based on tethering proteins, rather than lipids, to electrode surfaces; the lipids are then allowed to self-assemble around the tethered proteins [110]. To this end, a recombinant membrane protein is engineered to bear a stretch of six consecutive histidine residues. A gold surface is then functionalized by attaching a molecule terminated with a nitrilotriacetic (NTA) moiety at one end and with a sulfhydryl group for anchoring to gold at the other end. Complexation of Ni^{2+} ions to both the NTA functionality and the histidines of the stretch causes the protein to be anchored to the gold surface from its solution in detergent, as shown in Figure 4.12. To retain full functional integrity, the membrane protein is incorporated into a lipid bilayer. For this purpose, the protein layer tethered to gold is mixed with detergent-destabilized lipid vesicles of DMPC. By removing the detergent with microporous biobeads, the tethered proteins are surrounded by lipid molecules that form a lipid bilayer around them, as verified by SPR and EIS; a water layer remains interposed between the lipid bilayer and the NTA moiety, acting as an ionic reservoir. At low surface densities of the redox protein, the bilayer does not effectively form, and protein aggregates are observed; on the other hand, at very high surface densities, very little lipid is able to intrude between the closely packed protein molecules [111]. In both cases, redox activity is low. Redox activity is preserved in the biomimetic membrane only at moderate surface coverages, in which a continuous lipid bilayer is present and the protein molecules are not forced to aggregate.

This approach has been adopted to investigate the function of COX from the proteobacterium *Rhodobacter sphaeroides* [110], the last enzyme in the respiratory electron transport chain of bacteria, located in the bacterial inner membrane. It receives one electron from each of four ferrocytochrome c molecules, located on the periplasmic side of the membrane, and transfers them to one oxygen molecule, converting it into two water molecules. In the process, it binds four protons from the cytoplasm to make water, and in addition translocates four protons from the cytoplasm to the periplasm, to establish a proton electrochemical potential difference across the membrane. In this ptBLM, the orientation of the protein with respect to the membrane normal depends on the location of the histidine stretch

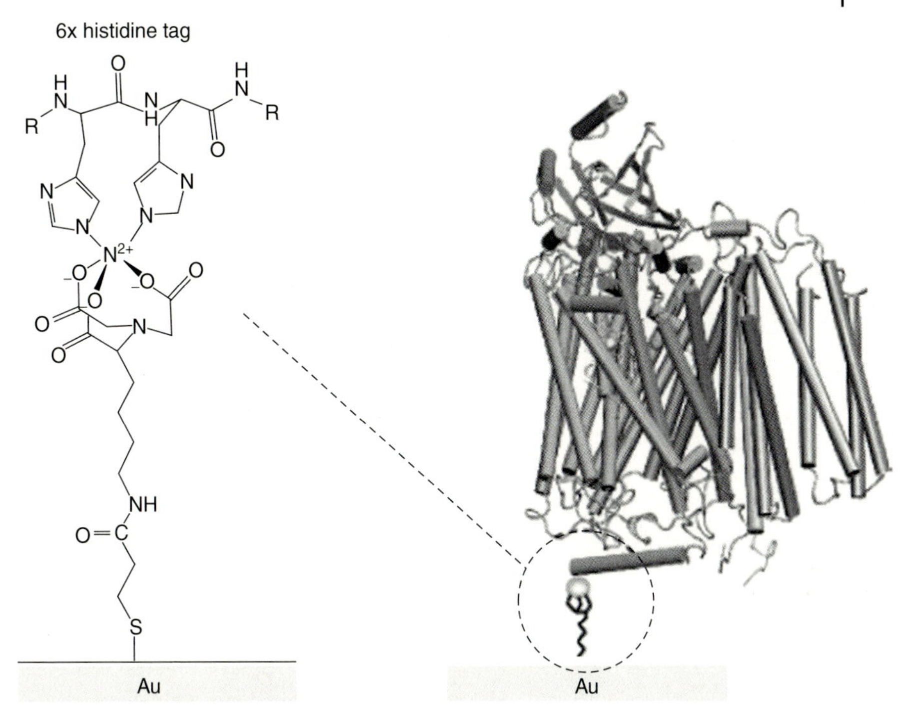

Figure 4.12 Adsorption of COX on Ni–NTA-modified gold surface via the His tag at the C-terminus of subunit I. Two nitrogen atoms of the imidazole rings from two of the histidines of the His stretch coordinate the Ni^{2+} ion. The coordinating histidine residues are not necessarily adjacent in the primary sequence, as drawn in the figure. Reprinted from [110] with kind permission from the American Chemical Society.

(His tag) within the protein. Two opposite orientations of the protein were investigated, either with the cytochrome c binding side pointing away from the electrode surface or directed toward the electrode, simply by engineering the His tag on the C-terminus of subunit SU I or SU II, respectively. The functional activity of COX was verified by cyclic voltammetry with both protein orientations. In this connection, it should be noted that electron transfer in COX occurs sequentially through the four redox centers Cu_A, heme a, heme a_3, and Cu_B, in the direction from the binding site of cytochrome c, located on the outer side of the bacterial membrane, to its inner side. With the cytochrome c binding site pointing away from the electrode surface, the primary electron acceptor, Cu_A, is far from the electrode surface. Hence, in the absence of cytochrome c, the cyclic voltammogram exhibits only a capacitive current. This indicates that COX is not electrically coupled to the electrode, and direct electron transfer does not take place. When the COX is oriented with the cytochrome c binding site pointing toward the electrode surface, the primary electron acceptor, Cu_A, is also oriented toward the electrode. In this case, the cyclic voltammogram in the absence of oxygen shows a single reduction peak at about −274 mV vs. NHE, due to the electroreduction of the enzyme, and a corresponding oxidation peak at about −209 mV vs. NHE [112, 113]. The peak currents increase linearly with the scan rate, denoting a surface-confined process. In the presence of oxygen, electrons transferred from the electrode to the redox centers of COX are irreversibly transferred to oxygen, leading to a notable increase of the reduction peak, which now lies at −202 mV vs. NHE, and to a continuous electron transfer. The absence of direct electron transfer and of proton electroreduction when COX is oriented with the cytochrome c binding site turned toward the solution confirms the orientation dependence both of direct electron transfer and of transmembrane proton transport.

4.6 Conclusions

The use of electrochemical techniques such as EIS, charge transient recordings, and cyclic voltammetry for the investigation of biological systems is becoming increasingly popular, just as is the application of the concepts of electrochemical kinetics and of the structure of electrified interfaces to the interpretation of the electrochemical response.

Much work is presently being done to realize biomembrane models consisting of a lipid bilayer anchored to a solid electrode through a hydrophilic spacer and satisfying those requirements of ruggedness, fluidity, and high electrical resistance that are necessary for the incorporation of integral proteins in a functionally active state. The capacitive currents resulting from the activation of ion pumps, transporters, ion channels, and channel-forming peptides incorporated in these biomembrane models can be analyzed over a broad potential range by electrochemical techniques, which are far less expensive than other techniques presently adopted.

The realization of these biomembrane models allows fundamental studies of the function of integral proteins. Biomimetic membranes are ideally suited to elucidate many problems in molecular membrane biology. This will open the way to the elucidation of structure–function relationships in ligand–receptor and protein–protein interactions. Moreover, the development of biomimetic systems that incorporate therapeutically or diagnostically important natural proteins will open the door to the realization of sensors targeting biological analytes. Many practical applications are foreseen for these sensors, such as the detection of drug candidates modulating the function of ion channels and pumps or targeting membrane receptors. In this respect, there is a strong need to develop novel, rapid, and highly sensitive methods for drug screening, capable of selecting and analyzing a huge number of compounds. At present, screening of pharmacologically active compounds follows traditional procedures that apply time-consuming ligand-binding studies and receptor-function tests separately. Thus, for instance, the function of ion channels and transporters is traditionally characterized in detail by patch-clamp studies, which investigate the proteins in their natural environment, the cellular membrane. These assays are tedious to perform and difficult to automate at high throughput, making the investigation of many samples difficult. The lack of knowledge about the different functions of these channels is due to a lack of specific inhibitors, which are unavailable due to the lack of efficient measuring systems. Present ligand-binding experiments identify only ligands to already known binding sites on the protein(s) of interest and neglect other potentially more interesting sites. Moreover, they cannot easily differentiate between agonists and antagonists. Thus, the direct, predominantly electrochemical determination of the function of ion channels and pumps in biomembrane models reconstituted from purified components addresses a strongly felt need for the development of new drug candidates or diagnostic test systems.

Acknowledgments

Thanks are due to Ente Cassa di Risparmio di Firenze for financial support to the authors' research on metal-supported biomimetic membranes. The technical support by Dr. Giovanni Aloisi in the preparation of the manuscript is gratefully acknowledged.

References

1 Schiller, S.M., Naumann, R., Lovejoy, K., Kunz, H., and Knoll, W. (2003) *Angewandte Chemie International Edition*, **42**, 208.

2 Becucci, L., Moncelli, M.R., Naumann, R., and Guidelli, R. (2005) *Journal of the American Chemical Society*, **127**, 13316.

3 Jeuken, L.J.C., Connell, S.D., Henderson, P.J.F., Gennis, R.B., Evans, S.D., and Bushby, R.J. (2006) *Journal of the American Chemical Society*, **128**, 1711.

4 Toby, A., Jenkins, A., Bushby, R.J., Boden, N., Evans, S.D., Knowles, P.F., Liu, Q., Miles, R.E., and Ogier, S.D. (1998) *Langmuir*, **14**, 4675.

5 Baumgart, M., Kreiter, M., Lauer, H., Naumann, R., Jung, G., Jonczyk, A., Offenhäusser, A., and Knoll, W. (2003) *Journal of Colloid and Interface Science*, **258**, 298.

6 (a) Knoll, W. (1998) *Annual Review of Physical Chemistry*, **49**, 569; (b) Knoll, W., Köper, I., Naumann, R., and Sinner, E.-K. (2008) *Electrochimica Acta*, **53**, 6680.

7 Plant, A.L. (1999) *Langmuir*, **15**, 5128.

8 Williams, L.M., Evans, S.D., Flynn, T.M., Marsh, A., Knowles, P.F., Bushby, R.J., and Boden, N. (1997) *Langmuir*, **13**, 751.

9 Williams, L.M., Evans, S.D., Flynn, T.M., Marsh, A., Knowles, P.F., Bushby, R.J., and Boden, N. (1997) *Supramolecular Science*, **4**, 513.

10 Kalb, E., Frey, S., and Tamm, L.K. (1992) *Biochimica et Biophysica Acta*, **1103**, 307.

11 Reimhult, E., Hook, F., and Kasemo, B. (2003) *Langmuir*, **19**, 1681.

12 Sackmann, E. (1996) *Science*, **271**, 43.

13 Keller, C.A., Glasmästar, K., Zhdanov, V.P., and Kasemo, B. (2000) *Physical Review Letters*, **84**, 5443.

14 Keller, C.A. and Kasemo, B. (1998) *Biophysical Journal*, **75**, 1397.

15 Lingler, S., Rubinstein, I., Knoll, W., and Offenhäusser, A. (1997) *Langmuir*, **13**, 7085.

16 Delahay, P. and Strassner, J.E. (1951) *Journal of the American Chemical Society*, **73**, 5219.

17 He, L., Robertson, J.W.F., Li, J., Kärcher, I., Schiller, S.M., Knoll, W., and Naumann, R. (2005) *Langmuir*, **21**, 11666.
18 Zebrowska, A. and Krysinski, P. (2004) *Langmuir*, **20**, 11127.
19 Puu, G. and Gustafson, I. (1997) *Biochimica et Biophysica Acta*, **1327**, 149.
20 Jass, J., Tjärnhage, T., and Puu, G. (2000) *Biophysical Journal*, **79**, 3153.
21 Naumann, R., Baumgart, T., Gräber, R., Jonczyk, A., Offenhäusser, A., and Knoll, W. (2002) *Biosensors & Bioelectronics*, **17**, 25.
22 Tamm, L.K. and McConnell, H.M. (1985) *Biophysical Journal*, **47**, 105.
23 McConnell, H.M., Watts, T.H., Weis, R.M., and Brian, A.A. (1986) *Biochimica et Biophysica Acta*, **864**, 95.
24 Zawisza, I., Bin, X., and Lipkowski, J. (2007) *Langmuir*, **23**, 5180.
25 Raguse, B., Braach-Maksvytis, V., Cornell, B.A., King, L.G., Osman, P.D.J., Pace, R.J., and Wieczorek, L. (1998) *Langmuir*, **14**, 648.
26 McGillivray, D.J., Valincius, G., Vanderah, D.J., Febo-Ayala, W., Woodward, J.T., Heinrich, F., Kasianowicz, J.J., and Lösche, M. (2007) *Biointerphases*, **2**, 21.
27 Richter, R.P., Mukhopadhyay, A. and Brisson, A. (2003) *Biophysical Journal*, **85**, 3035.
28 Axelrod, D., Koppel, D.E., Schlessinger, J., Elson, E., and Webb, W.W. (1976) *Biophysical Journal*, **16**, 1055.
29 Parikh, A.N. (2008) *Biointerphases*, **3**, FA22.
30 Becucci, L., Martinuzzi, S., Monetti, E., Mercatelli, R., Quercioli, F., Battistel, S., and Guidelli, R. (2010) *Soft Matter*, **6**, 2733.
31 Brian, A.A. and McConnel, H.M. (1984) *Proceedings of the National Academy of Sciences of the USA*, **81**, 6159.
32 Mossman, K.D., Campi, G., Groves, J.T., and Dustin, M.L. (2005) *Science*, **310**, 1191.
33 Boissiere, C., Brubach, J.B., Mermet, A., de Marzi, G., Bourgaux, C., Prouzet, E., and Roy, P. (2002) *Journal of Physical Chemistry B*, **106**, 1032.
34 Mashl, R.J., Joseph, S., Aluru, N.R., and Jakobsson, E. (2003) *Nano Letters*, **3**, 589.
35 Richter, R.P., Maury, N., and Brisson, A.R. (2005) *Langmuir*, **21**, 299.
36 Lenz, P., Ajo-Franklin, C.M., and Boxer, S.G. (2004) *Langmuir*, **20**, 11092.
37 Schmitt, J., Danner, B., and Bayerl, T.M. (2001) *Langmuir*, **17**, 244.
38 Dodd, C.E., Johnson, B.R.G., Jeuken, L.J.C., Bugg, T.D.H., Bushby, J., and Evans, S.D. (2008) *Biointerphases*, **3**, FA59.
39 Jackson, B.L., Nye, J.A., and Groves, J.T. (2008) *Langmuir*, **24**, 6189.
40 Zawisza, I., Lachenwitzer, A., Zamlynny, V., Horswell, S.L., Goddard, J.P., and Lipkowski, J. (2003) *Biophysical Journal*, **85**, 4055.
41 Burgess, L., Li, M., Horswell, S.L., Szymanski, G., Lipkowski, J., Majewski, J., and Satija, S. (2004) *Biophysical Journal*, **86**, 1763.
42 Bin, X., Horswell, S.L., and Lipkowski, J. (2005) *Biophysical Journal*, **89**, 592.
43 Xu, S., Szymanski, G., and Lipkowsky, J. (2004) *Journal of the American Chemical Society*, **126**, 12276.
44 Purrucker, O., Hillebrandt, H., Adlkofer, K., and Tanaka, M. (2001) *Electrochimica Acta*, **47**, 791.
45 Gritsch, S., Nollert, P., Jähnig, F., and Sackmann, E. (1998) *Langmuir*, **14**, 3118.
46 Steinem, C., Janshoff, A., Galla, H.-J., and Sieber, M. (1997) *Bioelectrochemistry and Bioenergetics*, **42**, 213.
47 Michalke, A., Schürholz, T., Galla, H.-J., and Steinem, C. (2001) *Langmuir*, **17**, 2251.
48 Becucci, L., Guidelli, R., Liu, Q., Bushby, R.J., and Evans, S.D. (2002) *Journal of Physical Chemistry B*, **106**, 10410.
49 Peggion, C., Formaggio, F., Toniolo, C., Becucci, L., Moncelli, M.R., and Guidelli, R. (2001) *Langmuir*, **17**, 6585.
50 Lang, H., Duschl, C., and Vogel, H. (1994) *Langmuir*, **10**, 197.
51 Lang, H., Duschl, C., Grätzel, M., and Vogel, H. (1992) *Thin Solid Films*, **210/211**, 818.
52 Boden, N., Bushby, R.J., Liu, Q., Evans, S.D., Jenkins, A.T.A., Knowles, P., and Miles, R.E. (1998) *Tetrahedron*, **54**, 11537.
53 Steinem, C., Janshoff, A., von dem Bruch, K., Reihs, K., Goossens, J., and

Galla, H.-J. (1998) *Bioelectrochemistry and Bioenergetics*, **45**, 17.

54 Lear, J.D., Wassermann, Z.D., and De Grado, W.D. (1988) *Science*, **240**, 1179.

55 Bunjes, N., Schmidt, E.K., Jonczyk, A., Rippmann, F., Beyer, D., Ringsdorf, H., Gräber, P., Knoll, W., and Naumann, R. (1997) *Langmuir*, **13**, 6188.

56 Naumann, R., Jonczyk, A., Kopp, R., van Esch, J., Ringsdorf, H., Knoll, W., and Gräber, P. (1995) *Angewandte Chemie (International Edition in English)*, **34**, 2056.

57 Naumann, R., Jonczyk, A., Hampel, C., Ringsdorf, H., Knoll, W., Bunjes, N., and Gräber, P. (1997) *Bioelectrochemistry and Bioenergetics*, **42**, 241.

58 Naumann, R., Schmidt, E.K., Jonczyk, A., Fendler, K., Kadenbach, B., Liebermann, T., Offenhäusser, A., and Knoll, W. (1999) *Biosensors & Bioelectronics*, **14**, 651.

59 Vockenroth, I.K., Ohm, C., Robertson, J.W.F., McGillivray, D.J., Lösche, M., and Köper, I. (2008) *Biointerphases*, **3**, FA68.

60 Leitch, J., Kunze, J., Goddard, J.D., Schwan, A.L., Faragher, R.J., Naumann, R., Knoll, W., Dutcher, J.R., and Lipkowski, J. (2009) *Langmuir*, **25**, 10354.

61 Naumann, R., Walz, D., Schiller, S.M., and Knoll, W. (2003) *Journal of Electroanalytical Chemistry*, **550–551**, 241.

62 Atanasov, V., Knorr, N., Duran, R.S., Ingebrandt, S., Offenhäusser, A., Knoll, W., and Koper, I. (2005) *Biophysical Journal*, **89**, 1780.

63 Vockenroth, I.K., Atanasova, P.P., Long, J.R., Jenkins, A.T.A., Knoll, W., and Köper, I. (2007) *Biochimica et Biophysica Acta*, **1768**, 1114.

64 Glazier, S.A., Vanderah, D.J., Plant, A.L., Bayley, H., Valincius, G., and Kasianowicz, J.J. (2000) *Langmuir*, **16**, 10428.

65 Vockenroth, I.K., Atanasova, P.P., Jenkins, A.T.A., and Köper, I. (2008) *Langmuir*, **24**, 496.

66 Roskamp, R.F., Vockenroth, I.K., Eisenmenger, N., Braunagel, J., and Köper, I. (2008) *ChemPhysChem*, **9**, 1920.

67 Robelek, R., Lemker, E.S., Wiltschi, B., Kirste, V., Naumann, R., Oesterhelt, D., and Sinner, E.-K. (2007) *Angewandte Chemie International Edition*, **46**, 605.

68 Leutenegger, M., Lasser, T., Sinner, E.-K., and Robelek, R. (2008) *Biointerphases*, **3**, FA136.

69 Schmidt, E.K., Liebermann, T., Kreiter, M., Jonczyk, A., Naumann, R., Offenhäusser, A., Neumann, E., Kukol, A., Maelicke, A., and Knoll, W. (1998) *Biosensors & Bioelectronics*, **13**, 585.

70 Andersson, M., Keizer, H.K., Zhu, C., Fine, D., Dodabalapur, A., and Duran, R.S. (2007) *Langmuir*, **23**, 2924.

71 Keizer, H.M., Dorvel, B.R., Andersson, M., Fine, D., Price, R.B., Long, J.R., Dodabalapur, A., Köper, I., Knoll, W., Anderson, P.A.V., and Duran, R.S. (2007) *ChemBioChem*, **8**, 1246.

72 Andersson, M., Okeyo, G., Wilson, D., Keizer, H., Moe, P., Blouunt, P., Fine, D., Dodabalapur, A., and Duran, R.S. (2008) *Biosensors & Bioelectronics*, **23**, 919.

73 Vockenroth, I.K., Fine, D., Dodobalapur, A., Jenkins, A.T.A., and Köper, I. (2008) *Electrochemistry Communications*, **10**, 323.

74 Becucci, L., Moncelli, M.R., and Guidelli, R. (2006) *Langmuir*, **22**, 1341.

75 Becucci, L., Carbone, M.V., Biagiotti, T., D'Amico, M., Olivotto, M., and Guidelli, R. (2008) *Journal of Physical Chemistry B*, **112**, 1315.

76 Becucci, L., Santucci, A., and Guidelli, R. (2007) *Journal of Physical Chemistry B*, **111**, 9814.

77 Becucci, L., and Guidelli, R. (2009) *Soft Matter*, **5**, 2294.

78 Becucci, L., Schwan, A.L., Sheepwash, E.E., and Guidelli, R. (2009) *Langmuir*, **25**, 1828.

79 Becucci, L., Innocenti, M., Salvietti, E., Rindi, A., Pasquini, I., Vassalli, M., Foresti, M.L., and Guidelli, R. (2008) *Electrochimica Acta*, **53**, 6372.

80 Grahame, D.C. (1954) *Journal of the American Chemical Society*, **76**, 4819.

81 Clavilier, J. and Nguyen Van Huong, C. (1973) *Journal of Electroanalytical Chemistry*, **41**, 193.

82 Becucci, L., Romero León, R., Moncelli, M.R., Rovero, P., and

Guidelli, R. (2006) *Langmuir*, **22**, 6644.
83 Becucci, L. and Guidelli, R. (2007) *Langmuir*, **23**, 5601.
84 (a) Trudeau, M.C., Warmke, J.W., Ganetzky, B., and Robertson, G.A. (1995) *Science*, **269**, 92; (b) Arcangeli, A., Bianchi, L., Becchetti, A., Faravelli, L., Coronnello, M., Mini, E., Olivotto, M., and Wanke, E. (1995) *Journal of Physiology*, **489**, 455.
85 Becucci, L., Guidelli, R., Karim, C.B., Thomas, D.D., and Veglia, G. (2007) *Biophysical Journal*, **93**, 1.
86 Becucci, L., Guidelli, R., Karim, C.B., Thomas, D.D., and Veglia, G. (2009) *Biophysical Journal*, **97**, 2693.
87 (a) Stefanova, H.I., East, J.M., and Lee, A.G. (1991) *Biochimica et Biophysica Acta*, **1064**, 321; (b) Stefanova, H.I., Jane, S.D., East, J.M., and Lee, A.G. (1991) *Biochimica et Biophysica Acta*, **1064**, 329.
88 (a) Robia, S.L., Campbell, K.S., Kelly, E.M., Hou, Z., Winters, D.L., and Thomas, D.D. (2007) *Circulation Research*, **101**, 1123; (b) Hou, Z., Kelly, E.M., and Robia, S.L. (2008) *Journal of Biological Chemistry*, **283**, 28996.
89 Becucci, L., Cembran, A., Karim, C.B., Thomas, D.D., Guidelli, R., Gao, J., and Veglia, G. (2009) *Biophysical Journal*, **96**, L60.
90 Becucci, L., D'Amico, M., Daniele, S., Olivotto, M., Pozzi, A., and Guidelli, R. (2010) *Bioelectrochemistry*, **78**, 176.
91 (a) Brown, D.A. and London, E. (1998) *Journal of Membrane Biology*, **164**, 103; (b) Brown, D.A. and London, E. (2000) *Journal of Biological Chemistry*, **275**, 17221.
92 Krishna, G., Schulte, J., Cornell, B.A., Pace, R., Wieczorek, L., and Osman, P.D. (2001) *Langmuir*, **17**, 4858.
93 Shon, Y.-S., Lee, S., Perry, S.S., and Lee, T.R. (2000) *Journal of the American Chemical Society*, **122**, 1278.
94 Chambers, R.C., Inman, C.E., and Hutchison, J.E. (2005) *Langmuir*, **21**, 4615.
95 Jeuken, L.J.C., Daskalakis, N.N., Han, X., Sheikh, K., Erbe, A., Bushby, R.J., and Evans, S.D. (2007) *Sensors & Actuators B*, **124**, 501.
96 Heyse, S., Ernst, O.P., Dienes, Z., Hofmann, K.P., and Vogel, H. (1998) *Biochemistry*, **37**, 507.
97 Heyse, S., Stora, T., Schmid, E., Lakey, J.H., and Vogel, H. (1998) *Biochimica et Biophysica Acta*, **1376**, 319.
98 (a) Tanaka, M. and Sackmann, E. (2005) *Nature*, **437**, 656; (b) Tanaka, M. and Sackmann, E. (2006) *Physica Status Solidi (a)*, **203**, 3452.
99 Wong, J.Y., Park, C.K., Seitz, M., and Israelachvili, J.N. (1999) *Biophysical Journal*, **77**, 1458.
100 Naumann, C.A., Prucker, O., Lehmann, T., Ruhe, J., Knoll, W., and Frank, C.W. (2002) *Biomacromolecules*, **3**, 27.
101 Wagner, M.L. and Tamm, L.K. (2000) *Biophysical Journal*, **79**, 1400.
102 Dietrich, C. and Tampé, R. (1995) *Biochimica et Biophysica Acta*, **1238**, 183.
103 Spinke, J., Yang, J., Wolf, H., Liley, M., Ringsdorf, H., and Knoll, W. (1992) *Biophysical Journal*, **63**, 1667.
104 Erdelen, C., Häussling, L., Naumann, R., Ringdorf, H., Wolf, H., and Yang, J. (1994) *Langmuir*, **10**, 1246.
105 Hillebrandt, H., Wiegand, G., Tanaka, M., and Sackmann, E. (1999) *Langmuir*, **15**, 8451.
106 Knoll, W., Naumann, R., Friedrich, M., Robertson, J.W.F., Lösche, M., Heinrich, F., McGillivray, D.J., Schuster, B., Gufler, P.C., Pum, D., and Sleytr, U.B. (2008) *Biointerphases*, **3**, FA125.
107 Gufter, P.C., Pum, D., Sleytr, U.B., and Schuster, B. (2004) *Biochimica et Biophysica Acta*, **1661**, 154.
108 Schuster, B., Pum, D., Sára, M., Braha, O., Bayley, H., and Sleytr, U.B. (2001) *Langmuir*, **17**, 499.
109 Schuster, B., Weigert, S., Pum, D., Sára, M., and Sleytr, U.B. (2003) *Langmuir*, **19**, 2392.
110 Ataka, K., Giess, F., Knoll, W., Naumann, R., Haber-Pohlmeier, S., Richter, B., and Heberle, J. (2004) *Journal of the American Chemical Society*, **126**, 16199.

111 Friedrich, M.G., Kirste, V.U., Zhu, J., Gennis, R.B., Knoll, W., and Naumann, R.L.C. (2008) *Journal of Physical Chemistry B*, **112**, 3193.

112 Friedrich, M.G., Robertson, J.W.F., Walz, D., Knoll, W., and Naumann, R.L.C. (2008) *Biophysical Journal*, **94**, 3698.

113 Naumann, R.L.C. and Knoll, W. (2008) *Biointerphases*, **3**, FA101.

5
Enzymatic Fuel Cells

Paul Kavanagh and Dónal Leech

5.1
Introduction

The rapid depletion of fossil fuels and the environmental concerns associated with their combustion have created a strong demand for new and improved technologies capable of harnessing energy from sustainable sources. The automobile industry's development of the hydrogen fuel cell as a "green" alternative to the traditional, polluting, combustion engine is one such initiative. Although hydrogen fuel cells are environmentally friendly (only producing water as a byproduct), major obstacles, such as cost and hydrogen production/storage issues, must be overcome before this technology could seriously compete in the marketplace.

An interesting spin-off from conventional fuel cell technology is the development of biofuel cells (BFCs) which utilize biological catalysts, in place of metal catalysts, to catalyze the reactions at the anode and cathode. Due to the versatile nature of the biocatalysts, BFCs are not limited to hydrogen as a fuel and can derive power from a wide range of organic substrates. Two types of BFCs are currently being explored for technological application: microbial fuel cells (MFCs) and enzymatic fuel cells (EFCs). MFCs use whole microorganisms, such as bacteria and algae, to produce and oxidize fuels (e.g., sugars and hydrogen) through fermentation or metabolism. This area is extensively reviewed elsewhere [1–5] and will not be covered in this chapter. EFCs use isolated enzymes, derived from microorganisms, to catalytically oxidize a specific fuel at the anode and reduce the oxidant, usually oxygen, at the cathode (shown schematically in Figure 5.1).

In 1964, Yahiro *et al.* reported the first EFC using glucose as a fuel and oxygen as an oxidant [6]. Shortly thereafter EFCs were investigated with specific applications in mind, such as using an EFC to power an artificial heart or to provide power to biomedical devices such as cardiac pacemakers [7]. It soon became clear that EFCs could not meet the energy and operational lifetime demands required for such devices and, for the next 25 years, this research area was largely abandoned. A re-emergence occurred in the late 1990s, partly driven by progress in (i) biosensor design and enzyme electrochemistry, particularly in terms of achieving high, and increasingly stable, current densities at modified electrodes and

Advances in Electrochemical Science and Engineering. Edited by Richard C. Alkire, Dieter M. Kolb, and Jacek Lipkowski

ISBN: 978-3-527-32885-7

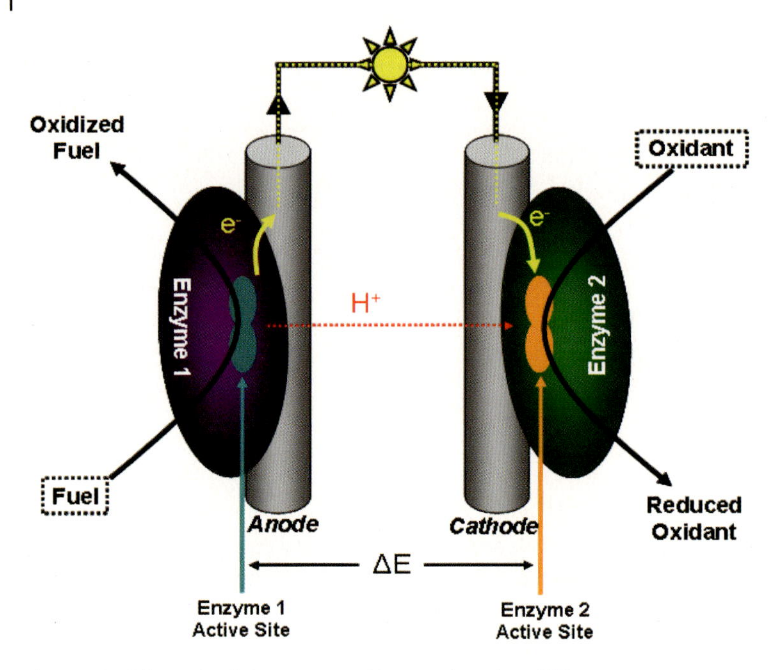

Figure 5.1 Schematic of the processes in a membraneless EFC.

(ii) microelectronics, where ever smaller and lower-energy-consuming devices are being manufactured. Such advances reignited interest in integrating EFCs with specific low-energy-demanding microelectronic biomedical devices [7].

As with the hydrogen fuel cell, EFCs must also compete with state-of-the-art batteries. Lithium–iodine batteries are currently used to power implantable devices, such as cardiac pacemakers, and can last for 5–6 years. Therefore, to compete in this market, a fully implantable EFC should have an operational lifetime of greater than 5 years. This seems quite a challenge considering that, to date, an implanted EFC generated intermittent power for 10 days [8]. One niche application which shows potential centers on the development of an integrated miniature EFC–sensor system for diabetes management. It is envisaged that these "semi-implantable" miniature EFCs will provide power for the lifetime of glucose sensors (typically less than a week) and be discarded after their first and only use [7], thereby eliminating the need for long-term stability. Other *in vivo* applications have been investigated, such as generating power from carbohydrates contained in plants [9]. Regardless of the intended application, the power densities, operational lifetime, and biocompatibility of such devices must be further investigated and improved upon before commercial applications can be realized. Indeed, whether or not EFCs will ever meet the operational and cost requirements to compete in the marketplace remains open to debate. Nonetheless fundamental EFC research will aid in providing new insights into redox enzyme structure–activity and stability in solution and at surfaces, surface and immobilization chemistry, and enzyme electrode electrochemistry and contribute to our fundamental understanding of catalysis, permitting future applications using biomimetics to be exploited.

This chapter is intended to provide a selective overview of the most recent developments in EFC research. Several excellent and comprehensive reviews focused on aspects of EFC research have appeared over the past few years [7, 10–16]. We focus here on developments that may lead to deployment of an implantable glucose-oxidizing, oxygen-reducing EFC, as an example of the extensive research on EFC bioelectrochemistry. Developments, focused on using other fuels, such as hydrogen, alcohols, or other sugars, or oxidants will not be considered in detail.

5.1.1 Enzymatic Fuel Cell Design

An EFC consists of two electrodes, anode and cathode, connected by an external load (shown schematically in Figure 5.1). In place of traditional nonselective metal catalysts, such as platinum, biological catalysts (enzymes) are used for fuel oxidation at the anode and oxidant reduction at the cathode. Judicious choice of enzymes allows such reactions to occur under relatively mild conditions (neutral pH, ambient temperature) compared to conventional fuel cells. In addition, the specificity of the enzyme reactions at the anode and cathode can eliminate the need for other components required for conventional fuel cells, such as a case and membrane. Due to the exclusion of such components, enzymatic fuel cells have the capacity to be miniaturized, and consequently micrometer-dimension membraneless EFCs have been developed [7]. In the simplest form, the difference between the formal redox potential (E^0) of the active site of the enzymes utilized for the anode and cathode determines the maximum voltage (ΔE) of the EFC. Ideally enzymes should possess the following qualities.

- A high specificity for fuel/oxidant.
- Retain good catalytic properties and high stability under cell operational conditions.
- Enzymes should not compete with each other for substrates (fuel and oxidant).
- Enzyme reactions should not generate any harmful side-products.

The chosen electrode material should be conductive and inert within the potential range of the cell. Materials composed of allotropes of carbon and, to a lesser extent, gold are most commonly used. The cell should be designed to minimize overpotentials due to kinetics, ohmic resistance, and mass transfer of fuel in order to maximize cell voltage (ΔE) and current (i) generation. In addition, all cell components should be mechanically stable within their operating environment.

5.1.2 Enzyme Electron Transfer

Redox enzymes typically consist of an apoenzyme (the protein component of an enzyme) and at least one coenzyme (a small nonproteinaceous electroactive

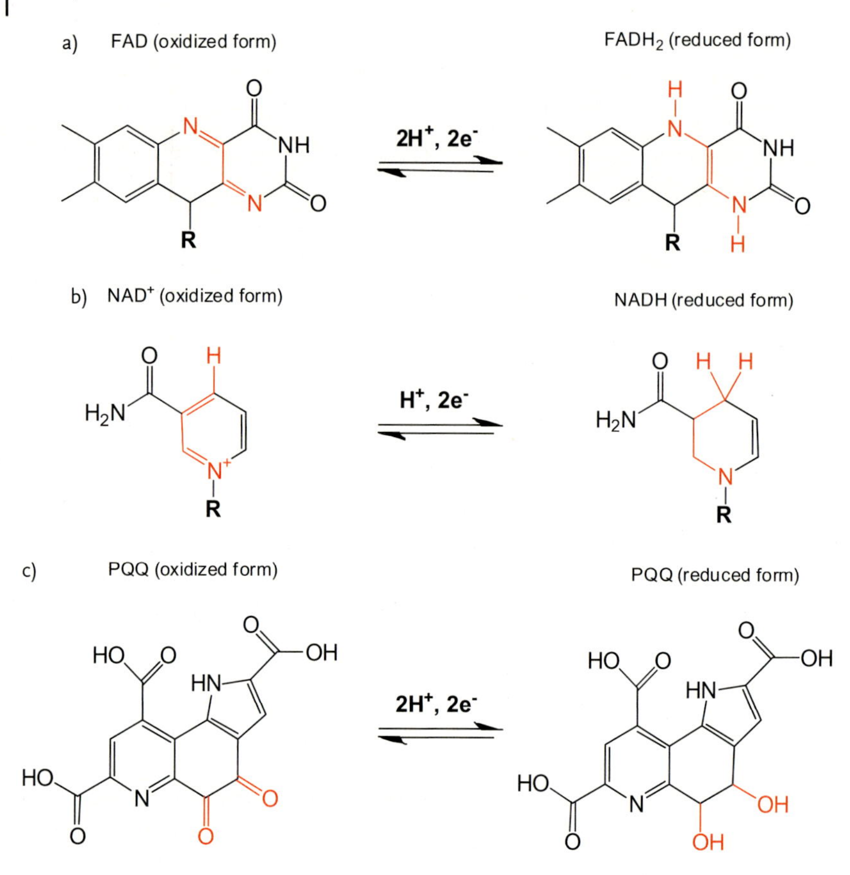

Figure 5.2 Enzyme cofactor structures and redox processes for: (a) FAD/$FADH_2$, (b) NAD^+/NADH, and (c) PQQ. R represents adenosine diphosphate.

molecule). The presence of the coenzyme is essential for facilitation of electron transfer between the enzyme and the substrate, which in EFCs can be the electrode surface. The coenzyme can be either tightly bound within the enzyme structure (coenzyme) or released from the enzyme's active site during the reaction (cofactor). Common coenzymes/cofactors include flavin adenine dinucleotide (FAD), nicotinamide adenine dinucleotide (NAD), and pyrroloquinoline quinone (PQQ) (Figure 5.2).

If the active site of the enzyme is located sufficiently close to the electrode surface electrons can be transferred directly from the enzyme to the electrode as depicted in Figure 5.3a. In the case of an anodic reaction, the electrode replaces the natural co-substrate (such as oxygen) as an electron acceptor. This process is known as direct electron transfer (DET), often categorized as "third-generation" enzyme electrodes in the biosensor literature, and is the most elegant and simplest method of bioelectrocatalysis between an enzyme active site and an electrode.

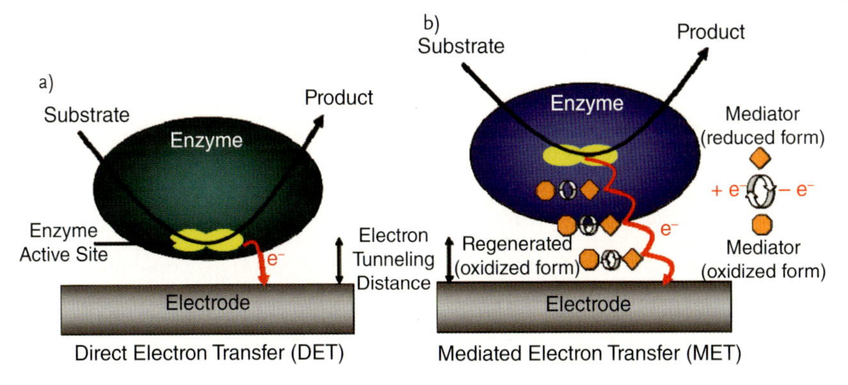

Figure 5.3 Schematics of (a) direct electron transfer (DET) and (b) mediated electron transfer (MET) between an enzyme active site and an electrode to oxidize a substrate.

Although DET has been reported for a wide range of enzymes [13, 17–19], several challenges need to be overcome to achieve significant rates of DET, leading to appreciable current densities, between active sites and solid electrode surfaces. For example, the rate of electron transfer is related exponentially to the distance of closest approach between an electron donor and acceptor [20], resulting in negligible rates for distances beyond 2 nm. This means that DET can only take place when an electrode is placed within this distance to an enzyme active site. In addition, for those enzymes that have active sites sufficiently exposed to permit DET to take place, correct orientation of each enzyme at the electrode surface is required to maintain the active site at the distance of closest approach. Finally, even when an electrode can approach sufficiently close to an active site to achieve DET, usually taking the place of a redox substrate/co-substrate, this may not necessarily lead to generation of a bioelectrocatalytic current for substrate electrolysis, as the electrode can block access to the active site of the co-substrate/substrate.

As an alternative to DET, small, artificial substrate/co-substrate electroactive molecules (mediators) can be used to shuttle electrons between the enzyme and the electrode (Figure 5.3b). This involves a process in which the enzyme takes part in the first redox reaction with the substrate and is re-oxidized or reduced by the mediator which in turn is regenerated, through a combination of physical diffusion and self-exchange, at the electrode surface. The mediator circulates continuously between the enzyme and the electrode, cycled between its oxidized and reduced forms, producing current. This process is known as mediated electron transfer (MET).

In MET, the thermodynamic redox potentials of the enzyme and the mediator should be accurately matched. The tuning of these potentials is of critical importance to EFC design as this will have a major bearing on cell voltage and catalytic current. When compared to the redox potential of the enzyme, the mediator should have a redox potential that is more positive for oxidative biocatalysis (at anode) and more negative for reductive biocatalysis (at cathode). For efficient electron transfer, it has been proposed that mediators and enzyme active sites should have a redox potential difference of approximately 50 mV, to provide a driving force for electron transfer [21]. More detailed studies suggest that, at least for biocatalytic oxygen reduction by a laccase enzyme within a redox hydrogel, an optimum compromise between driving force and current is obtained at 170 mV difference in redox potentials between mediator and enzyme active site [22]. Other considerations in design and selection of mediators include the following.

- The mediator dimensions should be sufficiently small to penetrate the protein structure and access the enzyme active site.
- The mediator should exhibit fast electron exchange rates with both the enzyme and the electrode to ensure that the mediator is not limited by electrode kinetics and to minimize competition with the enzyme natural substrate, if present.
- The mediator should be chemically stable in both the oxidized and reduced forms, and exhibit rapid diffusion and/or self-exchange, to provide stable and efficient electron transport through a three-dimensional layer.
- The mediator should present a redox potential sufficiently removed from all other electroactive species to avoid interferences and unwanted catalysis.
- The mediator and the mediated reaction should be insensitive to changes in pH or ionic strength.
- The mediator should be easily immobilized, to enable fabrication of devices.

Mediators can be either present in solution or immobilized at electrode surfaces. If contained in solution, separation of anodic and cathodic chambers is necessary to maintain a potential difference (ΔE) between the electrodes and to prevent crossover reactions. For eliminating the need for case and membrane all catalytic components (i.e., enzyme and mediator) should be immobilized at the electrode surfaces. The mediator and enzyme can be immobilized by adsorption onto electrodes, entrapment behind or within membranes and gels, linkage functionalization to the electrode surface, or integration into a polymer layer. Immobilization of mediator and enzyme is essential for cell miniaturization, and consequently EFC prototypes intended for *in vivo* applications tend to follow this approach. Therefore the following sections of this chapter will concentrate predominately on DET and MET for anodes and cathodes in immobilized systems for current/power generation, with some reference to solution-phase catalysis.

Comparison of current densities, and the potential for onset and for maximum current, is hampered when attempts are made, as will be undertaken in the following sections, to report on results from diverse sources, as there is as yet poor adherence to normalization of such data to standard or accepted conditions. The variability in experimental approaches includes differences, often unspecified, in specific surface areas of the electrodes used, the control of mass transport of fuel/oxygen to the surface, the coverage and activity of the enzyme on the surface, the stability of the system, the operational (and optimal) conditions (pH, ionic strength, buffer and chloride ion concentration, and temperature), and the method used to estimate current density (amperometry at fixed potential, cyclic voltammetry at various scan rates or in a BFC circuit) or power density (fixed load, potentiostatic or galvanostatic control). Where possible all potentials here are reported vs. Ag/AgCl (3 M NaCl) reference electrode. Recent focus on the need to adopt a standardized platform for testing of BFC electrodes, mediators, and configurations, to allow exchange and comparison of results, and to help understand contributions to improve bioelectrocatalysis at electrodes is to be welcomed [22–25].

5.2 Bioanodes for Glucose Oxidation

As in conventional fuel cells, the fuel is oxidized at the anode to release electrons and protons. The selection of the biocatalyst (redox enzyme) will determine the fuel consumed. Redox enzymes, oxidoreductases, are classified according to the nature of reactions that they catalyze. Of these, *oxidases*, which use oxygen as natural electron acceptor, and *dehydrogenases*, which transfer hydrogen to either NAD/NADP or a flavin, are most widely used. Comparison of current densities, and the potential for maximum current, is hampered by variability in the conditions used in each research report. Nonetheless, an attempt is made in Table 5.1 to compare the operational performance of bioanodes using a range of redox enzymes and electron transfer strategies (DET and MET) and under a diverse set of conditions. As recent EFC development has been predominately driven by potential *in vivo* applications, most research has focused on using glucose as a fuel due to its relatively high concentration in blood (about 6–8 mM). Glucose oxidase (GOx) from *Aspergillus niger* is the most widely used enzyme for glucose oxidation in EFCs owing to its remarkable stability, substrate selectivity, high electron turnover rate, and commercial availability. This enzyme catalyzes the oxidation of β-D-glucose to gluconolactone while reducing oxygen (as the natural electron acceptor) to hydrogen peroxide:

$$\text{Glucose} + \text{GOx–FAD} \rightarrow \text{Gluconolactone} + \text{GOx–FADH}_2 \tag{5.1}$$

$$\text{GOx–FADH}_2 + O_2 \rightarrow \text{GOx–FAD} + H_2O_2 \tag{5.2}$$

Aspergillus niger GOx is a dimer, composed of two identical 80 kDa subunits, each containing a tightly bound FAD cofactor (Figure 5.2a) which has a redox potential of −0.32 V (vs. Ag/AgCl) at pH = 7 [26]. The tightly bound FAD center is deeply buried (about 1.5 nm) within the protein structure, thus inhibiting DET at smooth electrode surfaces. DET to GOx has, however, been reported to be observed when using carbon nanotubes (CNTs) to molecularly "wire" the FAD active site [27–31]. However, the characteristic redox peaks of FAD are largely masked by background currents. In addition, the presence of CNTs within the protein structure may block substrate access to the enzyme active site and thus deactivate the enzyme. The most persuasive report of DET to a fully active GOx has been presented by Mano

Table 5.1 A comparison of the performance of bioanodes.

Sugar oxidizing enzyme	Surface	Mediator	Cell conditions	Potential (V)[a]	Current density ($mA\,cm^{-2}$) (approx.)[b]	Ref.
Aspergillus niger GOx	Crosslinking to glassy carbon electrode via ultraviolet irradiation	Ferrocene–benzophenone-modified polydimethylacrylamide polymer	pH = 5.2, 37 °C, N_2 saturated atmosphere, 1000 rpm rotation	0.45		[33]
			15 mM glucose		0.7	
			125 mM glucose		1.2	
Aspergillus niger GOx	Epoxide crosslinking to glassy carbon electrode	Ferrocene-modified linear polyethyleneimine polymer	pH = 7.4, 37 °C, 100 mM glucose, stirred solution	0.24	2.0	[34]
		Dimethylferrocene-modified linear polyethyleneimine polymer		0.13	2.1	
Aspergillus niger GOx	Epoxide crosslinking to carbon fiber electrode	PVI–[Os(4,4′-dimethyl-2,2′-bipyridine)$_2$Cl]$^{+/2+}$ coacrylamide	pH = 5, 23 °C, 15 mM glucose	0.2	0.2	[35]
Aspergillus niger GOx	Epoxide crosslinking to carbon fiber electrode	Polyvinypyridine–[Os(4,4′-dimethoxy-2,2′-bipyridine)$_2$Cl]$^{+/2+}$	pH = 7.4, 37.5 °C, 15 mM glucose	0	0.6	[36]
Aspergillus niger GOx	Epoxide crosslinking to carbon fiber electrode	PVI–[Os(4,4′-diamino-2,2′-bipyridine)$_2$Cl]$^{+/2+}$	pH = 7.4, 37.5 °C, 0.14 M NaCl, 15 mM glucose	−0.05	0.2	[37]
Aspergillus niger GOx	Epoxide crosslinking to carbon fiber electrode	Polyvinylpyridine–[Os(*N*,*N*′-dimethyl-2,2′-biimidazole)$_3$Cl]$^{2+/3+}$	pH = 7.2, 37.5 °C, argon saturated atmosphere 15 mM glucose	−0.1	1.3	[38]
Penicillium pinophilum GOx	Epoxide crosslinking to carbon fiber electrode	Polyvinylpyridine–[Os(*N*,*N*′-dimethyl-2,2′-biimidazole)$_3$Cl]$^{2+/3+}$	pH = 5, 37 °C, 32 mM glucose, 500 rpm rotation	−0.1	0.8	[39]

Aspergillus niger GOx (deglycosylated)	Adsorption to glassy carbon electrode	No mediator	pH = 7.4, 45 mM glucose, 500 rpm rotation	−0.2	0.2	[32]
Bacillus sp. GDH (NAD dependent), diaphorase	Adsorption onto a glassy carbon electrode	Vitamin K_3-modified poly(L-lysine)	pH = 7	0.2		[40]
			10 mM glucose		0.2	
	Adsorption onto a carbon fiber electrode		100 mM glucose	0.6	0.7	
			40 mM glucose		4	
GDH (NAD dependent), diaphorase	Adsorption onto a Ketjen black modified glassy carbon electrode	Vitamin K_3-modified poly(L-lysine)	pH = 7, 37 °C, 1 mM NAD, 1000 rpm rotation	0		[41]
			20 mM glucose		0.8	
			30 mM glucose		2	
GDH (NAD dependent)	Glutaraldehyde crosslinking to poly(methylene blue)–SWCNT composite, deposited on glassy carbon electrode	Poly(methylene blue)	pH = 6, 20 °C. 10 mM NAD	0.1		[42]
			15 mM glucose		0.1	
			90 mM glucose		0.3	
Acinetobacter calcoaceticus GDH (PQQ dependent)	Epoxide crosslinking to glassy carbon electrode	PVI–[Os(4,4′-dimethyl-2,2′ -bipyridine)$_2$Cl]$^{+/2+}$	pH = 5, 25 °C, 50 mM glucose, 500 rpm rotation	0.2	1.2	[43]
Acinetobacter calcoaceticus GDH (PQQ dependent) mutant	Glutaraldehyde crosslinking to carbon paste electrode	1-Methoxyphenazine methosulfate	pH = 7, 25 °C, 20 mM glucose	0	0.02	[44]
Gluconobacter sp. FDH	Adsorption onto a Ketjen black-modified carbon paper electrode	No mediator	pH = 5, 25 °C 200 mM fructose, stirred solution	0.4	4	[45]

(Continued)

Table 5.1 (*Continued*)

Sugar oxidizing enzyme	Surface	Mediator	Cell conditions	Potential (V)[a)]	Current density ($mA\,cm^{-2}$) (approx.)[b)]	Ref.
Gluconobacter sp. FDH	Alkanethiol-modified gold nanoparticles deposited on carbon fiber paper	No mediator	pH = 5, 25 °C 200 mM fructose	0.4	14	[46]
Gluconobacter industrius FDH	Adsorption to cellulose–MWCNT-modified glassy carbon electrode	No mediator	pH = 5, 15 mM fructose	0.3	0.8	[47]
Agaricus meleagris PDH	Epoxide crosslinking to graphite rod electrodes	Polyvinylpyridine–[Os(*N*,*N*′-dimethyl-2,2′ -biimidazole)3Cl]$^{2+/3+}$	pH = 8.5, 25 °C, 3 mM glucose, flow rate 1 ml/min	−0.1	0.02	[48]
		PVI–[Os(4,4′-dimethoxy-2,2′-bipyridine)$_2$Cl]$^{+/2+}$		−0.1	0.15	
		PVI–[Os(4,4′-dimethyl-2,2′ -bipyridine)$_2$Cl]$^{+/2+}$		0.1	0.15	
		PVI–[Os(2,2′ -bipyridine)$_2$Cl]$^{+/2+}$		0.15	0.16	
Phanerochaete sordida CDH	Epoxide crosslinking to SWCNTs deposited on pyrolytic graphite electrode	No mediator	pH = 3.5, 25 °C, 100 mM lactose	0.0	0.06	[49]
		Polyvinylpyridine–[Os(*N*,*N*′-dimethyl-2,2′ -biimidazole)$_3$Cl]$^{2+/3+}$		0.0	0.6	

a) Potential at which a steady-state current for oxidation is observed, corrected to the Ag/AgCl (3 M NaCl) reference.
b) Current densities are for the projected two-dimensional geometric underlying electrode area.

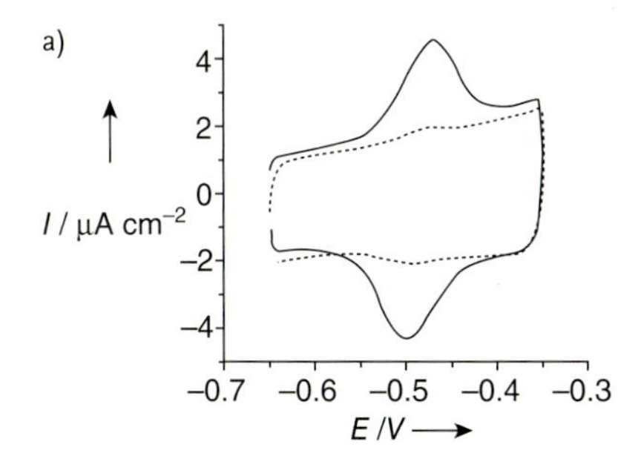

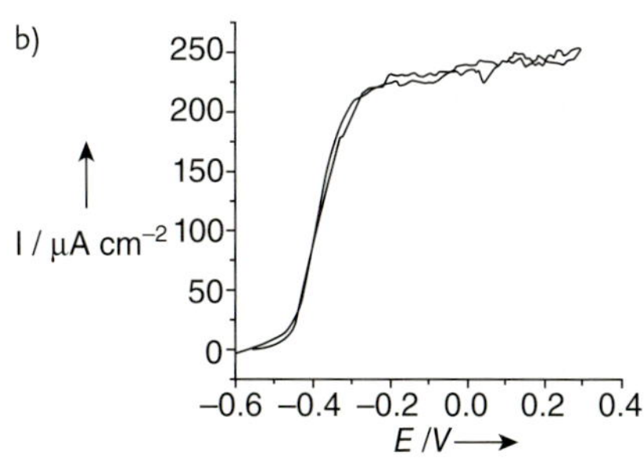

Figure 5.4 (a) Cyclic voltammograms of GOx (dotted curve) and dGOx (solid cirve) adsorbed on glassy carbon electrodes, 20 mM phosphate buffer, pH = 7.4, 37 °C, scan rate of 20 mV s^{-1}, argon atmosphere. (b) Direct electrooxidation of glucose (45 mM) on a monolayer of deglycosylated *Aspergillus niger* GOx adsorbed on a glassy carbon electrode (20 mM phosphate buffer, pH = 7.4, scan rate of 5 mV s^{-1}, 500 rpm, argon atmosphere). From [32] with permission from Wiley-VCH.

and coworkers [32] who report a redox couple in cyclic voltammograms at −0.49 V vs. Ag/AgCl, attributed to the FAD/$FADH_2$ redox process within a deglycosolated GOx (dGOx) adsorbed on a glassy carbon electrode (Figure 5.4a), although this redox potential is more negative than that observed by others for GOx–FAD/$FADH_2$. In the presence of 45 mM glucose a current density of 235 μA cm^{-2} at −0.2 V vs. Ag/AgCl is observed indicating that catalytic activity of the enzyme is retained (Figure 5.4b).

Despite the simplicity of DET, current densities are usually limited as only monolayer coverage is possible and enzymes may be electrically insulated due to incorrect orientation on the electrode surface. The simplest way to overcome these limitations is to immobilize the enzyme within a polymeric mediator matrix on the electrode surface. This enables multilayer enzyme coverages and, if the mediator can access the enzyme active site, the orientation of enzyme becomes irrelevant.

Much of the early research on MET to GOx stems from the development of enzyme electrodes for “second-generation” MET glucose sensing [50]. This involved replacing oxygen, used in “first-generation” glucose biosensors, with a synthetic electron acceptor to improve sensitivity and accuracy for determination of glucose concentrations in blood. Cass *et al.* [51] first demonstrated the use of ferrocene derivatives as mediators co-immobilized with GOx at a pyrolytic graphite electrode. This led to numerous studies examining ferrocene–GOx electron transfer reactions [52–56] including a comprehensive study by Forrow *et al.* [56] investigating the influence of ferrocene derivative structure on GOx mediation. Although adequate for single-use glucose biosensing, ferrocene derivatives are not particularly well suited as anodic mediators in EFCs due to their relatively high redox potentials (about 0.35 V vs. Ag/AgCl), and the instability of their oxidized form. The high redox potentials contribute to large thermodynamic losses in an EFC as shown by Liu *et al.* [57] using ferrocene monocarboxylic acid as an anodic mediator of GOx. Bunte *et al.* [33] recently described an EFC anode based on films of a ferrocene–benzophenone-modified polymer crosslinked with GOx via ultraviolet irradiation on glassy carbon. Although a high current density was achieved (about $1.2\,\text{mA}\,\text{cm}^{-2}$ at 0.45 V vs. Ag/AgCl), the potential remains relatively high for practical use in an assembled EFC. A more appropriate bioanode was developed by Meredith *et al.* [34] using GOx crosslinked with dimethylferrocene-modified polyethyleneimine polymer on a glassy carbon electrode, to provide about $2\,\text{mA}\,\text{cm}^{-2}$ at 0.24 V vs. Ag/AgCl at pH = 7.4. The electron-donating alkyl groups not only shifted the redox potential of the ferrocene but also resulted in better stability compared to unsubstituted ferrocene–polymers, reflected by superior power retention in assembled EFCs.

To date, osmium redox polymers are the most successful mediators for GOx in terms of current density and stability. As with ferrocenes, much of the early research concerning osmium redox polymers as mediators for GOx centered on their use in enzyme electrodes for glucose sensing [58]. Heller and coworkers [59, 60] first reported the use of the redox polymer $[Os(2,2'\text{-bipyridine})_2(\text{polyvinylimidazole})_{10}Cl]^{+/2+}$, depicted in Figure 5.5a, as an efficient mediator for glucose oxidation by GOx when co-immobilized with GOx at a glassy carbon electrode. This class of water-soluble redox polymer has the advantage of forming, when crosslinked with enzymes, hydrogel films on electrodes, permitting ingress and egress of substrate, solvent, and counter-ions. Shifting the Os(II/III) redox potential from +0.25 V vs. Ag/AgCl to lower, more suitable, values can be achieved through introduction of electron-donating functional groups on the bipyridyl ligands coordinated to the osmium redox center. De Lumley-Woodyear *et al.* [61] applied this approach to lower the redox potential by replacing the 2,2′-bipyridine

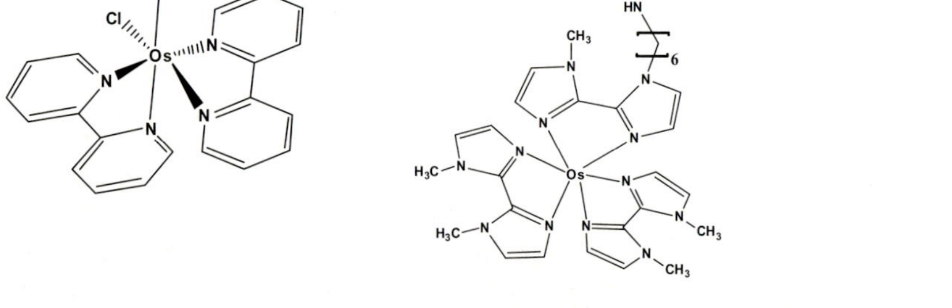

Figure 5.5 Structure proposed for the osmium redox polymers (a) $[Os(2,2'\text{-bipyridine})_2(PVI)_{10}Cl]^+$ and (b) polyvinylpyridine–$[Os(N,N'\text{-dimethyl-}2,2'\text{-biimidazole})_3Cl]^{2+}$.

ligands of osmium with 4,4′-dimethyl-2,2′-bipyridine, resulting in biocatalytic oxidation of glucose at about +0.1 V vs. Ag/AgCl [35, 61]. Further refinement using 4,4′-dimethoxy-2,2′-bipyridine [36] and 4,4′-diamino-2,2′-bipyridine [37] as ligands yielded redox polymers with potentials of −0.1 V and −0.15 V vs. Ag/AgCl, respectively. Hydrogels formed using the latter polymer crosslinked with GOx on carbon fiber electrodes yielded current densities of about 200 μA cm^{-2} at about −0.1 V vs. Ag/AgCl in solutions containing 15 mM glucose [37]. Electron transport within these redox hydrogels is controlled by the process of self-exchange between the oxidized and reduced osmium centers, which requires segmental chain motion of the polymer backbone to bring the centers close enough to enable exchange to occur, and can limit the current generated for glucose oxidation. This process may be characterized as diffusional, and an apparent charge (electron) transport diffusion coefficient can be estimated using voltammetry. The apparent electron transport diffusion coefficient, D_{app}, for a polyvinylimidazole (PVI)–[Os(4,4′-diamino-2,2′-bipyridine)$_2$(PVI)Cl]$^{+/2+}$ hydrogel was estimated as 1.2×10^{-9} cm^2 s^{-1} [62]. Subsequently, an improved D_{app} of 5.8×10^{-6} cm^2 s^{-1} for electron transport through an osmium-based hydrogel was achieved by Mao *et al.* [62] using the novel redox polymer polyvinylpyridine–[Os(*N,N*′-dimethyl-2,2′-biimidazole)$_3$Cl]$^{2+/3+}$ (Figure 5.5b). The increase in D_{app} was attributed to the 13-atom-long flexible spacer arm situated between the polyvinylpyridine polymer backbone and an alkyl functional group of an [Os(*N,N*′-dialkylated-2,2′-biimidazole)$_3$]$^{2+/3+}$ redox center, which increases the rate of electron transferring self-exchange collisions between the osmium centers. A current density of 1.5 mA cm^{-2} at −0.1 V vs. Ag/AgCl was obtained at a carbon fiber electrode coated with crosslinked films of GOx and this polymer in solutions containing 15 mM glucose (pH = 7.4) in phosphate buffered saline at 37 °C [38]. With a view to improving current generation at lower glucose concentrations, Mano [39] replaced *A. niger* GOx (K_m = 33 mM for glucose) [63] with *Penicillum pinophilum* GOx (K_m = 6.2 mM for glucose) [64] as biocatalyst in this hydrogel. Although improved current densities were achieved at pH = 5 using hydrogels containing *P. pinophilum* GOx at low glucose concentrations, operation of an EFC under physiological conditions would be hindered by the instability of *P. pinophilum* GOx at temperatures above 40 °C in the range of pH = 7–8.

Lowering the K_m value of a fuel-oxidizing enzyme is one of the many challenges faced when utilizing natural wild-type enzymes for EFC applications. Other issues, depending on the intended application, include enzyme immobilization and orientation, temperature and pH stability, salt inhibition, and substrate specificity. Such limitations may be overcome through protein engineering [65–68]. For example, recent reports by Schwaneberg and coworkers [66–68] describe a medium-throughput screening system for improving the properties of *A. niger* GOx through directed protein evolution. Using this method several mutants were obtained which displayed improved pH and thermal resistance, and higher activity for β-D-glucose oxidation compared to wild-type *A. niger* GOx. Despite these improvements, the natural reaction of GOx with oxygen (Eq. 5.2) represents a significant drawback for inclusion of this enzyme at the anode as oxygen competes with the electrode, or artificial electron acceptor, for electrons and in so doing can produce harmful hydrogen peroxide. In addition, oxygen consumption at the anode will reduce the concentration of oxidant available to the cathode in a membraneless EFC.

This has led to an increased focus on the use of enzymes which oxidize fuels yet do not produce hydrogen peroxide, such as the dehydrogenases. The majority of dehydrogenases contain a loosely bound NAD^+ cofactor (Figure 5.2b) which acts as a two-electron and one-proton carrier. The NAD^+ cofactor ($E^0 = -0.52$ V vs. Ag/AgCl at pH = 7) [69] itself is not an effective mediator due to the large overpotential required for direct oxidation at solid electrodes and lack of reversibility of the NAD^+/NADH redox process. Several reports describe appreciable current densities for oxidation of glucose catalyzed by NAD^+-dependent glucose dehydrogenase (GDH) at low overpotentials (>1 mA cm^{-2} at about 0.1 V vs. Ag/AgCl) using diaphorase, which catalyzes NADH oxidation, coupled to reduction of vitamin K_3 as mediator, all adsorbed at high-surface-area carbon electrodes [40–42, 70]. Yan *et al.* [42] employed polymethylene blue as mediator for NAD^+-dependent GDH, adsorbed at single-walled carbon nanotube (SWCNT) electrodes, generating currents of 0.33 mA cm^{-2} at 0.1 V vs. Ag/AgCl in buffered solutions (pH = 6) containing 60 mM glucose. Tsujimura *et al.* [43] first reported the use of a PQQ-dependent GDH at the anode of an EFC operating at physiological conditions. The PQQ cofactor (Figure 5.2c), which has a redox potential of −0.13 V vs. Ag/AgCl at pH = 7 [71], is tightly bound to the enzyme and is thus more practical than its NAD^+-dependent counterpart. However, the wild-type PQQ-dependent GDH used is unstable, severely impeding its performance at physiological conditions [70]. Sode and coworkers produced several PQQ-dependent GDH mutants which show increased thermal stability [44] and improved DET properties [44, 71, 72] for use in EFCs operating under physiological conditions. In addition, Durand *et al.* [73] recently reported a soluble PQQ–GDH mutant, obtained by site-directed mutagenesis, which displayed twofold higher activity towards glucose oxidation compared to wild-type GDH.

In addition to glucose, other saccharides, such as fructose, have been studied as potential fuels. Fructose dehydrogenase (FDH), for example, has been used in a number of studies for oxidation via DET of fructose at carbon black-modified [45, 46] and CNT-modified [47] electrodes. Kamitaka *et al.* [45] observed current densities of about 4 mA cm^{-2} at carbon paper electrodes modified with FDH in solutions containing 200 mM fructose at pH = 5. Gorton and coworkers have published extensively on the use of cellobiose dehydrogenases (CDHs) in biosensors and as putative bioanodes in EFCs [17, 49, 74–79]. CDH, which contains a flavin and heme domain, connected by a flexible linker region, is capable of both DET and MET, as depicted in Figure 5.6. DET occurs through intramolecular electron transfer from the substrate-oxidizing flavin to the heme domain followed by DET from the heme to the electrode surface [49]. Alternatively, electron transfer from the flavin domain can be facilitated through mediation by an appropriate mediator such as an osmium redox complex [49]. SWCNTs on graphite electrodes modified with CDH sourced from *Phanerochaete sordida* produced DET and MET current densities of 0.06 and 0.6 mA cm^{-2}, respectively, in solutions containing 100 mM

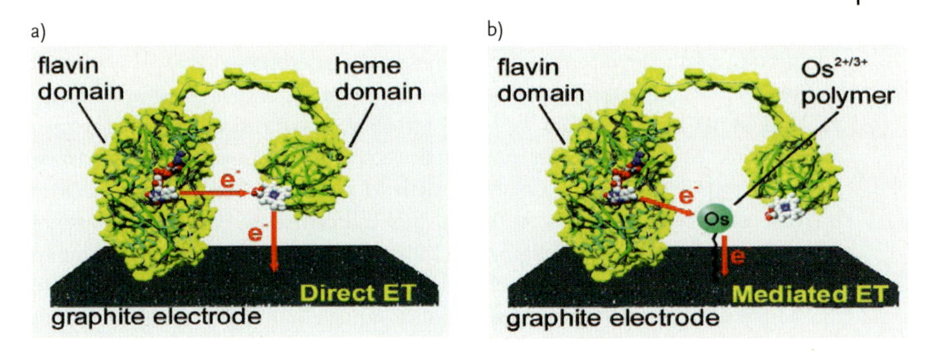

Figure 5.6 Schematic showing (a) DET and (b) MET for CDH. From [49] with permission from the American Chemical Society.

lactose [49]. Recent research has focused on production of CDHs that can efficiently oxidize glucose, instead of lactose, under physiological conditions, for application as glucose-oxidizing bioanodes in EFC configurations [17, 79]. Bioanode coulombic efficiencies can be greatly increased through combination of rationally selected enzymes to provide for more extensive (deeper) oxidation of the biofuel [80]. This was first demonstrated in an EFC anode by Palmore *et al.* [81] using a mixture of solution-phase dehydrogenases to catalyze the successive oxidation of methanol to carbon dioxide. Subsequently, Minteer and coworkers employed multiple-enzyme cascades for full or partial oxidation of ethanol [82, 83], pyruvate [84], and glycerol [85] in EFC assemblies. Recently, modified electrodes combining CDH and pyranose dehydrogenase (PDH) have been shown to be capable of extracting up to six electrons from one molecule of glucose [86].

5.3 Biocathodes

Several studies focused on understanding the factors that contribute to provide power in BFCs utilize hybrid systems, combining anodes that exploit the advantages of a biological catalyst with oxygen reduction reaction (ORR) catalysts used in traditional fuel cell research, such as platinum, platinum alloys, and other metal alloys [84, 85]. Such hybrid biological fuel cells, however, still require an ion-exchange membrane to separate the anolyte from the catholyte, and to efficiently transfer charge, usually protons, to maintain electro-neutrality and sustain the proton-coupled reduction of oxygen to peroxide or water. This is necessary as the ORR catalysts are not selective towards oxygen reduction alone, and may also oxidize the fuel, and/or be passivated by products of the fuel oxidation pathway. In addition the ORR at platinum at relatively low temperature and neutral pH has been shown to occur less efficiently than at electrodes modified with selected biological catalysts [87]. Stability, inhibition, and modulation of the biocatalytic processes remain, however, major problems for technological advances in the adoption of prototype biocatalytic cathodes on an industrial scale.

The theoretical thermodynamic reduction potential for oxygen to water is +1.23 V vs. NHE at pH = 0, or +0.82 V vs. NHE at pH = 7. The direct reduction of oxygen at metal-free carbon electrodes in neutral electrolytes is, however, hampered by slow electrode kinetics [88]. Catalysts, such as platinum, are therefore used to increase electrode kinetics in traditional fuel cell cathodes. However, the use of expensive, and nonselective, platinum catalysts is less compatible with operation of miniaturized membraneless fuel cells for portable systems or for application to *in vivo* power generation. An additional disadvantage of oxygen reduction at both metal-free carbon and platinum electrodes is that reduction occurs, at neutral pH, via a two-electron process to produce peroxide, a toxic reactive oxygen species. Recent research on the ORR has focused on replacement of platinum catalysts by less expensive metal alloys and co-catalysts, with some degree of success, beyond the scope of this chapter. The biological catalysts most widely studied for application to ORR in BFCs are a class of enzymes termed multiple copper oxidases (MCOs) that include laccases (EC 1.10.3.2), ascorbate oxidases (EC 1.10.3.3), bilirubin oxidases (EC 1.3.3.5), and others, that can reduce oxygen directly to water without apparent release of a peroxide intermediate. The first report of a MCO in a fuel cell for reduction of oxygen [89] used a solution-phase laccase isolated from *Pyricularia oryzae* and 2,2′-azinobis(3-ethylbenzothiazoline-6-sulfonate) (ABTS) as a diffusional mediator.

The MCOs generally contain at least four copper atoms in their complex active site, classified according to the electron paramagnetic resonance signals [19, 90–92]. One of these copper atoms, termed type 1 (T1), responsible for the blue color of copper proteins and enzymes, contains copper coordinated to at least two histidines and is the substrate oxidizing site of the MCOs. This substrate oxidation is coupled through intramolecular electron transfer via a conserved cysteine–histidine bridge to a trinuclear cluster of copper, consisting of one type 2 (T2) and two type 3 (T3) coordinated copper, approximately 1.3 nm distal to the T1 site, that reduces oxygen directly to water. While the exact detailed mechanism of oxygen reduction has yet to be clarified, recent computational, spectroscopic, and electrochemical studies on wild-type and mutant MCOs [92] indicate that oxygen binds to the trinuclear copper cluster of the reduced MCO, to form a peroxide intermediate, followed by intramolecular electron transfer and O–O bond cleavage to yield a fully oxidized, native intermediate, which reverts to the reduced form by accepting four electrons consecutively from the substrate with release of two molecules of water, as depicted in Figure 5.7.

Laccases are the most widely studied of the MCOs for application as biological ORR catalysts. Laccases, classed as polyphenol oxidases, catalyze the oxidation of diphenols, amines, and some inorganic ions as substrates, coupled to the reduction of oxygen to water. The substrate binding pocket at the T1 copper is rather open, resulting in the broad substrate range for the laccases, and this substrate range can be extended by the use of mediators [93], resulting in technological application of laccases in a range of applications, from biosensors [94–97], biob-

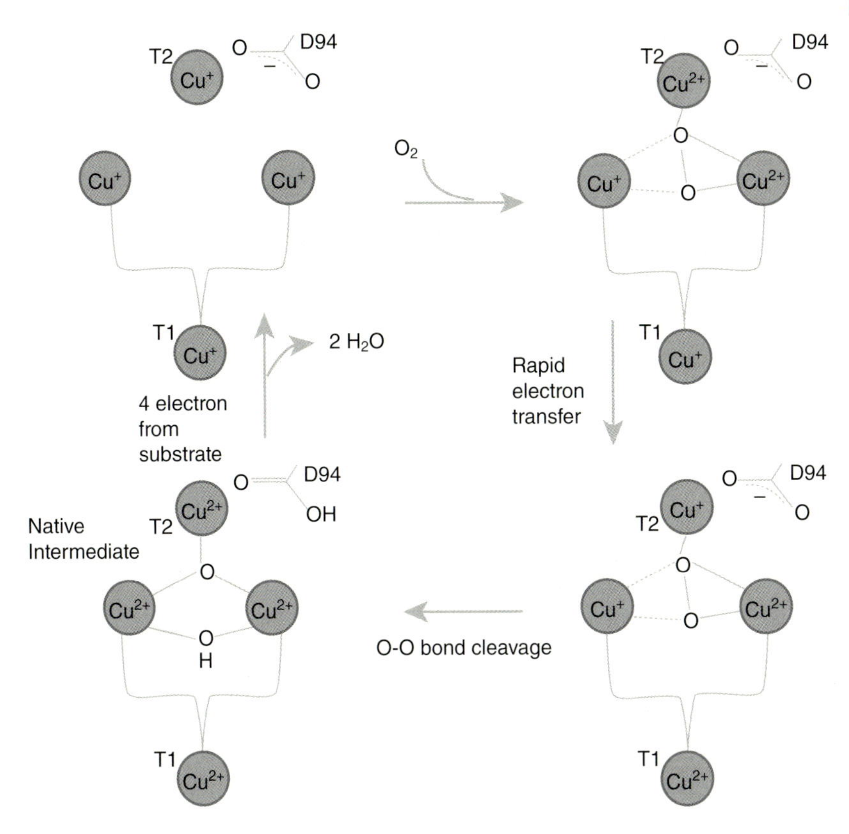

Figure 5.7 Proposed mechanism for reduction of oxygen to water at MCO active sites. Adapted from [92] with permission from the American Chemical Society.

leaching, and biodegradation [98–101], to EFCs. Laccase was first isolated by Yoshida [102] in 1883 from tree lacquer of *Rhus vernicifera,* with laccases subsequently isolated from plant, fungal, or, more recently, bacterial and insect sources [103]. Interest in the use of laccases as ORR catalysts stems from the reported low overpotential for ORR of a range of fungal laccases compared to that of platinum. The low overpotential for ORR has been ascribed to the relatively high reduction potential of their T1 copper sites, the capacity of the MCOs to utilize multi-atomic reaction sites, the influence of residues close to the active site on the reaction mechanism, and the structural flexibility of both the catalyst protein shell and the Cu–Cu interatomic distances of the active site, in comparison to simpler inorganic metal clusters [92, 103, 104]. The fungal laccases, in general, have higher T1 site redox potential, about 0.58 V vs. Ag/AgCl, than the plant laccases (e.g., 0.23 V vs. Ag/AgCl for the *R. vernicifera* laccase), with ligand coordination geometry around the copper atom, and in particular the presence of weakly axially coordinated residues, contributing to the differences. The MCOs have also been shown to transfer electrons directly with electrode materials, allowing estimation of the T1 redox potential using voltammetry, and possible correlation with structure and activity.

DET to a laccase was first reported by Yaropolov's group for enzymes adsorbed onto carbon electrodes [105]. Subsequent studies have investigated DET to the copper sites of MCOs leading to their classification, as suggested by Shleev *et al.* [19, 106], into separate groups, based upon the redox potential of the T1 site. The plant, bacterial, and some fungal laccases have a low T1 potential of about +0.23 V vs. Ag/AgCl; bilirubin oxidases (BODs) and some fungal laccases possess T1 sites of middle potential between +0.27 and +0.51 V vs. Ag/AgCl; and fungal laccases, especially isolated from *Trametes* sources, possess high T1 potentials of about +0.58 V vs. Ag/AgCl. Initial studies of MCOs as ORR catalysts, not surprisingly therefore, focused on the use of these *Trametes* laccases.

Unfortunately, the high-redox-potential fungal laccases are inhibited by hydroxyl ions, with maximal activity centered around pH = 4–5, and, to a lesser extent, by chloride ions [107]. This limits their use in neutral pH or saline conditions, for example as biocathodes in a putative implantable EFC. This low activity of most fungal laccases under pseudo-physiological conditions has led to increased focus on MCOs that are more active under those conditions. These include BODs [108], certain fungal laccases, such as that isolated from *Melanocarpus albomyces* [109–111], bacterial MCOs such as the copper efflux oxidase (CueO) isolated from the periplasmic space of *Escherichia coli* [112–116], CotA laccase isolated from *Bacillus subtilis* [117, 118], and small laccase from *Streptomyces coelicolor* (SLAC) [119, 120]. The first report on a BOD-based biocatalytic oxygen cathode focused on homogeneous ABTS-mediated reduction of oxygen at carbon felt electrodes using a BOD from *Myrothecium verrucaria* in phosphate buffer (pH = 7.0) [109]. However, the potential of the T1 site of these MCOs is of low to middle classification only, with activity at neutral pH for the ORR apparently being compromised by a decrease in the onset of the MCO-catalyzed ORR potential [108, 121].

DET to MCOs, for bioelectrocatalytic reduction of oxygen, is of immense interest to biocathode research and development, as this usually enables ORR onset at the T1 copper reduction potential, which can be close to the thermodynamic potential for oxygen reduction for the high-redox-potential laccases. For the reasons outlined above, DET to MCOs is surface specific. DET for the ORR is reported to occur through the four-electron reduction of oxygen to water, with the T1 copper site as the proposed linkage between electrode and MCO, for MCOs physically adsorbed at spectroscopic graphite and carbon aerogels [114, 122–124]. However, DET for the ORR is thought to occur through a different mechanism for MCOs on bare gold [123, 125, 126]. The earliest studies, focused on examining the possibility for DET to MCOs, were conducted by adsorption of fungal and plant laccases on carbon black [127] or spectroscopic graphite electrodes [128] of high surface areas, providing proof of DET through bioelectrocatalytic oxygen reduction at low overpotentials at MCO-modified surfaces compared to bare carbon electrodes. Comparison of current densities, and the potential for maximum ORR current, is hampered, however, by variability in the conditions used in each research report. Nonetheless, an attempt at a limited compilation of the more recent configurations of MCO-based biocathodes using DET for the ORR can be made, as summarized Table 5.2.

Table 5.2 A comparison of the performance of biocathodes for the ORR, based on DET to MCOs.

MCO	Surface	Conditions for ORR	Potential (V)[a]	Current density ($mA\,cm^{-2}$)[b]	Ref.
Rhus vernicifera laccase	EDC coupling to a mercaptopropionic acid-treated gold electrode	pH = 7.0, 21 °C, aeration of cell	0.1	0.080	[129]
Myrothecium verrucaria BOD	Adsorption to spectroscopic graphite electrode	pH = 4.0	0.3	0.03	[128]
Myrothecium verrucaria BOD	Adsorption onto a polylysine-modified plastic formed carbon electrode (high edge plane density)	pH = 7.0, 25 °C, aeration of cell, 1400 rpm rotation	0.25	0.85	[121]
Trametes versicolor laccase	Glutaraldehyde crosslinking to MWCNT, solubilized using cellulose derivatives, on glassy carbon electrode	pH = 6.0, aeration of cell	0.45	0.06	[130]
Myrothecium verrucaria BOD		pH = 7.0, aeration of cell	0.3	0.03	
Myrothecium verrucaria BOD	EDC coupling to MWCNT, deposited on glassy carbon electrode	pH = 7.0, aeration of cell	0.33	0.04	[131]
Trametes sp. laccases	Adsorption to carbon aerogel-modified carbon paper electrode	pH = 5.0, 25 °C, aeration of cell, 3000 rpm rotation	0.4	6.5	[124]
Myrothecium verrucaria BOD		pH = 7.0, 25 °C, aeration of cell, 2000 rpm rotation	0.3	3.5	
Rhus vernicifera laccase	Adsorption to highly oriented pyrolytic graphite	pH = 5.0, 25 °C, aeration of cell	0.1	0.052	[45]
Trametes sp. laccase			0.3	0.168	
Myrothecium verrucaria BOD	Adsorption to carbon aerogel-modified carbon paper		0	0.205	
Trametes sp. laccase			0.4	3.5	
Trametes hirsuta laccase	EDC coupling to aminophenyl-modified low density graphite electrode	pH = 5.1, 27 °C	0.2	0.5	[122]

(Continued)

Table 5.2 (Continued)

MCO	Surface	Conditions for ORR	Potential (V)[a]	Current density ($mA\,cm^{-2}$)[b]	Ref.
Myrothecium verrucaria BOD	EDC coupling to organic carboxylates (PQQ best) adsorbed on MWCNT, deposited on gold electrode	pH = 7.0, aeration of cell unstirred	0.25	0.5	[125]
		1300 rpm stirring		1.6	
E. coli CueO	Adsorption onto a Ketjen black-modified glassy carbon electrode	pH = 5.0, 25 °C, aeration of cell, 10 000 rpm rotation	0	12	[114]
		pH = 7.0, 25 °C, aeration of cell, 1000 rpm rotation	0	4	
Cerrena unicolor laccase	Adsoption onto CNT-loaded glassy carbon electrode with Nafion binder	pH = 5.2, 22 °C, aeration of cell	0.2	0.058	[132]
				0.11	
E. coli CueO	Adsorption onto a Ketjen black–polytetrafluoroethylene binder-modified carbon paper electrode	pH = 5.0, 25 °C, air diffusion configuration	−0.2	20	[116]
Cerrena unicolor laccase	Entrapment within carbon nanoparticle-loaded sol–gel on indium tin oxide electrode	pH = 4.8, 22 °C, aeration of cell	0.2	0.08	[133]
Myrothecium verrucaria BOD	Entrapment within carbon nanoparticle loaded sol-gel	22 °C, aeration of cell	0.2		[134]
		pH = 5		0.23	
		pH = 6		0.255	
		pH = 7		0.13	
Trametes hirsuta laccase	Adsorption to a gold nanoparticle dispersion on glassy carbon electrode	pH = 4.0	0.5	0.005	[135]

a) Potential at which a steady-state current for the ORR is observed, corrected to the Ag/AgCl (3 M NaCl) reference.
b) Current densities are for the projected two-dimensional geometric underlying electrode area.

Addition of conducting particles and tubes possessing nanometric dimensions has been postulated to improve DET for ORR by orienting the MCOs in a matrix of conductive electrode material that can approach the T1 copper site. For example, increased current density for ORR is observed for *Trametes* sp. laccase adsorbed onto three-dimensional carbon aerogel-modified electrodes, compared to that observed when the enzyme is adsorbed onto highly oriented pyrolytic graphite [45]. Addition of CNTs [124, 125, 131, 134], and carbon [114, 124, 133] or gold nanoparticles [135] to an electrode matrix, either drop-coated or by incorporation into a sol–gel [135], has also yielded substantial currents for the ORR when MCOs are either adsorbed or coupled to the nanostructures. Other approaches for achieving increased current density focus on retaining the MCOs in a hydrogel close to the active electrode surface, for example by using a polylysine matrix, resulting in a reaction-layer model where the enzyme, BOD in this case, can diffuse to and from the electrode [121]. An elegant approach for achieving improved DET current density for the ORR using a laccase, isolated from *Pycnoporus cinnabarinus*, proposed retention to a pyrolytic graphite electrode surface via a surface layer of anthracene, introduced via *in situ* diazotization of anthracene amine followed by electroreduction of the anthracene-diazonium salt. It was proposed that the anthracene orients and retains the laccase through interaction with a hydrophobic pocket close to the T1 active site, permitting DET, yielding a current density that is twice that for laccase adsorbed to an unmodified electrode. More importantly, this study reported that laccase on the anthracene-modified electrode maintained activity for ORR at levels exceeding 50% of the initial activity for over eight weeks, when the electrodes were stored hydrated at 4 °C between experiments (Figure 5.8) [136].

More recently, isolation of a MCO from the periplasmic space of *E. coli*, proposed to have a role in copper regulation [112], has led to its use as an ORR biocatalyst [113–116]. This CueO, while requiring a fifth copper to yield oxidase activity in the presence of mediators such as ABTS or phenylenediamine, has been shown to be a highly efficient catalyst for the ORR through DET at nanostructured carbon surfaces [114]. These surfaces are prepared by addition of a slurry of Ketjen black particles and binder to glassy carbon electrodes, and yield current densities exceeding 12 $mA\,cm^{-2}$ at pH = 5.0 for the ORR. This, however, comes at a thermodynamic cost, as steady-state ORR current is observed at a potential of 0 V vs. Ag/AgCl, approximately 0.4 V more negative than the potential for steady-state ORR at electrodes modified with fungal laccases.

An approach focused on fabrication of nanostructured three-dimensional electrodes and introduction of surface modifications for tethering/retention in an optimal orientation of the MCOs to permit DET to the T1 site from the electrode shows great promise for the production of biocathode prototypes for application to EFCs. A systematic study of such electrodes modified with each of the MCOs available, reporting on their activity for ORR, using DET, under defined conditions of pH, mass transport, and temperature is not yet available, and would be a valuable contribution to advance the technological application of EFCs. A welcome recent focus is normalization of ORR, based on DET to *Trametes versicolor* adsorbed on porous carbon-based electrode materials, to electrode volume and to electrode

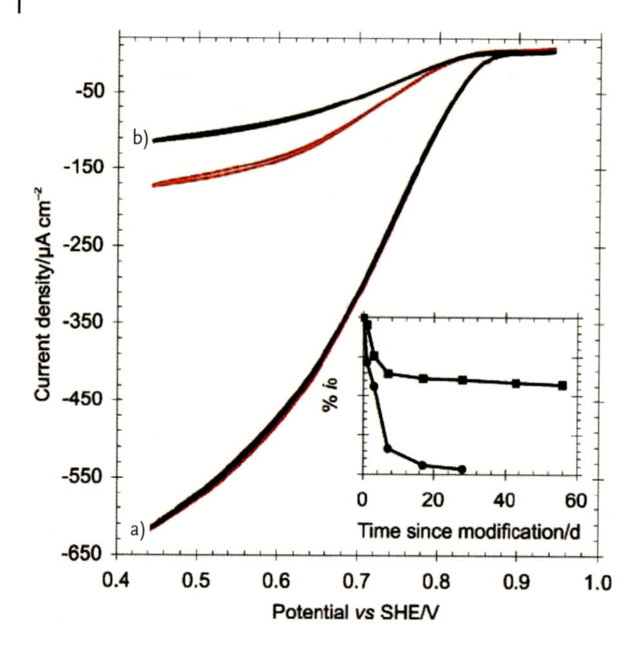

Figure 5.8 Cyclic voltammogram (5 mV s^{-1}, 2500 rpm) recorded for laccase adsorbed to (a) anthracene-modified and (b) unmodified pyrolytic graphite electrodes immediately following laccase deposition (black) and upon exchange of electrolyte (red). The inset shows long-term change in current density recorded at 0.44 V vs. SHE for modified (squares) and unmodified (circles) electrode, relative to the first wave with no enzyme in solution. Adapted from [136] with permission from the Royal Society of Chemistry.

Brunauer–Emmet–Teller (BET) estimated surface areas [23]. For example, from Figure 5.9, graphite felt electrodes show poor volume-normalized ORR current density compared to carbon nanofibers and multiwalled carbon nanotube (MWCNT)-based electrodes. However, the results also reveal that CNTs and porous carbon tubes exhibit dramatically lower ORR current densities when normalized to BET surface area, while graphite felt electrodes perform better, perhaps indicative of agglomeration of the carbon tubes, preventing enzyme adsorption over the entire area. Further research on methods to permit dispersion of nanotubes, while retaining electrical conductivity and adsorption of enzymes oriented for DET, is warranted.

Recent research on biocathode design has focused on gas-diffusion layers at the cathode to afford air-breathing biocathode configurations. Such configurations can overcome the limitation on current imposed by the availability of oxygen to the cathode due to its low solubility and diffusion coefficient in the aqueous electrolytes used heretofore. For example, CueO was adsorbed onto porous carbon paper, modified with carbon nanoparticles (Ketjen black) using hydrophobic binders, to yield biocathodes that could operate in air, permitting oxygen diffusion from air to the particles with minimal loss of electrolyte, as illustrated schematically in

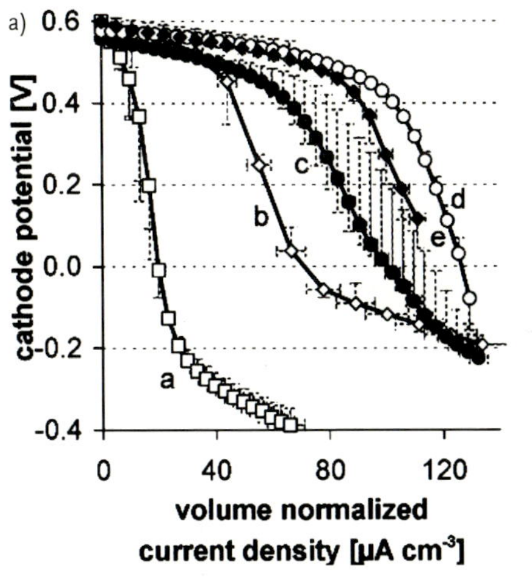

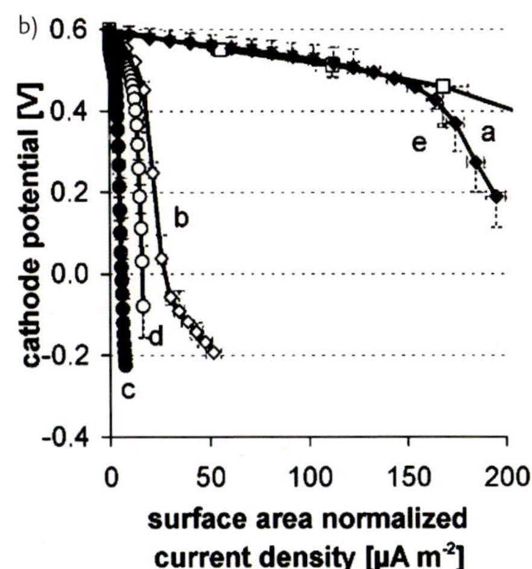

Figure 5.9 Cathode polarization plots for electrodes with laccase adsorbed to graphite felt (curves a), porous carbon tubes (curves b), SWCNTs (curves c), MWCNTs (curves d), and carbon nanofibers (curves e). (a) Comparison of cathode performance with a current density normalized to volume; (b) comparison of cathode performance with a current density normalized to BET surface area. From reference [23] with permission from Elsevier.

Figure 5.10 [116]. Such biocathodes yielded current densities in the $mA\,cm^{-2}$ range. However, the use of CueO as a biocatalyst meant that steady-state currents for ORR were only observed at potentials more negative than −0.2 V vs. Ag/AgCl. It was also suggested that the current was limited by supply of protons, as the current was dependent on buffer concentration in the electrolyte. Use of a 1.0 M citrate electrolyte (pH = 5.0), resulted in a steady-state current density of $20\,mA\,cm^{-2}$, the highest yet reported for the ORR at a biocathode.

Advances in nanostructured conducting materials for DET have resulted in impressive current densities for the ORR, and application of these three-dimensional materials to DET from MCOs other than CueO may provide biocathodes with the characteristics suitable for an implantable EFC. While a DET approach using MCOs can provide for ORR at potentials approaching the thermodynamic reduction potential for oxygen, the current density achievable in this approach still relies upon intimate contact, and correct orientation, of the MCO to a conducting surface. Use of a mediator, capable of close interaction with the T1 site of the MCOs, and with a redox potential tailored to permit rapid electron transfer to the T1 site, can eliminate the requirement for direct contact in the correct orientation between MCO and electrode, and offer the possibility of a three-dimensional biocatalytic reaction layer on electrodes for higher ORR current densities.

Initial reports on mediated ORR using MCOs focused on assessing the capability for ORR using solution-phase mediators with solution-phase MCOs [89].

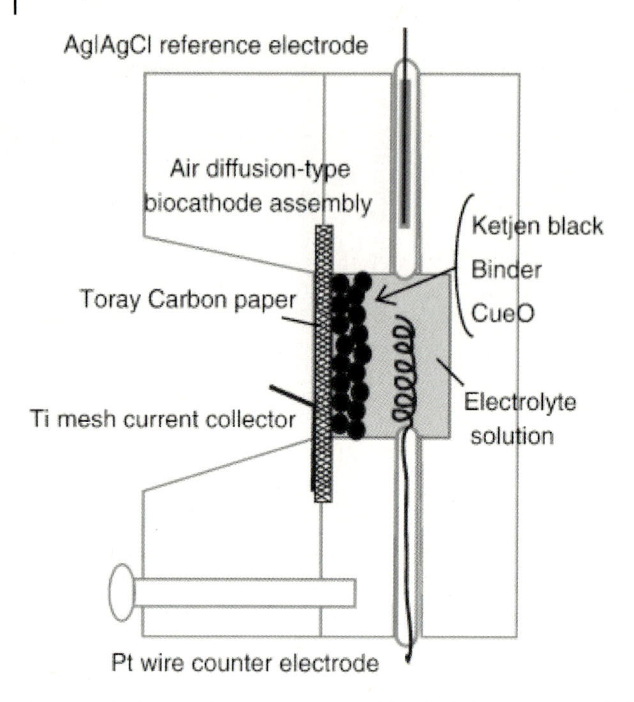

Figure 5.10 Schematic representation of the air diffusion biocathode half-cell based on CueO adsorbed on carbon particle-modified carbon paper electrode. Adapted from [116] with permission from Elsevier.

Immobilization of both mediator and MCO on relatively inert (to ORR) carbon-based electrodes can provide biocathodes for miniaturized membraneless BFCs. Trudeau *et al.* [96] were first to report on oxygen reduction by films of immobilized mediator and laccase, formed by crosslinking a *Trametes versicolor* laccase to the osmium-based redox polymer in Figure 5.5a, physisorbed on carbon electrodes. Although the laccase biocatalytic electrode for ORR was developed for detection of modulators of enzyme activity, steady-state current densities of greater than $125\,\mu A\,cm^{-2}$ were achieved at potentials of about +0.15 V vs. Ag/AgCl in oxygen-sparged acetate buffered solutions (pH = 4.5) [137]. Further improvements in the performance of redox polymer hydrogel-based biocathodes for the ORR were achieved by tailoring, through synthesis, the structure of the osmium ligands to yield redox polymers matching closely the redox potential of the T1 site of the MCOs. For example, substitution of the chloride ligand of a $[Os(4,4'\text{-dimethyl-}2,2'\text{-bipyridine})(2,2'{:}6',2''\text{-terpyridine})Cl]^+$ complex with imidazole units of PVI yields a redox polymer that may be co-immobilized with laccase from *Coriolus hirsutus* on carbon cloth fiber. The resulting biocathode can operate at $mA\,cm^{-2}$ current densities at +0.42 V vs. Ag/AgCl for the ORR in buffer of pH = 5, 37 °C, when rotated at 4000 rpm [138]. Co-immobilization of the redox polymer described above with a laccase from *Pleurotus ostreatus,* which has been reported to retain a high level of activity at pH = 7, yielded biocathodes capable of ORR in 0.1 M NaCl solution (pH = 7) at 37 °C [139]. Insertion of a more flexible tether between the polymer backbone and the redox complex, in an approach adopted for glucose-oxidizing anodes, is reported to yield improved performance, in terms of lower overpotentials and higher current densities, for mediated ORR in redox polymer hydrogels. This approach was implemented by coupling of a $[Os(4,4'\text{-dimethyl-}2,2'\text{-bipyridine})_2(4\text{-aminomethyl-}4'\text{-methyl-}2,2'\text{-bipyridine})]^+$ complex to a quaternized polyvinylpyridine polymer. The resulting redox polymer showed a twofold improvement in current density, and a decrease in overpotential, for the ORR at pH = 5, when co-immobilized with a fungal laccase at carbon fiber electrodes, compared to films prepared using a redox polymer without the tether [95]. Other studies have focused on matching the redox polymer potential to that reported for the T1 site of BODs, to overcome the reported chloride and pH sensitivity of the fungal laccases. A redox polymer prepared by substitution of one of the chloride ligands of a $Os(4,4'\text{-dichloro-}2,2'\text{-bipyridine})_2Cl_2$ complex with imidazole units of a copolymer of PVI and polyacrylamide can provide ORR current densities of $0.7\,mA\,cm^{-2}$ at +0.3 V vs. Ag/AgCl in unstirred phosphate buffered saline at 37 °C when co-immobilized with *M. verrucaria* BOD on carbon cloth electrodes [140]. Use of a *Trachyderma tsunodae* BOD, claimed to have a T1 redox potential of +0.44 V vs. Ag/AgCl, yields current densities of $6.25\,mA\,cm^{-2}$ for the ORR at +0.3 V vs. Ag/AgCl with electrodes rotated at 4000 rpm in oxygenated phosphate buffered saline at 37 °C [141]. Further purification of the enzyme yielded films providing current densities of $9.5\,mA\,cm^{-2}$ under these conditions [142]. More recent reports have focused on ORR using mediation to laccases that are active in neutral electrolytes. For example, SLAC produced higher current densities for the ORR at pH = 7 compared to that obtained with a *Trametes versicolor* laccase, when co-immobilized with the redox polymer in Figure 5.5a on carbon electrodes [120]. The use of a redox polymer with lower redox potential than those used with BOD or high-potential fungal laccases is required, as the reported redox potential of SLAC is about +0.3 V vs. Ag/AgCl. Nonetheless, current densities of $mA\,cm^{-2}$ are achieved for the ORR at 0.1 V vs. Ag/AgCl in electrolyte at pH = 7 for films of SLAC and the redox polymer adsorbed to high-surface-area carbon composite electrodes with mass transport aided by rotation of the electrode. The recently reported crystal structure for this enzyme, indicating a trimeric form containing 12 copper atoms [143], points to its unique characteristics over other MCOs, and may provide a route for tailoring of its structure to yield variants of increased stability and redox potential at neutral pH. An alternative laccase active in neutral electrolyte, isolated from *Melanocarpus albomyces*, can catalyze the ORR under pseudo-physiological conditions when co-immobilized with the redox polymer in Figure 5.5a on smooth glassy carbon electrodes, displaying current density of $0.38\,mA\,cm^{-2}$, but again only at lower redox potentials than that observed for the more acidophilic fungal laccases [110].

A series of recent reports have highlighted the improvement in current density that can be achieved by use of electrodeposition for formation of redox hydrogels, moving from relatively smooth surfaces to structured roughened surfaces [144], and by systematic variation in the structure of the redox complex and polymer backbone using osmium-based redox polymers as mediators [22, 144–148]. For example, electrodeposition, via electrochemically induced ligand exchange of labile chloride ligands of an osmium complex for imidazole units in a polymer, to provide crosslinked *Trametes versicolor*-containing redox hydrogels on carbon surfaces yielded higher specific activity for the ORR compared to chemically crosslinked hydrogels, postulated to be a result of thinner, more homogeneous film formation [144]. Using porous carbon fiber paper electrodes, with mass transport controlled using electrode rotation, ORR current densities of greater than $10\,mA\,cm^{-2}$ for both electrodeposited and chemically crosslinked *Trametes versicolor*-containing redox hydrogels, at 0.2 V vs. Ag/AgCl, could be obtained. Alternatively, incorporation of both functional groups for coordination to redox complexes and functional groups to provide for electrodeposition via protonation or de-protonation induced by local pH alteration at electrode surfaces by electrolysis of water provides a facile means for control of redox hydrogel formation [146, 147]. Initial reports using this procedure focused on an $Os(2,2'\text{-bipyridine})_2Cl_2$ complex, ligand-exchanged to an imdazolyl group on an electrodeposition polymer that also incorporated carboxylic acid groups, permitting co-deposition with BOD, by protonation in acidic electrolyte. ORR activity was observed at 0 V vs. Ag/AgCl, with local ORR over the biofilm visualized using scanning electrochemical microscopy [146]. Further refinement of the redox polymer illustrated in Figure 5.11 provided for a biofilm containing *Trametes hirsuta* laccase on glassy carbon electrodes that yielded ORR current densities of $0.3\,mA\,cm^{-2}$ at 0.4 V vs. Ag/AgCl in oxygen-saturated buffer (pH = 4.0) [147].

A systematic approach using *Trametes versicolor* laccase co-immobilized with osmium-based redox polymers has highlighted the importance of redox complex loading [144, 145], and of the redox potential of the redox polymer [22], on ORR current density. Increasing ORR current density is observed with increased difference in redox potential between the T1 site of laccase and redox polymer, but maximum power from an EFC is predicted to result using a redox polymer of redox potential 0.17 V more negative than the T1 site of laccase. It would be inter-

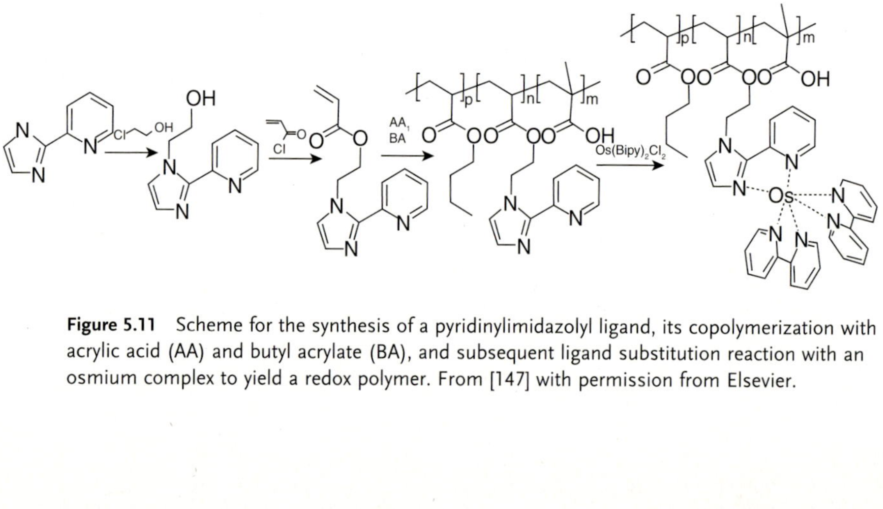

Figure 5.11 Scheme for the synthesis of a pyridinylimidazolyl ligand, its copolymerization with acrylic acid (AA) and butyl acrylate (BA), and subsequent ligand substitution reaction with an osmium complex to yield a redox polymer. From [147] with permission from Elsevier.

esting to extend this study to other MCOs, and to electrodeposited redox hydrogels. Indeed, in a preliminary study we reported a comparison of ORR currents observed for three redox polymers co-immobilized with BOD and *Trametes hirsuta* and *Melanocarpus albomyces* laccases [148], with highest ORR current density observed for redox hydrogels of BOD and the redox polymer in Figure 5.5a.

5.4 Assembled Biofuel Cells

We review in this section progress focused on development of devices capable of power generation through oxidation of glucose and reduction of oxygen. Studies aimed at device implementation as an implanted system to harvest energy from the body's own fuel under physiological conditions should focus on oxygen and glucose concentrations of about 10^{-4} M and about 6 mM, respectively, in solutions containing about 150 mM chloride, buffered to a pH of 7.4, thermostatically controlled at 37 °C. The velocity of blood in blood vessels is of the order of 1 to 10 cm s^{-1}. Unfortunately, to date, few studies have been undertaken in a configuration which permits exploration of EFC response under these conditions, usually because of limitations in condition choice due to biocatalyst sensitivity to one or other of the variables. Other body fluids or configurations may be considered, for instance EFCs using saliva, tears, or sweat or using high-energy-density fuels such as methanol or carbohydrates in devices for portable power.

Prototype EFC devices have been reported for both implanted power and portable power generation that point to technological application of devices, and provide a foundation to target improvements in device configuration and performance. These prototypes, to date, are based on demonstration of the concept using EFCs assembled from two chambers, separated by a membrane, containing diffusible biocatalysts and mediators adsorbed onto high-surface-area carbon [8, 40]. For example, mediated biocatalysis of glucose oxidation and oxygen reduction using vitamin K_3, diaphorase, NADH, and GDH in the anolyte and ferricyanide and BOD in the catholyte separated by cellophane membranes can provide power at pH = 7.0, in a demonstration unit for powering portable devices [40]. An air-breathing configuration adopted to enhance oxygen supply to the cathode resulted in a single EFC, operated as a "biobattery" with glucose introduced as a bolus at 0.4 M level, delivering maximum power density of 1.45 mW cm^{-2} at 0.3 V. A parallel stack structure achieved a power output of 50 mW from a 40 cm^3 volume that was expanded to provide over 100 mW from 80 cm^3 volume, when two such systems were connected in series, providing sufficient power to operate a toy car or portable music player over at least two hours. Another important demonstration of the potential application of EFCs focused on evaluating the performance of EFCs implanted in the retroperitoneal space of a rat [8]. The EFC was based on mediated biocatalysis of glucose oxidation and oxygen reduction using ubiquinone, catalase, and GOx confined behind a dialysis membrane at a high-surface-area carbon anode and quinone and a polyphenol oxidase (a tyrosinase) at a similar cathode, both electrodes then wrapped in an expanded polytetrafluoroethylene membrane. Maximum specific power density of 24.4 μW cm^{-3} was obtained at 0.13 V, with stable production of 2 μW of power observed over several hours, for the implanted system. Although little detail is provided on the performance of anode or cathode individually the platform proposed is amenable to simple interchanging of biocatalysts and mediator to help evaluate relative performances of a range of implanted EFCs.

An innovative alternative approach to the use of membranes in constructing prototype EFC assemblies for powering portable devices focuses on the use of either concentric porous tubes of graphite [149, 150] or of delivery of fuel and oxidant via laminar flow within a microchannel, to help prevent interference of fuel or oxidant in the cathodic or anodic reaction [41, 151–155]. The concentric glucose–oxygen EFC utilizes the outer surface of a porous carbon tube as cathode, and the inner surface of a second, larger diameter, porous tube encasing the first tube as anode, with delivery of oxygenated buffer through the inner tube bore while maintaining the rest of the system in deoxygenated 10 mM glucose buffer (pH = 7.4) [149]. Stability of the system was enhanced by grafting BOD and GOx at cathode and anode precoated with an electropolymerized layer of polyaminopropylpyrrole, and by trapping mediators, ABTS and 8-hydroxyquinoline-5-sulfonic acid, at cathode and anode using a diepoxide reagent, to yield a system providing maximum power of 20 μW cm^{-2} at 0.20 V in 10 mM glucose solutions over a 45-day period when used intermittently [150].

Microfluidic EFCs are designed to operate using laminar flow delivery of almost parallel streams of fuel and oxidant within the microchannel, preventing crossover and mixing [151–155]. For example, delivery in a polydimethylsiloxane-fabricated Y-shaped microchannel of separate streams of glucose anolyte (pH = 7) and oxygen in catholyte (pH = 3) for oxidation by GOx–Fe(CN)$_6$ [3], and reduction by laccase–ABTS at gold anodes and cathodes, respectively, produced a maximum power density of 110 μW cm^{-2} at 0.3 V at room temperature [154]. Further optimization of electrode and channel size, and spacings, led to smaller ohmic drop and depletion zone within the system providing power density up to 0.55 mW cm^{-2} at 0.3 V [155]. Of course, operation of microfluidic pumps and the requirement to add mediator and enzyme to the channel streams mitigate against adoption of microfluidic EFCs for energy harvesting to power portable electronic devices in this configuration. Progress depends upon immobilization and stabilization of enzymes, and mediators if required, onto channel electrodes, as demonstrated within microchannels fueled with 5 mM glucose and 1 mM NAD$^+$, oxidized at high-surface-area carbon anodes modified by co-adsorption of vitamin K$_3$-modified polylysine, diaphorase, and NAD-dependent GDH. This anode was coupled to oxygen reduction at a platinum cathode, to provide 32 μW cm^{-2} at 0.29 V [41]. The same anode has been coupled to a cathode, constructed by adsorption of BOD onto high-surface-area carbon for the ORR via DET, to examine the effect of electrode configuration on cell performance. Results indicate that placement of the cathode upstream from the anode can protect the anode from oxygen and lead to an increase in maximum cell current [153]. Prior to the appearance of these demon-

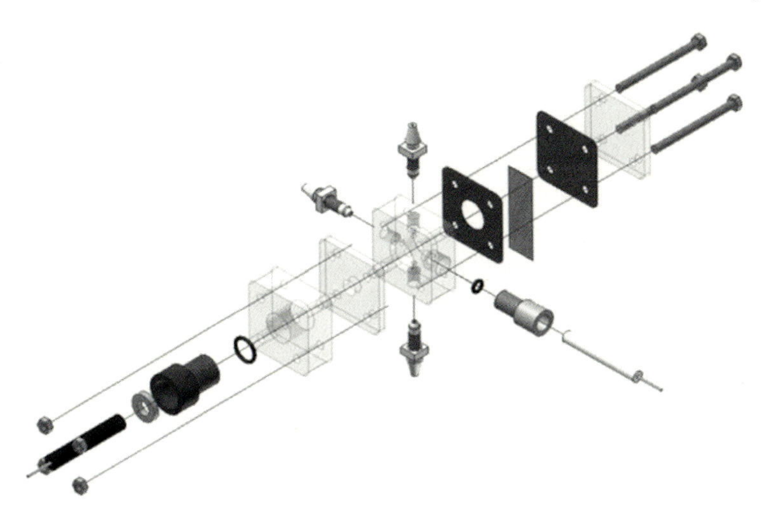

Figure 5.12 Three-dimensional drawing of the modular stack half-cell showing the central reaction chamber with reference electrode inlet (bottom right) and solution filling ports, the working electrode plate with inlet for the glassy carbon electrode (bottom left), and the counter electrode plate separated by gasket seal, all held leak-free together using bolts (top right). From [25] with permission from Wiley.

stration devices, a range of reports on the assembly, and the performance, of BFCs have appeared. These reports, with some notable exceptions, mostly focus on demonstration of the capability to generate power under a defined condition (pH, temperature, biocatalyst, surface, etc.), usually with little consideration of the stability of such power or of the consequences for powering devices. Comparison of power output from such assemblies under the range of conditions utilized thus becomes, as is evident from previous comparison of bioanode and biocathode performance, problematic. A welcome contribution to methodologies for comparison of bioelectrocatalytic fuel cell electrodes is the use of a standardized fuel cell set-up [25]. The set-up is actually a half-cell, anode, with a platinum mesh as counter electrode, assembled, as presented in Figure 5.12, to permit flow presentation of fuel to a glassy carbon anode. The reproducibility of the glassy carbon anode surface area was verified using voltammetry of a solution-phase ferricyanide redox probe. Reproducibility of voltammetry for deposited polymethylene green and for the voltammetric response of the deposited film to NADH oxidation is presented, with a view for use as a bioanode for NADH-dependent dehydrogenase reactions. Results from the use of such a standardized methodology should prove invaluable for comparison of surfaces, mediators, and enzymes in bioanodes and BFCs, and are eagerly anticipated!

A series of reports, focused on systematic approaches to optimizing choice of biocatalyst and either redox polymer-based mediator and/or carbon-based support, to provide for a glucose–oxygen EFC are noteworthy for the wide-ranging contribution to technological advances. This series of studies, based on initial reports of redox polymer-mediated biosensors for glucose [58–61] and oxygen [96, 137], focused on systematic selection and optimization of biocatalyst, mediator, and surface for anode and cathode, as detailed in the previous sections, and their combination to provide prototype EFCs. For example, initial demonstration of power density of 137 μW cm^{-2} at 0.37 V in 37 °C aerated buffer at pH = 5 and chloride-free medium containing 15 mM glucose was provided by a cell constructed using carbon fiber electrodes, modified by crosslinking GOx to [Os(4,4′-dimethyl-2,2′-bipyridine)$_2$(PVI)Cl]$^+$ and laccase (*Coriolus hirsitus*) to [Os(4,4′-dimethyl-2,2′-bipyridine)$_2$(2,2′,6′,2″-terpyridine)]$^{2+}$ tethered to a PVI polymer [35]. Systematic improvements in anodic redox polymer, by inclusion of a flexible tether to improve charge transport diffusion, and a more appropriate redox potential, provided power density of 268 μW cm^{-2} at 0.78 V [156]. When the same refinement procedure was followed for the cathode [87], a power density of 350 μW cm^{-2} at 0.88 V was obtained, in buffer pH = 5, a cell voltage only 0.3 V lower than that of the reversible cell voltage for glucose–oxygen.

Recent reports have highlighted improvement in power densities for assembled EFCs using osmium redox polymer mediators by purification of GOx [157], or replacement of GOx isolated from *Aspergillus niger* with GOx isolated from *Penicillium pinophilum*, used in the anode [39]. For example, a biofuel using purified *A. niger* GOx at the anode provided twice the power density than obtained using the "as-received" GOx, when operated using a BOD cathode in 15 mM glucose in phosphate buffered saline (pH = 7.4) [157]. Use of GOx isolated from *P. pinophilum* versus that from *A. niger* at a redox polymer anode coupled to a laccase cathode yielded a threefold increase in power density, to 0.28 mW cm^{-2} at 0.88 V, at glucose concentrations of 5 mM, approaching those in serum [39]. Unfortunately, the instability of the *P. pinophilum* GOx in neutral electrolyte precluded operation under physiological pH conditions. In addition, most fungal laccases lose activity in neutral electrolyte, and undergo inhibition by chloride. In an EFC we have developed, a maximum power density of 16 μW cm^{-2} at a cell voltage of 0.25 V can nonetheless be obtained in pseudo-physiological conditions of 10 mM glucose in phosphate buffer (pH = 7.4) at 37 °C, for graphite electrodes with an *A. niger* GOx crosslinked to [Os(4,4′-diamino-2,2′-bipyridine)$_2$(PVI)Cl]$^+$ as anode and *Trametes versicolor* laccase crosslinked to [Os(1,10-phenanthroline)$_2$(PVI)$_2$]$^{2+}$ as cathode. A maximum power density of 40 μW cm^{-2} was observed for this cell at pH = 5.5 where laccase is more active [158]. As discussed previously, replacement of acidophilic fungal laccases with biocatalysts that are active towards the ORR in neutral electrolyte can provide for improved ORR current densities, but not necessarily power output. An initial GOx–BOD fuel cell operating in pseudo-physiological conditions, with appropriate selection of the redox polymer structure to reflect the decreased T1 copper redox potential in BOD compared to fungal laccase, yielded a power density of 50 μW cm^{-2} at 0.5 V. In an illustration of the trade-off between cell voltage and maximum power, use of a redox polymer for the anode with a more positive redox potential increased maximum power density to 244 μW cm^{-2} but at the expense of a lower cell voltage of 0.36 V [36]. Again, inclusion of a flexible tether to improve charge transport diffusion for the anodic redox polymer improved anodic current density, producing a power density of 430 μW cm^{-2} at a cell voltage of 0.52 V [159]. One week of continuous operation of this cell at 0.52 V, resulted in an approximately 6% loss in power density per day. The operation of this miniaturized carbon fiber-based EFC in a grape, chosen for its high glucose content (>30 mM), was demonstrated, with power density of 47 μW cm^{-2} observed when the cell is placed in the oxygen-deficient grape center, whereas 240 μW cm^{-2} is obtained when the cell is placed near the grape skin, with a cell voltage of 0.52 V for both cases [9]. Further improvement in output for this EFC was achieved by increasing the redox site density of the anode redox polymer, resulting in a glucose flux-limited current density increase of 20% and an overpotential decrease of 50 mV [160]. This optimized EFC operated at +0.60 V with a 480 μW cm^{-2} power density in phosphate buffer of pH = 7.2 containing 0.1 M NaCl, 15 mM glucose at 37.5 °C, losing about 8% of its power each day of operation.

Increased power output of these redox polymer-mediated EFCs can be obtained by use of three-dimensional conducting materials as supports for the biocatalytic films, provided mass transport does not become rate limiting. This has been demonstrated by Gao *et al.* [161] through a comparison of carbon fiber-based electrodes with carbon microwire electrodes, formed via particle coagulation spinning of CNTs. Maximum power density of 0.74 mW cm^{-2} at 0.57 V was obtained for the miniaturized membraneless EFC in an unstirred phosphate buffer saline solution (pH = 7.2) under air, containing 15 mM glucose at 37 °C, more than ten times that obtained for the carbon fiber-based system.

5.5 Conclusions and Future Outlook

Bioelectrocatalytic studies have proven that enzymes as biocatalysts in EFC electrodes can perform better than conventional catalysts, providing advantages in terms of rates of reaction and specificity. This advantage can be transformed into provision of power with low overpotentials, and with ease of miniaturization, because removal of membranes and casings can be undertaken due to enzyme specificity, providing for niche opportunities for EFCs in the production of power using implanted or portable, miniaturized, devices. Challenges, nonetheless, remain in order to advance the knowledge gained over the past decade of research into technological products. Issues that require further attention relate to stabilization of power output of assembled cells and to increasing power output. Power output may be improved through use of optimized biocatalyst (and mediator if appropriate) on three-dimensional surfaces and deeper oxidation of fuels using multiple enzymes and enzyme cascades. An additional consideration is the cell voltage required to provide power to microelectronic devices. Cell voltages of greater than 0.5 V can be delivered under certain conditions, but most studies to date only report voltages of 0.2–0.3 V, requiring therefore stacking of cells to power electronic devices.

A systematic comparison of EFC benchmark figures of merit on stability, as was the case for reporting of power density benchmarks, is hampered by the lack of a defined measure of stability in each study. Many studies simply report on storage stability using either intermittent or single-shot measurement of current or power, with ill-defined storage conditions provided. Future studies should therefore focus on reporting operational stability of assembled cells under a defined load, in defined conditions, as is the case in traditional fuel cell research. Stability of power output is affected by enzyme deactivation and inhibition and by desorption/leaching of enzyme, and mediator if present, from the electrode surfaces. Recent reports highlight, for example, the effect of other endogenous serum components on EFC operation [17, 162–164]. For example, a glucose–oxygen EFC, based on DET for CDH (*Corynascus thermophilus*) oxidation of glucose and BOD reduction of oxygen, both enzymes simply adsorbed onto spectroscopic graphite rod electrodes, results in an operational half-life of more than six hours in buffer but only less than two hours in serum [17]. Others have reported that oxidation of uric acid in the presence of oxygen destabilizes BOD, leading to a decrease in performance of a BOD biocathode to only a few hours in urate-containing serum, compared to over a week in buffer [164]. Blocking access of urate to the enzyme can be achieved by, for example, coating the biocathode with a cubic-phase lyotropic liquid crystal that has channels modified using an anionic lipid, to increase mechanical strength and reduce urate permeation [165]. Desorption of enzyme and/or mediator from the electrode surface can also lead to a decrease in EFC performance, and attempts to prevent this range from entrapment behind or within membranes and crosslinking of films, to modification of electrode surface chemistry to provide for anchoring sites for chemical tethering of layers to the surface. For example, we have recently reported on improved stability of layers of redox complexes [166, 167], GOx [168], and MCOs in redox hydrogels [148] by chemical crosslinking of the layer to surface functional groups on carbon electrodes. The functional groups are introduced using *in situ* diazotization of an aryl amine, followed by electroreduction of the aryldiazonium to carbon–carbon couple to the electrode. Such an approach, combined with crosslinking and overcoating with additional membranes, could provide for EFCs with increased stability in serum.

EFCs operating on fuels other than glucose show promise as devices for powering portable electronics, and recent reports highlight advances in prototype design of cells using either fructose [45–47] or ethanol [169] as fuel in the anolyte, and either FDH or alcohol dehydrogenase as biocatalyst entrapped in three-dimensional conducting electrodes to provide for DET to the anode, in a membraneless EFC configuration. Maximum power densities of 0.85 mW cm^{-2} [45] and 0.126 mW cm^{-2} [47] at a cell voltage of about 0.4 V are observed in the fructose studies using either carbon black or MWCNTs as the three-dimensional electrode, and a fungal laccase or BOD as biocathode element, respectively, again making it difficult to compare performance.

The use of enzyme cascades, in an effort to extract more than two electrons from the fuel, as is the case for GOx oxidation of glucose, is a promising direction for improving power output for portable devices [80–86, 170]. Since the initial report, demonstrating complete oxidation of methanol to carbon dioxide based on NAD^+/NADH regeneration in the anolyte by diaphorase and benzylviologen coupled to three dehydrogenases [81], advances have focused on efforts to mimic metabolic pathways by combining multiple enzymes in the anolyte. For example, a mimic of the citric acid cycle, based on immobilization of dehydrogenases at, and electrocatalytic oxidation of NADH on, poly(methylene green)-modified carbon electrodes, provides for an ethanol–air (platinum-based) fuel cell that provides 8.7-fold more power compared to using a single alcohol dehydrogenase system in a Nafion-separated system [170]. Mediatorless, DET, glycerol oxidation using PQQ-dependent dehydrogenases and an oxalate oxidase co-immobilized on CNT-coated carbon paper can also be used to provide a glycerol–air fuel cell with a maximum power density of $1.32\,mW\,cm^{-2}$ in 100 mM glycerol solutions [85].

In conclusion, despite over a decade of increased attention, in the most recent phase of interest in EFC research, performance of operating prototype devices remains limited in terms of power output and stability. At a fundamental level, insight into enzyme electron transfer and power production can be gained by systematic studies on integration of surfaces, catalysts, substrates, co-substrates, and artificial substrates (mediators). Unfortunately, wide-scale comparison between devices (anodes, cathodes, assemblies), in terms of operating conditions, performance benchmarks, and stability benchmarks, is hampered by a lack of conformity to standardized procedures for testing, and reporting of data. In addition, there is scope for greater understanding of the role that each component plays in an integrated system to provide for an EFC, by comparison of performance to models. Relatively few examples exist to date of detailed studies comparing response to models, to enable identification of limiting factors in reaction kinetics, mass transport, and system design. Niche application areas for assembled EFCs, such as self-powered miniaturized disposable sensors [171], replaceable power for implantable sensors [7, 21, 172], or boosting power to electronic devices in remote locations using energy-dense fuels [45–47, 170], will emerge. Ultimately, the understanding gained from biocatalyst-based research may also lead to development of better biomimetics, mimicking not only the catalytic active site of a biocatalyst, but also the binding site and the structural flexibility provided by the protein assembly surrounding the active site, to provide for improved "small-molecule" catalysts in more traditional fuel cell systems, providing high power densities at low cost.

Acknowledgments

Funding by Science Foundation Ireland, through a Charles Parsons Energy Research Award, and by the EU (FP6 and FP7) is gratefully acknowledged.

References

1 Franks, A.E. and Nevin, K.P. (2010) *Energies*, **3**, 899–919.
2 Logan, B.E. (2009) *Nature Reviews Microbiology*, **7**, 375–381.
3 Kim, B.H., Chang, I.S., and Gadd, G.M. (2007) *Applied Microbiology and Biotechnology*, **76**, 485–494.
4 Lovley, D.R. (2006) *Current Opinion in Biotechnology*, **17**, 327–332.
5 Rabaey, K. and Verstraete, W. (2005) *Trends in Biotechnology*, **23**, 291–298.
6 Yahiro, A.T., Lee, S.M., and Kimble, D.O. (1964) *Biochimica et Biophysica Acta*, **88**, 375–383.
7 Heller, A. (2004) *Physical Chemistry Chemical Physics*, **6**, 209–216 and references therein.
8 Cinquin, P., Gondran, C., Giroud, F., Mazabrard, S., Pellissier, A., Boucher, F., Alcaraz, J.P., Gorgy, K., Lenouvel, F., Mathe, S., Porcu, P., and Cosnier, S. (2010) *PLoS ONE*, **5**, e10476.
9 Mano, N., Mao, F., and Heller, A. (2003) *Journal of the American Chemical Society*, **125**, 6588–6594.
10 Calabrese Barton, S., Gallaway, J., and Atanassov, P. (2004) *Chemical Reviews*, **104**, 4867–4886.
11 Davis, F. and Higson, S.P.J. (2007) *Biosensors & Bioelectronics*, **22**, 1224–1235.
12 Willner, I., Yan, Y.M., Willner, B., and Tel-Vered, R. (2009) *Fuel Cells*, **9**, 7–24.
13 Cracknell, J.A., Vincent, K.A., and Armstrong, F.A. (2008) *Chemical Reviews*, **108**, 2439–2461.
14 Cooney, M.J., Svoboda, V., Lau, C., Martin, G., and Minteer, S.D. (2008) *Energy & Environmental Science*, **1**, 320–337.
15 Ivanov, I., Vidakovic-Koch, T., and Sundmacher, K. (2010) *Energies*, **3**, 803–846.
16 Osman, M.H., Shah, A.A., and Walsh, F.C. (2011) *Biosensors & Bioelectronics*, **26**, 3087–3102.
17 Ludwig, R., Harreither, W., Tasca, F., and Gorton, L. (2010) *ChemPhysChem*, **11**, 2674–2697.
18 Leger, C., and Bertrand, P. (2008) *Chemical Reviews*, **108**, 2379–2438.
19 Shleev, S., Tkac, J., Christenson, A., Ruzgas, T., Yaropolov, A.I., Whittaker, J.W., and Gorton, L. (2005) *Biosensors & Bioelectronics*, **20**, 2517–2554.
20 Marcus, R.A. and Sutin, N. (1985) *Biochimica et Biophysica Acta*, **3**, 265–322.
21 Heller, A. (2005) *AIChE Journal*, **51**, 1054–1066.
22 Gallaway, J.W. and Calabrese Barton, S.A. (2008) *Journal of the American Chemical Society*, **130**, 8527–8536.
23 Rubenwolf, S., Strohmeier, O., Kloke, A., Kerzenmacher, S., Zengerle, R., and von Stetten, F. (2010) *Biosensors & Bioelectronics*, **26**, 841–845.
24 Kerzenmacher, S., Mutschler, K., Kräling, U., Baumer, H., Ducrée, J., Zengerle, R., and von Stetten, F. (2009) *Journal of Applied Electrochemistry*, **39**, 1477–1485.
25 Svoboda, V., Cooney, M., Liaw, B.Y., Minteer, S., Piles, E., Lehnert, D., Calabrese Barton, S., Rincon, R., and Atanassov, P. (2008) *Electroanalysis*, **20**, 1099–1109.
26 Stankovich, M.T., Schopfer, L.M., and Massey, V. (1978) *Journal of Biological Chemistry*, **253**, 4971–4979.
27 Guiseppi-Elie, A., Lei, C., and Baughman, R.H. (2002) *Nanotechnology*, **13**, 559–564.
28 Cai, C.X., and Chen, J. (2004) *Analytical Biochemistry*, **332**, 75–83.
29 Liu, Y., Wang, M., Zhao, F., Xu, Z., and Dong, S. (2005) *Biosensors & Bioelectronics*, **21**, 984–988.
30 Ivnitski, D., Branch, B., Atanassov, P., and Apblett, C. (2006) *Electrochemistry Communications*, **8**, 1204–1210.
31 Vaze, A., Hussain, N., Tang, C., Leech, D., and Rusling, J. (2009) *Electrochemistry Communications*, **11**, 2004–2007.
32 Courjean, O., Gao, F., and Mano, N. (2009) *Angewandte Chemie International Edition*, **48**, 5897–5899.
33 Bunte, C., Prucker, O., König, T., and Rühe, J. (2010) *Langmuir*, **26**, 6019–6027.
34 Meredith, M.T., Kao, D.Y., Hickey, D., Schmidtke, D.W., and Glatzhofera, D.T.

(2011) *Journal of the Electrochemical Society*, **158**, B166–B174.

35 Chen, T., Calabrese Barton, S., Binyamin, G., Gao, Z., Zhang, Y., Kim, H.H., and Heller, A. (2001) *Journal of the American Chemical Society*, **123**, 8630–8863.

36 Mano, N. and Heller, A. (2003) *Journal of the Electrochemical Society*, **150**, A1136–A1138.

37 Kim, H.H., Mano, N., Zhang, Y., and Heller, A. (2003) *Journal of the Electrochemical Society*, **150**, A209–A213.

38 Mano, N., Mao, F., and Heller, A. (2004) *Chemical Communications*, 2116–2117.

39 Mano, N. (2008) *Chemical Communications*, 2221–2223.

40 Sakai, H., Nakagawa, T., Tokita, Y., Hatazawa, T., Ikeda, T., Tsujimura, S., and Kano, K. (2009) *Energy & Environmental Science*, **2**, 133–138.

41 Togo, M., Takamura, A., Asai, T., Kaji, H., and Nishizawa, M. (2007) *Electrochimica Acta*, **52**, 4669–4674.

42 Yan, Y., Zheng, W., Su, L., and Mao, L. (2006) *Advanced Materials*, **18**, 2639–2643.

43 Tsujimura, S., Kano, K., and Ikeda, T. (2002) *Electrochemistry*, **70**, 940–942.

44 Yuhashi, N., Tomiyama, M., Okuda, J., Igarashi, S., Ikebukuro, K., and Sode, K. (2005) *Biosensors & Bioelectronics*, **20**, 2145–2150.

45 Kamitaka, Y., Tsujimura, S., Setoyama, N., Kajino, T., and Kano, K. (2007) *Physical Chemistry Chemical Physics*, **9**, 1793–1801.

46 Murata, K., Suzuki, M., Kajiya, K., Nakamura, N., and Ohno, H. (2009) *Electrochemistry Communications*, **11**, 668–671.

47 Wu, X., Zhao, F., Varcoe, J.R., Thumser, A.E., Avignone-Rossa, C., and Slade, R.C.T. (2009) *Biosensors & Bioelectronics*, **25**, 326–331.

48 Nadeem Zafar, M., Tasca, F., Boland, S., Kujawa, M., Patel, I., Peterbauer, C.K., Leech, D., and Gorton, L. (2010) *Bioelectrochemistry*, **80**, 38–42.

49 Tasca, F., Gorton, L., Harreither, W., Haltrich, D., Ludwig, R., and Nöll, G. (2009) *Analytical Chemistry*, **81**, 2791–2798.

50 Wang, J. (2008) *Chemical Reviews*, **108**, 814–825.

51 Cass, A.E.G., Davis, G., Francis, G.D., Hill, H.A.O., Aston, W.J., Higgins, I.J., Plotkin, E.V., Scott, L.D.L., and Turner, A.P.F. (1984) *Analytical Chemistry*, **56**, 667–671.

52 Green, M.J. and Hill, H.A.O. (1986) *Journal of the Chemical Society, Faraday Transactions*, **1** (82), 1237–1243.

53 Frew, J.E. and Hill, H.A.O. (1987) *Analytical Chemistry*, **59**, 933A–944A.

54 Schumann, W., Ohara, T.J., Schmidt, H.L., and Heller, A. (1991) *Journal of the American Chemical Society*, **113**, 1394–1397.

55 Anicet, A., Anne, N., Moiroux, J., and Savéant, J.M. (1998) *Journal of the American Chemical Society*, **120**, 7115–7116.

56 Forrow, N.J., Sanghera, G.S., and Walters, S.J. (2002) *Journal of the Chemical Society, Dalton Transactions*, 3187–3194.

57 Liu, Y., Wang, M.K., Zhao, F., Liu, B.F., and Dong, S.J. (2005) *Chemistry–a European Journal*, **11**, 4970–4974.

58 Gregg, B.A. and Heller, A. (1991) *Journal of Physical Chemistry*, **95**, 5976–5980.

59 Ohara, T.J., Rajagopalan, R., and Heller, A. (1993) *Analytical Chemistry*, **65**, 3512–3517.

60 Ohara, T., Rajagopalan, R., and Heller, A. (1994) *Analytical Chemistry*, **66**, 2451–2457.

61 de Lumley-Woodyear, T., Rocca, P., Lindsay, J., Dror, Y., Freeman, A., and Heller, A. (1995) *Analytical Chemistry*, **67**, 1332–1133.

62 Mao, F., Mano, N., and Heller, A. (2003) *Journal of the American Chemical Society*, **125**, 4951–4957.

63 Swoboda, B.E.P. and Massey, V. (1965) *Journal of Biological Chemistry*, **240**, 2209–2222.

64 Rando, D., Kohring, G.W., and Giffhorn, F. (1997) *Applied Microbiology and Biotechnology*, **48**, 34–40.

65 Caruana, D.J. and Howorka, S. (2010) *Molecular Biosystems*, **6**, 548–1556.

66 Güven, G., Prodanovic, R., and Schwaneberg, U. (2010) *Electroanalysis*, **22**, 765–775.

67 Zhu, Z., Wang, M., Gautam, A., Nazor, J., Momeu, C., Prodanovic, R., and Schwaneberg, U. (2007) *Biotechnology Journal*, **2**, 241–248.

68 Gorton, L. and Dominguez, E. (2002) Electrochemistry of NAD(P)+/NAD(P)H, in *Encyclopedia of Electrochemistry: vol. 9: Bioelectrochemistry* (ed. G.S. Wilson), Wiley-VCH Verlag GmbH, Weinheim, pp. 67–143.

69 Sato, F., Togo, M., Kamrul Islam, M., Matsue, T., Kosuge, J., Fukasaku, N., Kurosawa, S., and Nishizawa, M. (2005) *Electrochemistry Communications*, **7**, 643–647.

70 Sato, A., Takagi K. Kano, K., Kato, N., Duine, J.A., and Ikeda, T. (2001) *Biochemical Journal*, **357**, 893–898.

71 Okuda, J., Yamazaki, T., Fukasawa, M., Kakehi, N., and Sode, K. (2007) *Analytical Letters*, **40**, 431–440.

72 Okuda-Shimazaki, J., Kakehi, N., Yamazaki, T., Tomiyama, M., and Sode, K. (2008) *Biotechnology Letters*, **30**, 1753–1758.

73 Durand, F., Stines-Chaumeil, C., Flexer, V., André, I., and Mano, N. (2010) *Biochemical and Biophysical Research Communications*, **402**, 750–754.

74 Tasca, F., Gorton, L., Harreither, W., Haltrich, D., Ludwig, R., and Nöll, G. (2008) *Journal of Physical Chemistry C*, **112**, 9956–9961.

75 Tasca, F., Gorton, L., Harreither, W., Haltrich, D., Ludwig, R., and Nöll, G. (2008) *Journal of Physical Chemistry C*, **112**, 13668–13673.

76 Coman, V., Vaz-Domínguez, C., Ludwig, R., Harreither, W., Haltrich, D., De Lacey, A.L., Ruzgas, T., Gorton, L., and Shleev, S. (2008) *Physical Chemistry Chemical Physics*, **10**, 6093–6096.

77 Stoica, L., Dimcheva, N., Ackermann, Y., Karnicka, K., Guschin, D.A., Kulesza, P.J., Rogalski, J., Haltrich, D., Ludwig, R., Gorton, L., and Schuhmann, W. (2009) *Fuel Cells*, **9**, 53–68.

78 Coman, V., Ludwig, R., Harreither, W., Haltrich, D., Gorton, L., Ruzgas, T., and Shleev, S. (2010) *Fuel Cells*, **10**, 9–16.

79 Harreither, W., Sygmund, C., Augustin, M., Narciso, M., Rabinovich, M.L., Gorton, L., Haltrich, D., and Ludwig, R. (2011) *Applied and Environmental Microbiology*, **77**, 1804–1815.

80 Sokic-Lazic, D., Arechederra, R.L., Treu, B.L., and Minteer, S.D. (2010) *Electroanalysis*, **22**, 757–764.

81 Palmore, G.T.R., Bertschy, H., Bergens, S.H., and Whitesides, G.M. (1998) *Journal of Electroanalytical Chemistry*, **443**, 155–161.

82 Akers, N.L., Moore, C.M., and Minteer, S.D. (2005) *Electrochimica Acta*, **50**, 2521–2525.

83 Topcagic, S. and Minteer, S.D. (2006) *Electrochimica Acta*, **51**, 2168–2172.

84 Sokic-Lazic, D. and Minteer, S.D. (2009) *Electrochemical and Solid-State Letters*, **12**, F26–F28.

85 Arechederra, R.L. and Minteer, S.D. (2009) *Fuel Cells*, **9**, 63–69.

86 Tasca, F., Gorton, L., Kujawa, M., Patel, I., Harreither, W., Clemens, K.P., Ludwig, R., and Nöll, G. (2010) *Biosensors & Bioelectronics*, **25**, 1710–1716.

87 Soukharev, V., Mano, N., and Heller, A. (2004) *Journal of the American Chemical Society*, **126**, 8368–8369.

88 Yang, H.H. and McCreery, R.L. (2000) *Journal of the Electrochemical Society*, **147**, 3420–3428.

89 Palmore, G.T.R. and Kim, H.-H. (1999) *Journal of Electroanalytical Chemistry*, **464**, 110–117.

90 Solomon, E.I., Chen, P., Metz, M., Lee, S.-K., and Palmer, A.E. (2001) *Angewandte Chemie International Edition*, **40**, 4570–4590.

91 Quintanar, L., Stoj, C., Taylor, A.B., Hart, P.J., Kosman, D.J., and Solomon, E.I. (2007) *Accounts of Chemical Research*, **40**, 445–452.

92 Augustine, A.J., Kjaergaard, C., Qayyum, M., Ziegler, L., Kosman, D.J., Hodgson, K.O., Hedman, B., and Solomon, E.I. (2010) *Journal of the American Chemical Society*, **132**, 6057–6067.

93 Bourbonnais, R., and Paice, M.G. (1990) *FEBS Letters*, **267**, 99–102.

94 Yaropolov, A.I., Skorobogat'ko, O.V., Vartanov, S.S., and Varfolomeyev, S.D. (1994) *Applied Biochemistry and Biotechnology*, **49**, 257–275.

95 Duran, N., Rosa, M.A., D'Annibale, A., and Gianfreda, L. (2002) *Enzyme and Microbial Technology*, **31**, 907–931.

96 Trudeau, F., Daigle, F., and Leech, D. (1997) *Analytical Chemistry*, **69**, 882–886.

97 Kuznetsov, B.A., Shumakovich, G.P., Koroleva, O.V., and Yaropolov, A.I. (2001) *Biosensors & Bioelectronics*, **16**, 73–84.

98 Bourbonnais, R., Paice, M.G., and Leech, D. (1998) *Biochimica et Biophysica Acta*, **1379**, 381–390.

99 Rochefort, D., Leech, D., and Bourbonnais, R. (2004) *Green Chemistry*, **6**, 14–24.

100 Call, H.P. and Mucke, I. (1997) *Journal of Biotechnology*, **53**, 163–202.

101 Torres, E., Bustos-Jaimes, I., and le Borgne, S. (2003) *Applied Catalysis B*, **46**, 1–15.

102 Yoshida, H. (1883) *Journal of the Chemical Society*, **43**, 472–486.

103 Rodgers, C.J., Blanford, C.F., Giddens, S.R., Skamnioti, P., Armstrong, F.A., and Gurr, S.J. (2010) *Trends in Biotechnology*, **28**, 63–72.

104 Kjaergaard, C.H., Rossmeis, J., and Nørskov, J.K. (2010) *Inorganic Chemistry*, **49**, 3567–3572.

105 Berezin, I.V., Bogdanovskaya, V.A., Varfolomeev, S.D., Tarasevich, M.R., and Yaropolov, A.I. (1978) *Doklady Akademii Nauk SSSR*, **240**, 615–618.

106 Shleev, S., Jarosz-Wilkolazka, A., Khalunina, A., Morozova, O., Yaropolov, A., Ruzgas, T., and Gorton, L. (2005) *Bioelectrochemistry*, **67**, 115–124.

107 See for example Xu, F. (2001) *Applied Biochemistry and Biotechnology*, **95**, 125–133.

108 Tsujimura, S., Tatsumi, H., Ogawa, J., Shimizu, S., Kano, K., and Ikeda, T. (2001) *Journal of Electroanalytical Chemistry*, **496**, 69–75.

109 Kiiskinen, L.-L., Viikari, L., and Kruus, K. (2002) *Applied Microbiology and Biotechnology*, **59**, 198–204.

110 Kavanagh, P., Jenkins, P., and Leech, D. (2008) *Electrochemistry Communications*, **10**, 970–972.

111 Kavanagh, P., Boland, S., Jenkins, P., and Leech, D. (2009) *Fuel Cells*, **9**, 79–84.

112 Rensing, C. and Grass, G. (2003) *FEMS Microbiology Review*, **27**, 197–213.

113 Kataoka, K., Sugiyama, R., Hirota, S., Inoue, M., Urata, K., Minagawa, Y., Seo, D., and Sakurai, T. (2009) *Journal of Biological Chemistry*, **284**, 14405–14413.

114 Tsujimura, S., Miura, Y., and Kano, K. (2008) *Electrochimica Acta*, **53**, 5716–5720.

115 Miura, Y., Tsujimura, S., Kurose, S., Kamitaka, Y., Kataoka, K., Sakurai, T., and Kano, K. (2009) *Fuel Cells*, **9**, 70–78.

116 Kontani, R., Tsujimura, S., and Kano, K. (2009) *Bioelectrochemistry*, **76**, 10–13.

117 Martins, L.O., Soares, C.M., Pereira, M.M., Teixeira, M., Costa, T., Jones, G.H., and Henriques, A.O. (2002) *Journal of Biological Chemistry*, **277**, 18849–18859.

118 Beneyton, T., El Harrak, A., Griffiths, A.D., Hellwig, P., and Taly, V. (2011) *Electrochemistry Communications*, **13**, 24–27.

119 Machczynski, M.C., Vijgenboom, E., Samyn, B., and Canters, G.W. (2004) *Protein Science*, **13**, 2388–2397.

120 Gallaway, J., Wheeldon, I., Rincon, R., Atanassov, P., Banta, S., and Calabrese Barton, S. (2008) *Biosensors & Bioelectronics*, **23**, 1229–1235.

121 Tsujimura, S., Kano, K., and Ikeda, T. (2005) *Journal of Electroanalytical Chemistry*, **576**, 113–120.

122 Vaz-Dominguez, C., Campuzano, S., Rudiger, O., Gorbachev, M.P.M., Shleev, S., Fernandez, V.M., and De Lacey, A.L. (2008) *Biosensors & Bioelectronics*, **24**, 531–537.

123 Ramírez, P., Mano, N., Andreu, R., Ruzgas, T., Heller, A., Gorton, L., and Shleev, S. (2008) *Biochimica et Biophysica Acta*, **1777**, 1364–1369.

124 Tsujimura, S., Kamitaka, Y., and Kano, K. (2007) *Fuel Cells*, **7**, 463–469.

125 Goebel, G., and Lisdat, F. (2008) *Electrochemistry Communications*, **10**, 1691–1694.

126 Schubert, K., Goebel, G., and Lisdat, F. (2009) *Electrochimica Acta*, **54**, 3033–3038.

127 Tarasevich, M.R., Yaropolov, A.I., Bogdanovskaya, V.A., and Varfolomeev, S.D. (1979) *Bioelectrochemistry and Bioenergetics*, **6**, 393–403.

128 Shleev, S., El Kasmi, A., Ruzgas, T., and Gorton, L. (2004) *Electrochemistry Communications*, **6**, 934–939.

129 Yaropolov, A.I., Kharybin, A.N., Emneus, J., MarkoVarga, G., and Gorton, L. (1996) *Bioelectrochemistry and Bioenergetics*, **40**, 49–57.

130 Zheng, W., Li, Q., Su, L., Yan, Y., Zhang, J., and Mao, L. (2006) *Electroanalysis*, **18**, 587–594.

131 Weigel, M.C., Tritscher, E., and Lisdat, F. (2007) *Electrochemistry Communications*, **9**, 689–693.

132 Stolarczyk, K., Nazaruk, E., Rogalski, J., and Bilewicz, R. (2008) *Electrochimica Acta*, **53**, 3983–3990.

133 Szot, K., Nogala, W., Niedziolka-Jönsson, J., Jönsson-Niedziolka, M., Marken, F., Rogalski, J., Nunes Kirchner, C., Wittstock, G., and Opallo, M. (2009) *Electrochimica Acta*, **54**, 4620–4625.

134 Nogala, W., Celebanska, A., Szot, K., Wittstock, G., and Opallo, M. (2010) *Electrochimica Acta*, **55**, 5719–5724.

135 Dagys, M., Haberska, K., Shleev, S., Arnebrant, T., Kulys, J., and Ruzgas, T. (2010) *Electrochemistry Communications*, **12**, 933–935.

136 Blanford, C.F., Heath, R.S., and Armstrong, F.A. (2007) *Chemical Communications*, 1710–1712.

137 Leech, D. and Feerick, K.O. (2000) *Electroanalysis*, **12**, 1339–1342.

138 Barton, S.C., Kim, H.-H., Binyamin, G., Zhang, Y., and Heller, A. (2001) *Journal of the American Chemical Society*, **123**, 5802–5803.

139 Barton, S.C., Kim, H.-H., Binyamin, G., Zhang, Y., and Heller, A. (2001) *Journal of Physical Chemistry B*, **105**, 11917–11921.

140 Mano, N., Kim, H.-H., Zhang, Y., and Heller, A. (2002) *Journal of the American Chemical Society*, **124**, 6480–6486.

141 Mano, N., Kim, H.-H., and Heller, A. (2002) *Journal of Physical Chemistry B*, **106**, 8842–8848.

142 Mano, N., Fernandez, J.L., Kim, Y., Shin, W., Bard, A.J., and Heller, A. (2003) *Journal of the American Chemical Society*, **125**, 15290–15291.

143 Skálová, T., Dohnálek, J., Østergaard, L.H., Østergaard, P.R.P., Kolenko Dušková, J., Štěpánková, A., and Hašek, J. (2009) *Journal of Molecular Biology*, **385**, 1165–1178.

144 Hudak, N.S., Gallaway, J.W., and Calabrese Barton, S. (2009) *Journal of Electroanalytical Chemistry*, **629**, 57–62.

145 Gallaway, J.W. and Calabrese Barton, S.A. (2009) *Journal of Electroanalytical Chemistry*, **626**, 149–155.

146 Karnicka, K., Eckhard, K., Guschin, D.A., Stoica, L., Kulesza, P.J., and Schuhmann, W. (2007) *Electrochemistry Communications*, **9**, 1998–2002.

147 Ackermann, Y., Guschin, D.A., Eckhard, K., Shleev, S., and Schuhmann, W. (2010) *Electrochemistry Communications*, **12**, 640–643.

148 Jenkins, P.A., Boland, S., Kavanagh, P., and Leech, D. (2009) *Bioelectrochemistry*, **76**, 162–168.

149 Habrioux, A., Merle, G., Servat, K., Kokoh, K.B., Innocent, C., Cretin, M., and Tingry, S. (2008) *Journal of Electroanalytical Chemistry*, **622**, 97–102.

150 Merle, G., Habrioux, A., Servat, K., Rolland, M., Innocent, C., Kokoh, K.B., and Tingry, S. (2009) *Electrochimica Acta*, **54**, 2998–3003.

151 Lim, K.G. and Palmore, G.T.R. (2007) *Biosensors & Bioelectronics*, **22**, 941–947.

152 Lee, J., Lim, K.G., Palmore, G.T.R., and Tripathi, A. (2007) *Analytical Chemistry*, **79**, 7301–7307.

153 Togo, M., Takamura, A., Asai, T., Kaji, H., and Nishizawa, M. (2008) *Journal of Power Sources*, **178**, 53–58.

154 Zebda, A., Renaud, L., Cretin, M., Pichot, F., Innocent, C., Ferrigno, R., and Tingry, S. (2009) *Electrochemistry Communications*, **11**, 592–595.

155 Zebda, A., Renaud, L., Cretin, M., Innocent, C., Ferrigno, R., and Tingry, S. (2010) *Sensors and Actuarors B*, **149**, 44–50.

156 Mano, N., Mao, F., Shin, W., Chen, T., and Heller, A. (2003) *Chemical Communications*, 518–519.

157 Gao, F., Courjean, O., and Mano, N. (2009) *Biosensors & Bioelectronics*, **25**, 356–361.

158 Barrière, F., Kavanagh, P., and Leech, D. (2006) *Electrochimica Acta*, **51**, 5187–5192.

159 Mano, N., Mao, F., and Heller, A. (2002) *Journal of the American Chemical Society*, **124**, 12962–12963.

160 Mano, N., Mao, F., and Heller, A. (2004) *ChemBioChem*, **5**, 1703–1705.

161 Gao, F., Viry, L., Maugey, M., Poulin, P., and Mano, N. (2010) *Nature Communications*, **1**, 2.

162 Kang, C., Shin, H., Zhang, Y.C., and Heller, A. (2004) *Bioelectrochemistry*, **65**, 83–88.

163 Kang, C., Shin, H., and Heller, A. (2006) *Bioelectrochemistry*, **68**, 22–26.

164 Shin, H., Kang, C., and Heller, A. (2007) *Electroanalysis*, **19**, 638–643.

165 Rowinski, P., Kang, C., Shin, H., and Heller, A. (2007) *Analytical Chemistry*, **79**, 1173–1180.

166 Boland, S., Barrière, F., and Leech, D. (2008) *Langmuir*, **24**, 6351–6358.

167 Boland, S., Foster, K., and Leech, D. (2009) *Electrochimica Acta*, **54**, 1986–1991.

168 Pellissier, M., Barrière, F., Downard, A.J., and Leech, D. (2008) *Electrochemistry Communications*, **10**, 835–838.

169 Deng, L., Shang, L., Wen, D., Zhaia, J., and Dong, S. (2010) *Biosensors & Bioelectronics*, **26**, 70–73.

170 Sokic-Lazic, D. and Minteer, S.D. (2008) *Biosensors & Bioelectronics*, **24**, 939–944.

171 Deng, L., Chen, C.G., Zhou, M., Guo, S., Wang, E., and Dong, S. (2010) *Analytical Chemistry*, **82**, 4283–4287.

172 Heller, A. (2006) *Analytical and Bioanalytical Chemistry*, **385**, 469–473.

6
Raman Spectroscopy of Biomolecules at Electrode Surfaces

Philip Bartlett and Sumeet Mahajan

6.1
Introduction

Surface-enhanced Raman spectroscopy (SERS) is a remarkable phenomenon. The effect was first reported by Fleischmann *et al.* over 35 years ago [1] in an experiment in which they recorded the Raman spectrum of pyridine at a roughened silver electrode surface. From the start the effect provoked controversy because of its large magnitude and unexpected nature. Raman scattering is inherently inefficient with only a small fraction of the photons, typically around 1 in 10^{10}, being Raman scattered, the rest, the vast majority, being elastically, or Rayleigh, scattered. Thus to observe clear spectra for a monolayer of pyridine molecules adsorbed at the silver solution interface was unprecedented. In their original experiments Fleischman *et al.* used an electrochemically roughened silver surface but even so the increased surface area alone could not account for the many orders of magnitude increase in the Raman signal. At the time Fleischman *et al.* were aware of the unusually high intensity of the Raman signals and careful to exclude the possibility of multilayer formation [2]. This first publication, with its surprising results, sparked the interest of other groups and it was followed a few years later by publications by Jeanmarie and Van Duyne [3] and by Albrecht *et al.* [4] confirming the effect and demonstrating that the anomalous intensity of the spectra (a factor or 10^5 to 10^6) was the result of some kind of hidden resonance or a new type of Raman enhancement [5]. The term "surface enhanced Raman spectroscopy" was coined by Van Duyne and coworkers soon after to describe the effect [5]. This started a great interest in the effect and a search for detailed understanding of it and its application, which continues to this day.

One of the problems of SERS is that it is relatively easy to achieve but difficult to control. This has led, and continues to lead, to publications that claim to have new surfaces that show SERS but which in reality offer no advantage over existing roughened surfaces, or publications that claim insight into the mechanism of the phenomenon but which are in fact readily explained within the basis of current understanding of SERS [6]. As a consequence a vast and somewhat confusing literature has built up around the subject over the past 30 years and this is a

Advances in Electrochemical Science and Engineering. Edited by Richard C. Alkire, Dieter M. Kolb, and Jacek Lipkowski

ISBN: 978-3-527-32885-7

problem for those who wish to apply the technique to address specific chemical or biochemical questions or to use it for analytical purposes.

One of the attractions of SERS is that it is ideally suited to *in situ* electrochemical measurements because of the very high surface selectivity of the technique and the fact that water is a very weak Raman scatterer. As a result SERS is able to provide vibrational spectra for molecules at electrode surfaces under realistic operating conditions. In this chapter we focus on the applications of SERS in bioelectrochemistry, focusing on studies carried out at electrode surfaces under potential control. There are a number of excellent reviews in the literature of various aspects of SERS covering both the general field [6–8], as well as more specific topics including the biochemical applications of SERS [9, 10], analytical prospects [11–13], surfaces for SERS [14–17], and electrochemical SERS [18–20]; and the reader is directed to these for specific details. In this chapter we begin with a brief overview of SERS and SERS surfaces before reviewing the particular applications in bioelectrochemistry.

6.2 Raman Spectroscopy

Raman scattering is an inelastic process in which the change in wavelength of the scattered photon is associated with the excitation (Stokes) or relaxation (anti-Stokes) of vibrational modes of the scattering molecule (Figure 6.1). Raman spectroscopy, like infrared spectroscopy, thus gives information about molecular structure because the different functional groups within molecules have different characteristic vibrational frequencies – a kind of "molecular fingerprint" for the

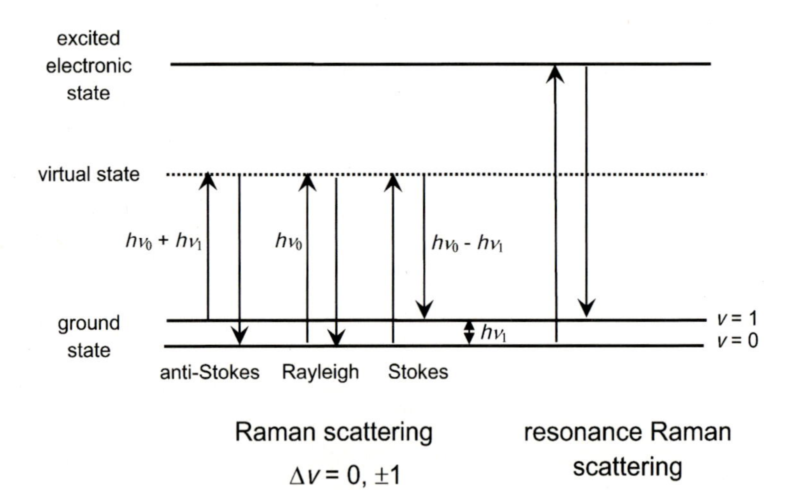

Figure 6.1 Spectroscopic transitions for Rayleigh scattering, Stokes and anti-Stokes Raman scattering, and resonance Raman scattering.

molecule. A Raman spectrum is thus a plot of scattered intensity as a function of the shift in energy, $h\nu_1$, with respect to the excitation energy, $h\nu_0$. As in infrared spectroscopy, only certain vibrational modes are active as determined by the selection rules. The virtual state, shown in Figure 6.1, can be thought of as a short-lived distortion of the electron density of the molecule brought about by the oscillating electric field of the exciting light.

The intensity for a transition from an initial state i to a final state f is given by [21, 22]

$$I_{\mathrm{f,i}} = \frac{8\pi e^4}{9c^4} \omega_0 \omega_{\mathrm{sc}}^3 \sum_{\rho,\sigma} \left| (\alpha_{\rho,\sigma})_{\mathrm{f,i}} \right|^2 I(\omega_0) \qquad (6.1)$$

where $I_{\mathrm{f,i}}$ is the number of scattered photons, $I(\omega_0)$ is the incident number of photons at the incident angular frequency ω_0, ω_{sc} is the angular frequency of the scattered light, e is the charge on the electron, c is the velocity of light, ρ is the polarization of the laser electric field at the molecule, σ is the direction of the Raman-scattered electric field at the molecule, and α is the transition polarizability tensor. From Eq. (6.1) we can see that the Raman scattering is first order in the intensity of the incident laser light. The primary selection rule for Raman spectroscopy states that only vibrations that change the polarizability of the molecule are Raman active; this is different from the primary selection rule for infrared spectroscopy which requires a change in the dipole for the vibrational mode to be active. Thus Raman spectra and infrared spectra are similar but not identical. In addition the vibrational selection rule requires $\Delta v = \pm 1$ so the Stokes and anti-Stokes shifts give information about the ground-state vibrational frequencies for the molecule (see Figure 6.1). Extensive compilations of infrared and Raman frequencies can be found in the literature. The intensity of Raman bands depends on the polarizability tensor, α (see Eq. (6.1)). Thus the intensities of Raman bands can be very different from the corresponding intensities in the infrared spectrum. Strong Raman bands are expected for molecules with extended π systems, for molecules with large electron-rich atoms (e.g., S–S bonds, C–Cl bonds, C–I bonds), and for molecules with multiple bonds (e.g., C=C, C≡N) as these are all easily polarized. By the same argument the Raman spectra of weakly polarizable bonds (C–H, C–O, C–C, etc.) will generally be weak. Thus water is a very weak Raman scatterer – a significant benefit for studies in aqueous solution.

Even so Raman scattering is a very weak process with typically only 1 in 10^{10} scattered photons being Raman scattered and the vast majority being elastically (or Rayleigh) scattered [23]. The cross sections for Raman scattering are very small, typically 6 to 8 orders of magnitude smaller than fluorescence cross sections [23] and, in the absence of resonance effects, decrease with the fourth power of the excitation wavelength. This wavelength dependence reflects the fact that light of shorter wavelength is scattered more efficiently. Another advantage of using shorter wavelengths for Raman measurements is that it is easier to detect light at shorter wavelength. However, for biological samples these advantages are offset by the fact that fluorescence becomes more likely and the risk of sample degradation at high laser intensity is increased. The commonly used wavelengths, reflecting the commonly available lasers, are 1064 and 785 nm in the near-infrared, and 632.8, 532, 514.5, and 488 nm in the visible.

When the laser photon energy is close to (pre-resonance) or matches (resonance) the energy difference between the ground state of a molecule and an excited vibronic state the intensity of Raman scattering is significantly increased. This is referred to as resonant Raman scattering. In this case the vibrations that show the resonant enhancement are those involving atoms or bands associated with the chromophore while Raman bands for parts of the molecule not associated with the chromophore are not enhanced. In biological systems this can be used to advantage to study specific parts of large redox proteins, for example the heme in cytochrome c [24]. For a comprehensive discussion of Raman spectroscopy, including the practical aspects, the reader is directed to the book by McCreery [23].

6.3 SERS and Surface-Enhanced Resonant Raman Spectroscopy

The key to the SERS effect, and to its utility as a tool to study processes *in situ* at electrode surfaces, is the large magnitude and short range of the surface enhancement. In the early days of the technique there was much discussion and disagreement over the origin and nature of this surface enhancement. It is now generally agreed that the enhancement is made up of two contributions: a chemical enhancement and an electromagnetic enhancement. Of the two the electromagnetic enhancement is the larger, contributing a factor of greater than 10^4 as compared to a contribution of typically no more than 10^2 for the chemical enhancement.

The chemical enhancement model [22] postulates a chemical bonding interaction between a molecule and a metal surface such that new electronic levels are produced by overlap of the molecular orbitals of the adsorbed molecule and the orbitals of the metal. Then, provided these new electronic levels are close to resonance with the exciting laser light, there will be an enhancement in the strength of the Raman scattering for the adsorbed molecule. Clearly this mechanism is very short range as it requires significant orbital overlap between the molecule and the metal surface (i.e., the molecules must be chemisorbed) and it will be highly specific to the precise structure of the molecule and choice of metal since it requires the appropriate matching of orbital energies and symmetries [25].

In contrast electromagnetic enhancement [6] relies on the intensification of the local electromagnetic field at the metal surface and the interaction of this localized electromagnetic field with the molecules close to, but not necessarily directly in contact with or chemisorbed at, the metal surface. The electromagnetic enhancement is of longer range than the chemical enhancement decaying over a distance of the order of 50 to 100 nm.

For electromagnetic enhancement the exciting laser light must first couple to the metal surface to produce a surface plasmon. A surface plasmon is an interaction between free surface charges and the electromagnetic field [26–28] (Figure

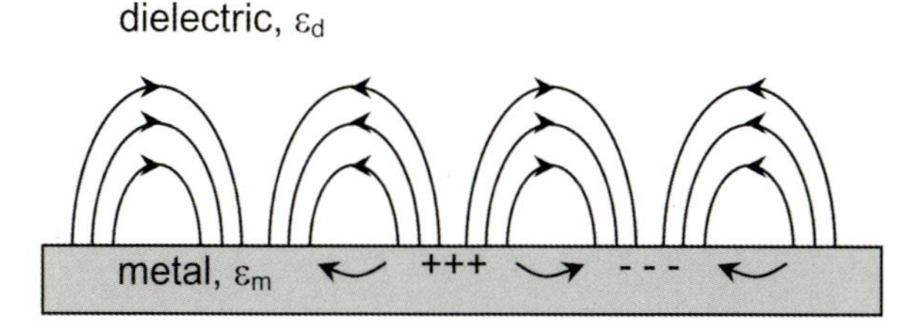

Figure 6.2 Combined electromagnetic wave and surface charge character of a surface plasmon.

6.2). Surface plasmons (also called surface plasmon polaritons) are characterized by a frequency-dependent surface wave vector

$$k_{sp} = \frac{\omega}{c}\left(\frac{\varepsilon_d \varepsilon_m}{\varepsilon_d + \varepsilon_m}\right)^{1/2} \tag{6.2}$$

where ω is the frequency of the electromagnetic field, c the velocity of light, and ε_d and ε_m the frequency-dependent permittivities of dielectric and metal, respectively. For surface plasmons to be possible ε_d and ε_m must have opposite signs. A consequence of the interaction between the electromagnetic field and the free surface charges is that the field perpendicular to the surface decays exponentially. Once the light is coupled to a flat surface as a surface plasmon it will propagate along the surface but will be slowly attenuated due to resistive losses in the metal. The greater the conductivity of the metal at the plasmon oscillation frequency the further the light will propagate. This depends on the complex permittivity of the metal. On a periodically structured or nanostructured surface, scattering of the surface plasmons can lead to the formation of standing waves or localized surface plasmons.

These surface plasmons can interact with the molecules close to, or at, the metal surface leading to Raman scattering which generates new surface plasmons shifted in energy by amounts corresponding to vibrational quanta for the molecules. The new surface plasmons then radiate light from the surface and can be detected by a spectrometer (Figure 6.3). The high magnitude of the electromagnetic contribution to the surface enhancement arises because there is a contribution from both the light coupling in to the surface and a contribution from the light coupling out from the surface [19, 29]:

$$I_{SERS} \propto \left|\frac{E_{local}(\omega_0)}{E_{inc}(\omega_0)}\right|^2 \left|\frac{E_{local}(\omega_{sc})}{E_{inc}(\omega_{sc})}\right|^2 \sum_{\rho,\sigma} \left|(\alpha_{\rho,\sigma})_{f,i}\right|^2 I(\omega_0) \tag{6.3}$$

where ω_0 is the incident and ω_{sc} the emitted frequency, and E_{inc} is the incident and E_{local} the local field strength. The first two terms describe the electromagnetic enhancement; the summation term contains the chemical enhancement contribution. As for the Raman effect itself, the SERS intensity is linear in the laser light intensity, $I(\omega_0)$.

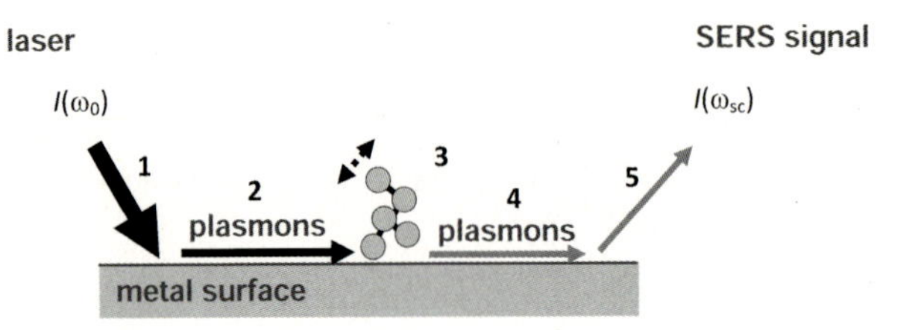

Figure 6.3 The SERS process. 1. The impinging laser light, ω_0, excites plasmons at the metal surface. 2. These plasmons convey optical energy into the molecule. 3. The molecule undergoes Raman scattering taking up a vibrational quantum of energy ($\omega_0 - \omega_{sc}$): Stokes scattering. 4. A plasmon at longer wavelength is produced. 5. This plasmon decays away into an emitted photon, ω_{sc}, which is detected in the spectrometer.

In both cases, coupling in and coupling out, the enhancement goes as the square of the local electromagnetic field and this is the origin of the very large enhancements that are possible in SERS – a 100-fold enhancement in the in-coupling and out-coupling electric fields at the metal surface will produce an electromagnetic contribution to the enhancement of 10^8. Often one sees in the literature the statement that the enhancement goes as the fourth power of the local electric field but this is actually a simplification [30–32], since, as shown by Eq. (6.3), the in- and out-coupling fields are at different frequencies. This has been clearly demonstrated experimentally in angle-resolved SERS measurements at gratings and structured surfaces where in-coupling and out-coupling are clearly separated [33–36].

In the literature there is often some confusion over the precise definition of the enhancement factor as measured experimentally and indeed there are several complications, and frequently assumptions, inherent in the estimation of enhancement factors. This is discussed in depth in a very useful article by Le Ru *et al.* [37].

The best-known metals for SERS are the coinage metals, Cu, Ag, and Au, but SERS has been observed on many other metals. The relative size of the enhancements for different metals depends on the wavelength-dependent optical properties of the metal [38]. In the visible region enhancements are strong on Cu, Au, and particularly Ag because of the optical properties of those metals. For other metals the effect is much less. Nevertheless there have been considerable efforts to extend the application of SERS to other metals, such as Pt, Pd, Rh, Ru, Fe, Co, Ni, etc., because of their importance in catalysis and electrochemistry [19, 39].

In addition to the Raman selection rules described above there are surface selection rules that apply for SERS because the process occurs close to metal surfaces [40–42]. The SERS surface selection rule predicts that the vibrational bands that have contributions from the Raman polarizability tensor component α_{zz}, where z is the surface normal, will be most intense with weaker contributions from vibrational bands which have contributions from α_{xz} and α_{yz}. This is essentially because the electric field of the exciting light is enhanced in the direction of the surface normal (Figure 6.2). The surface selection rule for Raman spectroscopy is more complex than that for infrared spectroscopy. Modes with the bond axis parallel to the surface often contain a substantial polarizability component in the direction of the surface normal and can, therefore, be SERS active. The surface selection rule means that the SERS spectrum is sensitive to the orientation of the molecule at the metal surface. For example, for an aromatic ring standing perpendicular to a metal surface the in-plane vibrational modes are expected to be much more intense in the SERS spectrum than the out-of-plane vibrational modes, whereas if the molecule is adsorbed flat on the surface this will change [43]. Pemberton *et al.* [44] have described a method to use SERS to determine the orientation of molecules with low symmetry using the relative intensities of vibrational modes for functional groups (such as $-CH_3$) which have multiple vibrational modes of known spatial relation. This approach has been used by Szafranski *et al.* [45, 46] to determine the tilt angle for aromatic thiols adsorbed at an Au electrode surface and to study the change in tilt angle with electrode potential.

The SERS enhancement falls quite sharply with distance from the surface; however, there is no generally agreed description for the distance dependence. This is probably because of a number of factors. First, it is quite difficult experimentally to establish the distance dependence because of the difficulty in unequivocally placing a probe molecule in the same orientation, at well-defined distances from a surface. Various approaches have been used including a polymer spacer layer [47], Langmuir–Blodgett layers [48] and multilayers [49, 50], self-assembled thiol monolayers [51, 52], and single-stranded DNA tethers [53, 54]. When using intervening layers there is the problem that this complicates the interpretation of the data since the intervening layer necessarily alters the plasmon field distribution and energy. Also, if a molecule is simply tethered to a surface by tethers of different length it is difficult to be certain that the molecule does not approach closer to the surface or change orientation [55]. Second, there are clearly differences in the distance dependence of the contributions from the chemical enhancement (which will be very short range) and the electromagnetic enhancement (expected to be longer range). Third, one can expect the distance dependence of the electromagnetic enhancement to depend on the surface geometry and to be different within a cavity or on a nanoparticle surface. There is, however, general consensus that the enhancement falls sharply with distance. Nevertheless, because the enhancement is so strong it is still possible to record spectra for molecules some tens of nanometers from a surface [54].

A further resonant contribution to the enhancement is possible when the molecule has an electronic transition in resonance, or close to resonance, with the exciting laser so that one obtains surface-enhanced resonant Raman scattering or surface-enhanced resonant Raman spectroscopy (SERRS). In these circumstances the resonant enhancement typically contributes an additional factor of between 100 and 1000 to the intensity of the signal and this makes SERRS a particularly attractive approach for analytical applications because of its extremely high sensitivity.

SERS and SERRS are powerful techniques for *in situ* electrochemical studies for several reasons. First, water is a weak Raman scatterer and is transparent in the visible region of the spectrum so it is relatively easy to carry out *in situ* measurements without interference from the solvent. Second, the SE(R)RS effect is highly surface selective so the spectra are obtained only from those molecules very close (typically 50 nm or less) to the electrode surface. Third, the technique gives molecular information about the species at the electrode surface including information about orientation and changes in molecular structure or bond strength.

In electrochemical SERS there can be a significant dependence, up to $100\,cm^{-1}\,V^{-1}$ or more, of the band frequencies on the electrode potential for adsorbates and molecules close to the electrode surface. This is referred to as the Stark effect or Stark tuning and can have contributions from the effect of the potential on metal–adsorbate bonding as well as from the purely electrostatic effect of the field at the electrode surface on the vibrational force constant [56]. For example, the Stark shifts seen for CO on Pt [56–58], CO on Ni [59], and NO on Pt [58] all have significant contributions from the effect of the electrode potential on the metal–adsorbate bonding as well as from purely electrostatic effects. When the functional group is remote from the metal surface, so that there are no direct bonding interactions, the Stark shift is entirely due to changes in the electric field in the double layer caused by changes in the electrode potential. In this case the magnitude of the Stark shift depends on the magnitude of the electric field and the dipole moment of the functional group so that groups with a large dipole such as CN are the most sensitive. This effect has been used very elegantly by Oklejas *et al.* to measure the interfacial electric fields in diffuse double layers using SERS [60, 61]. They used mixed monolayers of alkanethiol and a longer alkanethiol with a terminal CN moiety on an Ag electrode (Figure 6.4) to position the CN probe at different positions within the double layer. In this case, for measurements in 10 mM $NaClO_4$, they observed a Stark tuning rate of about $10\,cm^{-1}\,V^{-1}$.

Excellent reviews of electrochemical SERS, including details of instrumentation and applications, can be found in the work of Tian and the group in Xiamen [18–20]. One of the key impediments to the wider application of electrochemical SERS has been the problem of generating suitable, stable, reproducible electrode surfaces that show strong SERS signals. This is discussed below.

6.4
Comparison of SE(R)RS and Fluorescence for Biological Studies

Fluorescence-based methods are the most commonly used for biological labeling and studies of biological interactions. They require the covalent attachment of a fluorescent label to the biological molecule of interest that can then used in a number of different measurement formats including fluorescence resonance energy transfer. However these fluorescence-based techniques are not always compatible with *in situ* electrochemical measurements because of the quenching of fluorescence by the electrode surface. In contrast SERRS is an attractive alternative approach that is gathering support for biological applications [15, 16, 20, 62–64] and which is compatible with *in situ* electrochemical measurements.

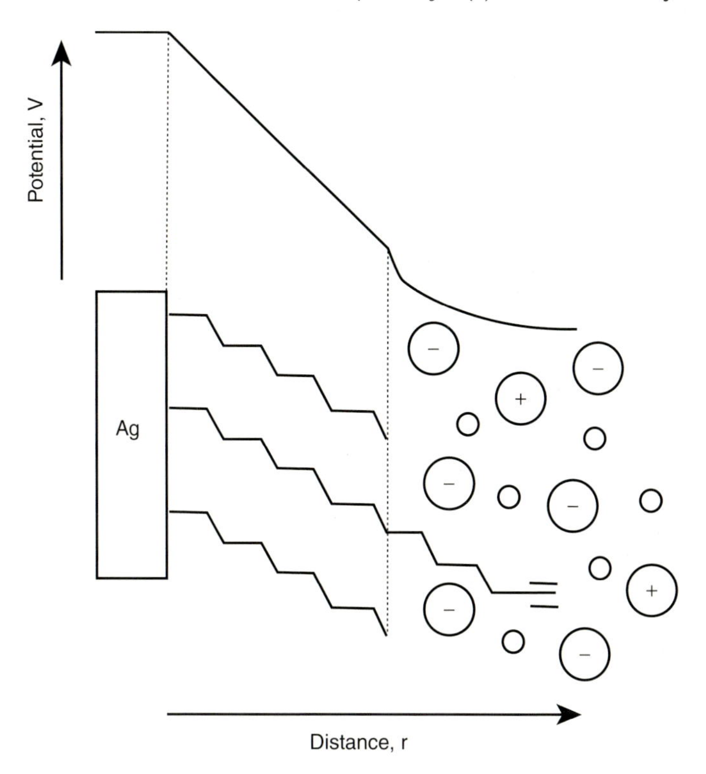

Figure 6.4 Schematic of the Ag–SAM–aqueous interface region showing the nitrile group extending into the double layer. Reprinted with permission from Oklejas, V., Sjostrom, C., and Harris, J.M. (2002) *Journal of the American Chemical Society*, **124**, 2408. Copyright 2002 American Chemical Society.

Both fluorescence and SERRS require labeling of the biological component, both are highly sensitive techniques, and both have been shown to be capable of single-molecule detection under favorable conditions [65–68]. Labels used for biological studies must fulfill several requirements [69]: they should not perturb the specificity of the biological recognition system under study; they should be robust and reproducible, stable on storage, in contact with the sample, and during measurement; they should give a linear response over the appropriate concentration range with high sensitivity; and the signal should be readily discernable from the background. These criteria are met by a wide range of Raman labels. Thus commonly available fluorescent labels can be used in SERRS and there are also a large number of Raman dyes [70–72] and nonfluorescent labels available [73]. This range of labels can be further extended by simple molecular modifications or the synthesis of new dyes with distinctive spectra [71, 74].

Compared to fluorescence, SERRS has some significant advantages. Unlike fluorescence-based methods, SERRS measurements are unaffected by quenching by oxygen or other species, are less sensitive to photobleaching, and are inherently more sensitive. For example, Sabatte *et al.* [75] showed that the limit of detection for SERRS in an immunoassay application was around 1000 times lower than for the corresponding fluorescence assay once account is taken of dilution effects. A similar improvement of three orders of magnitude in the limit of detection for SERRS over fluorescence was reported by Faulds *et al.* for a DNA assay [76].

For multiplex measurements, when compared to fluorescence, SERRS also offers significant advantages. In multiplex measurements fluorescence has the disadvantage that the electronic spectra produced are broad (typically 50 to 100 nm full width at half maximum) and therefore overlap so that the technique is limited to the simultaneous measurement of around four dye labels [69, 77]. In contrast, SERRS uses the vibrational Raman spectrum of the label as a spectroscopic molecular fingerprint. As a result the information content of the spectra is much higher and, because the vibrational bands are much narrower (about 1 nm full width at half maximum), spectral overlap is much less of a problem. Thus using SERRS it is possible to readily identify the components of a mixture without extensive separation procedures [78] and it has been estimated in the literature that simultaneous measurement with up to 30 SE(R)RS labels should be possible [79].

SE(R)RS is well suited as a technique for the detection and discrimination of DNA due to its high sensitivity and advantages over fluorescence [63]. The first applications of SE(R)RS to DNA detection by Vo-Dinh's group used roughed silver surfaces to achieve surface enhancement [80–82] but this approach suffers from problems of reproducibility. Using silver colloids Faulds *et al.* have demonstrated the quantitative detection of dye-labeled oligonucleotides by SERRS [83].

There are several examples of multiplexed SERRS measurements as applied to DNA assays [69, 70, 73, 77, 84–89] reported in the literature and a few examples of immunoassays. Thus Woo *et al.* [90] described a triplex immunoassay for the detection of bronchioalveolar stem cells, Cui *et al.* [91] used two dye labels to carry out a duplex sandwich immunoassay, Jun *et al.* [92] reported a triplex immunoassay using SERS, and Lutz *et al.* [93] described a quadruplex platelet activity state plate binding assay.

Finally, the instrumentation required for SE(R)RS measurements is not necessarily large or complex. Indeed, commercial, handheld spectrometers are available. In essence the technique requires a laser for excitation of the surface, a filter (notch or edge) to exclude the Raleigh-scattered radiation, a monochromator, and a detector to collect the Raman-scattered light [19, 20, 23].

6.5 Surfaces for SERS

Early studies of SERS were plagued by irreproducibility. The studies relied on electrochemically roughened coinage metals (Ag, Au, or Cu) with ill-defined and unstable surfaces. As a result the intensity of the spectra varied by orders of magnitude from place to place on the surface, from one surface to another, and over time. Thus although SERS was in principle, since it gives molecular structure-specific information with exceptionally high surface selectivity, ideally suited for *in situ* electrochemical studies it was dogged by problems. There was no way to be certain the spectra came from molecules in typical surface environments at the surface, as opposed to reflecting the properties of a small subset of molecules in an atypical environment; the method was restricted to studies of Ag, Au, and Cu electrodes and it was impossible to make quantitative measurements because of the irreproducibility and instability of the substrates.

This led to a lot of attention being focused on ways of roughening electrode surfaces, including potential- or current-controlled oxidation–reduction cycles, chemical etching, electrodeposition and template synthesis, and the development of optimized recipes for roughening different surfaces [11, 18–20]. It also led to many attempts to characterize these roughened surfaces down to the nanoscale, and a particular focus on generating "hotspots" – those regions on the surface where the random nanostructure was, by chance, such that there was a high surface enhancement. In turn this led to the interest in using colloids and in particular aggregates of colloids in an attempt to generate hotspots for SERS. While these colloidal aggregates are useful for solution studies it remains a challenge to control the aggregation process and use them in quantitative measurements [94]. This focus on the role of hotspots continues to the present where it is still common to read in papers that hotspots are essential for SERS even though this is clearly not the case. For *in situ* electrochemical studies relying on hotspots to give reliable SERS is not a productive approach; rather, the key to developing substrates for electrochemical SERS must be to make controlled surfaces where the nanostructure is stable and robust.

A lot of the published work on electrochemical SERS uses roughened Ag or Au electrodes because these give very large SERS enhancements. However, Ag and Au are not the only electrode materials of interest and consequently significant efforts have been made, particularly by Tian's group, to extend SERS to other transition metal surfaces [18, 19, 39]. The problem here is that, in the visible region of the spectrum, the enhancements that can be achieved on metals such as Pt or Pd are orders of magnitude less than for Ag or Au because of the optical properties of the metals. One way to overcome this is by roughening the metals or by synthesizing shaped nanoparticles of the metals that can be assembled on an electrode surface [19, 20, 95, 96]. An alternative approach, pioneered by Weaver's group [97–99], is the so-called borrowed SERS technique [18]. In borrowed SERS a thin (one to five atoms thick) overlayer of the metal of interest is formed on top of a SERS-active surface prepared from Ag or more usually Au. The technique can also be used with core–shell nanoparticles which have a metal shell over a SERS-active core [100]. The metal overlayer must be very thin otherwise it damps the plasmons in the underlying SERS-active surface and prevents the electromagnetic enhancement of the SERS. This obviously causes a problem because it is difficult to form perfectly pinhole-free overlayers of a metal just a few atoms thick. This has been addressed by Weaver's group using a "pinhole free" deposition technique [101–103]. It is also a potential problem because one cannot always be sure that the electrochemical properties of a very thin overlayer have not been altered from those of the bulk metal.

6.6 Plasmonic Surfaces

One of the key developments in SE(R)RS over the last 10 or so years has been the development of structured surfaces designed to achieve strong SERS enhancements by controlling the plasmonics of the surface and thus the electromagnetic enhancement. This is a very active area of research which overlaps with work in the wider fields of nanophotonics and optical metamaterials [104, 105]. A number of excellent reviews of the area have been published in the last few years [11, 15–17, 26–28, 106].

The plasmon resonances of metallic nanostructures can be modeled using classical electromagnetism and analytical solutions are available for simple highly symmetrical cases such as spherical particles. For more complex nanostructures it is necessary to use numerical methods such as the discrete dipole approximation [107], finite difference time domain methods [108, 109], or boundary element techniques [110]. However, these calculations are complex and computer intensive so that it is not possible to accurately model extended nanostructures. An important approach to understanding the plasmonics of these more complex nanostructures is the concept of plasmon hybridization [111–113]. The idea of plasmon hybridization exploits the fact that there are similarities between the way plasmons behave and the behavior of electrons in atomic orbitals so that, in the same way that atomic orbitals can combine to form new molecular orbits if the energies and symmetries are correct, plasmons in complex nanostructures can be envisaged as being made up from combinations of interacting plasmons. This is illustrated in Figure 6.5 for the example of a nanoshell particle. In the plasmon hybridization model the plasmon resonance of the nanoshell is built up by combining the plasmon resonances for a nanosphere and for a nanocavity. For these two simple, symmetric structures Mie theory gives the plasmon frequencies [111] for the sphere as

$$\omega_{S,l} = \omega_B \sqrt{\frac{l}{2l+1}} \tag{6.4}$$

and for the cavity as

$$\omega_{C,l} = \omega_B \sqrt{\frac{l+1}{2l+1}} \tag{6.5}$$

where l is the angular quantum number and ω_B is the bulk plasmon frequency.

Provided the plasmon resonances of the component structures are of the same symmetry and similar energy they can be combined together to form hybrid plasmon resonances in the same way that molecular orbital theory is used to combine together atomic orbitals to form bonding and antibonding molecular orbitals. As shown in Figure 6.5 the $l = 1$ modes for the cavity and sphere mix to

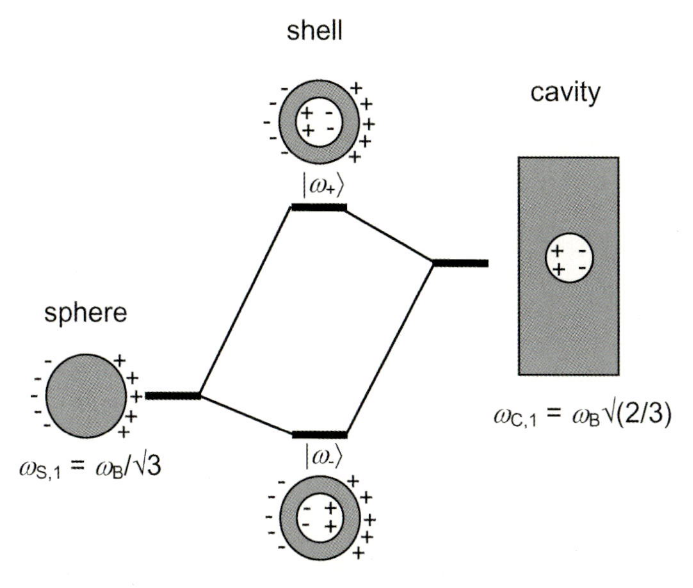

Figure 6.5 Energy level diagram for plasmon hybridization in a metal nanoshell by combination of the plasmons for a sphere and a cavity. The two nanoshell plasmons are symmetrically and antisymmetrically coupled and have energies given by Eq. (6.6).

produce two new modes with the higher energy ω_+ mode corresponding to the antisymmetric coupling (this would be the antibonding orbital in molecular orbital theory) and the lower energy ω_- mode corresponding to the symmetric coupling. The plasmon frequencies for these hybridized plasmons are given by [112]

$$\omega_{l\pm}^2 = \frac{\omega_B^2}{2}\left[1 \pm \frac{1}{2l+1}\sqrt{1+4l(l+1)\left(\frac{a}{b}\right)^{2l+1}}\right] \tag{6.6}$$

where a is the inner and b the outer radius of the nanoshell.

This plasmon hybridization approach [114] has been used very effectively to describe the plasmonics of a number of core–shell [112, 115, 116] and other structures [114, 117, 118] including effects on extended nanostructured surfaces [119–121].

For electrochemical SE(R)RS measurements it is necessary to have a continuous conducting surface to use as the electrode and this rules out the use of many nanoparticle and core–shell-type structures used for SE(R)RS studies. In the next section we concentrate on electrode structures for SE(R)RS.

6.7 SERS Surfaces for Electrochemistry

Over the last few years there has been an increasing move away from roughened electrode surfaces for electrochemical SE(R)RS with all their inherent limitations towards structured surfaces which show large, reproducible, repeatable surface enhancements [19, 20].

Knoll *et al.* [36] carried out a systematic study of SERS for Langmuir–Blodgett films of cadmium arachidate with different numbers of monolayers on Ag gratings. They clearly demonstrated, by a careful study of the angle dependence of the spectra, that the enhancement was caused by coupling with surface plasmons on the grating [122] and that it depended on both coupling of light in and coupling of the Raman scattered light out from the grating (see Figure 6.3). Subsequently Baltog *et al.* [34, 123] demonstrated a 10^4 enhancement of SERS of copper phthalocyanine on an Ag grating due to the resonant excitation of delocalized surface plasmons. Again they showed that the Raman-scattered light emerges at particular angles corresponding to resonance of the outgoing light with the grating and different from the angle for resonance of the incoming laser light. Essentially similar conclusions were reached by Kahl and Voges [35]. These initial studies on gratings pointed the way toward developing structured surfaces for SERS which can give surface enhancements without high surface roughness or randomly distributed hotspots.

There are several ways to structure surfaces on a scale close to the wavelength of light in order to generate surfaces with strong SERS enhancements. These can be divided into top-down methods, such as electron beam lithography, focused ion beam milling, and soft lithography, or bottom-up methods that rely on self-assembly to build up the structures. Electron beam lithography is an attractive approach because it allows a high degree of control over the structures that can be made, but it is an expensive approach requiring access to specialized equipment and is unsuited to large-scale fabrication.

Structures made by top-down methods are very useful in exploring the SERS effect and the role of surface plasmons [124]. One structure which has attracted attention because of its interesting plasmonic properties [118] is the nanohole array. Brolo *et al.* [125] used focused ion beam milling to fabricate square arrays of sub-wavelength holes on an Au film with periodicities of 560, 590, and 620 nm. They found enhancements in the SERRS for oxazine at the surface when irradiated at 633 nm. Anema *et al.* [126] carried out similar experiments for oxazine on square arrays of 133 nm diameter nanoholes in Cu. Their arrays had periodicities between 360 and 740 nm and were made in 144 nm thick Cu films on glass. The highest SERRS intensity, with an estimated enhancement factor of 10^4, was observed for the array with a periodicity of 578 nm when using 633 nm irradiation. These results were found to be consistent with calculations of the electric field intensity as a function of the array periodicity. Reilly *et al.* investigated SERS on arrays of 200 nm diameter holes in a 50 nm Ag film as a function of lattice spacing [127]. They concluded, on the basis of quantitative measurements for a nonresonant system, that the enhancement originated from two multiplicative contributions: a contribution from field localization near the aperture edges and a contribution from the surface roughness between the apertures. They concluded that the aperture contribution was a factor of around 600 and that from surface roughness around 10^5, giving a total enhancement of 6×10^7 in the optimal case.

Im *et al.* [128] have used optical lithography combined with atomic layer deposition to form vertically oriented metal–dielectric–metal nanogap structures designed to achieve significant field enhancements with strong localization within the gap. The method was used to fabricate a number of different nanogap structures and it was shown that the local enhancement increased as the size of the nanogap decreased from 20 to 5 nm. Dinish *et al.* [129] used deep-ultraviolet lithography to fabricate nanogaps between circular Au–Ag structures and obtained SERS intensities slightly greater than those for metal film over nanosphere (MFON) structures (see below) for 2-napthalenethiol. Jackel *et al.* [130] used electron beam lithography to fabricate bowtie structures (two Au nanotriangles oriented tip to tip with a nominal gap of 35 nm) on an indium tin oxide-coated fused silica substrate and used the structure to study electrochemical SERS of *p*-mercaptoaniline at individual bowtie structures. They found substantial changes in the shapes of the bowties caused by two electrochemical cycles between −0.4 and +0.4 V vs. Ag/AgCl in 0.1 M $NaClO_4$ (Figure 6.6).

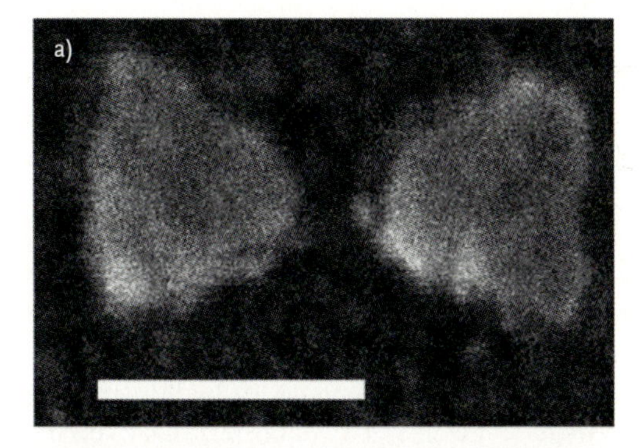

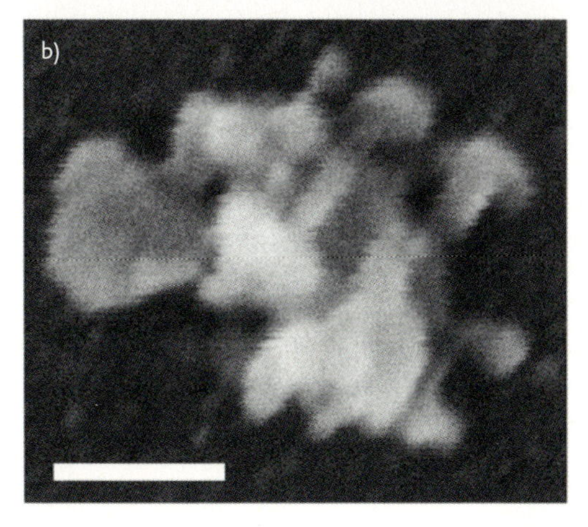

Figure 6.6 Scanning electron microscopy images of Au nano-bowtie structures (a) before and (b) after cycling in sodium perchlorate solution. The scale bars are 100 nm. Adapted from Jackel, F., Kinkhabwala, A.A., and Moerner, W.E. (2007) *Chemical Physics Letters*, **446**, 339, with permission from Elsevier.

Figure 6.7 Scanning electron microscopy image of part of a Klarite® substrate showing some of the Au-coated inverted pyramidal pits. Reprinted from Ignat, T., Munoz, R., Irina, K. *et al.* (2009) *Superlattices and Microstructures*, **46**, 451, with permission from Elsevier.

Lithography and anisotropic etching of (100)Si has also been used as a way to produce structures for SERS. In this case the etching produces pyramidal pits in the surface with a fixed apex pit angle of 70.5 °. These are then coated with Au by evaporation to produce a nanostructured surface (Figure 6.7). In this case the optical properties of the surface depend on the geometry of the pits (their spacing and the size of the opening or aperture). These structures are commercially available under the trade name Klarite® and have been used for SERS measurements by several authors including studies with molecularly imprinted polymers [131], supramolecular host molecules [132], squarines [74], and dye-labeled DNA [133, 134]. Meyer *et al.* [135] have compared the SERS of rhodamine 6G on Klarite® and several other surfaces. This type of structure has been fabricated by Ohta and Yagi [136] and used for electrochemical SERS measurements with pyridine in 0.1 M KCl between −0.06 V and +0.84 V vs. RHE.

Templated self-assembly approaches have the significant advantage over top-down approaches of being more generally accessible, since they do not require clean-room access and access to specialized expensive equipment, and are amenable to scale-up. In particular, methods based on nanosphere lithography and the assembly of colloidal templates have been very successfully applied to fabricate SE(R)RS substrates. There are a number of variations on the approach, producing several different types of structure (Figure 6.8), but all begin with the assembly of a closed packed array of uniform spherical particles to act as the template or mask. This approach originates with the work of Deckman and Dunsmuir on "natural lithography" [137].

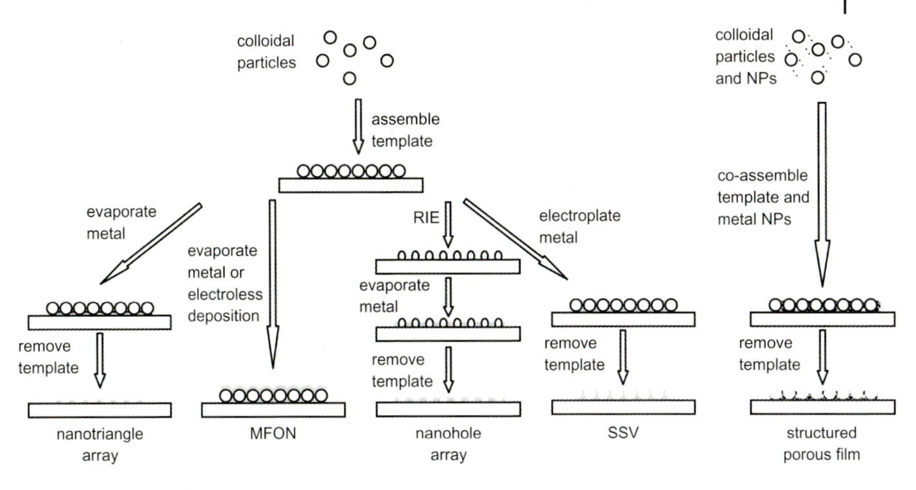

Figure 6.8 Different examples of colloidal templating, or "natural lithography," used to make SERS-active surfaces starting from uniform spherical colloidal particles, usually of polystyrene or silica. (NPs, nanoparticles; RIE, reactive ion etching.)

The most commonly used template particles are either polystyrene or silica spheres. These can be synthesized or obtained commercially and are available in a range of sizes (from about 100 nm upwards) with a narrow coefficient of variation (<2%) in diameter. The size variation is important because the more uniform the size of the particles the more regular the packing. The particles can be assembled into templates in a variety of ways, including evaporation of the solvent [138], convective self-assembly [139–141], Langmuir–Blodgett deposition [142, 143], or by withdrawing the substrate slowly from a suspension of the colloidal particles [144]. The optimum methods for the assembly of the silica and polystyrene particles differ because of the differences in their densities and hence sedimentation rates. Once assembled on the surface the particles act as a template to define the deposition.

These templates have been used in various different ways to make SE(R)RS substrates. Van Duyne and coworkers [106, 145] used evaporation of metal (typically Ag or Au) through the template. The metal evaporation occurs by line of sight so only the areas of the substrate exposed between the template spheres become coated by metal. Thus when the template is removed the substrate is left covered in an array of metal triangles oriented tip to tip (Figure 6.9) like an ordered array of the bowtie structures [130] discussed above. By varying the size of the template spheres, the amount of metal deposited and by using templates of one or of two monolayers of spheres and by varying the angle of metal deposition, they were able to make a range of structures with different sizes and geometries [146–148]. These surfaces have been used extensively by Van Duyne's group for localized surface plasmon studies [146–154] and as SE(R)RS substrates [145, 155, 156]. The plasmonics and SERS on these surfaces are dependent on the geometry [157, 158]

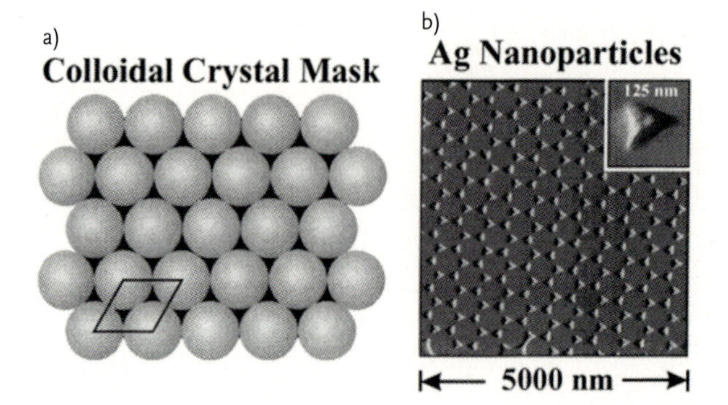

Figure 6.9 (a) Schematic of a single-layer colloidal template. The dark areas show the exposed parts of the underlying surface. (b) Representative ambient contact mode atomic force microscopy image of the array of nanotriangles formed by thermal evaporation of 48 nm of Ag through a single-layer colloidal template (sphere diameter 542 nm) followed by removal of the template by sonication in methylene chloride for 3 min. Reprinted with permission from Haynes, C.L. and Van Duyne, R.P. (2001) *Journal of Physical Chemistry B*, **105**, 5599. Copyright 2001 American Chemical Society.

and dominated by the small gap regions between the triangular structures. Haynes and Van Duyne have measured enhancement factors in excess of 10^8 for SERS of benzenethiol on these Ag structures and in excess of 7×10^9 for SERRS [155], although there was a significant amount of scatter in the measurements. As far as we are aware these triangular structures have not been used for electrochemical SERS, although such structures have been made on an indium tin oxide surface [159].

If a thicker metal film is evaporated so that it totally covers the template spheres a two-dimensional grating structure is formed [145]. This is referred to by Van Duyne and coworkers as a MFON structure [160] (Figure 6.10). Evaporated Ag film over nanosphere (AgFON) or Au film over nanosphere (AuFON) surfaces give large SERS enhancements [156, 161, 162], although data from Im *et al.* [128] indicate that the SERS comes from randomly distributed hotspots across the surface. Bantz and Haynes [163] have described an alternative approach to the preparation of AgFON surfaces that uses electroless plating of Ag to form the metal film rather than evaporation. The resulting surfaces are less topographically homogeneous but still exhibit strong surface enhancement. Van Duyne's group have used MFON surfaces to make biosensors [164–166] and have shown that for AgFON structures made with silica templates the thermal stability is good [167]. These MFON structures can be used for electrochemical SERS as there is now a continuous conducting surface. Van Duyne's group have used AuFON and AgFON surfaces for electrochemical SE(R)RS and surface-enhanced hyper-Raman scattering studies [156, 168]. For example, Dick *et al.* have studied the distance and orientation dependence of electron transfer to cytochrome c at thiol-coated AgFON electrodes [169]. A significant advantage of this type of structured surface over roughened

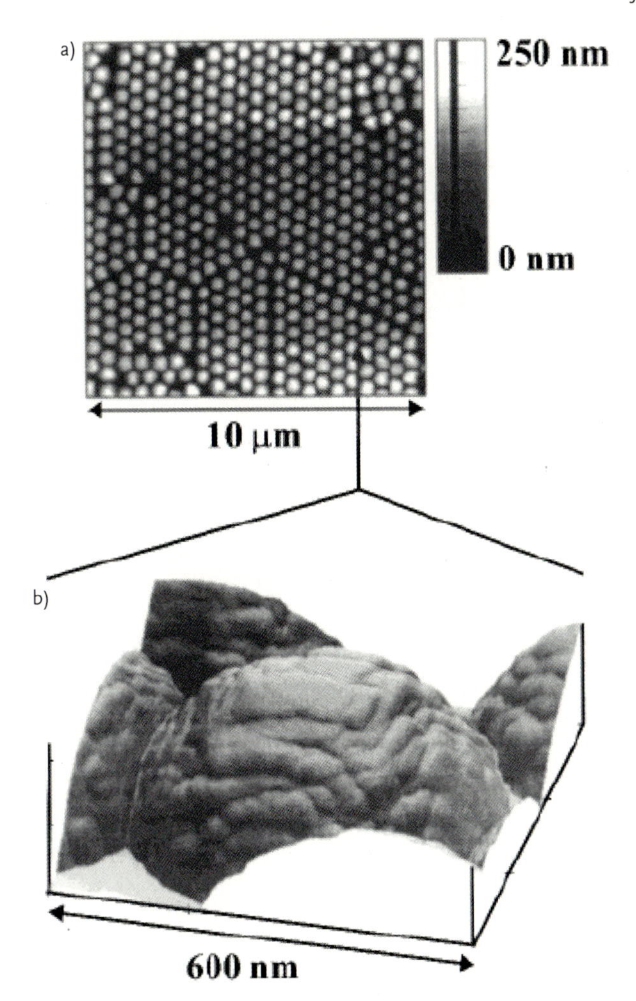

Figure 6.10 Ambient contact mode atomic force microscopy images of an AgFON structure. The structure was made by depositing 200 nm of Ag over 542 nm diameter polystyrene spheres. (a) The array of spheres and (b) image of one Ag-coated sphere showing the substructure roughness. Reprinted with permission from Dick, L.A., McFarland, A.D., Haynes, C.L., and Van Duyne, R.P. (2002) *Journal of Physical Chemistry B*, **106**, 853. Copyright 2002 American Chemical Society.

electrode surfaces is the significantly greater stability, so that the electrode can be cycled to extremely negative potentials (−1.3 V vs. Ag/AgCl) without loss of SERS activity [160]. AgFON surfaces have also been used for borrowed SERS by coating with a thin layer of Pt [170].

Comparison of the enhancement factors between triangular nanoparticle arrays formed by nanosphere lithography and the MFON structures [8] shows that although the enhancement factor is about 10 times larger for the triangular

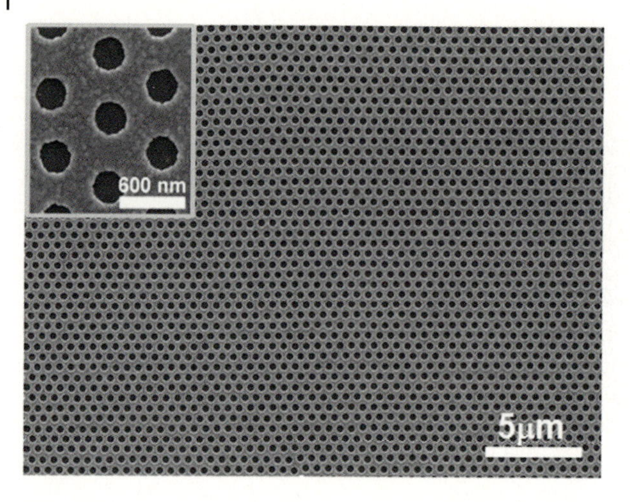

Figure 6.11 Scanning electron micrographs of a single-crystalline hexagonal nanohole array formed from a template of 600 nm polystyrene spheres by size reduction using reactive ion etching followed by deposition of 50 nm of Ag deposition and removal of the template with tape and solvent cleaning. Reprinted with permission from Lee, S.H., Bantz, K.C., Lindquist, N.C. *et al.* (2009) *Langmuir*, **25**, 13685. Copyright 2009 American Chemical Society.

nanoparticle array, the SERS signals are larger for the MFON structures because there is a larger active area covered by molecules.

In a variation of the nanosphere lithography approach Lee *et al.* [138] used reactive ion etching to reduce the size of polystyrene spheres before evaporation of Ag onto the structure. Removal of the spheres then produces a nanohole array (Figure 6.11) similar to that fabricated by focused ion beam milling. For benzenethiol Lee *et al.* obtained enhancement factors of up to 8×10^5 for these nanohole structures.

Self-assembled templates of colloidal particles have also been used to prepare metallic structures where the voids between the template spheres are filled with metal. Kubo *et al.* [144] used a dipping method to produce metal-coated colloidal crystal films by first depositing monodisperse 300 nm silica spheres onto a glass substrate in the presence of 10 nm Ag nanoparticles. This produced, after calcining, a structure in which the Ag nanoparticles were immobilized on the silica spheres. SERS signals of *p*-toluenethiol of this structure were 40 times stronger than for a flat Ag film and 3 times larger than for electrochemically roughened Ag with good reproducibility from place to place on the surface.

Velev and coworkers [140, 171] used a similar approach of co-assembling colloidal particles with Au nanoparticles. In this case they removed the colloidal template to produce a porous Au inverse opal structure (Figure 6.12). They found evidence for SERS activity on these structures but concluded that the enhancements were due to the nanoscale surface roughness stemming from the discrete aggregated Au particles [140].

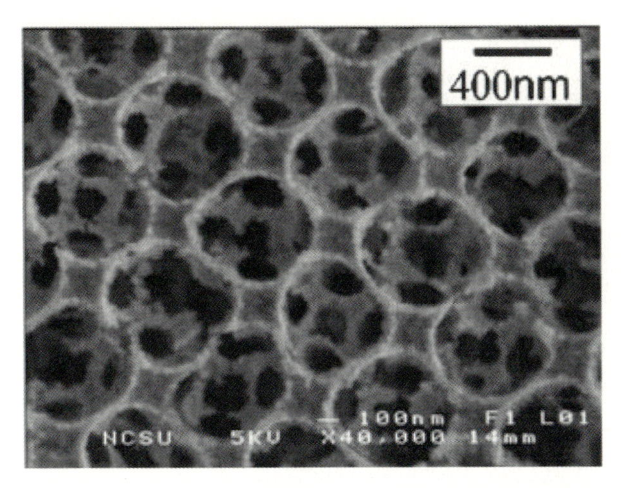

Figure 6.12 Scanning electron microscopy image of a mesoporous Au film made by co-assembly of colloidal template particles and Au nanoparticles following removal of the template. From Kuncicky, D.M., Prevo, B.G., and Velev, O.D. (2006) *Journal of Materials Chemistry*, **16**, 1207. Reproduced by permission of the Royal Society of Chemistry.

Lu *et al.* [172] used a template assembled from uniform silica spheres to produce both an ordered macroporous Au–Ag nanostructure and an ordered hollow Au–Ag nanostructured film by electroless deposition. Both films showed SERS activity but were rather rough on the nanoscale and the authors attributed the surface enhancement to the presence of interconnected nanostructured aggregates and nanoscale roughness.

The problem with the nanoparticle infiltration and electroless deposition approaches to fill the colloidal templates with metal as described above is that it is very difficult, if not impossible, to accurately and systematically control the structure and the deposited metal is rough on the nanoscale and often porous so that the surface area is poorly defined. These problems can be overcome by using electrodeposition through the template. Electrodeposition is a volume-filling method and the thickness of the electrodeposited metal film is readily controlled [173]. The resulting films have very low nanoscale roughness and their structure can be readily varied. This approach has been developed at the University of Southampton over the last 10 years. The surfaces are prepared by electrodeposition of a metal, typically Au, through a close-packed monolayer of uniform polystyrene spheres assembled on a flat conducting surface [141, 173]. After deposition of the metal the polystyrene spheres are removed by dissolution in an organic solvent to leave a film with a regular array of sphere segment voids (SSVs) (Figure 6.13). The diameter of the SSVs is controlled by the choice of the polystyrene spheres (typically around 400–900 nm in diameter) and the thickness of the metal film is controlled by the charge passed during deposition (typically 30 to 80% of the sphere diameter) [174].

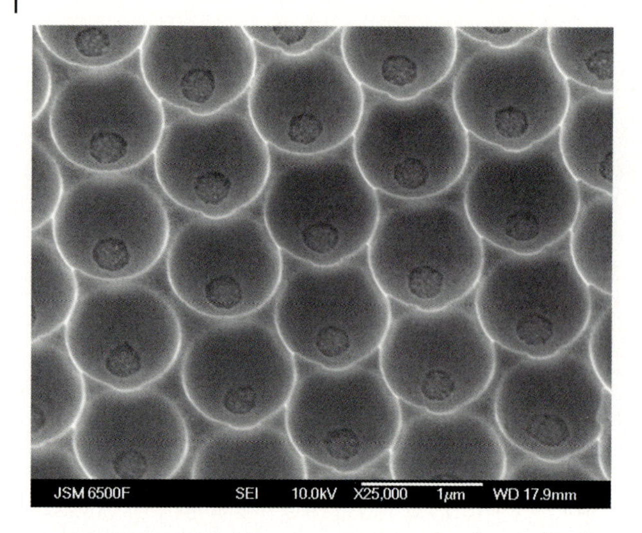

Figure 6.13 Scanning electron microscopy image of an Au SSV structure. Note the smooth electroplated metal walls and top surface. The rough circular areas at the bottom of each cavity are the evaporated Au substrate.

These SSV surfaces show strong reproducible SE(R)RS [175–177] with enhancement factors of 2.7×10^7 for benzenethiol on an Au SSV surface with typically about 10% variation from place to place across the surface [178]. By careful analysis of the reflection spectra of these nanostructured metal surfaces as a function of sphere diameter, film thickness, type of metal, light polarization, azimuthal angle, and angle of incidence it is possible to build up a detailed picture of the plasmonics on these structured surfaces and to use this knowledge to design surfaces for particular applications [33, 110, 120, 121, 174, 178–182]. These studies reveal that there are both Bragg-type propagating and Mie-type localized plasmons on the SSV surfaces and that these plasmons couple together [121].

This detailed understanding of the localized electromagnetic field on SSV metal surfaces due to the plasmonic resonances makes these surfaces ideal platforms for SE(R)RS [175, 178, 183]. Thus we have been able to design SSV surfaces suitable for SERS measurement in the ultraviolet [184], in the visible [175, 178, 183], and in the near-infrared [176] regions of the spectrum by varying the sphere diameter and film thickness. We have also been able to produce films of Pt and Pd, metals which do not show strong enhancements because of their fundamental optical properties, with surface enhancements equal to the highest reported values [185] for these metals. Our understanding of the detailed plasmonic properties of the SSV surfaces has enabled us to clearly demonstrate, by making angle-resolved measurements, the importance both of coupling light from the laser into the surface and of coupling the Raman-scattered light out to the detector [33]. We have demonstrated the relationship between the SERS intensity and the plasmonics of the surfaces [186], we have shown that the contribution of resonant enhancement on the SSV surfaces is around three orders of magnitude [177], and we have examined in detail the origins of the broad SERS background and how it arises on the SSV surfaces [187].

SSV surfaces are ideally suited for electrochemical SE(R)RS studies because of their low surface roughness, high surface enhancement, and good stability. They have been used for electrochemical SE(R)RS studies of pyridine [175], flavin [183, 188], adenine [184], β-thioglucose [189], and an osmium redox hydrogel [188], and for discrimination of DNA mutations [190, 191]. Recent work by Jose *et al.* has also demonstrated enhanced fluorescence on SSV surfaces [192].

6.8 Tip-Enhanced Raman Spectroscopy

It is appropriate to conclude this part of the chapter, before going on to review the literature on SE(R)RS of biomolecules at electrode surfaces, by briefly describing tip-enhanced Raman spectroscopy (TERS) since this rapidly developing technique offers the potential for studies at molecular resolution. In TERS a metal nanoparticle or metalized tip (usually Ag or Au) with an apex diameter of about 25 nm is illuminated by a laser as it is scanned across the surface (Figure 6.14). The tip is used to locally amplify and confine the electromagnetic field, in effect creating a local hotspot which can be scanned across the surface. The first examples of this approach were reported in 2000 [193–195]. Since then the approach has been

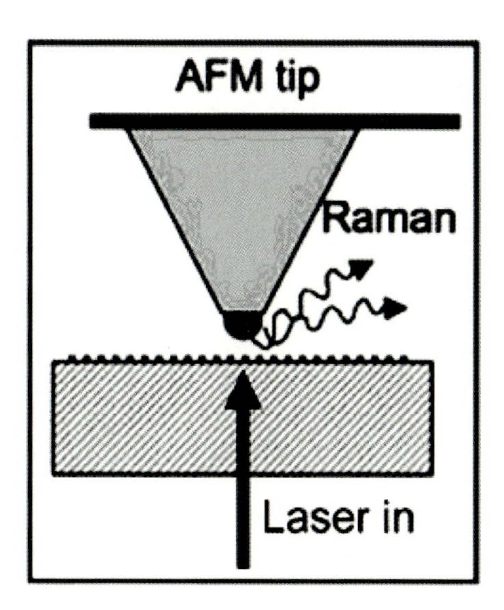

Figure 6.14 Schematic of a modified atomic force microscopy (AFM) tip used for TERS. Reprinted from Yeo, B.-S., Stadler, J., Schmid, T. *et al.* (2009) *Chemical Physics Letters*, **472**, 1. Copyright (2009), with permission from Elsevier.

applied to study a number of biomolecules at surfaces including oxidized glutathione on Au [196], histidine on Ag [197], cytochrome c [198], adenine [199, 200], and other DNA nucleotides and bases [201, 202], and RNA on mica [203]. For further details of TERS the reader is referred to several recent reviews [204–207].

The idea of TERS is to have a localized probe that, ideally, stimulates SERS but without perturbing the chemical system under study. This concept has been taken forward by Tian's group in recent work on silica- or alumina-coated Au nanoparticles in so-called shell-isolated nanoparticle-enhanced Raman spectroscopy, or SHINERS [208]. The idea is that the thin (a few nanometers) silica or alumina coating renders the particles inert so that they can be scattered over the surface of interest to generate SERS signals without perturbing the surface chemistry, rather like a large random array of TERS tips.

6.9 SE(R)RS of Biomolecules

As mentioned earlier in this chapter, water is a very weak Raman scatterer and therefore does not obscure SERS signals from biomolecules in aqueous solution. In addition SERS is very well suited as an *in situ* technique for monitoring changes in surface chemistry, including the adsorption and orientation of molecules. Thus SERS is an ideal complement to bioelectrochemical studies and provides surface-selective, molecule-specific information. Frequently it is possible to increase SERS intensities by manipulating adsorption of biomolecules through changes in the electrode potentials and this represents a further, significant advantage.

In the remainder of this chapter we review the applications of SE(R)RS in studies of biomolecules at electrode surfaces. We review SE(R)RS as applied in bioelectrochemistry, concentrating attention on studies where the interfacial potential is under direct electrochemical control. In general, and for the sake of brevity, we exclude any detailed discussion of SE(R)RS studies of biomolecules at colloidal nanoparticles where the interfacial potential is undefined; for accounts of these studies the reader is directed to some recent reviews [209–214].

6.9.1 DNA Bases, Nucleotides, and Their Derivatives

Nucleic acids and their constituents were among the first biomolecules to be studied on electrodes with SERS. Among the first reports on nucleic acid components studied on roughened Ag electrodes are those on adenine and its nucleotide derivatives by Koglin and coworkers [215, 216]. Their work showed that surface Raman spectra were stronger than their solution-based counterparts and they reported a detailed spectroelectrochemical study [216] where the intensity of the peaks was found to be strongly dependent on the applied potential and on the preparation conditions for the substrate. A strong peak at around 740 cm^{-1} was visible for all adenine derivatives including adenosine, adenosine monophosphate

(AMP), and adenosine triphosphate (ATP) in both the studies, and a band at 248 cm^{-1} was assigned to the Ag–phosphate vibration. In general the intensities of all peaks were found to decrease with increasing negative potentials for all of the nucleotide derivatives. Spectra of other nucleotide bases and of adenine, recorded on Ag electrode surfaces, were reported by Koglin *et al.* [217]. The SERS spectra of adenine, guanine, cytosine, and thymine adsorbed on a polycrystalline Ag electrode surface showed prominent SERS bands at 728, 648, 798, and 782 cm^{-1}, respectively, corresponding to their ring-breathing modes (Figure 6.15). The

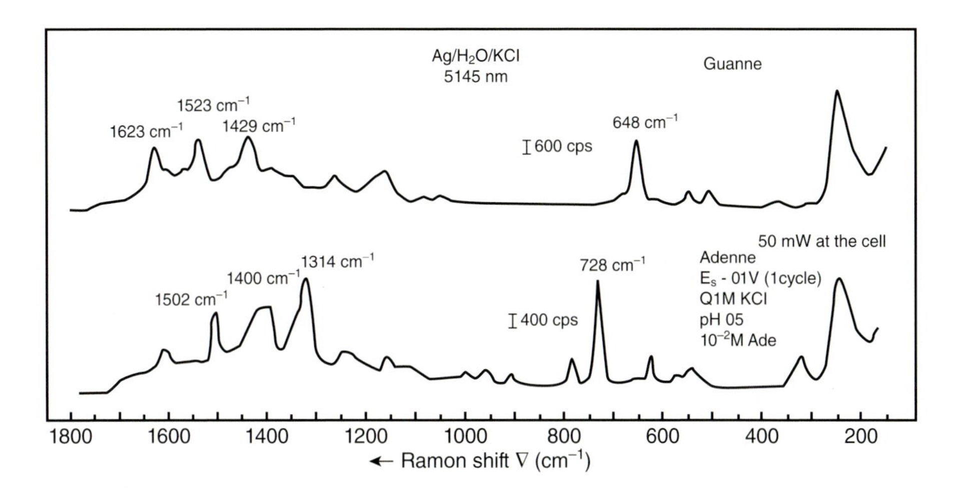

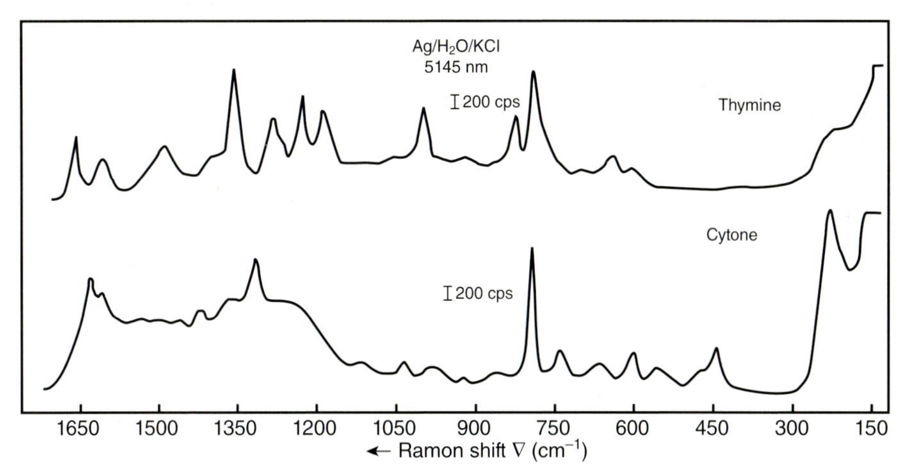

Figure 6.15 SERS spectra of protonated guanine, adenine, thymine, and cytosine (0.3 M HCl; 1×10^{-2} M guanine/adenine; pH = 0.5; electrode potential vs. sat. Ag/AgCl electrode, −0.1 V; electrode area, 0.2 cm^2; laser power at the cell, 100 mW). The Ag electrode was activated by one triangular voltage sweep between −0.1 and +0.2 V at a sweep rate of 50 mV s^{-1}. Reprinted from Koglin, E., Séquaris, J.M., and Valenta, P. (1982) *Journal of Molecular Structure*, **79**, 185. Copyright (1982), with permission from Elsevier.

spectral band positions are sensitive to pH and shift quite a lot in acidic media for the purine bases (adenine and guanine) compared to their position at pH = 7. Another study was carried out on DNA bases by Otto *et al.* in which they found typical differences between the normal Raman and SERS spectra [218]. They also saw changes in the SERS spectra with changes in potential but they did not study this in detail. The differences in shape and intensity of the carbonyl stretching vibrations are related to the different modes of adsorption with respect to the surface. The SERS spectra of DNA bases on surfaces are dependent on the adsorption process which, in turn, can be controlled by potential but is also clearly dependent on co-adsorption of ions [219] and the relative strength of the adsorbate–surface interaction as indicated by the heat of desorption as studied on Au surfaces [220].

SERS spectra are sensitive to proximity to the surface and the orientation of molecules, surface preparation, ions in solution, pH, and electrochemical potential, among other factors; these can lead to differences in the spectra reported for the same compound, giving rise to apparent contradictions in the literature. A case in point is the SERS spectra of adenine reported on Ag electrodes with both flat [216] as well as perpendicular [221] orientations being postulated. Proper assignment of each of the bands is extremely important in predicting orientations and this can be achieved by numerical simulations and comparison to experimental spectra [222]. Using electromagnetic theory-based surface selection rules, Giese and McNaughton conclude that on Ag colloids adenine adopts a more perpendicular orientation while on electrodes it has a more tilted orientation. That this is the case for Au surfaces has been reported by Kundu *et al.* for nanoshell surfaces where they have studied adenine, AMP, and polyadenine [223]. They have also carried out density functional theory (DFT) calculations which support the experimental conclusion that the adenine adsorbs "end-on" with the exocyclic NH_2 oriented almost normal to the surface.

Similarly the Raman and surface-enhanced Raman spectra of guanine and uracil have been recorded under a range of experimental conditions, including surface potential, and their spectra predicted by DFT calculations at the (B3LYP/6-31^{++}G(d,p)) level [224, 225]. Sardo *et al.* have recently studied a chemotherapeutic agent, 5-fluorouracil, on Ag nanostructures including electrodes [226]. They also carried out DFT calculations to deduce the positions of the main bands in the Raman spectra and studied the potential dependence on an Ag electrode to probe the charge transfer mechanism. Based on their calculations, and comparison with experimental spectra, they were able predict that the N_1 deprotonated anion is responsible for the spectral features. These studies highlight the importance of theoretical work in understanding the orientation of species at electrode surfaces and the possible adsorption behavior of neutral and charged molecules.

Not only can changes in orientation be studied and rationalized using SERS measurements but also the relative proportion of adsorbates can be determined. Watanabe and coworkers have studied potential-dependent changes in the spectra of adenosine, cytidine, and their mixtures among other derivatives of these nucleosides on Ag surfaces [221]. Cytidine displays weaker intensities than adenosine and they show very different potential dependence. Although it is clear that the relative proportions can be worked out by appropriate calibration, careful control of the conditions (such as the electrode potential) used when recording SERS spectra is necessary for quantification of the respective amounts.

Apart from adenine, much SERS work has been carried on thymine under electrochemical control. Cunha *et al.* studied the potential-dependent behavior of thymine at Ag electrodes [227]. Their detailed study demonstrates that determination of the potential of zero charge (PZC) is important for understanding the adsorption behavior of biomolecules at electrodes. Furthermore, changes in the spectra with potential can be rationalized on the basis of changes in adsorption behavior depending on whether one is positive or negative of the PZC. Drawing on the assignments from Aroca and Bujalski [228] the possible changes in orientation of thymine at potentials positive and negative of the PZC were deduced [227]. However, the most striking evidence seems to be the change of the low-frequency band at $216\,cm^{-1}$ (Ag–O) to $226\,cm^{-1}$ (Ag–N) on going from −0.3 to +0.6 V vs. Ag/AgCl/KCl(sat.), suggesting that the thymine adsorbs through O_8 in the first case and a ring nitrogen in the second (Figure 6.16a). Based on other potential-dependent changes, a model for adsorption of thymine on Ag was presented, as shown in Figure 6.16b. Another SERS study on thymine was carried out on a roughened Cu electrode by Shang *et al.* [229]; in this case potential-dependent signals were observed. The surface attachment (chelation) models were verified with DFT calculations leading to the conclusion that thymine molecules are likely to be adsorbed through N_1 and O_7 (through the O=C–N moiety). Besides changes in intensities and the relative heights of different peaks as a function of potential, SERS peaks also undergo shifts in position with electrode potential (Stark shift; see Section 6.3). The extent of this Stark shift is often related to the charge transfer

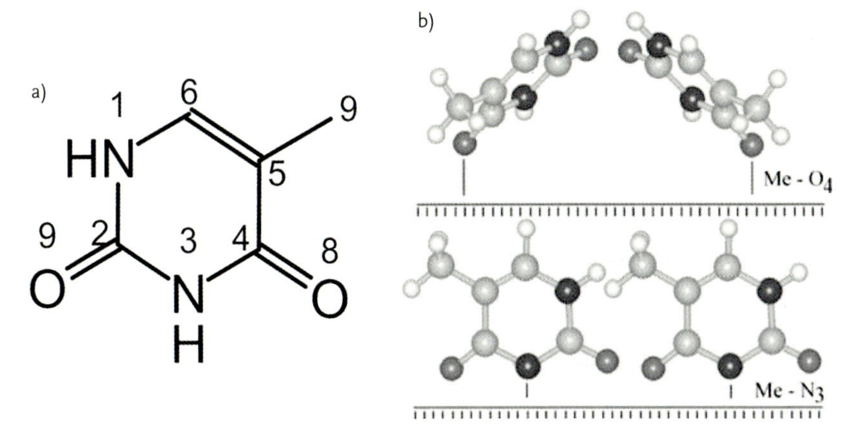

Figure 6.16 (a) Molecular structure of thymine. (b) Adsorption model of thymine on an Ag electrode. With kind permission from Springer Science+Business Media: Cunha, F., Garcia, J.R., Nart, F.C. *et al.* (2003) *Journal of Solid State Electrochemistry*, **7**, 576, figure 5.

mechanism and can be indicative of chemical interaction with the surface. In one such study using ultraviolet excitation, Hao and Fang found that the C=O bending and ring-breathing modes of thymine at 618 and 755 cm^{-1}, respectively, underwent a blue shift (to higher frequencies) of about 3 to 19 cm^{-1} on going from −0.1 to −1.3 V vs. Ag/AgCl on an Au electrode [230]. Recently Cui *et al.*, in their study of co-adsorption with perchlorate, found that thymine showed the weakest ability to co-adsorb on Au electrodes among DNA bases although the protonated forms of the bases were found to be more easily adsorbed at negative potentials [219]. These studies clearly demonstrate the importance of considering the chemical nature of the molecule under study and show that SERS is a sensitive probe of the nature of the interaction at the electrode interface.

6.9.2
DNA and Nucleic Acids

Most of the early work on SERS of nucleic acids themselves was carried out with colloidal nanoparticles [231–233]. However, DNA has also been extensively studied by SERS in tandem with electrochemistry. Early reports consisted of qualitative observation of the SERS spectra of nucleic acids and their possible conformational changes [215, 234, 235]. However, a proper study of DNA from calf thymus and the effect of adsorption on electrodes was reported by Brabec and Niki [236]; in this case they primarily observed adenine peaks in the spectra. Barhoumi *et al.* also found, in their SERS study of Au nanoshell substrates, that thermally uncoiled single-stranded DNA (ssDNA) sequences gave spectra dominated by the adenine and that thermal pretreatment resulted in better reproducibility as compared to the untreated oligonucleotides [237]. Furthermore, Brabec and Niki were able to prove that double stranded DNA (dsDNA) could be adsorbed on electrode surfaces without denaturation [236]. Interestingly they observed a change in the relative intensity of the adenine band (as shown in Figure 6.17) on denaturation of dsDNA on the surface suggesting that the helix–coil transition could be followed using SERS. It is pertinent to point out that Barhoumi *et al.* did not observe an increase in the adenine signal on thermal denaturation of dsDNA for the model sequences on their Au surfaces [237]. Nevertheless, spectra showing that the conformational rigidity of a DNA duplex is maintained on an Ag surface upon adsorption were provided by Koglin and Sequaris [238]. That the conformation of even a triplex oligonucleotide is preserved on adsorption on Ag electrode surfaces was shown by Fang *et al.* [239]; by carrying out potential-dependent SERS measurements they concluded that positive potentials were more conducive to the adsorption of the negatively charged triplex. At negative potentials (around and below the PZC of −0.8 V vs. Ag/AgCl) there was a decrease in the intensity of the peaks due to possible desorption.

Monolayers of nucleic acids are an important element in many putative biosensors. Some studies, mainly carried out over the last decade, have reported nucleic acid monolayers on surfaces. For example, monolayers of ssDNA and dsDNA were studied by recording both *ex situ* and *in situ* SERS spectra on Au electrodes [240].

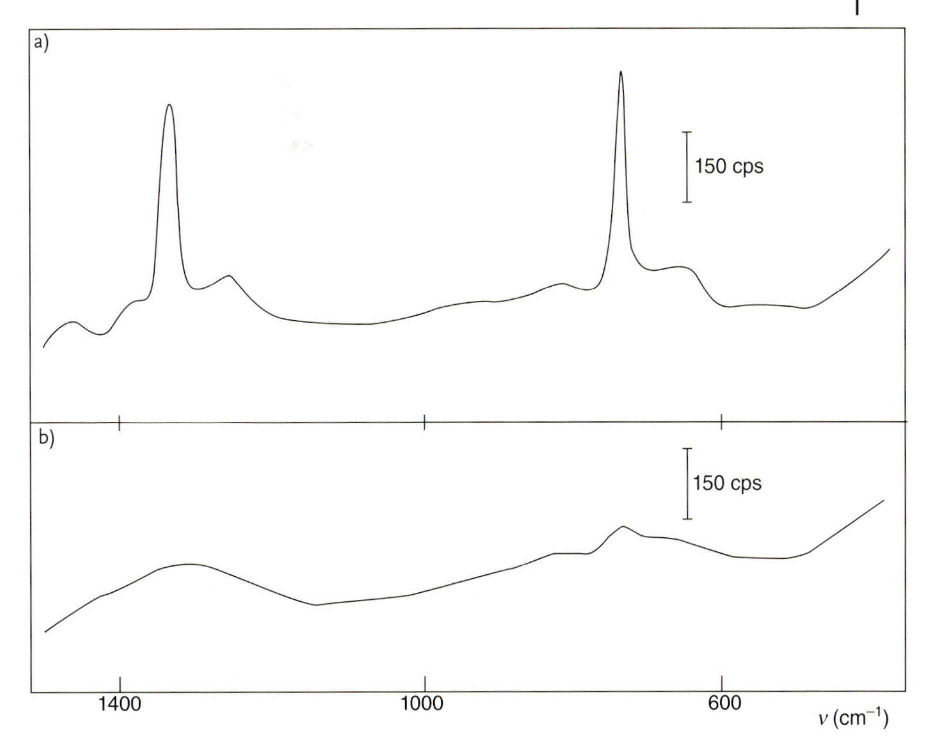

Figure 6.17 Surface Raman spectra of (a) thermally denatured DNA and (b) native DNA adsorbed at an Ag electrode polarized to −0.1 V in 0.1 M KCl with 1 mM phosphate buffer (pH = 7.0). DNA concentration was 0.2 mg/mL. Waiting time at −0.1 V was approx. 40 min. Reprinted from Brabec, V. and Niki, K. (1985) *Biophysical Chemistry*, **23**, 63, with permission from Elsevier.

In this work Dong *et al.* suggest that the duplexes undergo a vertical to horizontal transition in orientation on going from negative to positive potential. Furthermore, it was demonstrated by electrochemical scanning tunneling microscopy and electrochemical SERS on Au electrodes that dsDNA formed highly ordered and compact monolayers stable over a wide range of potentials [241]. The dsDNA was essentially arranged as rods parallel to each other and a change in orientation occurred only at positive potentials around +0.8 V vs. SCE. In contrast ssDNA adopted a coiled-like conformation.

Work on surfaces has proceeded more in the direction of developing sensing applications by using hybridization to surface-bound oligonucleotides which act as "probes" for solution "targets" with complementary sequences. Vo-Dinh and colleagues were the first to demonstrate this concept of a "SERS gene probe" using cresyl violet-labeled sequences [242]. Instead of signals from the nucleic acid, the stronger signature peaks of the label are detected. This strategy has been further developed for applications [79] including that for HIV [82] and cancer (BRCA-1 gene implicated in breast cancer) [81, 243] detection. This concept of hybridization sensors is very attractive for genomic analysis and incorporation onto DNA chips. Recent work has integrated on-chip DNA analysis by SERS with microfluidics wherein the platform consists of a structured surface on which probe oligonucleotides are immobilized [244]. In this regard an interesting study was reported by Huh *et al.* who used electrokinetically controlled microwells on a microfluidic device to increase the concentration of DNA for SERS detection of single nucleotide polymorphisms [245]. Although the SERS signals generated are nanoparticle based, the detection strategy is unique employing ligation at the mismatch site (called the ligation detection reaction) to bring the nanoparticle close to a dye label for detection.

Another way to develop DNA-based sensors is to utilize the reverse of hybridization, that is, the duplex denaturation process. The ability to controllably denature dsDNA on electrode surfaces and to monitoring this transition by SERS have been utilized by Mahajan *et al.* to distinguish between mutations [191, 246]. Although labeled oligonucleotides were used, the method relies on the concept of monitoring the helix–coil transition–denaturation by SERS as alluded to by Brabec and Niki [236]. The denaturation process can itself be induced thermally or electrochemically resulting in "melting" profiles which are dependent on the sequences and the number of mismatches, and thus allow distinction between healthy and mutated gene sequences [246]. The electrochemical denaturation option is particularly attractive for integration with gene chips for increased addressability, miniaturization, and rapid analysis. This has been applied for discriminating between the wild-type and triplet deletion (ΔF508 mutation) sequences in solutions containing the unpurified polymerase chain reaction products as shown in Figure 6.18. The electrochemically induced melting method, called "SERS-*E*melting," has been further extended to analyze short tandem repeats, which are commonly used to determine genetic profiles in forensic applications [190].

Recent work has focused on improving SERS labels for hybridization detection and the process of hybridization. Use of peptide nucleic acids as uncharged probes for DNA detection [247] offers some advantages in studying interactions with neutral species. New anthraquinone derivatives as electrochemical labels (redox indicators) for *in situ* SERS monitoring have been reported by Kowalczyk *et al.* [248]. An interesting approach to label-free detection for DNA hybridization has been reported by Halas and coworkers where they use SERS on nanoshell surfaces [249]. By replacing all the occurrences of adenine in the DNA probe sequence with 2-aminopurine, an artificial adenine substitution that preserves the hybridization characteristics, they were able to use the strong SERS band of adenine itself at 736 cm^{-1} as a marker for hybridization. Although it is clear that great progress has been made in nucleic acid detection and in developing sensing applications, gene detection, and in particular quantification, remains an issue. There is some indication that this issue could be addressed, for example, by internal referencing using what is referred to as the "gold-plasmon" band in Au nanoparticle films [250]; however, this is still an active area of research for the entire SERS community.

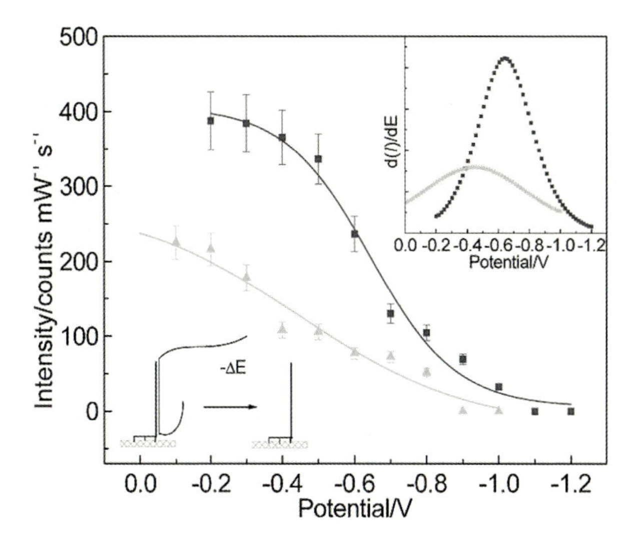

Figure 6.18 Resonant SERS-melting profiles of the polymerase chain reaction products for wild type (squares) and ΔF508 mutation (triangles) using the $1347\,cm^{-1}$ SERRS band for Cy5. A schematic of the dehybridization process of the polymerase chain reaction products is shown. The first derivatives of the melting profiles are shown in the inset. The mutation has a melting potential of −0.44 V, while that of the wild type is −0.64 V vs. SCE. Spectra were acquired with a single static scan of 2 s exposure at 1.5 mW laser power in 10 mM Tris buffer (pH = 7). Reprinted with permission from Mahajan, S., Richardson, J., Brown, T., and Bartlett, P.N. (2008) *Journal of the American Chemical Society*, **130**, 15589. Copyright 2008 American Chemical Society.

6.9.3
Amino Acids and Peptides

For historical reasons most of the work on amino acids is on electrochemically roughened electrodes. Nevertheless these studies provide important information on the adsorption and orientation of amino acids at interfaces. Phenylalanine (Phe), tyrosine (Tyr), and tryptophan (Trp) are by far the most studied amino acids using SERS. Phe, Tyr, and Trp was first studied by Nabiev *et al.* on Ag electrodes in 1981 [251]. In that work they showed detection at a sensitive level and clear differences between the spectra. While there are other studies of these amino acids that discuss the orientation and peak assignments on Ag colloids these quite often contradict each other and in some cases completely different spectra are reported for the same compound. In view of this probably the most comprehensive study of amino acids is that by Stewart and Fredericks in which 19 amino acids were studied on roughened Ag electrodes, their spectra analyzed, and molecular orientations proposed [252]. Owing to the large shift of the bands at 721 and $620\,cm^{-1}$, attributed, respectively, to the deformation and wagging modes of the carboxylate

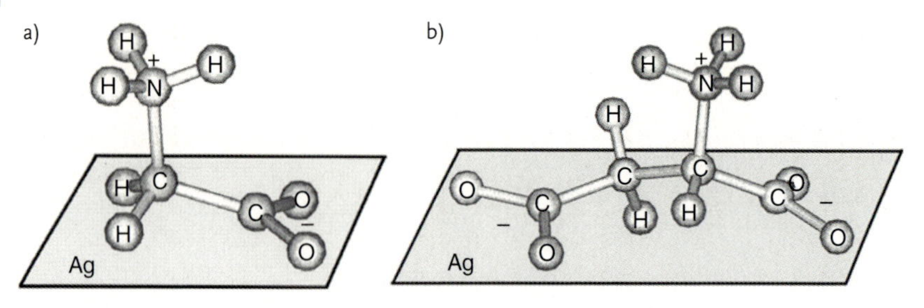

Figure 6.19 Proposed orientation of (a) glycine and (b) aspartic acid on a roughened Ag surface. Reprinted from Stewart, S. and Fredericks, P.M. (1999) *Spectrochimica Acta A*, **55**, 1641. Copyright 1999, with permission from Elsevier.

group, in the SERS spectrum as compared to the normal Raman spectrum the authors concluded that glycine adsorbs through the carboxylate group (Figure 6.19) which exists in its ionized form with the amino group pointing away from the surface. Suh and Moskovits have also studied the SERS of glycine, although with Ag colloids [41]. There is little similarity between their the spectra and those obtained by Stewarts and Fredericks [252]. The mode of adsorption is also different in the case of colloids, with the authors concluding that both the amino and carboxylic groups are attached to the surface since strong peaks for the C–N and carboxyl (–COO^-) vibrations at 1032 and 1384 cm^{-1} were observed. It is pertinent to point out that the same vibration is assigned to 1413 cm^{-1} in the study of Fredericks and Stewart [252]. This highlights the fact that SERS spectra depend strongly on the adsorption mode and hence on the nature of the surface, pH, and surface preparation, so that comparison between spectra and between experiments needs to be done carefully, not only when comparing results between nanostructured surfaces and colloids but also between different colloid preparations.

Nevertheless, we will try to briefly summarize the key features of SERS for amino acids as observed on Ag electrode surfaces primarily based on the study by Fredericks and Stewart [252] but referring to other authors where relevant. The strongest bands are obtained from amino acids with aromatic side chains such as Phe, Tyr, and Trp, each of which is adsorbed through the carboxyl group with the aromatic ring perpendicular to the surface. The carboxy bands shift in SERS on adsorption with respect to solution Raman, and the absence of C=O peaks above 1650 cm^{-1} clearly suggests that it is in its ionized form. The acidic amino acids, aspartic and glutamic acids, are structurally quite similar yet their spectra are distinguishable. Carboxylic acid peaks are more intense than for the others because they each have two carboxyl groups and both attach to the Ag surface (Figure 6.19). Similarly although glutamine and asparginine are quite similar, their spectra allow differentiation. Broadly the amide C=O bond is seen along with two types of NH_2 vibration which are shifted compared to the solution Raman spectra. The amide deprotonated form appears at 1100 and 1116 cm^{-1} and shifts more with respect to normal Raman, indicating chemical interaction of the amide bond, through the nitrogen lone pair, with the surface. The amino acids with aliphatic side chains, namely glycine, alanine, leucine, valine, isoleucine, and proline, can be distinguished based on their aliphatic C–C vibrations in the region around 850 to 950 cm^{-1}. As expected, with increasing length of the aliphatic side chain the amine terminus moves farther away from the surface and therefore the relative peak intensity for it decreases, except for proline. The alcohol-containing amino acids show no significant difference as compared to their aliphatic counterparts and show no evidence for interaction of the hydroxyl group with the Ag surface.

With sulfur-containing side chains, as in cysteine and methionine, the Ag–S interaction dominates. Cysteine adsorption on Ag has been studied separately and its potential dependence observed by Watanabe and Maeda [253]. While the adsorption of cysteine could be followed by observing the disappearance of the S–H band at 2576 cm^{-1}, the shift (650 to 670 cm^{-1}) and broadening of the C–S stretching peak indicated the formation of a strong Ag–S bond. Furthermore, although the adsorbed cysteine could not be oxidized to disulfide, cystine (the dimeric form of cysteine in which the two cysteines are linked by a disulfide bond) adsorption could be reversibly cycled between the disulfide and monothiolate forms by varying the potential between +0.3 and −0.3 V vs. Ag/AgCl (Figure 6.20).

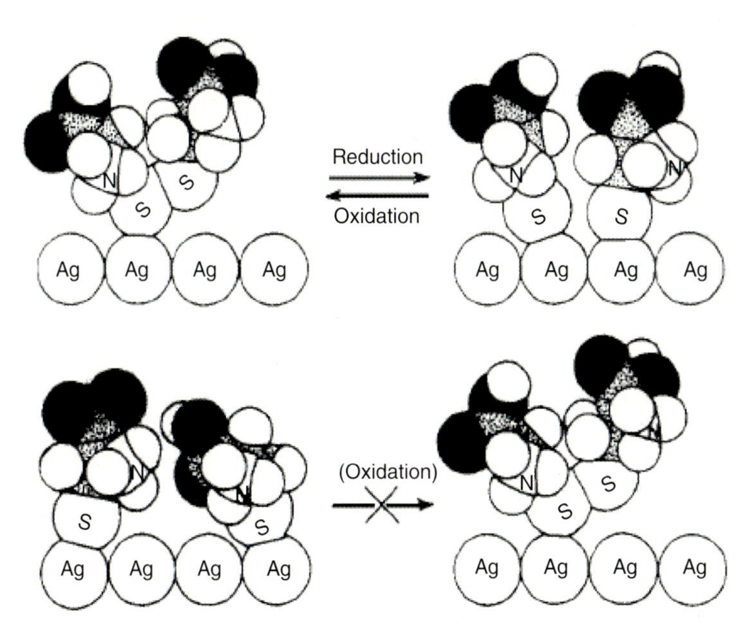

Figure 6.20 Surface adsorption layer model to account for experimental observations. The dotted, black, and white spheres represent carbon, oxygen, and hydrogen atoms, respectively. The atoms are scaled referring to the Ag–Ag distance (2.88 Å) for the Ag(111) plane together with known van der Waals radii and interatomic distances. Reprinted with permission from Watanabe, T. and Maeda, H. (1989) *Journal of Physical Chemistry*, **93**, 3258. Copyright 1989 American Chemical Society.

Adsorption and oxidation of glycine, serine, and threonine were studied on roughened Pt and Au electrode surfaces using SERS [254]. In alkaline solutions (0.1 M NaOH) dissociative adsorption was found to take place with a peak appearing at $2040\,cm^{-1}$ on Pt which is stable over a wide range of potentials. It was assigned to a CN vibration. On Au this peak appears only at positive potentials, resulting from oxidation of the amino acids to give cyanide which is then adsorbed on the surface. The dissociation of glycine is easier and starts in the hydrogen potential region while for serine and threonine dissociation occurs along with the oxidation of CHOH groups at positive potentials.

Among the basic amino acids, SERS of histidine has been studied on a roughened Cu electrode [255] in water and D_2O solutions. Depending on the pH of the solution histidine exists in five different ionic forms Eq. (6.7), which can be distinguished by the shift in Raman/SERS peak positions as studied by Martusevicius *et al.* [255]:

$$H_4His^{2+} \xleftrightarrow{pK_a=1.8} H_3His^{+} \xleftrightarrow{pK_a=6.0} H_2His^{0} \xleftrightarrow{pK_a=9.1} HHis^{-} \xleftrightarrow{pK_a=14.0} His^{2-} \quad (6.7)$$

Correspondingly, different adsorption states of histidine were observed depending on pH and electrode potential [255]. In acidic solutions at pH = 1.2 the imidazole ring of the adsorbed histidine remains protonated and is not involved in chemical coordination with the surface. The anion in the solution, SO_4^{2-} or Cl^-, also seems to play a role. At pH = 3.1 three different adsorption states of histidine are observed depending on the potential. It adsorbs with the protonated imidazole ring oriented perpendicular to the surface at potentials more positive than −0.2 V vs. NHE. At neutral pH histidine is adsorbed through the deprotonated nitrogen atom of the imidazole ring and the carboxyl group at potentials positive of −0.2 V vs. NHE. At more negative potentials the interaction is only through the amino group or the imidazole ring. In alkaline solutions at pH = 11.9 histidine is adsorbed on the Cu surface through the neutral imidazole ring. However, Stewart and Fredericks find that for histidine at neutral pH on an Ag surface the interaction is primarily through the carboxyl group [252]. The existence of chemically distinct states on Ag substrates has also been reported in the TERS study by Deckert-Gaudig and Deckert, although the zwitterionic state (His^0) was the dominant one [197]. It is postulated to be adsorbed with its neutral imidazole ring and with the carboxyl moiety bound to the Ag surface with an upright geometry predicted for the ring.

Similar results have been obtained for dipeptides and tripeptides studied on Ag surfaces, with the adsorption primarily through the carboxyl terminus and the SERS spectra dominated by the residue nearest the surface [256]. With an increase in chain length the strength of the amide bands increased while that of carboxyl peaks decreased. Although good-quality and distinguishable spectra could be obtained, it is evident that SERS is not effective for determining the secondary structure of peptides and proteins as the spectra were mostly dominated by the amino acid side chains. In cysteine-containing dipeptides the spectral signature of the aromatic amino acids dominated the spectra as studied on Au nanoshell

SERS substrates [257]. The dipeptide data were used as the basis set for predicting and understanding the spectral characteristics of a cell-penetrating peptide, penetratin, which, although much longer (19 amino acids), still overwhelmingly displayed only aromatic signatures.

6.9.4 Proteins and Enzymes

Studying proteins and enzymes with SE(R)RS is not simple especially if information regarding the native state is required. This is because their secondary and tertiary structure can be easily distorted near an electrode surface. Thus significant care, in terms of the correct functionalization of the surface, is required if the native structure is to be preserved. However, the variation in the structure of a protein at different potentials at the electrode, that is, with changes in the structure of the double layer, is itself a question of great interest. For the purposes of our review we will distinguish between proteins in terms of their electroactivity. Thus we first discuss redox-active proteins and enzymes, which have been widely studied in bioelectrochemistry, and after that move on to nonelectroactive proteins and enzymes.

6.9.4.1 Redox Proteins

Cytochromes, and in particular cytochrome c (Cyt c), are among the best studied redox proteins by electrochemical SE(R)RS. They have metallocenters which shuttle electrons by reversible oxidation–reduction and thus carry electrons in photosynthetic and respiratory processes. In aerobic organisms Cyt c transfers electrons to the membrane-bound enzyme complex cytochrome c oxidase (CcO or COX) resulting in reduction of oxygen to water. In mitochondria, Cyt c transfers electrons between two inner membrane-bound proteins, cytochrome c reductase (CcR) and CcO by interaction at surface binding sites on these two redox partners [258]. Cyt c immobilized on an electrode surface ("bare" or functionalized) provides a model system wherein changes with electrode potential can be studied by SERS in an environment free from spectroscopic interference from other entities in the CcO complex. Nonresonant SERS studies on Cyt c such as that by Niaura *et al.* [259] are relatively rare. The redox transition, monitored by SERS at 633 nm on roughened Ag electrodes by following the band at 742 cm^{-1}, was found to be −0.11 V vs. SHE largely in agreement with the formal potential of −0.17 V vs. SHE. Application of potentials more negative than −0.7 V vs. SHE led to irreversible changes in the spectra and the appearance of amide III vibrational modes (corresponding to the protein backbone) suggesting large conformational changes and possibly unfolding. They also observed bands due to the aromatic residues of the protein in their spectra.

Cyt c has a heme-based redox center (Figure 6.21a) which plays the key role in the electron transfer reaction. The heme moiety (based on the porphyrin structure) has electronic absorptions around 400 nm (called the "Soret" band) and a weaker band around 530 nm (called the "Q" band). Thus experiments can be carried out

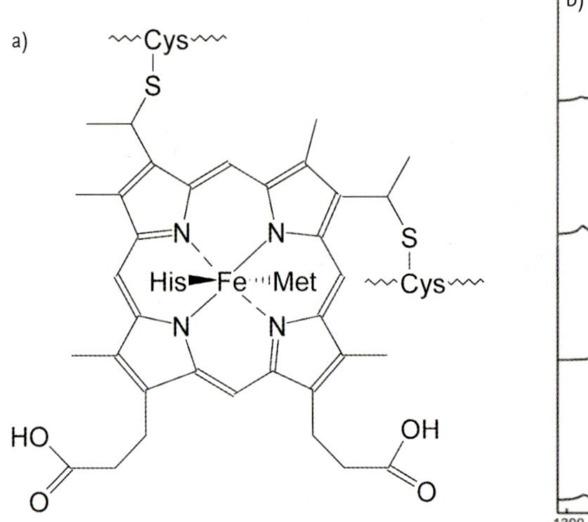

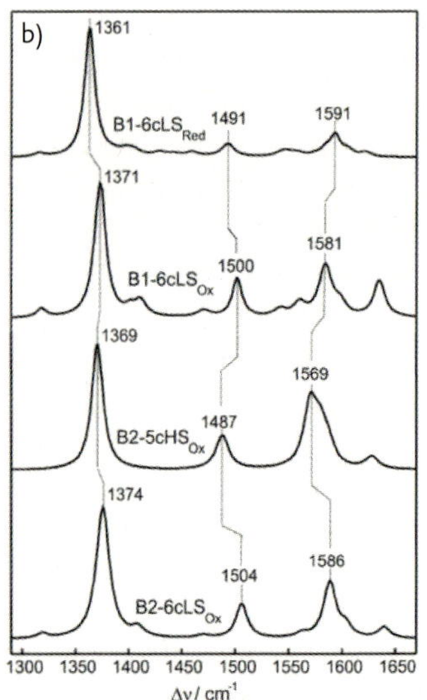

Figure 6.21 (a) Structure of heme center in cytochromes. (b) Resonant Raman component spectra of native (B1) and non-native (B2) forms Cyt c. From Murgida, D.H. and Hildebrandt, P. (2008) *Chemical Society Reviews*, **37**, 937. Reproduced by permission of the Royal Society of Chemistry.

under different excitation conditions resulting in selective, resonant enhancement of different signature vibration (symmetric or nontotally symmetric) modes of the heme moiety [260]. The peak positions also give information on the spin and oxidation states of the heme unit [261]. The marker band region from 1200 to 1700 cm^{-1} has typical peak positions which can be used to characterize the metallocenter–protein complex. A complete assignment of the resonance Raman peaks in Cyt c under various conditions has been published [262]. In particular ν_4, corresponding to C–N stretching vibration, is extremely sensitive to the redox state because the Fe center interacts strongly with the porphyrin core [24]. Furthermore, Cyt c has been shown to exist in its native hexa-coordinated low-spin or 6cLS (termed B1) state, where the central Fe–porphyrin is stabilized by His 18 and Met 80 axial ligands, as well as in a non-native state (termed B2), which lacks the Met 80 ligand, by *in situ* SERRS study on Ag electrodes [263–265]. Owing to bands in the marker band region which are sensitive to the spin state of the heme (see Figure 6.21b) complete characterization the state of Cyt c as well as its evolution as a function of redox potential is possible by SERRS.

SERRS studies on electrodes can not only probe at a very sensitive level but can also selectively probe the heme group giving electronic and conformation information, without interference from rest of the protein. In one of the first of such studies, Cotton *et al.* used roughened Ag electrodes to record the potential-dependent SERRS of Cyt c and myoglobin [266]. They were able to observe the shift in the marker band peaks at two different potentials, –0.2 V and –0.6 V vs. SCE, corresponding to a ferric low-spin and ferrous low-spin state, respectively. It was subsequently shown by Hilderbrandt and Stockburger that direct adsorption onto Ag induces some change in the spin state of the Cyt c as monitored by SERRS, indicating structural modification as compared to the native state [267]. With SERRS monitoring of the heme moiety of Cyt c at the electrode surface it was also shown that at low temperature the low-spin state was dominant while at room temperature a mixture of high- and low-spin states occurred [267]. Although the spin transitions remained reversible with temperature, indicating no denaturation at the electrode surface, electronic interaction did exist between the surface and the heme, possibly through the charged ligands, leading to a different crystal field splitting energy compared to the native state. By comparing the relative intensities the authors also concluded that the orientation of the heme–porphyrin ought to be perpendicular to the surface rather than parallel as found for free porphyrin molecules. The non-native conformational states of Cyt c have also been spectroscopically characterized by SERRS [265]. The non-native state B2 can itself consists of different sub-states depending on whether the vacant site is occupied by a water molecule or by a histidine ligand (His-33 or His-26) contributed by another peptide segment [265, 268]. Thus studying Cyt c by surface-enhanced Raman techniques on electrode surfaces is very attractive as it can give information regarding conformation and redox potential [269, 270]. It has been proposed that not only are redox transitions related to conformational changes in the protein but also that they may be even controlled by them [271, 272]. In this respect it can be said that the redox-controlled (and reversible) formation of the B2 state at the electrode might provide insight into deciphering the mechanism of electron transfer mediated by changes in conformation. For example, based on the observed spin state a model for the biochemical mechanism of interprotein electron transfer by Cyt c between the CcR and CcO was initially proposed [267] and subsequently refined [24]. A schematic of the electron transfer reaction between Cyt c and CcO is shown in Figure 6.22.

Electrochemical studies in combination with SE(R)RS can afford dynamic information regarding conformational change and coupled electron transfer processes. In this context the technique of time-resolved SERRS as developed by Hildebrandt and coworkers to obtain insight into the electron transfer kinetics of heme proteins [273–275] should be mentioned. The technique consists of a rapid potential step initiating the process coupled with synchronous monitoring of the SE(R)RS spectra after a specified delay time (δ, the probe interval). To improve signal-to-noise ratio the process has to be repeated several times using the same potential step; the transient behavior is built up by repeating the process for different probe times. Analysis of such kinetic data affords the heterogeneous electron transfer constant (k_{ET}). However, the need to repeat the potential steps limits the application of the technique to systems that are stable and chemically completely reversible. Subsequent improvements in the technique, including a better cell design with a rotating disc electrode to reduce photodegradation [276], and in the measurement technique by using two-color time-resolved SERS, have been

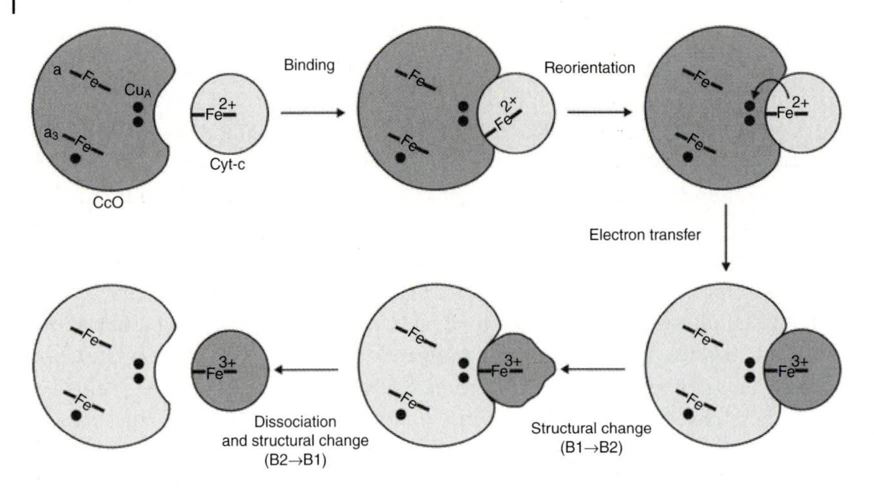

Figure 6.22 Schematic of the electron transfer reaction between Cyt c and CcO. From Murgida, D.H. and Hildebrandt, P. (2008) *Chemical Society Reviews*, **37**, 937. Reproduced by permission of the Royal Society of Chemistry.

implemented so that real-time kinetic information is generated [277] providing new insights into protein electron transfer dynamics. This technique of time-resolved SERRS has been successfully applied to study Cyt c as well as to a variety other heme proteins immobilized at electrodes [278–282].

For proteins, it is clear that the surface functionalization has a critical affect on the SERRS intensities observed due to the difference in orientation that it can induce and due to the affect on the electrochemistry. Self-assembled monolayers (SAMs) of thiols can provide a twofold advantage by (i) preventing denaturation and (ii) defining the orientation by either electrostatic or covalent binding. This particular aspect has been utilized elegantly by Dick *et al.* [169] to use SERRS to study heterogeneous electron transfer. By varying the chain length of carboxyl-terminated alkanethiols on AgFON substrates (see Section 6.7) they showed that the electrochemistry, as followed by SERRS, was most reversible for mercapto-hexanoic acid and became completely irreversible for mercaptoundecanoic acid. In the latter case the SERRS signals were 10 times weaker. A similar SERRS study by Murgida and coworkers has shown that, by controlling the chain length of carboxyl-terminated thiols on the surface, the formation of a non-native state of Cyt c can be controlled. Apparently the non-native redox state can be varied from 76% to 0% on going from C_2 to C_{11} SAMs [264] clearly suggesting the role of the electric field in determining the conformation of the redox center of the protein. Furthermore, close to the surface (for C_2 and C_3 SAMs) the interfacial redox process is found to be coupled to proton transfer as studied by monitoring the kinetic isotope effect with SERRS [263]. Obviously this proton coupling is also distance dependent and decreases with increasing distance from the surface. The study clearly demonstrates that the redox processes of proteins at electrode surfaces, and presumably other biomolecules, may not be entirely controlled by the electron transfer step, warranting careful interpretation of the data. Indeed, improved electrochemical performance, as demonstrated by cyclic voltammetry and SERS [283], can be achieved by the correct functionalization of the electrode such as by using 1,4-dithiane. Nevertheless the utility of these thiol-modified surfaces in understanding the functional aspects of protein complexes was established in the study by Dick *et al.* where their SERS-based titration (binding) curves gave association constants similar to that for the cytochrome c peroxidase–Cyt c complex [169]. Again stressing the role of functionalization in studying electron transfer rates, Grosserueschkamp *et al.* [276] used their time-resolved SERRS technique to study the electron transfer rates for Cyt c at mercaptoethanol-modified Ag electrodes. They obtained oxidative and reductive electron transfer rates of 46 and 84 s^{-1}, respectively, at least an order of magnitude lower than that found on Au electrodes. Recently Sezer *et al.* showed that the SERRS signals for Cyt c can be improved by using covalent attachment to Ag–(C_{11}–NH_2)–Au. In this case they obtained signals 15 times larger than those for Ag alone when using their multilayer structures [284]; however, the electron transfer deviated from Nernstian behavior upon covalent attachment while for electrostatically bound Cyt c fast long-distance electron transfer rates (49 s^{-1}) were observed.

One of the key advantages of SE(R)RS is the orientation information that it can provide as a function of the electrode potential. This can be crucial in understanding the functioning of redox proteins. As discussed above, Murgida and coworkers, among others, have studied the interaction of Cyt c on bare or SAM-coated electrodes. Studies of the conformational changes on interaction with other biomolecules coated on electrodes are an interesting prospect. A SERRS study by Jiang *et al.*, in conjunction with circular dichroism measurements, indicated that interaction with DNA does not affect the secondary structure of Cyt c, although the heme microenvironment was disturbed at the electrode–electrolyte interface [285, 286]. Cyt c adsorbed at a DNA-modified metal electrode showed voltage-dependent orientation resulting in non-Nernstian behavior at high electrode potentials.

Among other classes of redox proteins [258] those possessing Fe_2–S_2 clusters and blue Cu metallocenters have been explored by SERS on electrodes. Putidaredoxin (Pdx) is a 12 kD protein with an Fe_2–S_2 redox center that performs two stepwise one-electron transfers from Pdx reductase to the terminal cytochrome mono-oxygenase in *Pseudomonas putida* PpG786. Reipa *et al.* studied the reduction–oxidation of Pdx on bare roughened Ag electrodes and obtained spectroscopic evidence for quasi-reversible charge transfer between −0.6 and −0.3 V vs. Ag/AgCl [287]. Azurin, another nonheme protein [258], has also been studied on an Au electrode by a combination of electrochemistry and SERRS. In that work it was demonstrated that the Cu site of the immobilized protein remained intact upon adsorption and underwent reversible reduction–oxidation [288].

6.9.4.2 Other Proteins

We now turn our attention to nonredox proteins. SERS spectra of IgG adsorbed on roughened Ag electrodes were obtained by Grabbe and Buck [289]. The spectra contained bands attributed to the amino acid residues that were in close contact with the surface but interestingly did not include any amide backbone vibrations or solution bands. Usually these amide bonds, the bonds making up the protein backbone, produce the strongest bands in the Raman spectra of proteins, often indicating denaturation. Nevertheless, the vibrational frequencies depend on the secondary structure and change in position and intensity depending on whether they are part of a β-pleated sheet or α-helix. Again only amino acid side residues were observed in SERS indicative of the mode of adsorption on the Ag electrode. Adsorption was greatest at the PZC and at lower ionic strengths. The authors used the variation in band width and the relative intensity of the peaks to describe the molecular rearrangements.

Spectra of lysozyme on an Ag electrode have been recorded. The authors did not see peaks corresponding to the amide I, amide III, and S–S bands [238, 290] although Hu *et al.* were able to observe these on colloids [291]. Thus although chemisorption seems to be the key to what may be observed from the spectra, the orientation of the adsorbed protein and its side chains can be inferred from analysis of the spectra. Similarly the SERS spectrum of bovine insulin has been observed with high enhancement factors by Chumanov *et al.* [290]. A further, detailed, electrochemical SERS study for bovine insulin was carried out by Reipa *et al.* on an Ag electrode over the potential range of −0.2 to −1.2 V vs. Ag/AgCl [292]. Conformational changes with potential can be seen and the rupture of the disulfide bonds takes place sequentially with increasing negative potentials leading to a decrease in the α-helix conformation. Recently Drachev *et al.* have reported a protein sensor based on nanostructured adaptive Ag films using SERS detection [293]. The sensor can differentiate at the sub-monolayer level between two insulin isomers, human insulin and its analog insulin lispro, that differ by only two neighboring amino acids.

6.9.4.3 **Enzymes**

SE(R)RS spectra of enzymes are essentially similar to those of proteins except in those cases where they possess a distinct site that is chromophoric. In the latter case heme-based enzymes have again been predominantly studied by SERRS. Enzymes can have multiple redox centers (or multiple chromophores) or are often intimately associated with a cofactor; careful interpretation of the spectra is required to take account of these effects. If the cofactor is chromophoric then it can turn out to be an advantage as SERRS monitoring can be used. For example, Holt and Cotton used the SERRS from associated flavin adenine dinucleotide (FAD) to study the catalytic activity of glucose oxidase (GOx) studied by SERRS and combined electrochemistry [294]. Although successful as a strategy, they observed unusual potential shifts which could be attributed to the formation of a flavin–Ag^+ complex. The results caution against the use of Ag surfaces in solution SERS, as H_2O_2, an enzymatic reaction product of GOx, oxidized the Ag ultimately leading to flavin–Ag^+ complex formation. In another study the same authors found that free flavin was an interferent when recording spectra from commercially available GOx [295]. Potential-dependent SERS spectra of alkaline phosphatase, horseradish peroxidase, and lactoperoxidase were obtained by Razumas and coworkers on Ag electrodes [296, 297]. They also found that alkaline phosphatase and lactoperoxidase form a complex with Ag ions. Nevertheless, they were able to interpret the conformation and orientation changes as a result of changes in potential on the basis of the changes in the amino acid signatures. In another study Grabbe and Buck observed the SERS spectrum of an anti-human immunoglobulin G alkaline phosphatase conjugate (anti-IgG) adsorbed on an Ag electrode as a function of potential and the ionic strength of the buffer [298]. SERS spectra were recorded at −0.5 and −0.8 V vs. SCE. Upon reduction at −0.5 V most of the peaks lost intensity. In particular, the peaks corresponding to tryptophan and cysteine residues including those corresponding to C–S ($675\,cm^{-1}$), S–S ($511\,cm^{-1}$), and Ag–S ($389\,cm^{-1}$) were affected.

Microperoxidase is the heme-containing peptide portion of Cyt c that retains peroxidase activity. Several microperoxidases are available with different numbers of amino acid residues. The conformation of microperoxidase-11 (the microperoxidase with 11 amino acids in the peptide) adsorbed on roughened Ag electrodes was studied using Fourier transform SERS and shown to be adsorbed via the α-helical polypeptide chain [299]. As expected the characteristic amide I and III bands for the protein backbone were the strongest. Similarly microperoxidase-8 was studied by combined SERRS and electrochemistry where it was shown that the heme existed in the penta-coordinated state and could bind cyanide as the sixth ligand [300].

Iron tetraphenylporphyrin was immobilized on an electrode coated with poly(γ-ethyl-L-glutamate) functionalized with imidazole pendant arms [301]. Striking similarities with the model Cyt c3 system, which is a tetraheme protein, were observed. To mimic and understand biological function proper immobilization is necessary without disrupting the active site. CcO suspended in a lipid membrane was attached through a histidine-tag linker to an electrode and studied by SERRS [302]. It was demonstrated that the structures of the active sites remained intact in the enzyme and that electron transfer from the electrode could take place. Recently, biomimetic immobilization of CcO on a metal electrode has been carried out and studied by a combination of fast-scan cyclic voltammetry and SERRS [303]. This allowed the kinetic analysis of all four redox centers in the enzyme providing insights in to how the proton translocation might be coupled to electron transfer in such complex systems. In another study where CcO, embedded in a phospholipid bilayer, was tethered to an Ag electrode via a histidine tag it was found that the intramolecular electron transfer and proton translocation were perturbed due to the effect of the interfacial electric field although the heme active site structures were preserved [304].

The importance of the functional layer on an electrode to the interaction of an enzyme with the surface is brought out by Kudelski's studies with Cu-containing tyrosinase (a phenol oxidase) [305] and laccase [306]. Using ω-functionalized thiols he showed that electron transfer was not prevented between the electrode and the enzyme. That the local environment of the enzyme is important has also been demonstrated in a recent study of human sulfite oxidase using SERRS and cyclic voltammetry [307]. The native heme structure of the cytochrome b5 domain and the functionality of the enzyme were preserved on the electrode surface. It was further found that the heterogeneous electron transfer rate was directly affected by the ionic strength of the buffer.

6.9.5 Membranes, Lipids, and Fatty Acids

Studies of membranes, lipids (including phospholipids), and other such amphiphilic biomolecules on electrodes are rather limited due to the redox inactivity of these materials. Nevertheless, a few reports exist of studies of bilayer membranes and lipid amphiphiles on SERS surfaces and are mentioned here as exemplar studies due to the unique insight they offer.

SERS of adsorbed dipalmitoylphosphatidylcholine (DPPC) lipids was studied on an Ag electrode [308]. The potential-dependent behavior showed that the ratio of trans and gauche structures of the acyl chain, corresponding to bands at 1105 and 1142 cm^{-1}, respectively, changed around −1.05 V vs. SCE. However, in a later paper Guo *et al.* suggest that the adsorption and the potential-dependent order–disorder phase transition involves the phosphocholine head group [309]. Hybrid bilayer membranes wherein the layer in contact with the surface is formed by a thiolipid or a long-chain alkanethiol are robust and useful biomimetic model interfaces [310, 311]. Such tethered bilayer membranes on Au display good blocking properties, are well oriented, and have "fluid-like" properties [311]. Leverette and Dluhy characterized hybrid bilayer films composed of a long-chain alkanethiol, 1-dodecanethiol, and the phospholipid DPPC on Ag island films [312]. The DPPC layer was deposited by the Langmuir–Blodgett technique over the dodecanethiol-functionalized SERS surface. While spectral features from the dodecanethiol were strong and readily identified, the changes observed in the spectrum after addition of DPPC were quite subtle. The ratio of the antisymmetric to symmetric methylene stretches (also called the order parameter) reflects both intermolecular lateral chain interactions as well as intramolecular (i.e., conformational) chain order, with a decrease indicative of intramolecular disordering of the monolayer [313]. By employing deuterated DPPC the C–H stretches were shifted to around 2100 cm^{-1} from 2900 cm^{-1} allowing insight into the order of the transferred Langmuir–Blodgett layer along with the immobilizing layer of dodecanethiol [312]. The order parameter, based on C–D peaks and the position of the symmetric stretch of deuterated methylene, indicated the as-transferred gel-like state of the film.

As noted above lipid bilayer films supported on electrodes represent an interesting bioelectrochemical interface where the potentials are of the same order as those of physiological systems, and can be readily modulated. Nevertheless, this area remains less studied except in the case of enzyme complexes such CcO where the hydrophobic environment of the lipids is necessary to maintain the integrity of the enzyme when immobilizing it on electrodes [304, 314]. Studies of the interaction of drugs with lipid membranes are important from many points of view, particularly considering the fact that nearly 50% of drug molecules have membrane proteins as their targets. Among the first such reports, the interaction of various anthracyclines was studied with asymmetrical planar supported bilayers on Ag surfaces [315]. Different anthracyclines showed different interaction behavior depending on the hydrophobic–hydrophilic balance in the lipid bilayers. Halas and coworkers have recently examined the intercalation of the drug ibuprofen with lipid bilayers on nanoshells [316]. Using SERS measurements at two different pH values and various concentrations the authors are able to conclude that the drug molecule is intercalated into the lipid layer using primarily the hydrophobic effect and that the membrane is disrupted at higher concentrations. Studies such as this would enormously benefit if carried out on continuous conducting SERS surfaces and lead to more insight besides being able to corroborate conclusions such as membrane disruption. Recently Millone *et al.* used dimyristoylphosphatidylcholine-supported bilayers on dithiothreitol-treated roughened Au electrodes for an *in situ* electrochemical SERS study of the incorporation of methylene blue and FAD into the lipid layer [317]. Spectroscopically the lipids do not show any significant signal in the 1200–1700 cm^{-1} region while the characteristic bands for methylene blue and FAD are clearly visible. Based on their results the authors conclude that methylene blue is able to penetrate the bilayer easily reaching the Au interface due to its lipophilic positively charged nature, while FAD, due to its negative charge, remains in the outer part of the bilayer and therefore cannot be electrochemically detected.

6.9.6 Metabolites and Other Small Molecules

6.9.6.1 Neurotransmitters

Anodic voltammetry has conventionally been applied for detection of neurotransmitters and is most commonly the final step at the end of a chromatographic elution even for *ex vivo* analysis. In general these electrophysiological techniques are limited to easily oxidizable molecules and are not free of interferences such as ascorbate. Hence, SERS presents a unique tool for their interference-free detection at a very sensitive level. Furthermore, although colloidal nanoparticles might provide a better alternative for *in cellulo* work [318], SERS at electrodes provides unique insight into adsorption behavior and can be ideal for *in vitro* detection methodologies serving as a means to enhance signals and increase discrimination between different neurotransmitters. Among the earliest work in this area is that by Morris and coworkers who reported the detection of various catecholamines on Ag electrodes [319]. Distinguishable spectra with good signal-to-noise ratio were obtained for dopamine, norepinephrine, 3-methoxytyramine, and epinine in phosphate buffer of pH = 7.2. The intensities were maximized when the electrodes were polarized around −0.9 V vs. SCE or below. The PZC of Ag under these conditions is −0.9 V, which clearly indicates that these neurotransmitters do not adsorb well on positively charged surfaces. Furthermore under these conditions no interference from ascorbate, glutathione, or acetylcholine was observed. The lowest detection limit for dopamine was found to be 0.3 μM, which is much lower than the physiological limit of 1 to 100 μM, with seemingly modest detectors and moderate laser powers. A polymer-coated electrode probe has been developed to show the application of neurochemical measurements with SERS in protein-rich matrices without fouling [320].

Another important neurotransmitter is histamine. It is also involved in several biochemical functions such as regulating physiological function in the gut. Its detection by conventional electrochemical means is difficult as it is not readily reducible and its oxidation coincides with that of water. SERS of histamine adsorbed on Ag electrodes was observed by Morris and coworkers who found potential-dependent intensities [321]. Due to the imidazole moiety some of the peaks are pH dependent due to the ionizable nitrogens. Histamine is adsorbed in its neutral form at all pH values at negative potentials. At potentials more positive than −0.5 V vs. Ag/AgCl, the cation is adsorbed in acidic media. The maximum intensities were obtained at −0.8 V vs. Ag/AgCl. In addition to orientation changes brought about by the change in electrochemical potential the adsorption is also clearly affected. Nevertheless, the analytical utility of SERS detection was aptly demonstrated by showing the ability to differentiate between histamine and its primary metabolite, methyl histamine.

Much more work has been undertaken with colloidal nanoparticles over the past decades for developing various protocols for detecting neurochemicals and their metabolites. Recently it has even been shown that this SERS strategy can be used to image dopamine and norepinephrine release in cells [318] demonstrating the potential of the technique. However, issues with fast detection for live cell work and selectivity still remain.

6.9.6.2 Nicotinamide Adenine Dinucleotide

Nicotinamide adenine dinucleotide (NAD^+) is an important co-enzyme which plays a significant role in many biological processes [258]. The molecular structure of NAD^+ is shown in Figure 6.23. Studies on NAD^+ with SERS-active Ag and Au electrodes have been carried out primarily to understand its absorption behavior and orientation. Involvement of adenine in adsorption was clear from initial experiments on electrodes and similarly the vertical orientation of the adenine moiety could be inferred [322]. Some differences were found in the SERS spectra of NAD^+ between Ag and Au electrodes. While adsorption on Ag was primarily through the exocyclic amino group and N_7 of the adenine, on Au the orientation changed from N_7 to N_1 at −0.25 V vs. SCE. It was further found by Xiao and Marwell that Fourier transfer SERS in the near-infrared region gave better signals on an Au electrode and significant modulation in intensity of various peaks with potential was observed [323]. However, the structural orientation and assignment of peaks for the nicotinamide moiety could only be determined after careful experimentation. The potential dependence of various peaks has also helped in assignment of the peaks and disentanglement of the nicotinamide bands from those of adenine. Only after careful experiments were the same authors able to correctly assign the ring vibration of nicotinamide at 1575 cm^{-1} which appeared at −0.4 V vs. SCE [324]. Furthermore, some variations in peak positions and observed

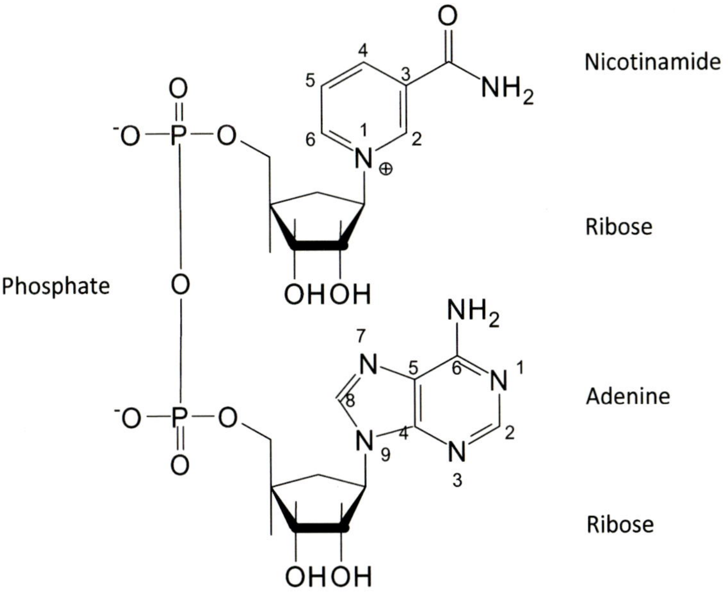

Figure 6.23 The structure of NAD^+.

spectra are apparent between different reports. For example, a 1340 cm^{-1} band assigned to the N_7–C_5 bond in the adenine moiety in NAD^+ is a strong feature in the spectra reported by Xiao *et al.* [324] while it is not observed in the study by Yang *et al.* [325]. This not only highlights the fact that reproducible SERS surfaces are essential for bioanalysis but also that careful consideration needs to be given to electrochemical potential, pH, and surface characteristics (adsorbed ions). Additionally, it is found that the spectra from NAD^+ changes with adsorption time as a full monolayer is formed [325]. The band at 735 cm^{-1}, assigned to in-plane NH_2 bending and a ring-breathing mode of adenine, appears only after 8 h, also suggesting that the adenine moiety might be changing its orientation from flat to vertical. Reorientation takes place from a flat adsorption mode for both the adenine and the nicotinamide moieties such that the final organized monolayer has the adenine vertically oriented while the ring and the carboxamide lie flat on the chemically roughened Ag surface.

6.9.6.3 **Flavin Adenine Dinucleotide**

Flavins are an important cofactor involved in many biological electron transfer processes [258]. They commonly occur as FAD, flavin mononucleotide, and riboflavin. For example, FAD is the cofactor in GOx. Cotton and Holt could even detect it at a very sensitive level of 10^{-10} M on a roughened Ag electrode by SERRS although in their study on GOx it was an interference [295]. Flavins can undergo oxidation–reduction in either a single two-electron transfer step or two one-electron transfer steps. The preference towards either of these mechanisms depends on the interaction with metal ions, pH, and the polarity of the matrix; SERS measurement made under electrochemical control can help to determine whether an intermediate state is involved. An unstable semiquinone state was detected on reduction of flavin in acidic media signifying that two one-electron transfers are involved in the overall two-electron reduction via this radical intermediate [326]. This was confirmed in the study by Abdelsalam *et al.* where they covalently attached the flavin moiety (isoalloxazine ring) to an Ag SSV surface [183]. The *in situ* electrochemical SERS spectra recorded at different potentials are shown in Figure 6.24. Additionally by quantitative analysis of their SERS data, based on a modified Nernst equation, they were able to estimate a redox potential, E', of −0.354 V vs. SCE and an interaction parameter, G, of −3.4, where the negative value of G indicates repulsive interactions between the immobilized molecules.

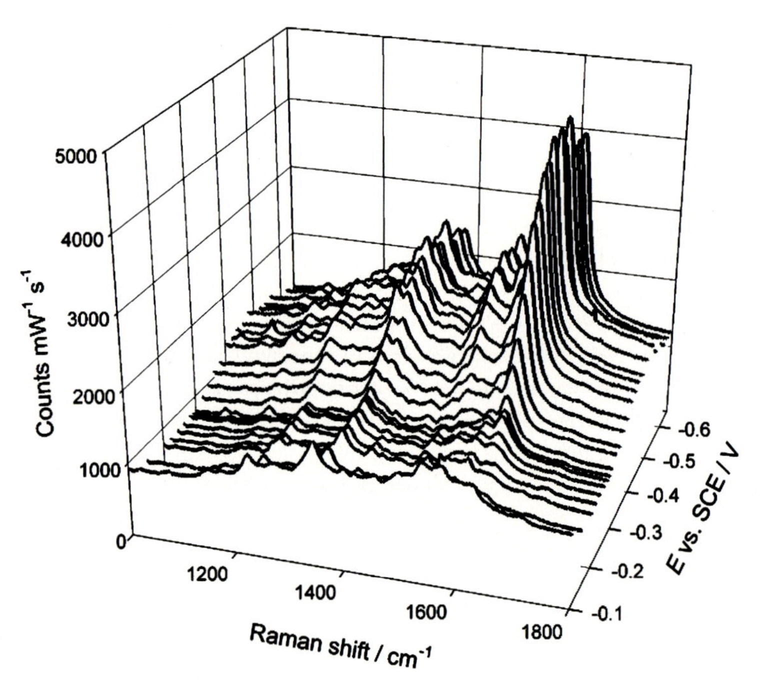

Figure 6.24 *In situ* electrochemical SERS spectra for flavin immobilized on an SSV Ag film produced using template spheres of 900 nm diameter and 540 nm thickness with a He–Ne laser (633 nm, single 10 s accumulation, 3 mW laser power, recorded in Tris buffer, pH = 7). The sample was purged with nitrogen for 20 min at the beginning of the experiment and then for 1 min between each measurement. Reprinted with permission from Abdelsalam, M., Bartlett, P.N., Russell, A.E. *et al.* (2008) *Langmuir*, **24**, 7018. Copyright 2008 American Chemical Society.

6.9.6.4 Bilirubin

Bilirubin is a lipid-soluble metabolite of hemoglobin that is transported through the bloodstream as an albumin complex to the liver where it is esterified for excretion in urine. It plays an important role in the pathology of many diseases. Its Raman spectrum has been difficult to observe because of its sensitivity to photoisomerization and decomposition. However, under conditions of SERS on Ag electrodes this is suppressed allowing potential-dependent spectra to be observed in its free form as well as complexed with cyclodextrins and albumin [327]. The band intensities are higher at more positive potentials implicating adsorption of anionic species on the electrode.

6.9.6.5 Glucose

Glucose is an important metabolite. However, its small Raman cross section makes it a difficult species to study by SERS. Approaches with colloidal nanoparticles, utilizing chemistry on the surface and solution to generate species with higher Raman cross sections such as using horseradish peroxidase and GOx co-immobilized on nanoparticles with *o*-phenylenediamine in solution to generate azoaniline on reaction with glucose, have been reported [328]. Detection of azoaniline, chosen because of its relatively high SERS cross section, enabled a dynamic range of 0.5 to 32 mM which is larger than the physiological range of 3.5 to 6.1 mM. Further work on glucose sensing has been carried out by Yonzon *et al.* using AgFON surfaces [166] using pre-concentration of the glucose molecules in alkanethiol monolayers to increase the sensitivity. Several improvements in sensor performance, such as increased temporal stability, could be achieved by switching to AuFON surfaces [165]. Recently Lipkowski and coworkers studied electrochemical SERS of β-thioglucose on Ag SSV surfaces (see Section 6.7) and compared them to electrochemically roughened electrodes [189]; the results for the SSV surfaces were found to be much more reproducible. *Ab initio* calculations were used to predict and carry out the assignments of the observed spectra. The results of this study showed that, under the experimental conditions, the β-thioglucose film anomerized into a mixture consisting of both α- and β-thioglucose anomers.

6.10 Conclusion

In this chapter we have provided an overview of the SE(R)RS technique and reviewed the literature in the context of bioelectrochemical studies. It has been impossible, due to the enormous amount of SERS literature, to be exhaustive and all-inclusive. Rather, we have tried to give an up-to-date account focusing on developments in the field of SERS of biomolecules and including progress made towards understanding the basic phenomena. The bulk of the historical literature covers roughened electrode surfaces but there is now an increasing move towards reproducible SERS surfaces. We believe that SERS is at an exciting stage where the ability to fabricate reproducible, stable SERS-active surfaces, and the growing understanding of the SERS phenomenon on these surfaces, will make possible much more quantitative measurements. These studies will not only help in advancing the understanding of interfacial behavior of biomolecules, which itself is important for biosensor and biofuel development, but will also, we believe, provide insights into the biological processes themselves. We hope that we have been successful in bringing out the advantages and novel insight that can be obtained by combining SE(R)RS with bioelectrochemistry.

References

1 Fleischmann, M., Hendra, P.J., and McQuillan, A.J. (1974) Raman spectra of pyridine adsorbed at a silver electrode. *Chemical Physics Letters*, **26**, 163–166.

2 McQuillan, A.J. (2009) The discovery of surface-enhanced Raman scattering. *Notes and Records of the Royal Society*, **63**, 105–109.

3 Jeanmarie, D.L. and Van Duyne, R.P. (1977) Surface Raman electrochemistry part I. Heterocyclic, aromatic and aliphatic amines adsorbed on the anodized silver electrode. *Journal of Electroanalytical Chemistry*, **84**, 1–20.

4 Albrecht, T., Li, W.W., Ulstrup, J., Haehnel, W., and Hildebrandt, P. (2005) Electrochemical and spectroscopic investigations of immobilized de novo designed heme proteins on metal electrodes. *ChemPhysChem*, **6**, 961–970.

5 Haynes, C.L., Yonzon, C.R., Zhang, X., and Van Duyne, R.P. (2005) Surface-enhanced Raman sensors: early history and the development of sensors for quantitative biowarfare agent and glucose detection. *Journal of Raman Spectroscopy*, **36**, 471–484.

6 Moskovits, M. (2005) Surface-enhanced Raman spectroscopy: a brief retrospective. *Journal of Raman Spectroscopy*, **36**, 485–496.

7 Haynes, C.L., McFarland, A.D., and Van Duyne, R.P. (2005) Surface-enhanced Raman spectroscopy. *Analytical Chemistry*, **77**, 338A–346A.

8 Stiles, P.L., Dieringer, J.A., Shah, N.C., and Van Duyne, R.P. (2008) Surface-enhanced Raman spectroscopy. *Annual Review of Analytical Chemistry*, **1**, 601–626.

9 Smith, W.E. (2008) Practical understanding and use of surface enhanced Raman scattering/surface enhanced resonance Raman scattering in chemical and biological analysis. *Chemical Society Reviews*, **37**, 955–964.

10 Hudson, S.D. and Chumanov, G. (2009) Bioanalytical applications of SERS (surface-enhanced Raman spectroscopy). *Analytical and Bioanalytical Chemistry*, **394**, 679–686.

11 Baker, G.A. and Moore, D.S. (2005) Progress in plasmonic engineering of surface-enhanced Raman-scattering substrates toward ultra-trace analysis. *Analytical and Bioanalytical Chemistry*, **382**, 1751–1770.

12 Efremov, E.V., Ariese, F., and Gooijer, C. (2008) Achievements in resonance Raman spectroscopy. *Analytica Chimica Acta*, **606**, 119–134.

13 Kudelski, A. (2008) Analytical applications of Raman spectroscopy. *Talanta*, **76**, 1–8.

14 Kudelski, A. (2009) Raman spectroscopy of surfaces. *Surface Science*, **603**, 1328–1334.

15 Banholzer, M.J., Millstone, J.E., Qin, L., and Mirkin, C.A. (2008) Rationally designed nanostructures for surface-enhanced Raman spectroscopy. *Chemical Society Reviews*, **37**, 886–897.

16 Lal, S., Grady, N.K., Kundu, J., Levin, C.S., Lassiter, J.B., and Halas, N.J. (2008) Tailoring plasmonic substrates for surface enhanced spectroscopies. *Chemical Society Reviews*, **37**, 898–911.

17 Lin, X.-M., Cui, Y., Xu, Y.-H., Ren, B., and Tian, Z.-Q. (2009) Surface-enhanced Raman spectroscopy: substrate-related issues. *Analytical and Bioanalytical Chemistry*, **394**, 1729–1745.

18 Tian, Z.-Q. and Ren, B. (2004) Adsorption and reaction at electrochemical interfaces as probed by surface-enhanced Raman spectroscopy. *Annual Review of Physical Chemistry*, **55**, 197–229.

19 Tian, Z.-Q., Ren, B., and Wu, D.-Y. (2002) Surface-enhanced Raman scattering: from noble to transition metals and from rough surfaces to ordered structures. *Journal of Physical Chemistry B*, **106**, 9463–9483.

20 Wu, D.-Y., Li, J.-F., Ren, B., and Tian, Z.-Q. (2008) Electrochemical surface-enhanced Raman spectroscopy of nanostructures. *Chemical Society Reviews*, **37**, 1025–1041.

21 Myers, A.B. and Mathies, R.A. (1987) Resonance Raman intensities: a probe of excited state structure and dynamics, in *Biological Applications of Raman Spectroscopy*, vol. 2 (ed. T.G. Spiro), John Wiley & Sons, Inc., New York, pp. 1–58.

22 Otto, A. (2005) The "chemical" (electronic) contribution to surface-enhanced Raman spectroscopy. *Journal of Raman Spectroscopy*, **36**, 497–509.

23 McCreery, R.L. (2000) *Raman Spectroscopy for Chemical Analysis*, vol. 157, Wiley, New York.

24 Murgida, D.H. and Hildebrandt, P. (2008) Disentangling interfacial redox processes of proteins by SERR spectroscopy. *Chemical Society Reviews*, **37**, 937–945.

25 Campion, A. and Kambhampati, P. (1998) Surface-enhanced Raman scattering. *Chemical Society Reviews*, **27**, 241–250.

26 Barnes, W.L., Dereux, A., and Ebbesen, T.W. (2003) Surface plasmon subwavelength optics. *Nature*, **424**, 824–830.

27 Genet, C. and Ebbesen, T.W. (2007) Light in tiny holes. *Nature*, **445**, 39–46.

28 Zayats, A.V., Smolyaninov, I.I., and Maradudin, A.A. (2005) Nano-optics of surface plasmon polaritons. *Physics Reports*, **408**, 131–314.

29 Otto, A., Mrozek, I., Grabhorn, H., and Akemann, W. (1992) Surface-enhanced Raman scattering. *Journal of Physics: Condensed Matter*, **4**, 1143–1212.

30 Etchegoin, P., Cohen, L.F., Hartigan, H., Brown, R.J.C., Milton, M.J.T., and Gallop, J.C. (2003) Electromagnetic contribution to surface enhanced Raman scattering revisited. *Journal of Chemical Physics*, **119**, 5281–5289.

31 Franzen, S. (2009) Intrinsic limitations of the $|E|^4$ dependence of the enhancement factor for surface-enhanced Raman scattering. *Journal of Physical Chemistry C*, **110**, 5912–5919.

32 Le Ru, E.C. and Etchegoin, P.G. (2006) Rigorous justification of the $|E|^4$ enhancement factor in surface enhanced Raman spectroscopy. *Chemical Physics Letters*, **423**, 63–66.

33 Baumberg, J.J., Kelf, T.A., Sugawara, Y., Cintra, S., Abdelsalam, M.E., Bartlett, P.N., and Russell, A.E. (2005) Angle-resolved surface-enhanced Raman scattering on metallic nanostructured plasmonic crystals. *Nano Letters*, **5**, 2262–2267.

34 Baltog, I., Primeau, N., Reinisch, R., and Coutaz, J.L. (1996) Observation of stimulated surface-enhanced Raman scattering through grating excitation of surface plasmons. *Journal of the Optical Society of America B*, **13**, 656–660.

35 Kahl, M. and Voges, E. (2000) Analysis of plasmon resonance and surface-enhanced Raman scattering on periodic silver structures. *Physical Review B*, **61**, 14078–14088.

36 Knoll, W., Philpott, M.R., Swalen, J.D., and Girlando, A. (1982) Surface plasmon enhanced Raman spectra of monolayer assemblies. *Journal of Chemical Physics*, **77**, 2254–2260.

37 Le Ru, E.C., Blackie, E., Meyer, M., and Etchegoin, P.G. (2007) Surface enhanced Raman scattering enhancement factors: a comprehensive study. *Journal of Physical Chemistry C*, **111**, 13794–13803.

38 Zeman, E.J. and Schatz, G.C. (1987) An accurate electromagnetic theory study of surface enhancement factors for Ag, Au, Cu, Li, Na, Al, Ga, In, Zn, and Cd. *Journal of Physical Chemistry*, **91**, 634–643.

39 Ren, B., Liu, G.-K., Lian, X.-B., Yang, Z.-L., and Tian, Z.-Q. (2007) Raman spectroscopy of transition metals. *Analytical and Bioanalytical Chemistry*, **388**, 29–45.

40 Moskovits, M. (1982) Surface selection rules. *Journal of Chemical Physics*, **77**, 4408–4416.

41 Suh, J.S. and Moskovits, M. (1986) Surface-enhanced Raman spectroscopy of amino acids and nucleotide bases adsorbed on silver. *Journal of the American Chemical Society*, **108**, 4711–4718.

42 Moskovits, M. and Suh, J.S. (1984) Surface selection rules for surface-enhanced Raman spectroscopy: calculations and application to the surface-enhanced Raman spectrum of phthalazine on silver. *Journal of Physical Chemistry*, **88**, 5526–5530.

43 Gao, X., Davies, J.P., and Weaver, M.J. (1990) A test of surface selection rules for surface-enhanced Raman scattering: the orientation of adsorbed benzene and monosubstituted benzenes on gold. *Journal of Physical Chemistry*, **94**, 6858–6864.

44 Pemberton, J.E., Bryant, M.A., Sobocinski, R.L., and Joa, S.L. (1992) A simple method for determination of orientation of adsorbed organics of low symmetry using surface-enhanced Raman scattering. *Journal of Physical Chemistry*, **96**, 3716–3782.

45 Szafranski, C.A., Tanner, W., Laibinis, P.E., and Garrell, R.L. (1998) Surface-enhanced Raman spectroscopy of aromatic thiols and disulfides on gold electrodes. *Langmuir*, **14**, 3570–3579.

46 Szafranski, C.A., Tanner, W., Laibinis, P.E., and Garrell, R.L. (1998) Surface-enhanced Raman spectroscopy of halogenated aromatic thiols on gold electrodes. *Langmuir*, **14**, 3580–3589.

47 Murray, C.A., and Allara, D.L. (1982) Measurement of the molecule–silver separation dependence of surface enhanced Raman scattering in multilayered structures. *Journal of Chemical Physics*, **76**, 1290–1303.

48 Aroca, R. and Guhathakurta-Ghosh, U. (1989) SERRS of Langmuir–Blodgett monolayers: spatial spectroscopic tuning. *Journal of the American Chemical Society*, **111**, 7681–7683.

49 Cotton, T.M., Uphaus, R.A., and Mobius, D. (1986) Distance dependence of surface-enhanced resonance Raman enhancement in Langmuir–Blodgett dye multiiayers. *Journal of Physical Chemistry*, **90**, 6071–6073.

50 Kovacs, G.J., Loutfy, R.O., Vincett, P.S., Jennings, C., and Aroca, R. (1986) Distance dependence of SERS enhancement factor from Langmuir–Blodgett monolayers on metal island films: evidence for the electromagnetic mechanism. *Langmuir*, **2**, 689–694.

51 Kennedy, B.J., Spaeth, S., Dickey, M., and Carron, K.T. (1999) Determination of the distance dependence and experimental effects for modified SERS substrates based on self-assembled monolayers formed using alkanethiols. *Journal of Physical Chemistry B*, **103**, 3640–3646.

52 Ye, Q., Fang, J., and Sun, L. (1997) Surface-enhanced Raman scattering from functionalized self-assembled monolayers. 2. Distance dependence of enhanced Raman scattering from an azobenzene terminal group. *Journal of Physical Chemistry B*, **101**, 8221–8224.

53 Lal, S., Grady, N.K., Goodrich, G.P., and Halas, N.J. (2006) Profiling the near field of a plasmonic nanoparticle with Raman-based molecular rulers. *Nano Letters*, **6**, 2338–2343.

54 Liu, F.M., Kollensperger, P.A., Green, M., Cass, A.E.G., and Cohen, L.F. (2006) A note on distance dependence in surface enhanced Raman spectroscopy. *Chemical Physics Letters*, **430**, 173–176.

55 Seballos, L., Zhang, J.Z., and Sutphen, R. (2005) Surface-enhanced Raman scattering detection of lysophosphatidic acid. *Analytical and Bioanalytical Chemistry*, **383**, 763–767.

56 Lambert, D.K. (1996) Vibrational Stark effect of adsorbates at electrochemical interfaces. *Electrochimica Acta*, **41**, 623–630.

57 Weaver, M.J., Williams, C.T., Zou, S., Chan, H.Y.H., and Takoudis, C.G. (1998) Surface potentials of metal–gas interfaces compared with analogous electrochemical systems as probed by

adsorbate vibrational frequencies. *Catalysis Letters*, **52**, 181–190.
58 Weaver, M.J., Zou, S., and Tang, C. (1999) A concerted assessment of potential-dependent vibrational frequencies for nitric oxide and carbon monoxide adlayers on low-index platinum-group surfaces in electrochemical compared with ultrahigh vacuum environments: structural and electrostatic implications. *Journal of Chemical Physics*, **111**, 368–381.
59 Lambert, D.K. (1988) Vibrational Stark effect of CO on Ni(100), and CO in the aqueous double layer: experiment, theory, and models. *Journal of Chemical Physics*, **89**, 3847–3860.
60 Oklejas, V., Sjostrom, C., and Harris, J.M. (2003) Surface-enhanced Raman scattering based vibrational stark effects as a spatial probe of interfacial electric fields in the diffuse double layer. *Journal of Physical Chemistry B*, **107**, 7788–7794.
61 Oklejas, V., Sjostrom, C., and Harris, J.M. (2002) SERS detection of the vibrational stark effect from nitrile-terminated SAMs to probe electric fields in the diffuse double-layer. *Journal of the American Chemical Society*, **124**, 2408–2409.
62 Bell, S.E.J. and Sirimuthu, N.M.S. (2008) Quantitative surface-enhanced Raman spectroscopy. *Chemical Society Reviews*, **37**, 1012–1024.
63 Graham, D. and Faulds, K. (2008) Quantitative SERRS for DNA sequence analysis. *Chemical Society Reviews*, **37**, 1042–1051.
64 Porter, M.D., Lipert, R.J., Siperko, L.M., Wang, G., and Narayanana, R. (2008) SERS as a bioassay platform: fundamentals, design, and applications. *Chemical Society Reviews*, **37**, 1001–1011.
65 Emory, S.R. and Nie, S. (1997) Probing single molecules and single nanoparticles by surface-enhanced Raman scattering. *Science*, **275**, 1102–1106.
66 Kneipp, K., Wang, Y., Kneipp, H., Perelman, L.T., Itzkan, I., Dasari, R.R., and Feld, M.S. (1997) Single molecule detection using surface-enhanced Raman scattering (SERS). *Physical Review Letters*, **78**, 1667–1670.
67 Li, H., Ying, L., Green, J.J., Balasubramanian, S., and Klenerman, D. (2003) Ultrasensitive coincidence fluorescence detection of single DNA molecules. *Analytical Chemistry*, **75**, 1664–1670.
68 Nie, S.M., Chiu, D.T., and Zare, R.N. (1995) Real-time detection of single-molecules in solution by confocal fluorescence microscopy. *Analytical Chemistry*, **61**, 2849–2857.
69 Faulds, K., Jarvis, R., Smith, W.E., Graham, D., and Goodacre, R. (2008) Multiplexed detection of six labelled oligonucleotides using surface enhanced resonance Raman scattering (SERRS). *Analyst*, **133**, 1505–1512.
70 Cao, Y.W.C., Jin, R.C., and Mirkin, C.A. (2002) Nanoparticles with Raman spectroscopic fingerprints for DNA and RNA detection. *Science*, **297**, 1536–1540.
71 Faulds, K., Fruk, L., Robson, D.C., Thompson, D.G., Enright, A., Smith, W.E., and Graham, D. (2006) A new approach for DNA detection by SERRS. *Faraday Discussions*, **132**, 261–268.
72 Stokes, R.J., Macaskill, A., Lundahl, P.J., Smith, W.E., Faulds, K., and Graham, D. (2007) Quantitative enhanced Raman scattering of labeled DNA from gold and silver nanoparticles. *Small*, **3**, 1593–1601.
73 Sun, L., Yu, C.X., and Irudayaraj, J. (2007) Surface-enhanced Raman scattering based nonfluorescent probe for multiplex DNA detection. *Analytical Chemistry*, **79**, 3981–3988.
74 Stokes, R.J., Ingram, A., Gallagher, J., Armstrong, D.R., Smith, W.E., and Graham, D. (2008) Squaraines as unique reporters for SERRS multiplexing. *Chemical Communications*, 567–569.
75 Sabatte, G., Keir, R., Lawlor, M., Black, M., Graham, D., and Smith, W.E. (2008) Comparison of surface-enhanced resonance Raman scattering and fluorescence for detection of a labeled antibody. *Analytical Chemistry*, **80**, 2351–2356.
76 Faulds, K., Barbagallo, R.P., Keer, J.T., Smith, W.E., and Graham, D. (2004) SERRS as a more sensitive technique for the detection of labelled

oligonucleotides compared to fluorescence. *Analyst*, **129**, 567–568.

77 MacAskill, A., Crawford, D., Graham, D., and Faulds, K. (2009) DNA sequence detection using surface-enhanced resonance Raman spectroscopy in a homogeneous multiplexed assay. *Analytical Chemistry*, **81**, 8134–8140.

78 Munro, C.H., Smith, W.E., and White, P.C. (1995) Qualitative and semiquantitative trace analysis of acidic monoazo dyes by surface-enhanced resonance Raman scattering. *Analyst*, **120**, 993–1003.

79 Vo-Dinh, T., Yan, F., and Wabuyele, M.B. (2005) Surface-enhanced Raman scattering for medical diagnostics and biological imaging. *Journal of Raman Spectroscopy*, **36**, 640–647.

80 Allain, L.R. and Vo-Dinh, T. (2002) Surface-enhanced Raman scattering detection of the breast cancer susceptibility gene BRCA1 using a silver-coated microarray platform. *Analytica Chimica Acta*, **469**, 149–154.

81 Culha, M., Stokes, D., Allain, L.R., and Vo-Dinh, T. (2003) Surface-enhanced Raman scattering substrate based on a self-assembled monolayer for use in gene diagnostics. *Analytical Chemistry*, **75**, 6196–6201.

82 Isola, N.R., Stokes, D.L., and Vo-Dinh, T. (1998) Surface enhanced Raman gene probe for HIV detection. *Analytical Chemistry*, **70**, 1352.

83 Faulds, K., Smith, W.E., and Graham, D. (2004) Evaluation of surface-enhanced resonance Raman scattering for quantitative DNA analysis. *Analytical Chemistry*, **76**, 412–417.

84 Cao, Y.C., Jin, R.C., Nam, J.M., Thaxton, C.S., and Mirkin, C.A. (2003) Raman dye-labeled nanoparticle probes for proteins. *Journal of the American Chemical Society*, **125**, 14676–14677.

85 Graham, D., Mallinder, B.J., Whitcombe, D., Watson, N.D., and Smith, W.E. (2002) Simple multiplex genotyping by surface-enhanced resonance Raman scattering. *Analytical Chemistry*, **74**, 1069–1074.

86 Lowe, A.J., Huh, Y.S., Strickland, A.D., Erickson, D., and Batt, C.A. (2010) Multiplex single nucleotide polymorphism genotyping utilizing ligase detection reaction coupled surface enhanced Raman spectroscopy. *Analytical Chemistry*, **82**, 5810–5814.

87 Sun, L. and Irudayaraj, J. (2009) PCR-free quantification of multiple splice variants in a cancer gene by surface-enhanced Raman spectroscopy. *Journal of Physical Chemistry B*, **113**, 14021–14025.

88 Sun, L., Yu, C.X., and Irudayaraj, J. (2008) Raman multiplexers for alternative gene splicing. *Analytical Chemistry*, **80**, 3342–3349.

89 Wang, H.N. and Vo-Dinh, T. (2009) Multiplex detection of breast cancer biomarkers using plasmonic molecular sentinel nanoprobes. *Nanotechnology*, **20**, 065101.

90 Woo, M.A., Lee, S.M., Kim, G., Baek, J., Noh, M.S., Kim, J.E., Park, S.J., Minai-Tehrani, A., Park, S.C., Seo, Y.T., Kim, Y.K., Lee, Y.S., Jeong, D.H., and Cho, M.H. (2009) Multiplex immunoassay using fluorescent-surface enhanced Raman spectroscopic dots for the detection of bronchioalveolar stem cells in murine lung. *Analytical Chemistry*, **81**, 1008–1015.

91 Cui, Y., Ren, B., Yao, J.L., Gu, R.A., and Tian, Z.Q. (2007) Multianalyte immunoassay based on surface-enhanced Raman spectroscopy. *Journal of Raman Spectroscopy*, **38**, 896–902.

92 Jun, B.H., Kim, J.H., Park, H., Kim, J.S., Yu, K.N., Lee, S.M., Choi, H., Kwak, S.Y., Kim, Y.K., Jeong, D.H., Cho, M.H., and Lee, Y.S. (2007) Surface-enhanced Raman spectroscopic-encoded beads for multiplex immunoassay. *Journal of Combinatorial Chemistry*, **9**, 237–244.

93 Lutz, B.R., Dentinger, C.E., Nguyen, L.N., Sun, L., Zhang, J.W., Allen, A.N., Chan, S., and Knudsen, B.S. (2008) Spectral analysis of multiplex Raman probe signatures. *ACS Nano*, **2**, 2306–2314.

94 Faulds, K., Littleford, R.E., Graham, D., Dent, G., and Smith, W.E. (2004) Comparison of surface-enhanced resonance Raman scattering from unaggregared and aggregated nanoparticles. *Analytical Chemistry*, **76**, 592–598.

95 Cui, L., Wang, A., Wu, D.-Y., Ren, B., and Tian, Z.-Q. (2008) Shaping and shelling Pt and Pd nanoparticles for ultraviolet laser excited surface-enhanced Raman scattering. *Journal of Physical Chemistry C*, **112**, 17618–17624.

96 Liu, Z., Yang, Z.-L., Cui, L., Ren, B., and Tian, Z.-Q. (2007) Electrochemically roughened palladium electrodes for surface-enhanced Raman spectroscopy: methodology, mechanism, and application. *Journal of Physical Chemistry C*, **111**, 1770–1775.

97 Leung, L.-W.H. and Weaver, M.J. (1987) Extending surface-enhanced Raman spectroscopy to transition-metal surfaces: carbon monoxide adsorption and electrooxidation on platinum- and palladium-coated gold electrodes. *Journal of the American Chemical Society*, **109**, 5113–5119.

98 Leung, L.-W.H. and Weaver, M.J. (1988) Adsorption and electrooxidation of carbon monoxide on rhodium- and ruthenium-coated gold electrodes as probed by surface-enhanced Raman spectroscopy. *Langmuir*, **4**, 1076–1083.

99 Wilke, T., Gao, X., Takoudis, C.G., and Weaver, M.J. (1991) Surface-enhanced Raman spectroscopy as a probe of adsorption at transition metal-high-pressure gas interfaces: NO, CO, and oxygen on platinum-, rhodium-, and ruthenium-coated gold. *Langmuir*, **7**, 714–721.

100 Tian, Z.-Q., Ren, B., Li, J.-F., and Yang, Z.-L. (2007) Expanding generality of surface-enhanced Raman spectroscopy with borrowing SERS activity strategy. *Chemical Communications*, 3514–3534.

101 Zou, S., and Weaver, M.J. (1998) Surface-enhanced Raman scattering on uniform transition-metal films: toward a versatile adsorbate vibrational strategy for solid-non vacuum interfaces? *Analytical Chemistry*, **70**, 2387–2395.

102 Zou, S., Williams, C.T., Chen, E.K.-Y., and Weaver, M.J. (1998) Probing molecular vibrations at catalytically significant interfaces: a new ubiquity of surface-enhanced Raman scattering. *Journal of the American Chemical Society*, **120**, 3811–3812.

103 Zou, S., Williams, C.T., Chen, E.K.-Y., and Weaver, M.J. (1998) Surface-enhanced Raman scattering as a ubiquitous vibrational probe of transition-metal interfaces: benzene and related chemisorbates on palladium and rhodium in aqueous solution. *Journal of Physical Chemistry B*, **102**, 9039–9049.

104 Halas, N.J. (2010) Plasmonics: an emerging field fostered by *Nano Letters*. *Nano Letters*, **10**, 3816–3822.

105 Zheludev, N.I. (2010) The road ahead for metamaterials. *Science*, **328**, 582–583.

106 Willets, K.A. and Van Duyne, R.P. (2007) Localized surface plasmon resonance spectroscopy and sensing. *Annual Review of Physical Chemistry*, **58**, 267–297.

107 Draine, B.T. and Flatau, P.J. (1994) Discrete-dipole approximation for scattering calculations. *Journal of the Optical Society of America A*, **11**, 1491–1499.

108 Futamata, M., Maruyama, Y., and Ishikawa, M. (2003) Local electric field and scattering cross section of Ag nanoparticles under surface plasmon resonance by finite difference time domain method. *Journal of Physical Chemistry B*, **107**, 7607–7617.

109 Oubre, C. and Nordlander, P. (2004) Optical properties of metallodielectric nanostructures calculated using the finite difference time domain method. *Journal of Physical Chemistry B*, **108**, 17740–17747.

110 Cole, R.M., Baumberg, J.J., Garcia de Abajo, F.J., Mahajan, S., Abdelsalam, M., and Bartlett, P.N. (2007) Understanding plasmons in nanoscale voids. *Nano Letters*, **7**, 2094–2100.

111 Prodan, E. and Nordlander, P. (2004) Plasmon hybridization in spherical nanoparticles. *Journal of Chemical Physics*, **120**, 5444–5454.

112 Prodan, E., Radloff, C., Halas, N.J., and Nordlander, P. (2003) A hybridization model for the plasmon response of complex nanostructures. *Science*, **302**, 419–422.

113 Wang, H., Brandl, D.W., Norlander, P., and Halas, N.J. (2007) Plasmonic nanostructures: artificial molecules. *Accounts of Chemical Research*, **40**, 53–62.

114 Brown, L.V., Sobhani, H., Lassiter, J.B., Norlander, P., and Halas, N.J. (2010) Heterodimers: plasmonic properties of mismatched nanoparticle pairs. *ACS Nano*, **4**, 819–832.

115 Prodan, E. and Nordlander, P. (2003) Structural tunability of the plasmon resonances in metallic nanoshells. *Nano Letters*, **3**, 543–547.

116 Bardhan, R., Grady, N.K., Ali, T., and Halas, N.J. (2010) Metallic nanoshells with semiconductor cores: optical characteristics modified by core medium properties. *ACS Nano*, **4**, 6169–6179.

117 Lassiter, J.B., Aizpurua, J., Hernandez, L.I., Brandl, D.W., Romero, I., Lal, S., Hafner, J.H., Nordlander, P., and Halas, N.J. (2008) Close encounters between two nanoshells. *Nano Letters*, **8**, 1212–1218.

118 Park, T.-H., Mirin, N., Lassiter, J.B., Nehl, C.L., Halas, N.J., and Peter Nordlander, P. (2008) Optical properties of a nanosized hole in a thin metallic film. *ACS Nano*, **2**, 25–32.

119 Teperik, T.V., Popov, V.V., de Abajo, F.J.G., Kelf, T.A., Sugawara, Y., Baumberg, J.J., Abdelsalem, M., and Bartlett, P.N. (2006) Mie plasmon enhanced diffraction of light from nanoporous metal surfaces. *Optics Express*, **14**, 11964–11971.

120 Kelf, T.A., Sugawara, Y., Baumberg, J.J., Abdelsalam, M., and Bartlett, P.N. (2005) Plasmonic band gaps and trapped plasmons on nanostructured metal surfaces. *Physical Review Letters*, **95**, 116802.

121 Kelf, T.A., Sugawara, Y., Cole, R.M., Baumberg, J.J., Abdelsalam, M.E., Cintra, S., Mahajan, S., Russell, A.E., and Bartlett, P.N. (2006) Localized and delocalized plasmons in metallic nanovoids. *Physical Review B*, **74**, 245415.

122 Barnes, W.L., Preist, T.W., Kitson, S.C., and Sambles, J.R. (1996) Physical origin of photonic energy gaps in the propagation of surface plasmons on gratings. *Physical Review B*, **54**, 6227–6244.

123 Baltog, I., Primeau, N., Reinisch, R., and Coutaz, J.L. (1995) Surface enhanced Raman scattering on silver grating: optimized antennalike gain of the stokes signal of 10^4. *Applied Physics Letters*, **66**, 1187–1189.

124 Henzie, J., Lee, J., Lee, M.H., Hasan, W., and Odom, T.W. (2009) Nanofabrication of plasmonic structures. *Annual Review of Physical Chemistry*, **60**, 147–165.

125 Brolo, A.G., Arctander, E., Gordon, R., Leathem, B., and Kavanagh, K.L. (2004) Nanohole-enhanced Raman scattering. *Nano Letters*, **4**, 2015–2018.

126 Anema, J.R., Brolo, A.G., Marthandam, P., and Gordon, R. (2008) Enhanced Raman scattering from nanoholes in a copper film. *Journal of Physical Chemistry C*, **112**, 17051–17055.

127 Reilly, T.H., Chang, S.-H., Corbman, J.D., Schatz, G.C., and Rowlen, K.L. (2007) Quantitative evaluation of plasmon enhanced Raman scattering from nanoaperture arrays. *Journal of Physical Chemistry C*, **111**, 1689–1694.

128 Im, H., Bantz, K.C., Lindquist, N.C., Haynes, C.L., and Oh, S.-H. (2010) Vertically oriented sub-10-nm plasmonic nanogap arrays. *Nano Letters*, **10**, 2231–2236.

129 Dinish, U.S., Yaw, F.C., Agarwal, A., and Olivo, M. (2011) Development of highly reproducible nanogap SERS substrates: comparative performance analysis and its application for glucose sensing. *Biosensors & Bioelectronics*, **26**, 1987–1992.

130 Jackel, F., Kinkhabwala, A.A., and Moerner, W.E. (2007) Gold bowtie nanoantennas for surface-enhanced Raman scattering under controlled electrochemical potential. *Chemical Physics Letters*, **446**, 339–343.

131 Kantarovich, K., Tsarfati, I., Gheber, L.A., Haupt, K., and Bar, I. (2009) Writing droplets of molecularly imprinted polymers by nano fountain pen and detecting their molecular interactions by surface-enhanced Raman scattering. *Analytical Chemistry*, **81**, 5686–5690.

132 Mahajan, S., Lee, T.-C., Biedermann, F., Hugall, J.T., Baumberg, J.J., and Scherman, O.A. (2010) Raman and SERS spectroscopy of cucurbit[n]urils.

Physical Chemistry Chemical Physics, **12**, 10429–10433.
133 Stokes, R.J., Dougan, J.A., and Graham, D. (2008) Dip-pen nanolithography and SERRS as synergic techniques. *Chemical Communications*, 5734–5736.
134 Stokes, R.J., Macaskill, A., Dougan, J.A., Hargreaves, P.G., Stanford, H.M., Smith, W.E., Faulds, K., and Graham, D. (2007) Highly sensitive detection of dye-labelled DNA using nanostructured gold surfaces. *Chemical Communications*, 2811–2813.
135 Meyer, S.A., Le Ru, E.C., and Etchegoin, P.G. (2010) Quantifying resonant Raman cross sections with SERS. *Journal of Physical Chemistry A*, **114**, 5515–5519.
136 Ohta, N., and Yagi, I. (2008) *In situ* surface-enhanced Raman scattering spectroscopic study of pyridine adsorbed on gold electrode surfaces comprised of plasmonic crystal structures. *Journal of Physical Chemistry C*, **112**, 17603–17610.
137 Deckman, H.W. and Dunsmuir, J.H. (1982) Natural lithography. *Applied Physics Letters*, **41**, 377–397.
138 Lee, S.H., Bantz, K.C., Lindquist, N.C., Oh, S.-H., and Haynes, C.L. (2009) Self-assembled plasmonic nanohole arrays. *Langmuir*, **25**, 13685–13693.
139 Ormonde, A.D., Hicks, E.C.M., Castillo, J., and Van Duyne, R.P. (2004) Nanosphere lithography: fabrication of large-area Ag nanoparticle arrays by convective self-assembly and their characterisation by scanning UV-visible extinction spectroscopy. *Langmuir*, **20**, 6927–6931.
140 Kuncicky, D.M., Prevo, B.G., and Velev, O.D. (2006) Controlled assembly of SERS substrates templated by colloidal crystal films. *Journal of Materials Chemistry*, **16**, 1207–1211.
141 Bartlett, P.N., Birkin, P.R., and Ghanem, M.A. (2000) Electrochemical deposition of macroporous platinum, palladium and cobalt films using polystyrene latex sphere templates. *Chemical Communications*, 1671–1672.
142 Bardosova, M., Pemble, M.E., Povey, I.M., and Tredgold, R.H. (2010) The Langmuir–Blodgett approach to making colloidal photonic crystals from silica spheres. *Advanced Materials*, **22**, 3104–3124.
143 Szamocki, R., Reculusa, S., Ravaine, S., Bartlett, P.N., Kuhn, A., and Hempelmann, R. (2006) Tailored mesostructuring and biofunctionalization of gold for increase electroactivity. *Angewandte Chemie International Edition*, **45**, 1317–1321.
144 Kubo, S., Gu, Z.-Z., Tryk, D.A., Ohko, Y., Sato, O., and Fujishima, A. (2002) Metal-coated colloidal crystal film as surface-enhanced Raman scattering substrate. *Langmuir*, **18**, 5043–5046.
145 Camden, J.P., Dieringer, J.A., Zhao, J., and Van Duyne, R.P. (2008) Controlled plasmonic nanostructures for surface-enhanced spectroscopy and sensing. *Accounts of Chemical Research*, **41**, 1653–1661.
146 Haynes, C.L., McFarland, A.D., Smith, M.T., Hulteen, J.C., and Van Duyne, R.P. (2002) Angle-resolved nanosphere lithography: manipulation of nanoparticle size, shape and interparticle spacing. *Journal of Physical Chemistry B*, **106**, 1898–1902.
147 Haynes, C.L. and Van Duyne, R.P. (2001) Nanosphere lithography; a versatile nanofabrication tool for studies of size-dependent nanoparticle optics. *Journal of Physical Chemistry B*, **105**, 5599–5611.
148 Hulteen, J.C., Treichel, D.A., Smith, M.T., Duval, M.L., Jensen, T.R., and Van Duyne, R.P. (1999) Nanosphere lithography: size-tunable silver nanoparticle and surface cluster arrays. *Journal of Physical Chemistry B*, **103**, 3854–3863.
149 Chan, G.H., Zhao, J., Hicks, E.M., Schatz, G.C., and Van Duyne, R.P. (2007) Plasmonic properties of copper nanoparticles fabricated by nanosphere lithography. *Nano Letters*, **7**, 1947–1952.
150 Haes, A.J., Hall, W.P., Chang, L., Klein, W.L., and Van Duyne, R.P. (2004) A localised surface plasmon resonance biosensor: first steps toward an assay for Alzheimer's disease. *Nano Letters*, **4**, 1029–1034.
151 Haes, A.J. and Van Duyne, R.P. (2002) A nanoscale optical biosensor: sensitivity and selectivity of an approach based on

localised surface plasmon resonance spectroscopy of triangular silver nanoparticles. *Journal of the American Chemical Society*, **124**, 10596–10604.

152 Haes, A.J., Zou, S., Schatz, G.C., and Van Duyne, R.P. (2004) Nanoscale optical biosensor: short range distance dependence of the localized surface plasmon resonance of noble metal nanoparticles. *Journal of Physical Chemistry B*, **108**, 6961–6968.

153 Jensen, T.R., Duval, M.L., Kelly, K.L., Lazarides, A.A., Schatz, G.C., and Van Duyne, R.P. (1999) Nanosphere lithography: effect of the external dielectric medium on the surface plasmon resonance spectrum of a periodic array of silver nanoparticles. *Journal of Physical Chemistry B*, **103**, 9846–9853.

154 Malinsky, M.D., Kelly, K.L., Schatz, G.C., and Van Duyne, R.P. (2001) Nanosphere lithography: effect of substrate on the localized surface plasmon resonance spectrum of silver nanoparticles. *Journal of Physical Chemistry B*, **105**, 2343–2350.

155 Haynes, C.L. and Van Duyne, R.P. (2003) Plasmon-sampled surface-enhanced Raman excitation spectroscopy. *Journal of Physical Chemistry B*, **107**, 7426–7433.

156 Van Duyne, R.P., Hulteen, J.C., and Treichel, D.A. (1993) Atomic force microscopy and surface-enhanced Raman spectroscopy. I. Ag island films and Ag film over polymer nanosphere surfaces supported on glass. *Journal of Chemical Physics*, **99**, 2101–2115.

157 Jensen, T.R., Malinsky, M.D., Haynes, C.L., and Van Duyne, R.P. (2000) Nanosphere lithography: tunable localized surface plasmon resonance spectra of silver nanoparticles. *Journal of Physical Chemistry B*, **104**, 10549–10556.

158 Jensen, T.R., Schatz, G.C., and Van Duyne, R.P. (1999) Nanosphere lithography: surface plasmon resonance spectrum of a periodic array of silver nanoparticles by ultraviolet-visible extinction spectroscopy and electrodynamic modeling. *Journal of Physical Chemistry B*, **103**, 2394–2401.

159 Zhang, X., Hicks, E.M., Zhao, J., Schatz, G.C., and Van Duyne, R.P. (2005) Electrochemical tuning of silver nanoparticles fabricated by nanosphere lithography. *Nano Letters*, **5**, 1503–1507.

160 Dick, L.A., McFarland, A.D., Haynes, C.L., and Van Duyne, R.P. (2002) Metal film over nanosphere (MFON) electrodes for surface-enhanced Raman spectroscopy (SERS): improvements in surface nanostructure stability and suppression of irreversible loss. *Journal of Physical Chemistry B*, **106**, 853–860.

161 Biggs, K.B., Camden, J.P., Anker, J.N., and Van Duyne, R.P. (2009) Surface-enhanced Raman spectroscopy of benzenethiol adsorbed from the gas phase onto silver film over nanosphere surfaces: determination of the sticking probability and detection limit time. *Journal of Physical Chemistry A*, **113**, 4581–4586.

162 Baia, M., Baia, L., and Astilean, S. (2005) Gold nanostructured films deposited on polystyrene colloidal crystal templates for surface-enhanced Raman spectroscopy. *Chemical Physics Letters*, **404**, 3–8.

163 Bantz, K.C. and Haynes, C.L. (2008) Surface-enhanced Raman scattering substrates fabricated using electroless plating on polymer-templated nanostructures. *Langmuir*, **24**, 5862–5867.

164 Shafer-Peltier, K.E., Haynes, C.L., Glucksberg, M.R., and Van Duyne, R.P. (2003) Toward a glucose biosensor based on surface-enhanced Raman scattering. *Journal of the American Chemical Society*, **125**, 588–593.

165 Stuart, D.A., Yonzon, C.R., Zhang, X., Lyandres, O., Shah, N.C., Glucksberg, M.R., Walsh, J.T., and Van Duyne, R.P. (2005) Glucose sensing using near-infrared surface-enhanced Raman spectroscopy: gold surfaces, 10-day stability, and improved accuracy. *Analytical Chemistry*, **77**, 4013–4019.

166 Yonzon, C.R., Haynes, C.L., Zhang, X., Walsh, J.T., and Van Duyne, R.P. (2004) A glucose biosensor based on surface-enhanced Raman scattering: improved partition layer, temporal stability, reversibility, and resistance to serum protein interference. *Analytical Chemistry*, **76**, 78–85.

167 Litorja, M.L., Haynes, C.L., Haes, A.J., Jensen, T.R., and Van Duyne, R.P. (2001) Surface-enhanced Raman scattering detected temperature programmed desorption: optical properties, nanostructure, and stability of silver film over SiO_2 nanosphere. *Journal of Physical Chemistry B*, **105**, 6907–6915.

168 Hulteen, J.C., Young, M.A., and Van Duyne, R.P. (2006) Surface-enhanced hyper-Raman scattering (SEHRS) on Ag film over nanosphere (FON) electrodes: surface symmetry of centrosymmetric adsorbates. *Langmuir*, **22**, 10354–10364.

169 Dick, L.A., Haes, A.J., and Van Duyne, R.P. (2000) Distance and orientation dependence of heterogeneous electron transfer: a surface-enhanced resonance Raman scattering study of cytochrome *c* bond to carboxylic acid terminated alkanethiols adsorbed on silver electrodes. *Journal of Physical Chemistry B*, **104**, 11752–11762.

170 Freunscht, P., Van Duyne, R.P., and Schneider, S. (1997) Surface-enhanced Raman spectroscopy of trans-stilbene adsorbed on platinum- or self-assembled monolayer-modified silver film over nanosphere surfaces. *Chemical Physics Letters*, **281**, 372–378.

171 Tessier, P.M., Velev, O.D., Kalambur, A.T., Rabolt, J.F., Lenhoff, A.M., and Kaler, E.W. (2000) Assembly of gold nanostructured films templated by colloidal crystals and use in surface-enhanced Raman spectroscopy. *Journal of the American Chemical Society*, **122**, 9554–9555.

172 Lu, L., Eychmuller, A., Kobayashi, A., Hirano, Y., Yoshida, K., Kikkawa, Y., Tawa, K., and Ozaki, Y. (2006) Designed fabrication of ordered porous Au/Ag nanostructured films for surface-enhanced Raman scattering substrates. *Langmuir*, **22**, 2605–2609.

173 Bartlett, P.N., Baumberg, J.J., Birkin, P.R., Ghanem, M.A., and Netti, M.C. (2002) Highly ordered macroporous gold and platinum films formed by electrochemical deposition through templates assembled from submicron diameter monodisperse polystyrene spheres. *Chemistry of Materials*, **14**, 2199–2208.

174 Bartlett, P.N., Baumberg, J.J., Coyle, S., and Abdelsalam, M.A. (2004) Optical properties of nanostructured metal films. *Faraday Discussions*, **125**, 117–132.

175 Abdelsalam, M.E., Bartlett, P.N., Baumberg, J.J., Cintra, S., Kelf, T.A., and Russell, A.E. (2005) Electrochemical SERS at a structured gold surface. *Electrochemistry Communications*, **7**, 740–744.

176 Mahajan, S., Abdelsalam, M., Sugawara, Y., Cintra, S., Russell, A., Baumberg, J., and Bartlett, P. (2007) Tuning plasmons on nano-structured substrates for NIR-SERS. *Physical Chemistry Chemical Physics*, **9**, 104–109.

177 Mahajan, S., Baumberg, J.J., Russell, A.E., and Bartlett, P.N. (2007) Reproducible SERRS from structured gold surfaces. *Physical Chemistry Chemical Physics*, **9**, 6016–6020.

178 Cintra, S., Abdelsalam, M.E., Bartlett, P.N., Baumberg, J.J., Kelf, T.A., Sugawara, Y., and Russell, A.E. (2006) Sculpted substrates for SERS. *Faraday Discussions*, **132**, 191–199.

179 Cole, R.M., Mahajan, S., Bartlett, P.N., and Baumberg, J.J. (2009) Engineering SERS via absorption control in novel hybrid Ni/Au nanovoids. *Optical Express*, **17**, 13298–13308.

180 Coyle, S., Netti, M.C., Baumberg, J.J., Ghanem, M.A., Birkin, P.R., Bartlett, P.N., and Whitaker, D.M. (2001) Confined surface plasmons in metallic nanocavites. *Physical Review Letters*, **87**, 176801.

181 Lacharmoise, P.D., Tognalli, N.G., Goñi, A.R., Alonso, M.I., Fainstein, A., Cole, R.M., Baumberg, J.J., Garcia de Abajo, J., and Bartlett, P.N. (2008) Imaging optical near fields at metallic nanoscale voids. *Physical Review B*, **78**, 125410.

182 Netti, M.C., Coyle, S., Baumberg, J.J., Ghanem, M.A., Birkin, P.R., Bartlett, P.N., and Whitaker, D.M. (2001) Confined surface plasmons in gold photonic nanocavities. *Advanced Materials*, **13**, 1368–1370.

183 Abdelsalam, M., Bartlett, P.N., Russell, A.E., Baumberg, J.J., Calvo, E.J.,

Tognalli, N.G., and Fainstein, A. (2008) Quantitative electrochemical SERS of flavin at a structured silver surface. *Langmuir*, **24**, 7018–7023.

184 Cui, L., Mahajan, S., Cole, R.M., Soares, B., Bartlett, P.N., Baumberg, J.J., Hayward, I.P., Ren, B., Russell, A.E., and Tian, Z.Q. (2009) UV SERS at well ordered Pd sphere segment void (SSV) nanostructures. *Physical Chemistry Chemical Physics*, **11**, 1023–1026.

185 Abdelsalam, M.E., Mahajan, S., Bartlett, P.N., Baumberg, J.J., and Russell, A.E. (2007) SERS at structured palladium and platinum surfaces. *Journal of the American Chemical Society*, **129**, 7399–7406.

186 Mahajan, S., Cole, R.M., Soares, B.F., Pelfrey, S.H., Russell, A.E., Baumberg, J.J., and Bartlett, P.N. (2009) Relating SERS intensity to specific plasmon modes on sphere segment void surfaces. *Journal of Physical Chemistry C*, **113**, 9284–9289.

187 Mahajan, S., Cole, R.M., Speed, J.D., Pelfrey, S.H., Russell, A.E., Bartlett, P.N., Barnett, S.M., and Baumberg, J.J. (2010) Understanding the surface-enhanced Raman spectroscopy "background". *Journal of Physical Chemistry C*, **114**, 7242–7250.

188 Tognalli, N.G., Scodeller, P., Flexer, V., Szamocki, R., Ricci, A., Tagliazucchi, M., Calvo, E.J., and Fainstein, A. (2009) Redox molecule based SERS sensors. *Physical Chemistry Chemical Physics*, **11**, 7412–7423.

189 Vezvaie, M., Brosseau, C.L., Goddard, J.D., and Lipkowski, J. (2010) SERS of beta-thioglucose adsorbed on nanostructured silver electrodes. *ChemPhysChem*, **11**, 1460–1467.

190 Corrigan, D.K., Gale, N., Brown, T., and Bartlett, P.N. (2010) Analysis of short tandem repeats (STRs) using SERS monitoring and electrochemical melting. *Angewandte Chemie International Edition*, **49**, 5917–5920.

191 Mahajan, S., Richardson, J., Ben Gaied, N., Zhao, Z., Brown, T., and Bartlett, P.N. (2009) The use of an electroactive marker as a SERS label in an E-melting mutation discrimination assay. *Electroanalysis*, **21**, 2190–2197.

192 Jose, B., Steffen, R., Neugebauer, U., Sheridan, E., Marthi, R., Forster, R.J., and Keyes, T.E. (2009) Emission enhancement within gold spherical nanocavity arrays. *Physical Chemistry Chemical Physics*, **11**, 10923–10933.

193 Stockle, R.M., Suh, Y.D., Deckert, V., and Zenobi, R. (2000) Nanoscale chemical analysis by tip-enhanced Raman spectroscopy. *Chemical Physics Letters*, **318**, 131–136.

194 Anderson, M.S. (2000) Locally enhanced Raman spectroscopy with an atomic force microscope. *Applied Physics Letters*, **76**, 3130–3132.

195 Hayazawa, N., Inouye, Y., Sekkat, Z., and Kawata, S. (2000) Metallized tip amplification of near-field Raman scattering. *Optics Communications*, **183**, 333–336.

196 Deckert-Gaudig, T., Bailo, E., and Deckert, V. (2009) Tip-enhanced Raman scattering (TERS) of oxidised glutathione on an ultraflat gold nanoplate. *Physical Chemistry Chemical Physics*, **11**, 7360–7362.

197 Deckert-Gaudig, T., and Deckert, V. (2009) Tip-enhanced Raman scattering studies of histidine on novel silver substrates. *Journal of Raman Spectroscopy*, **40**, 1446–1451.

198 Yeo, B.S., Madler, S., Schmid, T., Zhang, W.H., and Zenobi, R. (2008) Tip-enhanced Raman spectroscopy can see more: the case of cytochrome *c*. *Journal of Physical Chemistry C*, **112**, 4867–4873.

199 Watanabe, H., Ishida, Y., Hayazawa, N., Inouye, Y., and Kawata, S. (2004) Tip-enhanced near-field Raman analysis of tip-pressurized adenine molecule. *Physical Review B*, **96**, 155418.

200 Ichimura, T., Watanabe, H., Morita, Y., Verma, P., Kawata, S., and Inouye, Y. (2007) Temporal fluctuation of tip-enhanced Raman spectra of adenine molecules. *Journal of Physical Chemistry C*, **111**, 9460–9464.

201 Rasmussen, A. and Deckert, V. (2006) Surface- and tip-enhanced Raman scattering of DNA components. *Journal of Raman Spectroscopy*, **37**, 311–317.

202 Domke, K.F., Zhang, D., and Pettinger, B. (2007) Tip-enhanced Raman spectra of picomole quantities of DNA nucleobases at Au(111). *Journal of the American Chemical Society*, **129**, 6708–6709.

203 Bailo, E. and Deckert, V. (2008) Tip-enhanced Raman spectroscopy of single RNA strands: towards a novel direct-sequencing method. *Angewandte Chemie International Edition*, **47**, 1658–1661.

204 Yeo, B.-S., Stadler, J., Schmid, T., Zenobi, R., and Zhang, W. (2009) Tip-enhanced Raman spectroscopy–its status, challenges and future directions. *Chemical Physics Letters*, **472**, 1–13.

205 Deckert-Gaudiga, T. and Deckert, V. (2010) Tip-enhanced Raman scattering (TERS) and high-resolution bio nano-analysis–a comparison. *Physical Chemistry Chemical Physics*, **12**, 12040–12049.

206 Pettinger, B. (2010) Single-molecule surface- and tip-enhanced Raman spectroscopy. *Molecular Physics*, **108**, 2039–2059.

207 Hartschuh, A. (2008) Tip-enhanced near-field optical microscopy. *Angewandte Chemie International Edition*, **47**, 8178–8191.

208 Li, J.-F., Huang, Y.F., Ding, Y., Yang, Z.L., Li, S.B., Zhou, X.S., Fan, F.R., Zhang, W., Zhou, Z.Y., Wu, D.Y., Ren, B., Wang, Z.L., and Tian, Z.Q. (2010) Shell-isolated nanoparticle-enhanced Raman spectroscopy. *Nature*, **464**, 392–394.

209 Alvarez-Puebla, R.A. and Liz-Marzan, L.M. (2010) SERS-based diagnosis and biodetection. *Small*, **6**, 604–610.

210 Schlucker, S. (2009) SERS microscopy: nanoparticle probes and biomedical applications. *ChemPhysChem*, **10**, 1344–1354.

211 Liu, S.Q. and Tang, Z.Y. (2010) Nanoparticle assemblies for biological and chemical sensing. *Journal of Materials Chemistry*, **20**, 24–35.

212 Lee, S.E. and Lee, L.P. (2010) Biomolecular plasmonics for quantitative biology and nanomedicine. *Current Opinion in Biotechnology*, **21**, 489–497.

213 Hu, J.A. and Zhang, C.Y. (2010) Surface-enhanced Raman scattering technology and its application to gene analysis. *Progress in Chemistry*, **22**, 1641–1647.

214 Graham, D. (2010) The next generation of advanced spectroscopy: surface enhanced Raman scattering from metal nanoparticles. *Angewandte Chemie International Edition*, **49**, 9325–9327.

215 Ervin, K.M., Koglin, E., Sequaris, J.M., Valenta, P., and Nürnberg, H.W. (1980) Surface enhanced Raman spectra of nucleic acid components adsorbed at a silver electrode. *Journal of Electroanalytical Chemistry*, **114**, 179–194.

216 Koglin, E., Sequaris, J.M., and Valenta, P. (1980) Surface Raman spectra of nucleic acid components adsorbed at a silver electrode. *Journal of Molecular Structure*, **60**, 421–425.

217 Koglin, E., Séquaris, J.M., and Valenta, P. (1982) Surface enhanced Raman spectroscopy of nucleic acid bases on Ag electrodes. *Journal of Molecular Structure*, **79**, 185–189.

218 Otto, C., van den Tweel, T.J.J., de Mul, F.F.M., and Greve, J. (1986) Surface-enhanced Raman spectroscopy of DNA bases. *Journal of Raman Spectroscopy*, **17**, 289–298.

219 Cui, L., Ren, B., and Tian, Z.Q. (2010) Surface-enhanced Raman spectroscopic study on the co-adsorption of DNA bases with perchlorate. *Acta Physico-Chimica Sinica*, **26**, 397–402.

220 Demers, L.M., Ostblom, M., Zhang, H., Jang, N.H., Liedberg, B., and Mirkin, C.A. (2002) Thermal desorption behavior and binding properties of DNA bases and nucleosides on gold. *Journal of the American Chemical Society*, **124**, 11248–11249.

221 Watanabe, T., Kawanami, O., Katoh, H., Honda, K., Nishimura, Y., and Tsuboi, M. (1985) SERS study of molecular adsorption–some nucleic acid bases on Ag electrodes. *Surface Science*, **158**, 341–351.

222 Giese, B. and McNaughton, D. (2002) Surface-enhanced Raman spectroscopic and density functional theory study of adenine adsorption to silver surfaces. *Journal of Physical Chemistry B*, **106**, 101–112.

223 Kundu, J., Neumann, O., Janesko, B.G., Zhang, D., Lal, S., Barhoumi, A., Scuseria, G.E., and Halas, N.J. (2009) Adenine– and adenosine monophosphate (AMP)–gold binding interactions studied by surface-enhanced Raman and infrared spectroscopies. *Journal of Physical Chemistry C*, **113**, 14390–14397.

224 Giese, B. and McNaughton, D. (2002) Density functional theoretical (DFT) and surface-enhanced Raman spectroscopic study of guanine and its alkylated derivatives: 2. Surface-enhanced Raman scattering on silver surfaces. *Physical Chemistry Chemical Physics*, **4**, 5171–5182.

225 Giese, B. and McNaughton, D. (2002) Surface-enhanced Raman spectroscopic study of uracil. The influence of the surface substrate, surface potential and pH. *Journal of Physical Chemistry B*, **106**, 1461–1470.

226 Sardo, M., Ruano, C., Castro, J.L., Lopez-Tocon, I., Soto, J., Ribeiro-Claro, P., and Otero, J.C. (2009) Surface-enhanced Raman scattering of 5-fluorouracil adsorbed on silver nanostructures. *Physical Chemistry Chemical Physics*, **11**, 7437–7443.

227 Cunha, F., Garcia, J.R., Nart, F.C., Corio, P., and Temperini, M. (2003) Surface enhanced Raman spectroscopy study of the potential dependence of thymine on silver electrodes. *Journal of Solid State Electrochemistry*, **7**, 576–581.

228 Aroca, R., and Bujalski, R. (1999) Surface enhanced vibrational spectra of thymine. *Vibrational Spectroscopy*, **19**, 11–21.

229 Shang, Z.G., Gao, Y., Jia, T.J., and Mo, Y.J. (2009) Vibrational modes study of thymine on the surface of copper electrode using SERS-measurement and the DFT method. *Journal of Molecular Structure*, **930**, 60–64.

230 Hao, Y. and Fang, Y. (2007) Ultraviolet Raman study of thymine on the Au electrode. *Spectrochimica Acta A*, **68**, 778–782.

231 Dou, X.M. and Ozaki, Y. (1999) Surface-enhanced Raman scattering of biological molecules on metal colloids: basic studies and applications to quantitative assay. *Reviews in Analytical Chemistry*, **18**, 285–321.

232 Kneipp, K. and Flemming, J. (1986) Surface enhanced Raman scattering (SERS) of nucleic acids adsorbed on colloidal silver particles. *Journal of Molecular Structure*, **145**, 173–179.

233 Kneipp, K., Pohle, W., and Fabian, H. (1991) Surface enhanced Raman spectroscopy on nucleic acids and related compounds adsorbed on colloidal silver particles. *Journal of Molecular Structure*, **244**, 183–192.

234 Sequaris, J.M., Koglin, E., Valenta, P., and Nurnberg, H.W. (1981) Surface-enhanced Raman-scattering (SERS) spectroscopy of nucleic-acids. *Berichte der Bunsengesellschaft fur Physikalische Chemie*, **85**, 512–513.

235 Valenta, P., Sequaris, J.M., Koglin, E., and Nurnberg, H.W. (1982) Study of adsorption and conformational changes of nucleic acids by voltammetry and by surface-enhanced Raman-scattering. *Journal of the Electrochemical Society*, **129**, C131–C131.

236 Brabec, V. and Niki, K. (1985) Raman scattering from nucleic acids adsorbed at a silver electrode. *Biophysical Chemistry*, **23**, 63–70.

237 Barhoumi, A., Zhang, D., Tam, F., and Halas, N.J. (2008) Surface-enhanced Raman spectroscopy of DNA. *Journal of the American Chemical Society*, **130**, 5523–5529.

238 Koglin, E. and Sequaris, J.M. (1986) Surface enhanced Raman-scattering of biomolecules. *Topics in Current Chemistry*, **134**, 1–57.

239 Fang, Y., Bai, C.L., Wang, T., Zhong, F.P., Tang, Y.Q., Lin, S.B., and Kan, L.S. (1996) Evidence for the conformational rigidity of triples $D(C^{+}T)_8$-$D(AG)_8$-$(CT)_8$ on silver electrode revealed by Fourier transform Raman scattering studies. *Journal of Molecular Structure*, **377**, 1–11.

240 Dong, L.Q., Zhou, J.Z., Wu, L.L., Dong, P., and Lin, Z.H. (2002) SERS studies of self-adsorbed DNA monolayer–characterization of adsorption orientation of oligonucleotide probes and their hybridized helices on gold substrate. *Chemical Physics Letters*, **354**, 458–465.

241 Zhang, R.Y., Pang, D.W., Zhang, Z.L., Yan, J.W., Yao, J.L., Tian, Z.Q., Mao, B.W., and Sun, S.G. (2002) Investigation of ordered ds-DNA monolayers on gold electrodes. *Journal of Physical Chemistry B*, **106**, 11233–11239.

242 Vo-Dinh, T., Houck, K., and Stokes, D.L. (1994) Surface-enhanced Raman gene probes. *Analytical Chemistry*, **66**, 3379–3383.

243 Pal, A., Isola, N.R., Alarie, J.P., Stokes, D.L., and Vo-Dinh, T. (2006) Synthesis and characterization of SERS gene probe for BRCA-1 (breast cancer). *Faraday Discussions*, **132**, 293–301.

244 Strelau, K.K., Kretschmer, R., Moller, R., Fritzsche, W., and Popp, J. (2010) SERS a tool for the analysis of DNA chips in a microfluidic platform. *Analytical and Bioanalytical Chemistry*, **396**, 1381–1384.

245 Huh, Y.S., Lowe, A.J., Strickland, A.D., Batt, C.A., and Erickson, D. (2009) Surface-enhanced Raman scattering based ligase detection reaction. *Journal of the American Chemical Society*, **131**, 2208–2213.

246 Mahajan, S., Richardson, J., Brown, T., and Bartlett, P.N. (2008) SERS-melting: a new method for discriminating mutations in DNA sequences. *Journal of the American Chemical Society*, **130**, 15589–15601.

247 Fang, C., Agarwal, A., Buddharaju, K.D., Khalid, N.M., Salim, S.M., Widjaja, E., Garland, M.V., Balasubramanian, N., and Kwong, D.L. (2008) DNA detection using nanostructured SERS substrates with rhodamine B as Raman label. *Biosensors & Bioelectronics*, **24**, 216–221.

248 Kowalczyk, A., Nowicka, A.M., Jurczakowski, R., Niedzialkowski, P., Ossowski, T., and Stojek, Z. (2010) New anthraquinone derivatives as electrochemical redox indicators for the visualization of the DNA hybridization process. *Electroanalysis*, **22**, 49–59.

249 Barhoumi, A. and Halas, N.J. (2010) Label-free detection of DNA hybridization using surface enhanced Raman spectroscopy. *Journal of the American Chemical Society*, **132**, 12792–12793.

250 Sun, L. and Irudayaraj, J. (2009) Quantitative surface-enhanced Raman for gene expression estimation. *Biophysical Journal*, **96**, 4709–4716.

251 Nabiev, I.R., Trakhanov, S.D., Efremov, E.S., Marinyuk, V.V., and Lasorenkomanevich, R.M. (1981) Surface enhanced Raman spectra of some biological molecules adsorbed at silver electrodes. *Bioorganic Khimica*, **7**, 941–945.

252 Stewart, S. and Fredericks, P.M. (1999) Surface-enhanced Raman spectroscopy of amino acids adsorbed on an electrochemically prepared silver surface. *Spectrochimica Acta A*, **55**, 1641–1660.

253 Watanabe, T. and Maeda, H. (1989) Adsorption-controlled redox activity. Surface-enhanced Raman investigation of cystine versus cysteine on silver electrodes. *Journal of Physical Chemistry*, **93**, 3258–3260.

254 Xiao, X.Y., Sun, S.G., Yao, J.L., Wu, Q.H., and Tian, Z.Q. (2002) Surface-enhanced Raman spectroscopic studies of dissociative adsorption of amino acids on platinum and gold electrodes in alkaline solutions. *Langmuir*, **18**, 6274–6279.

255 Martusevicius, S., Niaura, G., Talaikyté, Z., and Razumas, V. (1996) Adsorption of histidine on copper surface as evidenced by surface-enhanced Raman scattering spectroscopy. *Vibrational Spectroscopy*, **10**, 271–280.

256 Stewart, S. and Fredericks, P.M. (1999) Surface-enhanced Raman spectroscopy of peptides and proteins adsorbed on an electrochemically prepared silver surface. *Spectrochimica Acta A*, **55**, 1615–1640.

257 Wei, F., Zhang, D.M., Halas, N.J., and Hartgerink, J.D. (2008) Aromatic amino acids providing characteristic motifs in the Raman and SERS spectroscopy of peptides. *Journal of Physical Chemistry B*, **112**, 9158–9164.

258 Bartlett, P.N. (2008) Bioenergetics and biological electron transport, in *Bioelectrochemistry. Fundamentals, Experimental Techniques Ans Applications* (ed. P.N. Bartlett), John Wiley & Sons, Ltd, Chichester, pp. 1–38.

259 Niaura, G., Gaigalas, A.K., and Vilker, V.L. (1996) Non-resonant SERS study of

the adsorption of cytochrome *c* on a silver electrode. *Journal of Electroanalytical Chemistry*, **416**, 167–178.

260 Spiro, T.G. (1975) Resonance Raman spectroscopic studies of heme proteins. *Biochimica et Biophysica Acta*, **416**, 169–189.

261 Spiro, T.G. and Strekas, T.C. (1974) Resonance Raman spectra of heme proteins. Effects of oxidation and spin state. *Journal of the American Chemical Society*, **96**, 338–345.

262 Hu, S., Morris, I.K., Singh, J.P., Smith, K.M., and Spiro, T.G. (1993) Complete assignment of cytochrome *c* resonance Raman spectra via enzymic reconstitution with isotopically labeled hemes. *Journal of the American Chemical Society*, **115**, 12446–12458.

263 Murgida, D.H. and Hildebrandt, P. (2001) Proton-coupled electron transfer of cytochrome *c*. *Journal of the American Chemical Society*, **123**, 4062–4068.

264 Murgida, D.H. and Hildebrandt, P. (2001) Heterogeneous electron transfer of cytochrome *c* on coated silver electrodes. Electric field effects on structure and redox potential. *Journal of Physical Chemistry B*, **105**, 1578–1586.

265 Oellerich, S., Wackerbarth, H., and Hildebrandt, P. (2002) Spectroscopic characterization of nonnative conformational states of cytochrome *c*. *Journal of Physical Chemistry B*, **106**, 6566–6580.

266 Cotton, T.M., Schultz, S.G., and Van Duyne, R.P. (1980) Surface-enhanced resonance Raman scattering from cytochrome-*c* and myoglobin adsorbed on a silver electrode. *Journal of the American Chemical Society*, **102**, 7960–7962.

267 Hildebrandt, P. and Stockburger, M. (1986) Surface-enhanced resonance Raman spectroscopy of cytochrome *c* at room and low temperatures. *Journal of Physical Chemistry*, **90**, 6017–6024.

268 Oellerich, S., Lecomte, S., Paternostre, M., Heimburg, T., and Hildebrandt, P. (2004) Peripheral and integral binding of cytochrome *c* to phospholipids vesicles. *Journal of Physical Chemistry B*, **108**, 3871–3878.

269 Hildebrandt, P. and Stockburger, M. (1989) Cytochrome *c* at charged interfaces. 1. Conformational and redox equilibria at the electrode/electrolyte interface probed by surface-enhanced resonance Raman spectroscopy. *Biochemistry*, **28**, 6710–6721.

270 Hildebrandt, P. and Stockburger, M. (1989) Cytochrome *c* at charged interfaces. 2. Complexes with negatively charged macromolecular systems studied by resonance Raman spectroscopy. *Biochemistry*, **28**, 6722–6728.

271 Döpner, S., Hildebrandt, P., Rosell, F.I., Mauk, A.G., Walter, M.V., Buse, G., and Soulimane, T. (1999) The structural and functional role of lysine residues in the binding domain of cytochrome *c* in the electron transfer to cytochrome *c* oxidase. *European Journal of Biochemistry*, **261**, 379–391.

272 Hildebrandt, P. (1991) Resonance Raman-spectroscopic studies of cytochrome-*c* at charged interfaces. *Journal of Molecular Structure*, **242**, 379–395.

273 Lecomte, S., Wackerbarth, H., Soulimane, T., Buse, G., and Hildebrandt, P. (1998) Time-resolved surface-enhanced resonance Raman spectroscopy for studying electron-transfer dynamics of heme proteins. *Journal of the American Chemical Society*, **120**, 7381–7382.

274 Wackerbarth, H., Klar, U., Gunther, W., and Hildebrandt, P. (1999) Novel time-resolved surface-enhanced (resonance) Raman spectroscopic technique for studying the dynamics of interfacial processes: application to the electron transfer reaction of cytochrome *c* at a silver electrode. *Applied Spectroscopy*, **53**, 283–291.

275 Naumann, H., Klare, J.P., Engelhard, M., Hildebrandt, P., and Murgida, D.H. (2006) Time-resolved methods in biophysics. 1. A novel pump and probe surface-enhanced resonance Raman approach for studying biological photoreceptors. *Photochemical & Photobiological Sciences*, **5**, 1103–1108.

276 Grosserueschkamp, M., Friedrich, M.G., Plum, M., Knoll, W., and Naumann,

R.L.C. (2009) Electron transfer kinetics of cytochrome *c* probed by time-resolved surface-enhanced resonance Raman spectroscopy. *Journal of Physical Chemistry B*, **113**, 2492–2497.

277 Kranich, A., Ly, H.K., Hildebrandt, P., and Murgida, D.H. (2008) Direct observation of the gating step in protein electron transfer: electric-field-controlled protein dynamics. *Journal of the American Chemical Society*, **130**, 9844–9848.

278 Murgida, D.H. and Hildebrandt, P. (2004) Electron-transfer processes of cytochrome *c* at interfaces. New insights by surface-enhanced resonance Raman spectroscopy. *Accounts of Chemical Research*, **37**, 854–861.

279 Murgida, D.H. and Hildebrandt, P. (2005) Redox and redox-coupled processes of heme proteins and enzymes at electrochemical interfaces. *Physical Chemistry Chemical Physics*, **7**, 3773–3784.

280 Bernad, S., Leygue, N., Korri-Youssoufi, H., and Lecomte, S. (2007) Kinetics of the electron transfer reaction of cytochrome *c*(552) adsorbed on biomimetic electrode studied by time-resolved surface-enhanced resonance Raman spectroscopy and electrochemistry. *European Biophysics Journal*, **36**, 1039–1048.

281 Feng, J.J., Murgida, D.H., Kuhlmann, U., Utesch, T., Mroginski, M.A., Hildebrandt, P., and Weidinger, I.M. (2008) Gated electron transfer of yeast iso-1 cytochrome *c* on self-assembled monolayer-coated electrodes. *Journal of Physical Chemistry B*, **112**, 15202–15211.

282 Kranich, A., Naumann, H., Molina-Heredia, F.P., Moore, H.J., Lee, T.R., Lecomte, S., de la Rosa, M.A., Hildebrandt, P., and Murgida, D.H. (2009) Gated electron transfer of cytochrome *c*(6) at biomimetic interfaces: a time-resolved SERR study. *Physical Chemistry Chemical Physics*, **11**, 7390–7397.

283 de Sousa, J.R., Batista, A.A., Diogenes, I.C.N., Andrade, G.F.S., Temperini, M.L.A., Lopes, L.G.F., and de Sousa Moreira, I. (2003) Characterization of a 1,4-dithiane gold self-assembled monolayer: an electrochemical sensor for the cyt-*c* redox process. *Journal of Electroanalytical Chemistry*, **543**, 93–99.

284 Sezer, M., Feng, J.J., Ly, H.K., Shen, Y.F., Nakanishi, T., Kuhlmann, U., Hildebrandt, P., Mohwald, H., and Weidinger, I.M. (2010) Multi-layer electron transfer across nanostructured Ag-SAM-Au-SAM junctions probed by surface enhanced Raman spectroscopy. *Physical Chemistry Chemical Physics*, **12**, 9822–9829.

285 Jiang, X., Wang, Y.L., Qu, X.H., and Dong, S.J. (2006) Surface-enhanced resonance Raman spectroscopy and spectroscopy study of redox-induced conformational equilibrium of cytochrome *c* adsorbed on DNA-modified metal electrode. *Biosensors & Bioelectronics*, **22**, 49–55.

286 Jiang, X., Zhang, Z., Bai, H., Qu, X., Jiang, J., Wang, E., and Dong, S. (2005) Effect of electrode surface microstructure on electron transfer induced conformation changes in cytochrome monitored by *in situ* UV and CD spectroelectrochemistry. *Spectrochimica Acta A*, **61**, 943–951.

287 Reipa, V., Gaigalas, A.K., Edwards, J.J., and Vilker, V.L. (1995) Surface-enhanced Raman spectroscopy (SERS) evidence of charge transfer between putidaredoxin and a silver electrode. *Journal of Electroanalytical Chemistry*, **395**, 299–303.

288 Gaigalas, A.K. and Niaura, G. (1997) Measurement of electron transfer rates between adsorbed azurin and a gold electrode modified with a hexanethiol layer. *Journal of Colloid and Interface Science*, **193**, 60–70.

289 Grabbe, E.S. and Buck, R.P. (1989) Surface enhanced Raman spectroscopic investigation of human immunoglobulin G adsorbed on a silver electrode. *Journal of the American Chemical Society*, **111**, 8362–8366.

290 Chumanov, G.D., Efremov, R.G., and Nabiev, I.R. (1990) Surface-enhanced Raman spectroscopy of biomolecules: I. Water-soluble proteins, dipeptides and amino acids. *Journal of Raman Spectroscopy*, **21**, 43–48.

291 Hu, J., Sheng, R.S., Xu, Z.S., and Zeng, Y. (1995) Surface enhanced Raman spectroscopy of lysozyme. *Spectrochimica Acta A*, **51**, 1087–1096.

292 Reipa, V., Gaigalas, A., and Abramowitz, S. (1993) Conformational alterations of bovine insulin adsorbed on a silver electrode. *Journal of Electroanalytical Chemistry*, **348**, 413–428.

293 Drachev, V.P., Thoreson, M.D., Khaliullin, E.N., Davisson, V.J., and Shalaev, V.M. (2004) Surface-enhanced Raman difference between human insulin and insulin lispro detected with adaptive nanostructures. *Journal of Physical Chemistry B*, **108**, 18046–18052.

294 Holt, R.E. and Cotton, T.M. (1989) Surface-enhanced resonance Raman and electrochemical investigation of glucose oxidase catalysis at a silver electrode. *Journal of the American Chemical Society*, **111**, 2815–2821.

295 Holt, R.E. and Cotton, T.M. (1987) Free flavin interference in surface enhanced resonance Raman-spectroscopy of glucose oxidase. *Journal of the American Chemical Society*, **109**, 1841–1845.

296 Razumas, V.J., Vidugiris, G.J.A., and Kulys, J.J. (1987) Giant Raman-scattering of alkaline-phosphatase, horseradish-peroxidase and lactoperoxidase on silver electrodes. *Biofizika*, **32**, 967–971.

297 Vidugiris, G.J.A., Gudavicius, A.V., Razumas, V.J., and Kulys, J.J. (1989) Structure–potential dependence of adsorbed enzymes and amino acids revealed by the surface enhanced Raman effect. *European Biophysis Journal*, **17**, 19–23.

298 Grabbe, E.S. and Buck, R.P. (1991) Evidence for a conformational change with potential for adsorbed anti-IGG alkaline-phosphatase conjugate at the silver electrode interface using SERS. *Journal of Electroanalytical Chemistry*, **308**, 227–237.

299 Wang, F., Zheng, J., Li, X., Ji, Y., Gao, Y., Xing, W., and Lu, T. (2003) Surface-enhanced Raman spectroscopy of microperoxidase-11 on roughed silver electrodes. *Journal of Electroanalytical Chemistry*, **545**, 123–128.

300 Lecomte, S., Ricoux, R., Mahy, J.P., and Korri-Youssoufi, H. (2004) Microperoxidase 8 adsorbed on a roughened silver electrode as a monomeric high-spin penta-coordinated species: characterization by SERR spectroscopy and electrochemistry. *Journal of Biological Inorganic Chemistry*, **9**, 850–858.

301 Bell, S.E.J., Devenney, M.D., Grimshaw, J., Hara, S., Rice, J.H., and Trocha-Grimshaw, J. (1998) Resonance Raman and surface-enhanced resonance Raman studies of polymer-modified electrodes which mimic heme enzymes. *Journal of the Chemical Society, Faraday Transactions*, **94**, 2955–2960.

302 Friedrich, M.G., Giess, F., Naumann, R., Knoll, W., Ataka, K., Heberle, J., Hrabakova, J., Murgida, D.H., and Hildebrandt, P. (2004) Active site structure and redox processes of cytochrome *c* oxidase immobilised in a novel biomimetic lipid membrane on an electrode. *Chemical Communications*, 2376–2377.

303 Friedrich, M.G., Robertson, J.W.F., Walz, D., Knoll, W., and Naumann, R.L.C. (2008) Electronic wiring of a multi-redox site membrane protein in a biomimetic surface architecture. *Biophysical Journal*, **94**, 3698–3705.

304 Hrabakova, J., Ataka, K., Heberle, J., Hildebrandt, P., and Murgida, D.H. (2006) Long distance electron transfer in cytochrome *c* oxidase immobilised on electrodes. A surface enhanced resonance Raman spectroscopic study. *Physical Chemistry Chemical Physics*, **8**, 759–766.

305 Kudelski, A. (2008) *In situ* SERS studies on the adsorption of tyrosinase on bare and alkanethiol-modified silver substrates. *Vibrational Spectroscopy*, **46**, 34–38.

306 Kudelski, A. (2006) Raman studies on the coverage integrity of monolayers formed on silver from various ω-functionalised alkanethiols. *Vibrational Spectroscopy*, **41**, 83–89.

307 Sezer, M., Spricigo, R., Utesch, T., Millo, D., Leimkuehler, S., Mroginski, M.A., Wollenberger, U., Hildebrandt, P., and Weidinger, I.M. (2010) Redox properties and catalytic activity of surface-bound human sulfite oxidase studied by a combined surface enhanced

resonance Raman spectroscopic and electrochemical approach. *Physical Chemistry Chemical Physics*, **12**, 7894–7903.

308 Guo, F.C., Chou, Y.C., and Wu, W.G. (1990) The surface enhanced Raman spectroscopic study of the ordering of lipid molecules deposited on the silver electrode. *Chinese Journal of Physics*, **28**, 173–180.

309 Guo, F.C., Chou, Y.C., Huang, W.N., and Wu, W.G. (1992) SERS study of the conformational transition of the ordering of lipid molecules deposited on a silver electrode. *Journal of Raman Spectroscopy*, **23**, 425–430.

310 Krysinski, P., Zebrowska, A., Michota, A., Bukowska, J., Becucci, L., and Moncelli, M.R. (2001) Tethered mono- and bilayer lipid membranes on Au and Hg. *Langmuir*, **17**, 3852–3857.

311 Krysinski, P., Zebrowska, A., Palys, B., and Lotowski, Z. (2002) Spectroscopic and electrochemical studies of bilayer lipid membranes tethered to the surface of gold. *Journal of the Electrochemical Society*, **149**, E189–E194.

312 Leverette, C.L. and Dluhy, R.A. (2004) Vibrational characterization of a planar-supported model bilayer system utilizing surface-enhanced Raman scattering (SERS) and infrared reflection-absorption spectroscopy (IRRAS). *Colloids and Surfaces A*, **243**, 157–167.

313 Leverette, C.L. and Dluhy, R.A. (2000) A novel fiber-optic interface for unenhanced external reflection Raman spectroscopy of supported monolayers. *Langmuir*, **16**, 3977–3983.

314 Friedrich, M.G., Plum, M.A., Santonicola, M.G., Kirste, V.U., Knoll, W., Ludwig, B., and Naumann, R.L.C. (2008) *In situ* monitoring of the catalytic activity of cytochrome *c* oxidase in a biomimetic architecture. *Biophysical Journal*, **95**, 1500–1510.

315 Heywang, C., Chazalet, M.S.P., Masson, M., and Bolard, J. (1998) Orientation of anthracyclines in lipid monolayers and planar asymmetrical bilayers: a surface-enhanced resonance Raman scattering study. *Biophysical Journal*, **75**, 2368–2381.

316 Levin, C.S., Kundu, J., Janesko, B.G., Scuseria, G.E., Raphael, R.M., and Halas, N.J. (2008) Interactions of ibuprofen with hybrid lipid bilayers probed by complementary surface-enhanced vibrational spectroscopies. *Journal of Physical Chemistry B*, **112**, 14168–14175.

317 Millone, M.A.D., Vela, M.E., Salvarezza, R.C., Creczynski-Pasa, T.B., Tognalli, N.G., and Fainstein, A. (2009) Phospholipid bilayers supported on thiolate-covered nanostructured gold: *in situ* Raman spectroscopy and electrochemistry of redox species. *ChemPhysChem*, **10**, 1927–1933.

318 Dijkstra, R.J., Scheenen, W.J.J.M., Dam, N., Roubos, E.W., and ter Meulen, J.J. (2007) Monitoring neurotransmitter release using surface-enhanced Raman spectroscopy. *Journal of Neuroscience Methods*, **159**, 43–50.

319 Lee, N.S., Hsieh, Y.Z., Paisley, R.F., and Morris, M.D. (1988) Surface-enhanced Raman spectroscopy of the catecholamine neurotransmitters and related compounds. *Analytical Chemistry*, **60**, 442–446.

320 McGlashen, M.L., Davis, K.L., and Morris, M.D. (1990) Surface-enhanced Raman scattering of dopamine at polymer-coated silver electrodes. *Analytical Chemistry*, **62**, 846–849.

321 Davis, K.L., McGlashen, M.L., and Morris, M.D. (1992) Surface-enhanced Raman scattering of histamine at silver electrodes. *Langmuir*, **8**, 1654–1658.

322 Taniguchi, I., Umekita, K., and Yasukouchi, K. (1986) Surface-enhanced Raman scattering of nicotinamide adenine dinucleotide (NAD^+) adsorbed on silver and gold electrodes. *Journal of Electroanalytical Chemistry*, **202**, 315–322.

323 Xiao, Y.-J. and Markwell, J.P. (1997) Potential dependence of the conformations of nicotinamide adenine dinuceletide on gold electrode determined by FT-near-IR-SERS. *Langmuir*, **13**, 7068–7074.

324 Xiao, Y.-J., Chen, Y.-F., and Gao, X.-X. (1999) Comparative study of the surface enhanced near infrared Raman spectra of adenine and NAD^+ on a gold electrode. *Spectrochimica Acta A*, **55**, 1209–1218.

325 Yang, H.F., Yang, Y., Liu, Z.M., Zhang, Z.R., Shen, G.L., and Yu, R.Q. (2004) Self-assembled monolayer of NAD at silver surface: a Raman mapping study. *Surface Science*, **551**, 1–8.

326 Xu, J., Birke, R.L., and Lombardi, J.R. (1987) Surface-enhanced Raman-spectroscopy from flavins adsorbed on a silver electrode–observation of the unstable semiquinone intermediate. *Journal of the American Chemical Society*, **109**, 5645–5649.

327 Hsieh, Y.Z., Lee, N.S., Sheng, R.S., and Morris, M.D. (1987) Surface-enhanced Raman spectroscopy of free and complexed bilirubin. *Langmuir*, **3**, 1141–1146.

328 Wu, Z.S., Zhou, G.Z., Jiang, J.H., Shen, G.L., and Yu, R.Q. (2006) Gold colloid–bienzyme conjugates for glucose detection utilizing surface-enhanced Raman scattering. *Talanta*, **70**, 533–539.

7
Membrane Electroporation in High Electric Fields

Rumiana Dimova

7.1
Introduction

The autonomy of a cell, the basic building unit of most living creatures, is ensured by a bounding membrane. The scaffold of this membrane is made of a double lipid layer, which is basically impermeable to all substances in the cellular environment except for water. The mechanical and rheological properties of the bilayer define the response of the membrane to external perturbations. The membrane "intolerance" towards letting solute molecules easily cross it creates the main obstacle in biomedical applications where drugs or genes have to be introduced into cells. One approach finding broad use nowadays in overcoming the barrier functions of membranes relies on the temporary bilayer perforation when exposed to strong electric fields. This phenomenon, called electroporation, is the main focus of this chapter.

The main topics to be covered build upon our knowledge of the mechanical and rheological properties of membranes and their response to perturbations. In the remaining parts of this introductory section, some of these properties are briefly described to set the basis for a discussion of the behavior and response of membranes to external forces. The following sections consider in detail the morphological changes and poration electric fields can induce in vesicles made of membranes in different phases, and the effects of media environment and various molecular inclusions in the lipid bilayer, that is, the specific membrane composition. Finally, some application aspects of the work are discussed.

7.1.1
Giant Vesicles as Model Membrane Systems

The field of membrane structure and characterization is attracting the attention of a growing number of researchers. The basic research in this area builds upon studies performed on the simplest and minimal systems mimicking cell membranes, namely model membranes. Examples of such model membranes are

Advances in Electrochemical Science and Engineering. Edited by Richard C. Alkire, Dieter M. Kolb, and Jacek Lipkowski

ISBN: 978-3-527-32885-7

lipid monolayers at the air–water interface, solid-supported bilayers, black lipid membranes, vesicles, and bilayer stacks. Among them, vesicles or liposomes are membrane "bubbles" formed by bending and closing up of a lipid bilayer. They are the most natural systems, because, in terms of shape and structure, they are closest to membranes of cells and cell organelles.

Various experimental techniques have been developed for preparing liposomes of different sizes (from nanometers to tens of micrometers) [1]. The largest ones, several tens of micrometers in size, are called "giant vesicles" [2] and are an extraordinarily convenient system for studying membrane behavior [3, 4]. They are well visible under an optical microscope using various enhancing techniques like phase contrast, differential interference contrast, or confocal and conventional fluorescence microscopy, the latter two being particularly useful in distinguishing domains on membranes (Figure 7.1). Thus, giant vesicles allow for direct manipulation and observation of membrane interactions and responses to external perturbations. On the contrary, working with conventional vesicles (a few hundreds of nanometers) usually involves the application of indirect methods and techniques for observation. In addition, their small sizes often raise questions about effects due to high membrane curvature when molecular interactions are considered. In contrast, giant vesicles, which have sizes in the micrometer range (i.e., comparable to the sizes of cells), and therefore have nearly zero membrane curvature, reflect the properties and behavior of cell plasma membranes.

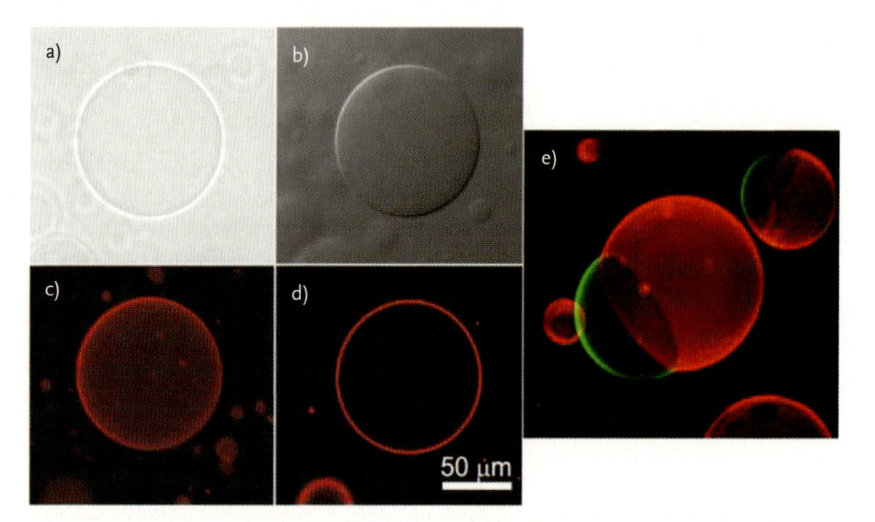

Figure 7.1 Snapshots of the same giant vesicles observed under different microscopy modes: (a) phase contrast; (b) differential interference contrast; (c) projection averaged confocal microscopy; (d) equatorial section confocal microscopy. Adapted from [3] by permission of IOP Publishing Ltd. (e) Confocal three-dimensional projection image of vesicles with immiscible fluid domains visualized with fluorescent dyes, which preferentially partition in one or the other lipid phase.

The two most popular techniques for the formation of giant vesicles (other available methods are summarized in [4]) are spontaneous swelling or gentle hydration, introduced by Reeves and Dowben [5] and electroswelling, introduced by Angelova and Dimitrov [6]; for a brief description of these two preparation protocols, see [3]. These two methods were further developed and improved by several groups (e.g., [4, 7–10]). Interestingly enough, the underlying mechanism of the electroformation protocol, which is based on exposing lipid layers to alternating electric fields, is still poorly understood even though widely used. Both protocols yield giant vesicles with sizes in the range of a few tens of micrometers.

Giant unilamellar vesicles (GUVs) are increasingly employed for quantitative characterization of the physicochemical properties of lipid membranes with various compositions, but also to study membrane-related processes like cell adhesion, phase separation and domain formation, protein sorting in lipid rafts, endo- and exocytosis, uptake of various molecules, and protein mobility, to mention just a few of the studied fields. Examples of using GUVs as simple model systems for unraveling certain physicochemical properties of biological membranes include lipid domain formation [11–13], mechanical and rheological properties of the entire vesicle [14] or of the membrane [3, 15, 16], lipid dynamics, membrane growth [17–20], membrane adhesion [21–25], wetting phenomena [26, 27], budding and fission [28–32], and membrane fusion [33–37]. Giant vesicles are also a very practical tool to study the response of membranes to external perturbations like hydrodynamic flows [38], locally applied forces [39], micromanipulation [40], and electric fields [41, 42]. This chapter focuses on the effects of strong electric pulses on model membranes as exhibited by the behavior of GUVs exposed to such pulses. The vesicle response will be interpreted in view of general concepts of membrane biophysics.

7.1.2 Mechanical and Rheological Properties of Lipid Bilayers

The physical properties of lipid bilayers are those that define their response to external perturbations. Knowing the mechanical and rheological characteristics of lipid membranes will prepare us to tackle problems related to stress induced in bilayers by electric fields and the phenomena that it triggers, for example, dynamics of vesicle and cell deformation, bilayer instability, electroporation, and electrofusion.

As a simple depiction of a lipid bilayer, one can consider it as a film or a slab, which may be curved, compressed or dilated, and sheared. At physiological temperatures most natural lipid membranes are fluid. Therefore, within this slab, the lipid molecules are free to move. Below the lipid phase transition temperature, single-component membranes crystallize. In this so called "gel" phase, the relative motion of lipids and membrane inclusions is principally hindered. The fluidity of the membrane and resistance to shear in the plane of the film are characterized by the shear viscosity, η_S (or the diffusion coefficient of the lipids). Typical values of η_S lie in the range 1×10^{-9}–$5 \times 10^{-9}\,\mathrm{N\,s\,m^{-1}}$ [16] for fluid membranes, but for gel-phase membranes divergence is observed [43]. One may equivalently define a viscosity η_D related to the dilation and compression of the membrane. The value of η_D is of the order of $3.5 \times 10^{-7}\,\mathrm{N\,s\,m^{-1}}$ [44].

Phospholipid membranes in the fluid phase are very soft: the energy required for their bending is comparable to the thermal energy. The bilayer bending rigidity, κ, which characterizes how easy it is to curve the lipid bilayer, is typically of the order of $0.9 \times 10^{-19}\,\mathrm{J}$ [45–47], which is equivalent to $20k_BT$, where k_B is the Boltzmann constant and T is the absolute temperature. Thus, fluid membranes fluctuate due to thermal noise. These fluctuations can be directly observed on tensionless giant vesicles under the microscope. They are the basis of the so-called fluctuation spectroscopy method used to measure membrane bending rigidity [48–55]. For gel-phase membranes, the bending rigidity increases significantly, and a few degrees below the main phase transition temperature it reaches values of the order of $(15–20) \times 10^{-19}\,\mathrm{J}$ (about $350k_BT$) [43, 56, 57].

Weak tensions applied to a fluid membrane smooth out bilayer undulations. At high tensions the membrane can be stretched leading to a change in the area per lipid molecule. The stretching elasticity modulus, K_a, characterizing this response is of the order of that of a rubber sheet with the same thickness (about 4 nm). Typical values of K_a for fluid membranes lie in the range $200–300\,\mathrm{mN\,m^{-1}}$ [47] and for gel-phase membranes can reach values of $850\,\mathrm{mN\,m^{-1}}$ [58]. Upon stretching, a lipid bilayer can sustain tensions up to several $\mathrm{mN\,m^{-1}}$. At a certain critical tension, also known as the lysis tension, σ_{lys}, the membrane ruptures. For fluid membranes, σ_{lys} is of the order of $5–10\,\mathrm{mN\,m^{-1}}$ [59, 60]. Note that the membrane tensile strength depends on the tension loading rate [61]. Membranes in the gel phase can sustain higher tensions and rupture at higher values of σ_{lys} [62].

After rupture or poration, the rearrangement of the lipids to close the bilayer sheet is energetically favorable because in this way the hydrophobic tails of the lipid molecules are shielded from exposure to water. The energy penalty of closing a hole in the membrane is described by the edge tension, γ, which is of the order of several piconewtons [63]. The edge tension plays a strong role in processes of pore stability and resealing as in electroporation, which is discussed in more detail in Section 7.5.1.

7.2
Electrodeformation and Electroporation of Membranes in the Fluid Phase

Vesicles exposed to electric fields deform. The response of GUVs to electric fields has been the subject of extensive investigation. When exposed to AC electric fields, as a stationary state they attain ellipsoidal shapes – prolate or oblate – depending on the field frequency and media conductivity [42, 64]. Initiated by the seminal work of Winterhalter and Helfrich [65], this effect has been considered theoretically [66–74] and experimentally [42, 64, 67, 75–77], whereby interesting dynamics and flows in the membrane and in the surrounding medium were observed

[78, 79]. This chapter mainly discusses the response of giant vesicles to short DC rectangular pulses with duration in the range 100 μs–5 ms. Parallels to the vesicle behavior in AC fields will be also drawn. In this section, only vesicles in the fluid phase will be considered. In Section 7.3, we will discuss the response of vesicles in the gel phase and compare it to that of fluid vesicles.

While vesicle deformation in AC fields concerns stationary shapes, DC pulses induce short-lived shape deformations. In different studies, the pulse duration has been typically varied from several microseconds to milliseconds, while studies on cells have investigated a much wider range of pulse durations – from tens of nanoseconds to milliseconds and even seconds [80], as discussed in other chapters of this book. Various pulse profiles, unipolar or bipolar, as well as trains of pulses have been also employed (e.g., [81, 82]). Because the application of both AC fields and DC pulses creates a transmembrane potential, vesicle deformations of similar nature are to be expected in both cases. However, the working field strength for DC pulses is usually higher by several orders of magnitude. Thus, the degree of deformation can be different.

Vesicle deformation induced by DC pulses has been studied theoretically [83–85]. The majority of experimental studies were preformed on small vesicles of hundreds of nanometers in size [86–88], but their size did not allow for direct observation of the deformation dynamics. Employing fast digital imaging, recently we succeeded in revealing vesicle deformation [89], whereby the vesicle response was recorded with high temporal resolution of up to 30 000 frames per second (fps), that is, acquiring an image every 33 μs. The GUVs were observed to deform into prolate ellipsoids during the pulse and subsequently relax back to their initial spherical shape.

The degree of deformation of an ellipsoidal vesicle can be characterized by the aspect ratio of the two principal radii, a and b (see the inset of Figure 7.2). For $a/b = 1$ the vesicle is a sphere. In the absence of poration, that is, relatively weak or short pulses, the relaxation can be described by a single exponential with a characteristic decay time, τ_1. Figure 7.2 gives one example of the response of a giant vesicle, which is initially spherical. The maximum deformation of this vesicle under the applied pulse conditions corresponds to about 10% change in the vesicle aspect ratio. The degree of vesicle deformation depends on the initial tension of the vesicle as well as on the excess area [90], the latter being defined as an excess compared to the area of a spherical vesicle with identical volume.

The typical decay time for the relaxation of nonporated vesicles, τ_1, is of the order of 100 μs. It is set by the relaxation of the membrane tension achieved at the end of the pulse. The membrane tension, σ_{el}, acquired during the pulse, also referred to as "electric tension," arises from the transmembrane potential, Ψ_m, built across the membrane during the pulse. Lipid membranes are impermeable to ions and, in the presence of an electric field, charges accumulate on both sides of the bilayer, which gives rise to this transmembrane potential [91]:

$$\Psi_m(t) = 1.5R|\cos\theta|E\left[1 - \exp\left(\frac{-t}{t_c}\right)\right] \tag{7.1}$$

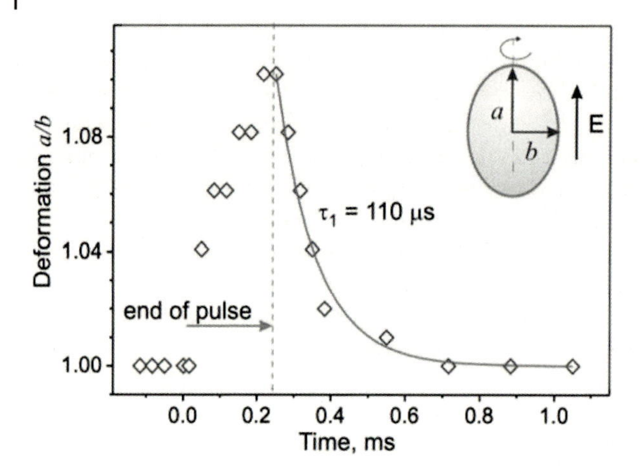

Figure 7.2 Time dependence of the degree of deformation, a/b, of a vesicle exposed to a square-wave DC pulse with field strength $E = 100\,\mathrm{kV\,m^{-1}}$ and pulse duration $t_p = 250\,\mu\mathrm{s}$. The solid curve is an exponential fit with a decay time τ_1 as indicated. The inset schematically illustrates the shape of the deformed vesicle and the principal radii. The dashed line indicates the end of the pulse. Adapted from [89] by permission of the Biophysical Society.

where R is the radius of a spherical vesicle, θ is the tilt angle between the electric field and the surface normal, t is time, and t_c is the charging time as defined by

$$t_c = RC_m\left(\frac{1}{\lambda_{in}} + \frac{1}{\lambda_{ex}}\right) \tag{7.2}$$

Here, λ_{in} and λ_{ex} are the conductivities of the solutions inside and outside the vesicle, respectively. Equations (7.1) and (7.2) are valid only for a nonconductive membrane.

The effective electrical tension, σ_{el}, induced by the transmembrane potential, Ψ_m, is defined by the Maxwell stress tensor [59, 89, 92]

$$\sigma_{el} = \varepsilon_m\left(\frac{h}{2h_e^2}\right)\Psi_m^2 \tag{7.3}$$

Here h is the total bilayer thickness ($\approx 4\,\mathrm{nm}$), h_e is the dielectric thickness ($\approx 2.8\,\mathrm{nm}$ for lecithin bilayers [93, 94]), and ε_m is the membrane permittivity ($\approx 2\varepsilon_0$, where ε_0 is the vacuum permittivity). For vesicles with some initial tension σ_0, the total tension reached during the pulse is simply

$$\sigma = \sigma_0 + \sigma_{el} \tag{7.4}$$

As mentioned above, the decay time for the relaxation of nonporated vesicles, τ_1, is defined by the total membrane tension at the end of the pulse. The tension relaxation requires relative displacement of the lipid molecules in the bilayer, whose intermolecular distance is increased during the pulse, that is, the membrane is stretched. This relative displacement is characterized by the dilational viscosity η_D as introduced in Section 7.1.2. Thus, τ_1 relates mainly to the relaxation of membrane stretching: $\tau_1 \sim \eta_D/\sigma$. For membrane tensions of the order of $5\,mN\,m^{-1}$ (which should be around the maximum tension before the membrane ruptures, $\sigma \approx \sigma_{lys}$) and for typical values of η_D, one obtains $\tau_1 \sim 100\,\mu s$, which corresponds to the experimentally measured value (see Figure 7.2).

Above some electroporation threshold, the transmembrane potential Ψ_m cannot be further increased, and can even decrease due to transport of ions across the membrane [91, 95]. The phenomenon of membrane electroporation can also be understood in terms of tension. If the total membrane tension exceeds the lysis tension σ_{lys}, the vesicle ruptures. This corresponds to building up a certain critical transmembrane potential, $\Psi_m = \Psi_c$. According to Eqs. (7.3) and (7.4), this poration potential Ψ_c depends on the initial membrane tension σ_0 as previously reported [59, 89, 90, 96, 97]. The critical transmembrane potential for cell membranes is about 1 V (e.g., [98, 99]).

Vesicle poration induced by DC pulses has been studied extensively, initially mainly on small vesicles [88, 100, 101]. Experiments on giant vesicles are of special relevance because their size allows for direct observation using optical microscopy [37, 89, 102–107]. The pores can reach various sizes depending on the location on the vesicle or cell surface (e.g., [99, 108] and work cited therein). For the case of plane parallel electrodes, the poration occurs predominantly in the area at the poles of the vesicle facing the electrodes. This is because the transmembrane potential attains its maximal value at the two poles as expressed by the angular dependence in Eq. (7.1). With optical microscopy, only pores that are of diameter larger than about $0.5\,\mu m$ can be resolved. We refer to them as macropores. The lifetime of macropores, τ_{pore}, observed in vesicles in the fluid state, varies with pore radius, r_{pore} [104], and depends on the membrane edge tension, γ, and the membrane dilatational viscosity: $\tau_{pore} \approx 2r_{pore}\eta_D/\gamma$. For phosphatidylcholine vesicles with low tension, the lifetime τ_{pore} is typically shorter than 30 ms [89]. The effect of membrane edge tension on pore resealing is discussed in detail in Section 7.5.1.

Figure 7.3a shows the time dependence of the deformation of vesicles in which macropores were observed. The maximum deformation is much higher than that observed for nonporated vesicles (compare with the aspect ratio a/b in Figure 7.2). The typical relaxation time is $\tau_2 \approx 7 \pm 3\,ms$. The relaxation process associated with τ_2 takes place during the time interval when pores are present (shaded region in Figure 7.3a; snapshots of a porated vesicle are shown in Figure 7.3b). Thus, τ_2 is determined by the closing of the pores: $\tau_2 \approx \eta_D r_{pore}/2\gamma$. The edge tension γ is of the order of 10^{-11} N [109, 110]. For a typical pore radius of $1\,\mu m$ one obtains $\tau_2 \approx 10\,ms$. When the vesicles have some excess area, the relaxation proceeds in two steps: a fast relaxation characterized by τ_2 and a second, longer, relaxation with decay time τ_3: $0.5\,s < \tau_3 < 3\,s$ [89]. This relaxation time is related to the presence of some excess area available for shape changes corresponding well to the experimentally measured value.

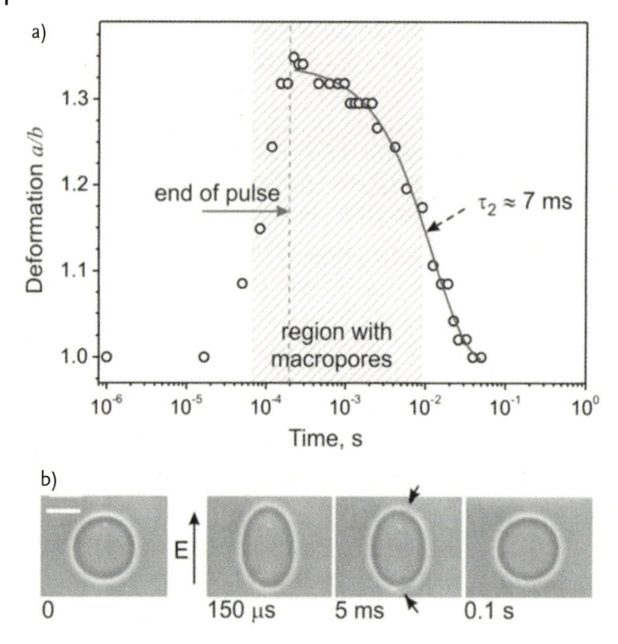

Figure 7.3 Deformation of macroporated phosphatidylcholine vesicles exposed to rectangular DC pulses with amplitude $E = 200\,\text{kV}\,\text{m}^{-1}$, and pulse duration $t_p = 200\,\mu\text{s}$. (a) Time dependence of the degree of deformation a/b of a vesicle. Time $t = 0$ was set as the beginning of the pulse. The dashed line indicates the end of the pulse. The relaxation of the vesicle is described by a single exponential fit (solid curve) with a decay time τ_2 as indicated. The shaded area indicates the time interval when macropores were optically detected. (b) Snapshots from another vesicle before, during, and after poration. The time after the beginning of the pulse is indicated below each micrograph. The scale bar corresponds to $15\,\mu\text{m}$. The arrows in the third image point to the porated zones, which are visualized as interruptions in the bright halo around the vesicle. Adapted from [89] by permission of the Biophysical Society.

7.3 Response of Gel-Phase Membranes

As discussed in Section 7.1.2, the mechanical and rheological properties of membranes in the gel phase differ significantly from those of fluid membranes. These differences introduce new features in the response of gel-phase membranes to electric fields. We compared the response to DC pulses of a vesicle in the fluid phase with that of a vesicle in the gel phase. The applied DC pulses were weak enough not to induce formation of macropores in the membranes. Figure 7.4a shows the deformation of the two vesicle types in response to DC pulses with duration of $300\,\mu\text{s}$. To achieve similar maximal degree of deformation in vesicles with comparable radii, stronger pulses have to be applied to the gel-phase vesicle as compared to the fluid one. Pulses with field strength of about $100\,\text{kV}\,\text{m}^{-1}$ do not produce optically detectable deformations in gel-phase vesicles, while strong

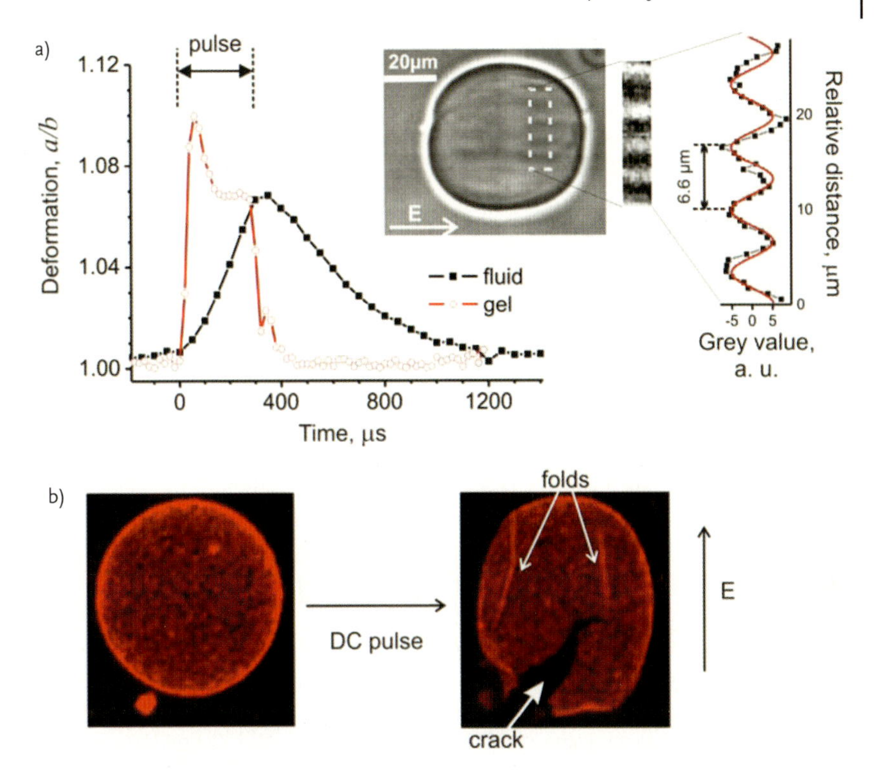

Figure 7.4 Deformation and electroporation of gel-phase vesicles. (a) Deformation response of a gel-phase vesicle with a radius of 22 µm, and a fluid-phase vesicle with a radius of 20 µm. The applied rectangular DC pulses were of duration of 300 µs as indicated. The field strength of the pulses was 500 and 80 kV m^{-1} for the gel and the fluid vesicle, respectively, which is below the corresponding poration thresholds for the two types of membranes. The gel-phase vesicle exhibits an intrapulse relaxation arising from wrinkling of the membrane. The vesicle wrinkling is visible in the inset snapshot recorded at time $t = 200$ µs. A magnified and enhanced section of the image indicated with a dashed rectangle is given to the right of the image, showing the membrane wrinkling parallel to the electric field direction. The gray value intensity from such a section is plotted and fitted with a sinusoidal function. The corresponding wavelength of the wrinkles is about 6.6 µm as indicated. (b) Electroporation of a gel-phase vesicle with radius 25 µm observed with confocal microscopy. Before the pulse, the vesicle has a spherical shape. After applying a pulse with field strength of 600 kV m^{-1} and duration of 300 µs, the vesicle cracks open and folds as indicated by the white arrows. The field direction is indicated with a vertical arrow. The second image was recorded a few seconds after the end of the pulse. A relatively large crack is visible in the vesicle as shown in the three-dimensional projection of the vesicle top part. Adapted from [111] by permission of the Royal Society of Chemistry.

pulses of about 500 kV m^{-1} applied to fluid-phase vesicles cause poration. The fluid vesicle gradually deforms and reaches maximal deformation (maximal aspect ratio a/b) at the end of the pulse as in Figure 7.1. The gel-phase vesicle responds significantly faster, and exhibits a relaxation with a decay time of about 50 μs already during the pulse. This unusual intrapulse relaxation was found to be due to wrinkling of the membrane [111] as shown with the inset in Figure 7.4a. Typical wavelengths of the wrinkles, Λ, lie in the range 5–8 μm and were found to obey laws for wrinkling of elastic sheets [112–114]: $\Lambda = 2(\pi^2 L^2 \kappa/\sigma)^{1/4}$, where L is the characteristic length of the system, which in our case is the vesicle size, $L \approx 2R$. The bending stiffness, κ, of membranes in the gel phase is orders of magnitude greater than that of fluid membranes (see Section 7.1.2), which is one of the reasons why wrinkling of fluid vesicles is not observed. Furthermore, fluid membranes have zero shear modulus and deform smoothly rather than exhibiting wrinkles.

The behavior of gel-phase vesicles exposed to stronger pulses above the poration threshold is also significantly different from that of porated fluid membranes. While pores in GUVs made of lipids in the fluid phase reseal within a few tens of milliseconds (Section 7.2), gel-phase giant vesicles exhibit long-living pores [42, 111]. The pores may resemble cracks on solid shells (Figure 7.4b), which remain open for minutes. Similar arrest was reported for mechanically induced pores in membranes below the main phase transition temperature [58]. Having in mind that the pore lifetime is defined by the membrane viscosity ($\tau_{pore} \approx 2r_{pore}\eta_D/\gamma$, see Section 7.2), it is easy to understand the long lifetime of pores in gel-phase membranes. The membrane viscosity diverges when the membrane crosses the main phase transition [43]. Thus, the resealing process is strongly suppressed. The irregular shape of the pores in gel-phase vesicles may be further indicative of the relatively low edge tension in such membranes.

Above we discussed the behavior of macroporated giant vesicles. Electroporated gel-phase vesicles with sizes around 100 nm were reported to reseal within milliseconds [100, 115]. For pores created in such vesicles, the pore radius, r_{pore}, should be in the nanometer range, yielding pore lifetimes, τ_{pore}, that are three orders of magnitude shorter than those of the micrometer-size pores in giant vesicles. Furthermore, the fields applied in references [100, 115] were of the order of 30 kV cm^{-1}. Such pulses induce Joule heating resulting in a temperature increase of up to 15 K [115], which can bring the membranes close to and even above the pretransition temperature of the investigated lipid and lead to a decrease in the membrane viscosity. Finally, recent coarse-grain simulations on gel-phase vesicles with similar sizes suggest that below the main phase transition temperature a fraction of the lipids in such highly curved membranes remain in the fluid phase [116], which may further facilitate pore closure. To summarize, the response of small and giant vesicles in the gel phase to DC pulses above the poration threshold differs significantly.

One additional feature that differentiates the response of gel-phase from fluid-phase membranes is the critical poration threshold. As demonstrated in Figure 7.4a, pulses that porate fluid vesicles lead only to deformation of gel-phase GUVs.

The critical transmembrane potential for the latter was found to be in the range 8–10 V [111]. This value is significantly higher than the critical potential of 1 V reported for fluid membranes [59, 98, 99]. Thus, membranes in the gel phase can sustain higher tensile stresses (see Eqs. (7.3) and (7.4)). This is also confirmed by micropipette aspiration experiments showing that fluid-phase dimyristoylphosphatidylcholine membranes undergo lysis at tensions around 2–3 $mN\,m^{-1}$, but when in the gel phase, the membranes rupture at tensions above 15 $mN\,m^{-1}$ [62]. It is important to note that the rupture process depends on the loading rate [61, 117, 118]. At high loading rates a membrane can sustain much higher tensions before it ruptures. Similar behavior was demonstrated by simulation studies, where fluid dipalmitoylphosphatidylcholine (DPPC) bilayers were shown to spontaneously rupture at tensions exceeding 90 $mN\,m^{-1}$ [119].

7.4 Effects of Membrane Inclusions and Media on the Response and Stability of Fluid Vesicles in Electric Fields

In the previous two sections we discussed the electrodeformation and electroporation of vesicles made of single-component membranes in water. In this section, we consider the effect of salt present in the solutions. The membrane response discussed above was based on data accumulated for vesicles made of phosphatidylcholines (PCs), the most abundant fraction of lipids in mammalian cells. PC membranes are neutral and predominantly located in the outer leaflet of the plasma membrane. The inner leaflet, as well as the bilayer of bacterial membranes, is rich in charged lipids. This raises the question as to whether the presence of such charged lipids would influence the vesicle behavior in electric fields. Cholesterol is also present at a large fraction in mammalian cell membranes. It is extensively involved in the dynamics and stability of raft-like domains in membranes [120]. In this section, apart from considering the response of vesicles in salt solutions, we describe aspects of the vesicle behavior of fluid-phase vesicles when two types of membrane inclusions are introduced, namely cholesterol and charged lipids.

7.4.1 Vesicles in Salt Solutions

In the presence of salt in a vesicle exterior, unusual shape changes are observed during an applied DC pulse [107]. The vesicles adopt spherocylindrical shapes (Figure 7.5) with lifetimes of the order of 1 ms. These deformations occur only in the presence of salt outside the vesicles, irrespective of their inner content (note that, in the absence of salt in the external solution, the vesicles deform only into prolates; see Figure 7.5a). When the solution conductivities inside and outside are identical, $\lambda_{in} \approx \lambda_{ex}$, vesicles with square cross section are observed (Figure 7.5d). For the case $\lambda_{in} < \lambda_{ex}$, the vesicles adopt disc-like shapes (Figure 7.5c), while in the

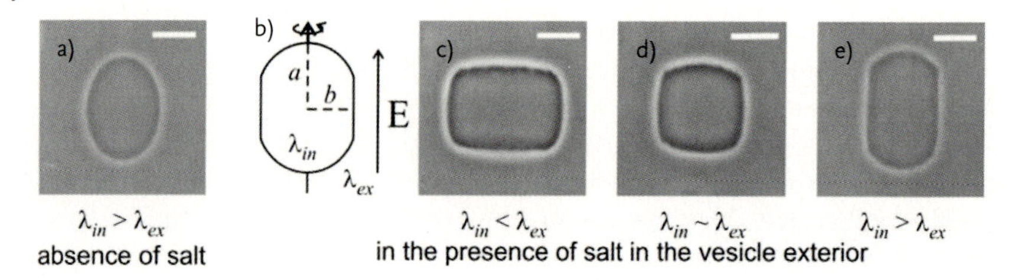

Figure 7.5 Deformation of vesicles in the absence and in the presence of salt at different conductivity conditions when subjected to DC pulses (200 kV m^{-1}, 200 μs). (a) In the absence of salt in the external solution, prolate deformation is observed. (b) Schematic of a cross section of a vesicle, which has adopted spherocylindrical deformation (a cylinder with spherical caps) when salt is present in the vesicle exterior. The field direction is indicated with an arrow. The presence of salt flattens the vesicle walls into (c) disc-like, (d) "square"-like, and (e) tube-like shapes. The scale bars correspond to 15 μm. Adapted from [107] by permission of the Biophysicsl Society.

opposite case, $\lambda_{in} > \lambda_{ex}$, they deform into long cylinders with rounded caps (Figure 7.5e).

Analogy can be drawn between the cylindrical shapes observed in the presence of salt at various conductivity conditions and the elliptical morphologies of vesicles exposed to AC fields. The beginning of Section 2 briefly discussed the latter shapes. In AC fields, vesicles attain different deformations: prolate (with the long principal radius oriented in the field direction) and oblate (with the long principal radius oriented perpendicular to the field direction). For a fixed field frequency in the range 10^4–10^6 Hz, the type of deformation depends on the solution conductivities [42, 64]. In particular, vesicles with lower internal conductivity ($\lambda_{in} < \lambda_{ex}$) adopt oblate shapes ($a/b < 1$), analogously to the disc-like deformation in Figure 7.5c, whereas vesicles with higher internal conductivity ($\lambda_{in} > \lambda_{ex}$) adopt prolate shapes ($a/b > 1$), analogously to the tube-like deformation shown in Figure 7.5e. However, in AC fields, no flattening of the vesicles is observed.

The cylindrical deformations shown in Figure 7.5 are nonequilibrium shapes and have a very short lifetime, which is why they have not been observed previously when standard video acquisition speed was accessible only. The flattening of the vesicle walls starts during the applied pulse and is observed throughout a period of about 1 ms. So far, cylindrical deformations have not been observed in studies on cells [121, 122]. However, the temporal resolution in those experiments (3.3 ms per image) was not high enough to detect such short-lived deformations.

The formation of the spherocylindrical shapes is not well understood. They are observed not only on lipid vesicles but also on polymersomes (vesicles made of diblock copolymers [123, 124]) [107]. Therefore, lipid-specific effects, for example partial head group charge and membrane thickness, as a possible cause for the observed cylindrical deformations are to be excluded. One possible explanation could be that ions flatten the equatorial zone of the deformed vesicle. During the pulse there is an inhomogeneity in the membrane tension due to the fact that the electric field is strongest at the poles of the vesicle, and almost zero close to the equator. The kinetic energy of the accelerated ions "hitting" the equatorial (tensionless) region of the vesicle is higher than the energy needed to bend the membrane, thus presumably leading to the observed deformation. To a certain extent, the effect of the ions could be considered as that of tiny particles subject to electrophoresis. Indeed, cylindrical deformations were observed also in quasi-salt-free solutions but in the presence of small negatively charged nanoparticles (40 nm in radius) in the vesicle exterior [42]. Thus, particle-driven flows may be yet another possible factor inducing membrane instability and giving rise to higher order modes of vesicle shape [125]. Another effect that could be considered is related to a local change in the spontaneous curvature of the bilayer due to the ion or particle asymmetry across the membrane [126]. During the pulse, local and transient accumulation of particles or ions in the membrane vicinity can occur inducing a change in the bilayer spontaneous curvature and driving the cylindrical deformations. Furthermore, another influencing factor might be an electrohydrodynamic instability caused by electric fields interacting with flat membranes, which was predicted to increase membrane roughness [127]. Finally, the flexoelectric properties of lipid membranes, first postulated by Petrov and coworkers [128–130], may also be involved as recently proposed [73, 131]. The interplay between surface ion concentration gradients combined with the overall ionic strength and bilayer material properties and tension could be expected to produce the observed cylindrical deformations.

7.4.2
Vesicles with Cholesterol-Doped Membranes

In order to understand the complex behavior of cellular membranes and their response to external perturbations like electric fields, one has to elucidate the basic mechanical properties of the lipid bilayer. The significant expansion in recent years of the field of membrane raft-like domain formation [11, 132, 133] imposes the compelling need for understanding the effect of lipid bilayer composition on membrane properties. Cholesterol, a ubiquitous species in eukaryotic membranes, is an important component in raft-like domains in cells and in vesicles, which motivates studies aimed at understanding its influence on the mechanical properties and stability of membranes.

A widely accepted view in the past regarded cholesterol as a stiffening agent in membranes. The conventional belief was that cholesterol orders the acyl chains in fluid membranes and increases the bending stiffness. This concept was supported by observations of lipids such as stearoylοleoylphosphatidylcholine (SOPC) [134, 135], dimyristoylphosphatidylcholine [53, 136], and palmitoyloleoylphosphatidylcholine (not only in mixtures with cholesterol but also with other sterols) [137]. However, as demonstrated recently [55, 138–140], the bending rigidity of membranes made of dioleoylphosphatidylcholine (DOPC) and cholesterol does not show any significant correlation with the cholesterol content. This suggests that the effect of cholesterol is not universal, but rather specific to the lipid architecture with respect to unsaturation and acyl chain length, and probably the lipid interfacial region.

The influence of cholesterol on other properties relevant for the stability and electroporation of membranes has not been studied extensively. The inverted-cone shape of this molecule should prevent it from locating at the rim of pores in membranes. Thus, the presence of cholesterol would require more energy to rearrange the lipids along the pore walls leading to an effective increase in the edge tension. Observations in this direction have been reported [141]. In [141] the pores were induced by strong illumination of giant vesicles inducing membrane stretching. Electroporation of giant vesicles and observation on the dynamics of pore closure also supported the notion of cholesterol increasing the edge tension [110]. The associated implication is that the lifetime of pores in such membranes is shorter (as defined above, the pore lifetime is inversely proportional to the edge tension; see Section 7.2). Note that both above-mentioned studies on the effect of cholesterol on pore stability [110, 141] were performed using DOPC membranes. It remains to be clarified whether the tendency will be preserved when other lipids with different degrees of saturation are considered.

Another peculiarity of DOPC–cholesterol GUVs is that they are destabilized and collapse when exposed to DC pulses at which cholesterol-free vesicles (e.g., made of egg phosphatidylcholine (egg PC) or pure DOPC) preserve their stability and only porate and reseal. The cholesterol-doped GUVs burst and disintegrated in a fashion reminiscent of that of charged membranes [142] (see Figure 7.6 for an example of a collapsing charged vesicle). As a quantitative characteristic of mem-

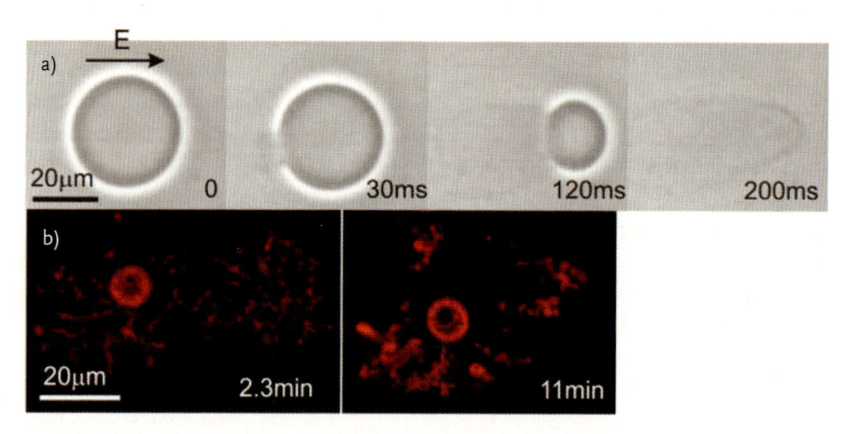

Figure 7.6 Bursting of charged vesicles subjected to electric pulses. The time after the beginning of the pulse is marked on each image. (a) Phase contrast microscopy snapshots from fast-camera observation of a vesicle in salt solution subjected to a pulse with field strength of $120\,kV\,m^{-1}$ and duration of $200\,\mu s$. The field direction is indicated in the first snapshot. The vesicle bursts and disintegrates. (b) Confocal microscopy cross sections of a vesicle that has been subjected to an electric pulse and after bursting has rearranged into a network of tubes and smaller vesicles. Adapted from [42] by permission of the Royal Society of Chemistry.

brane stability in electric fields one can consider the critical transmembrane potential, Ψ_c, at which poration occurs. Following [80], at the moment of maximally expanded pore, one can define the critical transmembrane potential as $\Psi_c = 1.5 RE\cos\theta_p$, where R is the initial vesicle radius before poration, E is the field magnitude, and θ_p is the inclination angle defining the location of the pore edge with respect to the vesicle center. This expression can be presented as $\Psi_c = 1.5E\left(R^2 - r_{pore}^2\right)^{1/2}$, where r_{pore} is the maximal pore radius. Thus, measuring r_{pore} immediately after applying a pulse with a field strength E provides a *rough* estimate for the critical transmembrane poration potential Ψ_c. While cholesterol-free membranes made of different lipids porate at similar values of Ψ_c of around 0.9 V, the addition of 17 mol% cholesterol to DOPC bilayers decreases the critical transmembrane potential to around 0.7 V, that is, destabilizes these membranes when exposed to an electric pulse [110]. This finding implies that cholesterol lowers the lysis tension of DOPC membranes. This observation is interesting and unexpected having in mind the somewhat opposite effect of cholesterol observed on SOPC giant vesicles [59, 103], where it was found to increase the critical poration threshold Ψ_c. On the contrary, DPPC vesicles doped with cholesterol appear to porate at transmembrane potentials lower than those of pure DPPC membranes [143]. Finally, for egg PC planar membranes [144] and vesicles [143] no effect of cholesterol on the critical permeabilization potential was observed.

The origin of the cholesterol-induced changes to the critical poration potential could be sought in its effect on altering the membrane conductivity justified in detail in [143]. This behavior may be related to lipid ordering, which may occur to a different extent depending on the type of lipid. In summary, cholesterol, which alters the hydrophobic core of the bilayer, affects the membrane stability in a different fashion depending on the specific molecular architecture of the lipid building the membrane.

7.4.3 Membranes with Charged Lipids

As discussed in Section 7.2, strong electric pulses applied to single-component giant vesicles made of zwitterionic lipids like PCs induce the formation of pores, which reseal within tens of milliseconds. When negatively charged lipids, like phosphatidylglycerol (PG) or phosphatidylserine, are present in a membrane a very different response of the vesicles can be observed, partially influenced by the medium conditions [142].

In buffered solutions containing ethylenediaminetetraacetic acid, PC:PG vesicles with molar ratios 9:1, 4:1, and 1:1 behave in the same way as pure PC vesicles, that is, the pulses induce opening of macropores with a diameter up to about 10 μm, which reseal within tens of milliseconds. In nonbuffered solution, the membranes with low fractions of charged lipids (9:1 and 4:1) retain this behavior, but for membrane composition of 1:1 PC:PG, the vesicles collapse and disintegrate after electroporation [142] (see Figure 7.6). Typically, one macropore forms and expands in the first 50–100 ms at a very high speed of approximately 1 mm s^{-1}.

The entire vesicle content is released seen as darker fluid in Figure 7.6. The bursting is followed by restructuring of the membrane into what seem to be interconnected bilayer fragments in the first seconds, and a tether-like structure in the first minute. Then the membrane stabilizes into interconnected micrometer-sized tubules and small vesicles. Similar behavior is observed for vesicles prepared from lipid extracts from the plasma membrane of red blood cells which also contain a fraction of charged lipids [142]. These observations suggest that vesicle bursting and membrane instability are related to the amount of charged lipid in the bilayer as theoretically predicted in earlier studies [145–147].

As mentioned above, observations of GUVs prepared from lipid extracts from the plasma membrane of red blood cells show a response similar to that of vesicles prepared from synthetic charged lipids. Thus, one would intuitively expect that cells should exhibit similar bursting behavior. However, cell membranes are subjected to internal mechanical constraints imposed by the cytoskeleton, which prevents their disintegration even if their membranes are prone to disruption when exposed to strong DC pulses. Instead, the pores in the cell membrane are stable for a long time [148] and can either lead to cell death by lysis or reseal depending on the media [149, 150]. The latter is the key to efficient electroporation-based protocols for drug or gene transfer in cells. The results discussed in this section suggest that membrane charge might be an important, but not yet well understood, regulating agent in these protocols.

7.5 Application of Vesicle Electroporation

In this section some application aspects of giant vesicle electroporation are considered. In particular, it will be demonstrated that creating macropores in GUVs and observing their closing dynamics can be successfully applied to the evaluation of material properties of membranes. While in Section 7.4.2 we saw that such experiments can be used to characterize membrane stability in terms of the critical poration potential Ψ_c, here we will find out how one can also evaluate the edge tension of porated membranes. In addition, another application based on electroporation, namely vesicle electrofusion, is introduced whereby the use of GUVs as microreactors suitable for the synthesis of nanoparticles is demonstrated.

7.5.1 Measuring Membrane Edge Tension from Vesicle Electroporation

Upon poration, the lipid molecules in a bilayer reorient so that their polar heads can line the pore walls and form a hydrophilic pore [151]. The energetic penalty per unit length for this reorganization is described by the edge tension, which emerges from the physicochemical properties and the amphiphilic nature of lipids. It also gives rise to a force driving the closure of transient pores, playing a crucial role in membrane resealing mechanisms taking place after physical protocols for drug delivery, such as sonoporation [152] or electroporation [153]. Being able to experimentally measure the edge tension is thus of significant interest for understanding various biological events and physicochemical processes in membranes. However, only a few experimental methods have been developed to directly assess this physical quantity. Among them, an elegant approach was based on rapid freezing of cells with a controlled time delay after electroporation, and examining the pores with electron microscopy [154, 155]. This method, however, provides a static picture of the porated membrane and there is a danger of ice crystal damage.

Only a few previous studies have employed GUVs for estimating edge tension. Observations of open cylindrical giant vesicles exposed to AC fields [109] provided an estimate for the edge tension, but this technique did not allow for good control over the system. In another work, vesicles were porated with an electric pulse, and the pores were kept open by externally adjusting the membrane tension with a micropipette [103]. Even though solid, this approach requires the use of sophisticated equipment like a set-up for vesicle micropipette aspiration. GUVs were also used in [141, 156], where the pore closure dynamics was analyzed in light of a theory developed earlier [44]. However, for the direct visualization of pore closure, the use of viscous glycerol solutions and fluorescent dyes in the membrane was required, both of which potentially influence the edge tension. (The theoretical approach in [44] was also applied to experiments where the membrane was disrupted by laser ablation in [157], but the results of the latter study seem questionable since laser ablation is associated with local evaporation; indeed, the images provided in this work suggest the formation of bubbles in the sample rather than pores in the membrane.)

A much simpler approach introduced recently is based on the electroporation of giant vesicles and observation of the pore closure dynamics with fast digital imaging [110, 158]. The analysis of the pore dynamics was based on the theoretical work of Brochard-Wyart *et al.* [44]. The process of pore closure was observed under phase contrast microscopy with a high-speed digital camera (the acquisition speed was typically above 1000 fps). In this way, the need to use viscous solutions to slow down the system dynamics was avoided and the application of fluorescent dyes to visualize the vesicles as in [104, 141, 156] was not necessary. Note that both glycerol and fluorescent markers may influence the measured value for the edge tension. Vesicle electroporation was induced by applying electric pulses of 5 ms duration and field strength in the range 20–80 kV m^{-1}. The pore dynamics typically consisted of four stages: growing, stabilization at some maximal pore radius, slow decrease in pore size, and fast closure (e.g., data in Figure 7.7). The third stage of slow pore closure is the one that can be used to determine the membrane edge tension applying the dependence derived in [44]:

$$R^2 \ln(r) = -\frac{2\gamma}{3\pi\eta} t + C \tag{7.5}$$

where R and r are the vesicle and pore radii, respectively, γ is the edge tension, η is the viscosity of the aqueous medium, t is time, and C is a constant depending

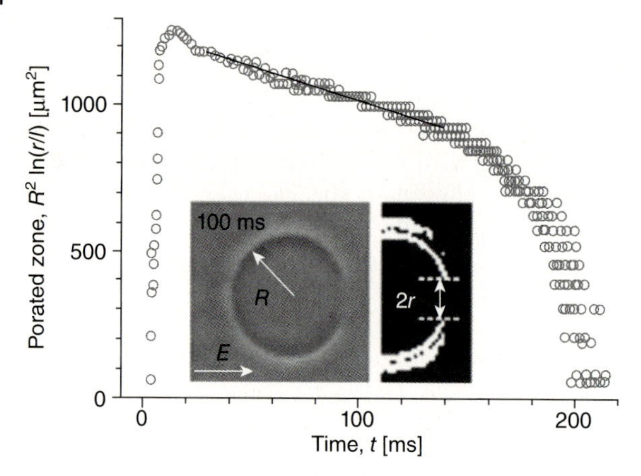

Figure 7.7 Evolution of the porated region in an egg PC vesicle as characterized by $R^2\ln(r/l)$ as a function of time t (see Eq. (7.5); note that to avoid plotting a dimensional value in the logarithmic term, we have introduced $l = 1\,\mu m$). The open circles are experimental data and the solid line is a linear fit, whose slope yields the edge tension γ. The inset shows a raw image (left) of a porated vesicle with a radius of 17 µm 100 ms after being exposed to an electric pulse with duration of 5 ms and amplitude of 50 kV m^{-1}. The field direction is indicated with an arrow. The right-hand side of the inset is an enhanced and processed image of the vesicle half facing the cathode. The inner white contour corresponds to the location of the membrane. The pore radius is schematically indicated.

on the maximal pore radius reached. Then, one only has to consider the linear part of $R^2\ln(r)$ as a function of time corresponding to the slow closure stage. Linear fit of this part is characterized by a slope a and the edge tension γ is estimated from the relation $\gamma = -(3/2)\pi\eta a$. Figure 7.7 illustrates the analysis performed on one egg PC vesicle. Measurements on many vesicles yielded an average value of $\gamma = 14.3\,\text{pN}$ for the edge tension of such membranes.

Using this approach, one can measure the edge tension in membranes of various compositions, thus characterizing the stability of pores in these membranes, and evaluate the effect of various inclusions. For example, it was found that the addition of cholesterol to DOPC membranes increases the edge tension confirming previously reported results [141]. The inverted-cone shape of cholesterol prevents it from locating at the rim of pores. Thus, the presence of cholesterol requires more energy to rearrange the lipids along the pore walls increasing the edge tension. Surprisingly, doping DOPC membranes with another cone-shaped type of lipid like dioleoylphosphatidylethanolamine (DOPE) was found to decrease the edge tension, that is, DOPE has a pore-stabilizing effect [110]. Presumably, the molecular architecture of phosphatidylethanolamine (PE)–lipids leading to their tendency to form an inverted hexagonal phase, which facilitates fusion and vesicle leakage (e.g., [159]), is also responsible for stabilizing pores. A plausible explanation for this behavior is also provided by the propensity of PE to form interlipid hydrogen bonds [160, 161], that is, inter-PE hydrogen bonding in the pore region can effectively stabilize pores.

As demonstrated above, the edge tension is a sensitive parameter, which effectively characterizes the stability of pores in membranes. Compiling a database for the effect of various types of membrane inclusions will be useful for understanding the lifetime of pores in membranes with more complex compositions, which is important for achieving control over medical applications for drug and gene delivery in cells.

7.5.2 Vesicle Electrofusion

The phenomenon of membrane electrofusion is of particular interest, because of its widespread use in cell biology and biotechnology (e.g., [162–164] and the references cited therein). The application of electrofusion to cells can lead to the creation of multinucleated viable cells with new properties (this phenomenon is also known as hybridization) (e.g., [164]). In addition, electroporation and electrofusion are often used to introduce molecules like proteins, foreign genes (plasmids), antibodies, and drugs into cells.

When a DC pulse is applied to a couple of fluid-phase vesicles, which are in contact and oriented in the direction of the field, electrofusion can be observed. Vesicle orientation (and even alignment into pearl chains) can be achieved by application of an AC field to a vesicle suspension. This phenomenon is also observed with cells [164, 165] and is due to dielectric screening of the field. When the suspension is dilute, two vesicles can be brought together via the AC field and aligned. A subsequent application of a DC pulse to such a vesicle couple can lead to fusion. The necessary condition is that poration is induced in the contact area between the two vesicles. The possible steps of the electrofusion of two membranes are schematically illustrated in Figure 7.8a. In Sections 7.5.2.1 and 7.5.2.2, consideration will be given to the fusion of vesicles with different membrane composition or different composition of the enclosed solutions.

7.5.2.1 Fusing Vesicles with Identical or Different Membrane Composition

Membrane fusion is a fast process. The time needed for the formation of a fusion neck can be rather short as demonstrated by electrophysiological methods applied to the fusion of small vesicles with cell membranes [166–169]. The time evolution of the observed membrane capacitance indicates that the formation of the fusion neck is presumably faster than 100 μs. Direct observation of the fusion of giant vesicles recently confirmed this finding and suggested that this time is even shorter [36, 37]. An example of a few snapshots taken from the electrofusion of two GUVs with identical membrane compositions and in the presence of salt is given in Figure 7.8b. The overall deformation of each vesicle as seen in the second snapshot corresponds to the spherocylindrical shapes as observed with individual vesicles in the presence of salt (see Section 7.4.1). From such micrographs, one can measure the fusion neck diameter, denoted by L in Figures 7.8a and 7.8b, and follow the dynamics of its expansion as shown in Figure 7.8c. From the data, two stages of the fusion process can be distinguished (note that the data are displayed in a semi-logarithmic plot): an early stage, which is very fast and with average

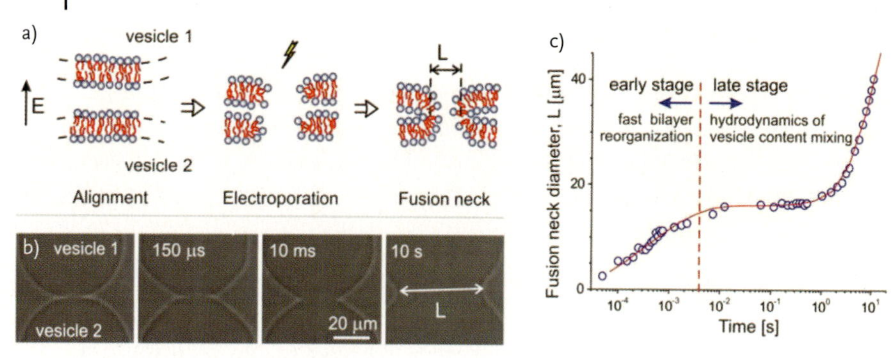

Figure 7.8 Membrane electrofusion. (a) Schematic of the possible steps of the electrofusion process: two lipid vesicles are brought into contact (only the membranes in the contact zone of the vesicles are sketched), followed by electroporation and formation of a fusion neck of diameter *L*. The pore sizes are not to scale but almost an order of magnitude larger than the bilayer thickness. (b) Micrographs from the electrofusion of a vesicle couple. Only segments of the vesicles are visible. The external solution contains 1 mM NaCl, which causes flattening of the vesicle walls in the second snapshot (see Section 7.4.1). The amplitude of the DC pulse was 240 kV m^{-1}, and its duration was 120 μs. The time after the beginning of the pulse is indicated on the snapshots. (c) Time evolution of the fusion neck diameter, *L*, formed between two vesicles with radii of about 15 μm. The solid curve is a guide to the eye. The vertical dashed line indicates the border between the two stages in the fusion dynamics. Adapted from [41] by permission of the Royal Society of Chemistry.

expansion velocity of about $2 \times 10^4\,\mu\text{m s}^{-1}$, followed by a later slower stage with an expansion rate that is orders of magnitude smaller (about $2\,\mu\text{m s}^{-1}$). The early stage is governed by fast relaxation of the membrane tension built during the pulse, whereby the dissipation occurs in the bilayer. Essentially, the driving forces here are the same as those responsible for the relaxation dynamics of nonporated vesicles (as characterized by τ_1 in Section 7.2). Thus, the characteristic time for this early stage of fusion, τ_{early}, can be expressed as $\tau_{\text{early}} \sim \eta_D/\sigma$, where the membrane tension σ should be close to the tension of rupture σ_{lys}. Thus for τ_{early} one obtains a value of 100 μs, which is in agreement with the experimental observations for the time needed to complete the early stage of fusion. Linear extrapolation of the data in the early stage predicts that the formation of a fusion neck with a diameter of about 10 nm should occur within a time period of about 250 ns [37]. It is quite remarkable that this time scale of the order of 200 ns was also obtained from computer simulations of a vesicle fusing with a tense membrane segment [169].

In the later stage of fusion, the neck expansion velocity slows down by more than two orders of magnitude. Here the dynamics is mainly governed by the displacement of the volume of fluid around the fusion neck between the fused vesicles. The restoring force is related to the bending elasticity of the lipid bilayer [36, 37].

Fusing two vesicles with membranes of different composition can provide a promising tool for studying raft-like domains in membranes [11, 12, 133, 170, 171].

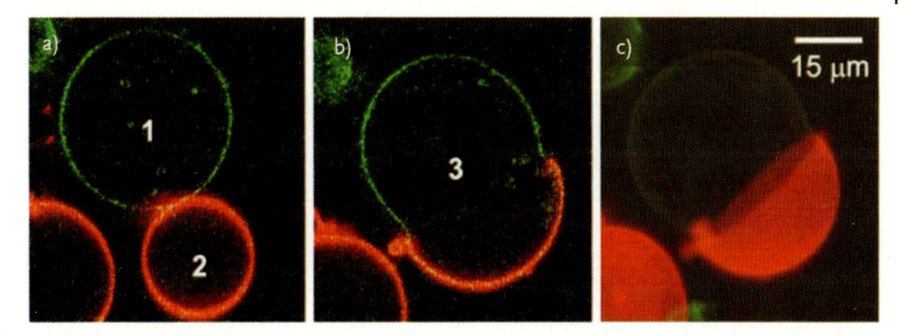

Figure 7.9 Creating a multidomain vesicle by electrofusion of two vesicles with different membrane composition as observed with fluorescence microscopy. (a, b) Images acquired with confocal microscopy scans nearly at the equatorial plane of the fusing vesicles. (a) Vesicle 1 is made of sphingomyelin and cholesterol (7 : 3) and labeled with one fluorescent dye (green). Vesicle 2 is composed of dioleoylphosphatidylcholine and cholesterol (8 : 2) and labeled with another fluorescent dye (red). (b) The two vesicles were subjected to an electric pulse ($220\,kV\,m^{-1}$, duration $300\,\mu s$) and fused to form vesicle 3. Because the lipids with this final membrane composition form immiscible fluid phases, the resulting vesicle has two domains. (c) A three-dimensional image projection of vesicle 3 with the two domains formed from vesicles 1 and 2. Adapted from [41] by permission of the Royal Society of Chemistry.

In particular, vesicle electrofusion is a very attractive experimental approach for producing multicomponent vesicles of well-defined composition [36]. One example for the fusion of two vesicles with different membrane composition is shown in Figure 7.9. To distinguish the vesicles according to their composition, two fluorescent markers have been used. In this particular example, one of the vesicles (vesicle 1) is composed of sphingomyelin and cholesterol in 7 : 3 molar ratio and labeled in green. The other vesicle (vesicle 2) is composed of DOPC and cholesterol in 8 : 2 molar ratio and labeled in red (with fluorescence microscopy, the two vesicle can be distinguished by their color). Thus, the membrane of the fused vesicle is a three-component one. At room temperature, this lipid mixture separates into two phases, liquid ordered (rich in sphingomyelin and cholesterol) and liquid disordered (rich in DOPC), which is why the final vesicle exhibits immiscible fluid domains. The exact composition of each of these domains is not well known because lipids may redistribute among the domains. However, from the domain area and the area of the initial vesicles before fusion, one can judge whether there is redistribution of cholesterol, and eventually calculate the actual domain composition.

7.5.2.2 **Vesicle Electrofusion: Employing Vesicles as Microreactors**

Vesicle fusion can be employed to scale down the interaction volume of a chemical reaction and reduce it to a few picoliters or less. In other words, fusion of two vesicles of different content can be used for the realization of a tiny microreactor [172–176]. The principle of fusion-mediated synthesis is simple: the starting reagents are separately loaded into different vesicles, and then the reaction is triggered by the fusion of these vesicles, which allows the mixing of their contents. The success of this approach is guaranteed by two important factors. First, the lipid membrane is impermeable to the reactants such as ions or macromolecules. Second, fusion can be initiated by a variety of fusogens such as membrane stress [169, 177, 178], ions or synthetic fusogenic molecules [37, 179–181], fusion proteins [182], or electric fields [36, 176]. Among these approaches, electrofusion becomes increasingly important because of its reliable, fast, and easy handling. An immediate benefit of this strategy is that precise temporal control of the synthesis process can be easily achieved.

Synthesis of nanoparticles in such microreactors is of particular interest. Indeed, cells and microorganisms themselves have been reported to have the unique ability to synthesize inorganic nanoparticles such as CdS, ZnS, gold, and silver [183–185]. The tentative interpretation of this observation is related to the mediation ability of specific molecules such as inorganic-binding peptides [186]. Nevertheless, the underlying processes are still not well understood at the molecular scale. Therefore, attempts to perform similar reactions in simplified artificial systems become very important towards detailed exploration of biological synthesis mechanisms and biomimetic fabrication [185].

Although nanoparticle synthesis in vesicle nanoreactors has been well documented [187–190], much effort is still needed to perform such synthesis in GUVs so as to elucidate the biological mechanism of nanoparticle synthesis in cells and construct novel functional "artificial cells" for advanced technological applications. Other processes in giant vesicles occurring in their enclosed compartments or at the membrane surface have also been studied [172, 174, 175, 191–194]. Furthermore, one great and exclusive advantage of GUVs is that giant vesicles and the corresponding products of their interaction can be visualized in real time under a light microscope, providing the potential possibility for on-line monitoring of material growth at micrometer and submicrometer scales.

Success in this direction was recently reported for the synthesis of CdS quantum-like nanoparticles [176]. The protocol applied is schematically illustrated in Figure 7.10a. Two vesicle populations loaded with reactant A ($CdCl_2$) or B (Na_2S) are mixed in A-, B-free isotonic solution. To be able to distinguish the two vesicle types from each other, the membranes of A-loaded and B-loaded vesicles are labeled by small fractions of red and green fluorescent dyes, respectively, added to the main lipid building the membrane. Similar to pearl-chain formation in cell suspensions, the two types of vesicles align along the direction of an exogenous AC field. This field-induced self-arrangement makes reactive vesicles match well for the CdS synthesis reaction: half of the aligned vesicle couples are A–B couples. After that, a strong DC pulse is applied (typically pulses of 50–200 $kV\,m^{-1}$ field strength and 150–300 μs duration suffice) to initiate vesicle fusion and, thus, the reaction between A and B. The product, in this case quantum-dot-like CdS nanoparticles (with sizes between 4 and 8 nm as determined from transmission electron microscopy [176]), is visualized under laser excitation as a fluorescent bright spot in the fusion zone (see the second snapshot in Figure 7.10b). The intensity from this spot increases gradually from 0 s (the starting time point of electrofusion), and

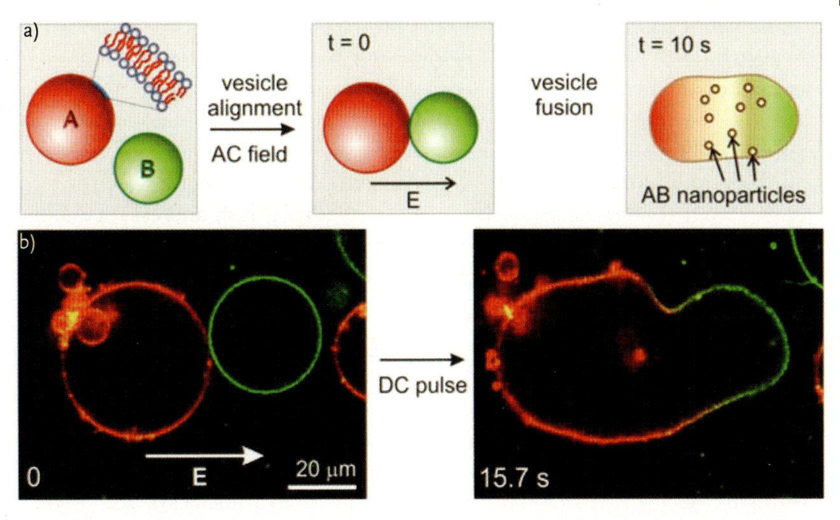

Figure 7.10 Electrofusion of giant vesicles as a method for nanoparticle synthesis. (a) Schematic of the electrofusion protocol. Two populations of vesicles containing reactant A or B are mixed (in A- and B-free environment) and subjected to an AC field to align them in the direction of the field and bring them close together. A DC pulse initiates the electrofusion of the two vesicles and the reaction between A and B proceeds to the formation of nanoparticles encapsulated in the fused vesicle. (b) Confocal scans of vesicles loaded with 0.3 mM Na_2S (red) and 0.3 mM $CdCl_2$ (green) undergoing fusion. The direction of the field is indicated in the first snapshot. After fusion (second snapshot), fluorescence from the product is detected in the interior of the fused vesicle. The time after applying the pulse is indicated on the micrographs. Adapted from [176] by permission of Wiley-VCH Verlag.

reaches maximum around 10 s. One can therefore infer that the reaction begins at the electrofusion point and quickly reaches a balance around 10 s. Obviously, this protocol provides us with a visualizing analytical tool to follow the reaction kinetics with high temporal sensitivity. This novel and facile method is especially suitable for the on-line monitoring of ultrafast physicochemical processes such as photosynthesis, enzyme catalysis, and photopolymerization, which usually require complex and abstracted spectroscopy techniques at present. These results also show that even without the mediation of biomacromolecules, nanoparticles can still be synthesized in biological compartments. This outcome provides a new insight in the developing research on biomineralization mechanisms.

7.6 Conclusions and Outlook

The issues addressed in this chapter demonstrate that cell-sized giant vesicles provide a very useful model for resolving various effects of electric fields on lipid membranes because vesicle dynamics can be directly observed with optical microscopy. It has been shown that the vesicle response to electric fields can be interpreted and understood considering the basic mechanical properties of membranes.

Until recently, the dynamics of vesicle relaxation and poration, which occur on microsecond time scales, has eluded direct observation because the temporal resolution of optical microscopy observations with analog video technology is in the range of milliseconds. The recent use of fast digital imaging has helped us to characterize membrane deformation and poration and also to discover new features of the membrane response arising from the presence of charged lipids or cholesterol in the membrane and nanoparticles in the surrounding media, and to compare the response of gel-phase membranes to fluid-phase ones. Due to this high temporal resolution, new shape deformations, such as spherocylindrical ones, have been detected. The observations of vesicle fusion revealed the presence of two stages of the fusion process. Finally, a novel application of membrane electrofusion was introduced, which allows the construction of vesicles with fluid domains.

In conclusion, the reported observations demonstrate that giant vesicles as biomimetic membrane compartments can be of significant help to advance fundamental knowledge about the complex behavior of cells and membranes in electric fields and can inspire novel practical applications.

Acknowledgments

In preparing this chapter, I have profited from numerous experimental contributions and insightful discussions with skillful members of my group, and current and former collaborators. The work in the laboratory on vesicles in electric fields was initiated together with Karin A. Riske. I would also like to thank Roland L. Knorr, Natalya Bezlyepkina, Thomas Portet, Margarita Staykova, Peng Yang, Said Aranda, Marie Domange Jordö, Benjamin Klasczyk, and D. Duda. I also wish to acknowledge the valuable insight I got from the enlightening discussions with theoreticians like Reinhard Lipowsky, Petia M. Vlahovska, Tetsuya Yamamoto, and Rubèn S. Gracià.

References

1 Lasic, D.D. (ed.) (1993) *Liposomes: From Physics to Applications*, Elsevier, Amsterdam.

2 Luisi, P.L. and Walde, P. (eds) (2000) *Giant Vesicles*, John Wiley & Sons, Ltd, Chichester.

3 Dimova, R., Aranda, S., Bezlyepkina, N., Nikolov, V., Riske, K.A., and Lipowsky, R. (2006) A practical guide to giant vesicles. Probing the membrane nanoregime via optical microscopy. *Journal of Physics: Condensed Matter*, **18** (28), S1151–S1176.

4 Walde, P., Cosentino, K., Engel, H., and Stano, P. (2010) Giant vesicles: preparations and applications. *ChemBioChem*, **11** (7), 848–865.

5 Reeves, J.P. and Dowben, R.M. (1969) Formation and properties of thin-walled phospholipid vesicles. *Journal of Cellular Physiology*, **73** (1), 49–60.
6 Angelova, M.I. and Dimitrov, D.S. (1986) Liposome electroformation. *Faraday Discussions*, **81**, 303–311.
7 Akashi, K., Miyata, H., Itoh, H., and Kinosita, K. (1996) Preparation of giant liposomes in physiological conditions and their characterization under an optical microscope. *Biophysical Journal*, **71** (6), 3242–3250.
8 Akashi, K., Miyata, H., Itoh, H., and Kinosita, K. (1998) Formation of giant liposomes promoted by divalent cations: critical role of electrostatic repulsion. *Biophysical Journal*, **74** (6), 2973–2982.
9 Montes, L.R., Alonso, A., Goni, F.M., and Bagatolli, L.A. (2007) Giant unilamellar vesicles electroformed from native membranes and organic lipid mixtures under physiological conditions. *Biophysical Journal*, **93**, 3548–3554.
10 Pott, T., Bouvrais, H., and Meleard, P. (2008) Giant unilamellar vesicle formation under physiologically relevant conditions. *Chemistry and Physics of Lipids*, **154** (2), 115–119.
11 Lipowsky, R., and Dimova, R. (2003) Domains in membranes and vesicles. *Journal of Physics: Condensed Matter*, **15** (1), S31–S45.
12 Baumgart, T., Hess, S.T., and Webb, W.W. (2003) Imaging coexisting fluid domains in biomembrane models coupling curvature and line tension. *Nature*, **425** (6960), 821–824.
13 Veatch, S.L. and Keller, S.L. (2002) Organization in lipid membranes containing cholesterol. *Physical Review Letters*, **89** (26), 4.
14 Vitkova, V., Mader, M., and Podgorski, T. (2004) Deformation of vesicles flowing through capillaries. *Europhysics Letters*, **68** (3), 398–404.
15 Meleard, P., Gerbeaud, C., Pott, T., and Mitov, M.D. (2000) Electromechanical properties of model membranes and giant vesicle deformation, in *Giant Vesicles* (eds P.L. Luisi and P. Walde), John Wiley & Sons, Ltd, Chichester, pp. 185–205.
16 Dimova, R., Dietrich, C., Hadjiisky, A., Danov, K., and Pouligny, B. (1999) Falling ball viscosimetry of giant vesicle membranes: finite-size effects. *European Physical Journal B*, **12**, 589.
17 Wick, R., Walde, P., and Luisi, P.L. (1995) Light-microscopic investigations of the autocatalytic self-reproduction of giant vesicles. *Journal of the American Chemical Society*, **117** (4), 1435–1436.
18 Menger, F.M. and Gabrielson, K. (1994) Chemically-induced birthing and foraging in vesicle systems. *Journal of the American Chemical Society*, **116** (4), 1567–1568.
19 Takakura, K., Toyota, T., and Sugawara, T. (2003) A novel system of self-reproducing giant vesicles. *Journal of the American Chemical Society*, **125** (27), 8134–8140.
20 Zhu, T.F. and Szostak, J.W. (2009) Coupled growth and division of model protocell membranes. *Journal of the American Chemical Society*, **131** (15), 5705–5713.
21 Evans, E. and Metcalfe, M. (1984) Free-energy potential for aggregation of giant, neutral lipid bilayer vesicles by van der waals attraction. *Biophysical Journal*, **46** (3), 423–426.
22 Evans, E. (1991) Entropy-driven tension in vesicle membranes and unbinding of adherent vesicles. *Langmuir*, **7** (9), 1900–1908.
23 Dietrich, C., Angelova, M.I., and Pouligny, B. (1997) Adhesion of latex spheres to giant phospholipid vesicles: statics and dynamics. *Journal de Physique II France*, **7**, 1651–1682.
24 Gruhn, T., Franke, T., Dimova, R., and Lipowsky, R. (2007) Novel method for measuring the adhesion energy of vesicles. *Langmuir*, **23** (10), 5423–5429.
25 Fang, N., Chan, V., and Wan, K.T. (2003) The effect of electrostatics on the contact mechanics of adherent phospholipid vesicles. *Colloids and Surfaces B*, **27** (1), 83–94.
26 Li, Y., Lipowsky, R., and Dimova, R. (2008) Transition from complete to partial wetting within membrane compartments. *Journal of the American Chemical Society*, **130** (37), 12252–12253.

27 Kusumaatmaja, H., Li, Y., Dimova, R., and Lipowsky, R. (2009) Intrinsic contact angle of aqueous phases at membranes and vesicles. *Physical Review Letters*, **103** (23), 238103–238104.

28 Kas, J. and Sackmann, E. (1991) Shape transitions and shape stability of giant phospholipid-vesicles in pure water induced by area-to-volume changes. *Biophysical Journal*, **60** (4), 825–844.

29 Dobereiner, H.G., Kas, J., Noppl, D., Sprenger, I., and Sackmann, E. (1993) Budding and fission of vesicles. *Biophysical Journal*, **65** (4), 1396–1403.

30 Peterlin, P., Arrigler, V., Kogej, K., Svetina, S., and Walde, P. (2009) Growth and shape transformations of giant phospholipid vesicles upon interaction with an aqueous oleic acid suspension. *Chemistry and Physics of Lipids*, **159** (2), 67–76.

31 Staneva, G., Seigneuret, M., Koumanov, K., Trugnan, G., and Angelova, M.I. (2005) Detergents induce raft-like domains budding and fission from giant unilamellar heterogeneous vesicles – a direct microscopy observation. *Chemistry and Physics of Lipids*, **136** (1), 55–66.

32 Inaoka, Y. and Yamazaki, M. (2007) Vesicle fission of giant unilamellar vesicles of liquid-ordered-phase membranes induced by amphiphiles with a single long hydrocarbon chain. *Langmuir*, **23** (2), 720–728.

33 Pantazatos, D.P. and MacDonald, R.C. (1999) Directly observed membrane fusion between oppositely charged phospholipid bilayers. *Journal of Membrane Biology*, **170** (1), 27–38.

34 Lei, G.H. and MacDonald, R.C. (2003) Lipid bilayer vesicle fusion: intermediates captured by high-speed microfluorescence spectroscopy. *Biophysical Journal*, **85** (3), 1585–1599.

35 Zhou, Y.F. and Yan, D.Y. (2005) Real-time membrane fusion of giant polymer vesicles. *Journal of the American Chemical Society*, **127** (30), 10468–10469.

36 Riske, K.A., Bezlyepkina, N., Lipowsky, R., and Dimova, R. (2006) Electrofusion of model lipid membranes viewed with high temporal resolution. *Biophysical Review Letters*, **1** (4), 387–400.

37 Haluska, C.K., Riske, K.A., Marchi-Artzner, V., Lehn, J.M., Lipowsky, R., and Dimova, R. (2006) Time scales of membrane fusion revealed by direct imaging of vesicle fusion with high temporal resolution. *Proceedings of the National Academy of Sciences of the USA*, **103** (43), 15841–15846.

38 Abkarian, M. and Viallat, A. (2008) Vesicles and red blood cells in shear flow. *Soft Matter*, **4** (4), 653–657.

39 Raphael, R. and Waugh, R. (1996) Accelerated interleaflet transport of phosphatidylcholine molecules in membranes under deformation. *Biophysical Journal*, **71**, 1374–1388.

40 Evans, E. and Rawicz, W. (1997) Elasticity of "fuzzy" biomembranes. *Physical Review Letters*, **79** (12), 2379–2382.

41 Dimova, R., Riske, K.A., Aranda, S., Bezlyepkina, N., Knorr, R.L., and Lipowsky, R. (2007) Giant vesicles in electric fields. *Soft Matter*, **3** (7), 817–827.

42 Dimova, R., Bezlyepkina, N., Jordo, M.D., Knorr, R.L., Riske, K.A., Staykova, M., Vlahovska, P.M., Yamamoto, T., Yang, P., and Lipowsky, R. (2009) Vesicles in electric fields: some novel aspects of membrane behavior. *Soft Matter*, **5** (17), 3201–3212.

43 Dimova, R., Pouligny, B., and Dietrich, C. (2000) Pretransitional effects in dimyristoylphosphatidylcholine vesicle membranes: optical dynamometry study. *Biophysical Journal*, **79** (1), 340–356.

44 Brochard-Wyart, F., de Gennes, P.G., and Sandre, O. (2000) Transient pores in stretched vesicles: role of leak-out. *Physica A*, **278** (1–2), 32–51.

45 Marsh, D. (2006) Elastic curvature constants of lipid monolayers and bilayers. *Chemistry and Physics of Lipids*, **144** (2), 146–159.

46 Seifert, U. and Lipowsky, R. (1995) Morphology of vesicles, in *Structure and Dynamics of Membranes (Handbook of Biological Physics)* (eds R. Lipowsky and E. Sackmann), Elsevier, Amsterdam, pp. 403–463.

47 Rawicz, W., Olbrich, K.C., McIntosh, T., Needham, D., and Evans, E. (2000) Effect of chain length and unsaturation

on elasticity of lipid bilayers. *Biophysical Journal*, **79** (1), 328–339.
48 Brochard, F. and Lennon, J.F. (1975) Frequency spectrum of flicker phenomenon in erythrocytes. *Journal of Physics*, **36** (11), 1035–1047.
49 Schneider, M.B., Jenkins, J.T., and Webb, W.W. (1984) Thermal fluctuations of large quasi-spherical bimolecular phospholipid-vesicles. *Journal of Physics*, **45** (9), 1457–1472.
50 Engelhardt, H., Duwe, H.P., and Sackmann, E. (1985) Bilayer bending elasticity measured by Fourier analysis of thermally excited surface undulations of flaccid vesicles. *Journal of Physics Letters*, **46** (8), L395–L400.
51 Bivas, I., Hanusse, P., Bothorel, P., Lalanne, J., and Aguerrechariol, O. (1987) An application of the optical microscopy to the determination of the curvature elastic-modulus of biological and model membranes. *Journal of Physics*, **48** (5), 855–867.
52 Faucon, J.F., Mitov, M.D., Meleard, P., Bivas, I., and Bothorel, P. (1989) Bending elasticity and thermal fluctuations of lipid membranes – theoretical and experimental requirements. *Journal of Physics*, **50** (17), 2389–2414.
53 Duwe, H.P., Kaes, J., and Sackmann, E. (1990) Bending elastic moduli of lipid bilayers – modulation by solutes. *Journal of Physics*, **51** (10), 945–962.
54 Henriksen, J.R., and Ipsen, J.H. (2002) Thermal undulations of quasi-spherical vesicles stabilized by gravity. *European Physical Journal E*, **9** (4), 365–374.
55 Gracia, R.S., Bezlyepkina, N., Knorr, R.L., Lipowsky, R., and Dimova, R. (2010) Effect of cholesterol on the rigidity of saturated and unsaturated membranes: fluctuation and electrodeformation analysis of giant vesicles. *Soft Matter*, **6** (7), 1472–1482.
56 Mecke, K.R., Charitat, T., and Graner, F. (2003) Fluctuating lipid bilayer in an arbitrary potential: theory and experimental determination of bending rigidity. *Langmuir*, **19** (6), 2080–2087.
57 Lee, C.H., Lin, W.C., and Wang, J.P. (2001) All-optical measurements of the bending rigidity of lipid-vesicle membranes across structural phase transitions. *Physical Review E*, **64** (2), 020901.
58 Needham, D. and Evans, E. (1988) Structure and mechanical properties of giant lipid (DMPC) vesicle bilayers from 20-degrees-C below to 10-degrees-C above the liquid-crystal crystalline phase transition at 24-degrees-C. *Biochemistry*, **27** (21), 8261–8269.
59 Needham, D. and Hochmuth, R.M. (1989) Electro-mechanical permeabilization of lipid vesicles – role of membrane tension and compressibility. *Biophysical Journal*, **55** (5), 1001–1009.
60 Olbrich, K., Rawicz, W., Needham, D., and Evans, E. (2000) Water permeability and mechanical strength of polyunsaturated lipid bilayers. *Biophysical Journal*, **79** (1), 321–327.
61 Evans, E., Heinrich, V., Ludwig, F., and Rawicz, W. (2003) Dynamic tension spectroscopy and strength of biomembranes. *Biophysical Journal*, **85** (4), 2342–2350.
62 Evans, E. and Needham, D. (1987) Physical properties of surfactant bilayer membranes – thermal transitions, elasticity, rigidity, cohesion, and colloidal interactions. *Journal of Physical Chemstry*, **91** (16), 4219–4228.
63 Boal, D. (ed.) (2002) *Mechanics of the Cell*, Cambridge University Press, Cambridge.
64 Aranda, S., Riske, K.A., Lipowsky, R., and Dimova, R. (2008) Morphological transitions of vesicles induced by alternating electric fields. *Biophysical Journal*, **95** (2), L19–L21.
65 Winterhalter, M. and Helfrich, W. (1988) Deformation of spherical vesicles by electric fields. *Journal of Colloid and Interface Science*, **122** (2), 583–586.
66 Hyuga, H., Kinosita, K., and Wakabayashi, N. (1991) Transient and steady-state deformations of a vesicle with an insulating membrane in response to step-function or alternating electric fields. *Japanese Journal of Applied Physics Part 1*, **30** (10), 2649–2656.
67 Mitov, M.D., Meleard, P., Winterhalter, M., Angelova, M.I., and Bothorel, P. (1993) Electric-field-dependent thermal

fluctuations of giant vesicles. *Physical Review E*, **48** (1), 628–631.

68 Hyuga, H., Kinosita, K., and Wakabayashi, N. (1993) Steady-state deformation of a vesicle in alternating electric fields. *Bioelectrochemistry and Bioenergetics*, **32** (1), 15–25.

69 Peterlin, P., Svetina, S., and Zeks, B. (2007) The prolate-to-oblate shape transition of phospholipid vesicles in response to frequency variation of an AC electric field can be explained by the dielectric anisotropy of a phospholipid bilayer. *Journal of Physics: Condensed Matter*, **19** (13), 136220.

70 Peterlin, P. (2010) Frequency-dependent electrodeformation of giant phospholipid vesicles in AC electric field. *Journal of Biological Physics*, **36**, 339–354.

71 Vlahovska, P.M., Gracia, R.S., Aranda-Espinoza, S., and Dimova, R. (2009) Electrohydrodynamic model of vesicle deformation in alternating electric fields. *Biophysical Journal*, **96** (12), 4789–4803.

72 Yamamoto, T., Aranda-Espinoza, S., Dimova, R., and Lipowsky, R. (2010) Stability of spherical vesicles in electric fields. *Langmuir*, **26**, 12390–12407.

73 Gao, L.T., Feng, X.Q., and Gao, H.J. (2009) A phase field method for simulating morphological evolution of vesicles in electric fields. *Journal of Computational Physics*, **228** (11), 4162–4181.

74 Gao, L.T., Liu, Y., Qin, Q.H., and Feng, X.Q. (2010) Morphological stability analysis of vesicles with mechanical-electrical coupling effects. *Acta Mechanica Sinica*, **26** (1), 5–11.

75 Niggemann, G., Kummrow, M., and Helfrich, W. (1995) The bending rigidity of phosphatidylcholine bilayers – dependences on experimental method, sample cell sealing and temperature. *Journal of Physics II*, **5** (3), 413–425.

76 Peterlin, P., Svetina, S., and Zeks, B. (2000) The frequency dependence of phospholipid vesicle shapes in an external electric field. *Pfluegers Archiv/European Journal of Physiology*, **439**, R139–R140.

77 Antonova, K., Vitkova, V., and Mitov, M.D. (2010) Deformation of giant vesicles in AC electric fields-Dependence of the prolate-to-oblate transition frequency on vesicle radius. *Europhysics Letters*, **89** (3), 38004.

78 Staykova, M., Lipowsky, R., and Dimova, R. (2008) Membrane flow patterns in multicomponent giant vesicles induced by alternating electric fields. *Soft Matter*, **4** (11), 2168–2171.

79 Lecuyer, S., Ristenpart, W.D., Vincent, O., and Stone, H.A. (2008) Electrohydrodynamic size stratification and flow separation of giant vesicles. *Applied Physics Letters*, **92** (10), 104105.

80 Neumann, E., Sowers, A.E., and Jordan, C. (eds) (1989) *Electroporation and Electrofusion in Cell Biology*, Plenum, New York.

81 Kotnik, T., Pucihar, G., Rebersek, M., Miklavcic, D., and Mir, L.M. (2003) Role of pulse shape in cell membrane electropermeabilization. *Biochimica et Biophysica Acta*, **1614** (2), 193–200.

82 Tekle, E., Astumian, R.D., and Chock, P.B. (1991) Electroporation by using bipolar oscillating electric-field – an improved method for DNA transfection of Nih 3t3 cells. *Proceedings of the National Academy of Sciences of the USA*, **88** (10), 4230–4234.

83 Hyuga, H., Kinosita, K., and Wakabayashi, N. (1991) Deformation of vesicles under the influence of strong electric-fields. *Japanese Journal of Applied Physics Part 1*, **30** (5), 1141–1148.

84 Hyuga, H., Kinosita, K., and Wakabayashi, N. (1991) Deformation of vesicles under the influence of strong electric fields. 2. *Japanese Journal of Applied Physics Part 1*, **30** (6), 1333–1335.

85 Sokirko, A., Pastushenko, V., Svetina, S., and Zeks, B. (1994) Deformation of a lipid vesicle in an electric-field – a theoretical study. *Bioelectrochemistry and Bioenergetics*, **34** (2), 101–107.

86 Neumann, E., Kakorin, S., and Toensing, K. (1998) Membrane electroporation and electromechanical deformation of vesicles and cells. *Faraday Discussions*, **111** (111), 111–125.

87 Griese, T., Kakorin, S., and Neumann, E. (2002) Conductometric and electrooptic relaxation spectrometry of lipid vesicle electroporation at high

fields. *Physical Chemistry Chemical Physics*, **4** (7), 1217–1227.

88 Kakorin, S. and Neumann, E. (2002) Electrooptical relaxation spectrometry of membrane electroporation in lipid vesicles. *Colloids and Surfaces A*, **209** (2–3), 147–165.

89 Riske, K.A. and Dimova, R. (2005) Electro-deformation and poration of giant vesicles viewed with high temporal resolution. *Biophysical Journal*, **88** (2), 1143–1155.

90 Riske, K.A. and Dimova, R. (2005) Timescales involved in electro-deformation, poration and fusion of giant vesicles resolved with fast digital imaging. *Biophysical Journal*, **88** (1), 241A–241A.

91 Kinosita, K., Ashikawa, I., Saita, N., Yoshimura, H., Itoh, H., Nagayama, K., and Ikegami, A. (1988) Electroporation of cell-membrane visualized under a pulsed-laser fluorescence microscope. *Biophysical Journal*, **53** (6), 1015–1019.

92 Abidor, I.G., Arakelyan, V.B., Chernomordik, L.V., Chizmadzhev, Y.A., Pastushenko, V.F., and Tarasevich, M.R. (1979) Electrical breakdown of bilayer lipid-membranes. 1. Main experimental facts and their qualitative discussion. *Bioelectrochemistry and Bioenergetics*, **6** (1), 37–52.

93 Simon, S.A. and McIntosh, T.J. (1986) Depth of water penetration into lipid bilayers. *Methods in Enzymology*, **127**, 511–521.

94 Nagle, J.F. and Tristram-Nagle, S. (2000) Structure of lipid bilayers. *Biochimica et Biophysica Acta*, **1469** (3), 159–195.

95 Hibino, M., Shigemori, M., Itoh, H., Nagayama, K., and Kinosita, K. (1991) Membrane conductance of an electroporated cell analyzed by submicrosecond imaging of transmembrane potential. *Biophysical Journal*, **59** (1), 209–220.

96 Akinlaja, J. and Sachs, F. (1998) The breakdown of cell membranes by electrical and mechanical stress. *Biophysical Journal*, **75** (1), 247–254.

97 Portet, T., Febrer, F.C.I., Escoffre, J.M., Favard, C., Rols, M.P., and Dean, D.S. (2009) Visualization of membrane loss during the shrinkage of giant vesicles under electropulsation. *Biophysical Journal*, **96** (10), 4109–4121.

98 Tsong, T.Y. (1991) Electroporation of cell-membranes. *Biophysical Journal*, **60** (2), 297–306.

99 Weaver, J.C. and Chizmadzhev, Y.A. (1996) Theory of electroporation: a review. *Bioelectrochemistry and Bioenergetics*, **41** (2), 135–160.

100 Teissie, J. and Tsong, T.Y. (1981) Electric-field induced transient pores in phospholipid-bilayer vesicles. *Biochemistry*, **20** (6), 1548–1554.

101 Glaser, R.W., Leikin, S.L., Chernomordik, L.V., Pastushenko, V.F., and Sokirko, A.I. (1988) Reversible electrical breakdown of lipid bilayers – formation and evolution of pores. *Biochimica et Biophysica Acta*, **940** (2), 275–287.

102 Kinosita, K.J., Hibino, M., Itoh, H., Shigemori, M., Hirano, K., Kirino, Y., and Hayakawa, T. (1992) Events of membrane elecroporation visualized on a time scale from microseconds to seconds, in *Guide to Electroporation and Electrofusion* (eds D.C. Chang, B.M. Chassy, J.A. Saunders, and A.E. Sowers), Academic Press, New York, pp. 29–46.

103 Zhelev, D.V. and Needham, D. (1993) Tension-stabilized pores in giant vesicles – determination of pore-size and pore line tension. *Biochimica et Biophysica Acta*, **1147** (1), 89–104.

104 Sandre, O., Moreaux, L., and Brochard-Wyart, F. (1999) Dynamics of transient pores in stretched vesicles. *Proceedings of the National Academy of Sciences of the USA*, **96** (19), 10591–10596.

105 Tekle, E., Astumian, R.D., Friauf, W.A., and Chock, P.B. (2001) Asymmetric pore distribution and loss of membrane lipid in electroporated DOPC vesicles. *Biophysical Journal*, **81** (2), 960–968.

106 Rodriguez, N., Cribier, S., and Pincet, F. (2006) Transition from long- to short-lived transient pores in giant vesicles in an aqueous medium. *Physical Review E*, **74** (6), 061902.

107 Riske, K.A. and Dimova, R. (2006) Electric pulses induce cylindrical deformations on giant vesicles in salt solutions. *Biophysical Journal*, **91** (5), 1778–1786.

108 Krassowska, W. and Filev, P.D. (2007) Modeling electroporation in a single cell. *Biophysical Journal*, **92** (2), 404–417.

109 Harbich, W. and Helfrich, W. (1979) Alignment and opening of giant lecithin vesicles by electric fields. *Zeitschrift fur Naturforschung A*, **34** (9), 1063–1065.

110 Portet, T. and Dimova, R. (2010) A new method for measuring edge tensions and stability of lipid bilayers: effect of membrane composition. *Biophysical Journal*, **99**, 3264–3273.

111 Knorr, R.L., Staykova, M., Gracia, R.S., and Dimova, R. (2010) Wrinkling and electroporation of giant vesicles in the gel phase. *Soft Matter*, **6** (9), 1990–1996.

112 Cerda, E. and Mahadevan, L. (2003) Geometry and physics of wrinkling. *Physical Review Letters*, **90** (7), 074302.

113 Cerda, E., Ravi-Chandar, K., and Mahadevan, L. (2002) Thin films – wrinkling of an elastic sheet under tension. *Nature*, **419** (6907), 579–580.

114 Finken, R. and Seifert, U. (2006) Wrinkling of microcapsules in shear flow. *Journal of Physics: Condensed Matter*, **18** (15), L185–L191.

115 Elmashak, E.M. and Tsong, T.Y. (1985) Ion selectivity of temperature-induced and electric-field induced pores in dipalmitoylphosphatidylcholine vesicles. *Biochemistry*, **24** (12), 2884–2888.

116 Risselada, H.J. and Marrink, S.J. (2009) The freezing process of small lipid vesicles at molecular resolution. *Soft Matter*, **5**, 4531–4541.

117 Evans, E. and Heinrich, V. (2003) Dynamic strength of fluid membranes. *Comptes Rendus Physique*, **4** (2), 265–274.

118 Boucher, P.A., Joos, B., Zuckermann, M.J., and Fournier, L. (2007) Pore formation in a lipid bilayer under a tension ramp: modeling the distribution of rupture tensions. *Biophysical Journal*, **92** (12), 4344–4355.

119 Leontiadou, H., Mark, A.E., and Marrink, S.J. (2004) Molecular dynamics simulations of hydrophilic pores in lipid bilayers. *Biophysical Journal*, **86** (4), 2156–2164.

120 Mukherjee, S. and Maxfield, F.R. (2004) Membrane domains. *Annual Review of Cell and Developmental Biology*, **20**, 839–866.

121 Gabriel, B. and Teissie, J. (1997) Direct observation in the millisecond time range of fluorescent molecule asymmetrical interaction with the electropermeabilized cell membrane. *Biophysical Journal*, **73** (5), 2630–2637.

122 Gabriel, B. and Teissie, J. (1999) Time courses of mammalian cell electropermeabilization observed by millisecond imaging of membrane property changes during the pulse. *Biophysical Journal*, **76** (4), 2158–2165.

123 Discher, B.M., Won, Y.Y., Ege, D.S., Lee, J.C.M., Bates, F.S., Discher, D.E., and Hammer, D.A. (1999) Polymersomes: tough vesicles made from diblock copolymers. *Science*, **284** (5417), 1143–1146.

124 Dimova, R., Seifert, U., Pouligny, B., Förster, S., and Döbereiner, H.-G. (2002) Hyperviscous diblock copolymer vesicles. *European Physical Journal B*, **7**, 241–250.

125 Kantsler, V., Segre, E., and Steinberg, V. (2007) Vesicle dynamics in time-dependent elongation flow: wrinkling instability. *Physical Review Letters*, **99** (17), 178102.

126 Lipowsky, R. and Dobereiner, H.G. (1998) Vesicles in contact with nanoparticles and colloids. *Europhysics Letters*, **43** (2), 219–225.

127 Sens, P. and Isambert, H. (2002) Undulation instability of lipid membranes under an electric field. *Physical Review Letters*, **88** (12), 128102.

128 Petrov, A.G. (1984) Flexoelectricity of lyotropics and biomembranes. *Nuovo Cimento della Societa Italiana di Fisica D*, **3** (1), 174–192.

129 Petrov, A.G. and Bivas, I. (1984) Elastic and flexoelectic aspects of out-of-plane fluctuations in biological and model membranes. *Progress in Surface Science*, **16** (4), 389–512.

130 Petrov, A.G. (2002) Flexoelectricity of model and living membranes. *Biochimica et Biophysica Acta*, **1561** (1), 1–25.

131 Gao, L.T., Feng, X.Q., Yin, Y.J., and Gao, H.J. (2008) An electromechanical liquid crystal model of vesicles. *Journal* of the Mechanica and Physics of Solids, **56** (9), 2844–2862.

132 Simons, K. and Ikonen, E. (1997) Functional rafts in cell membranes. *Nature*, **387** (6633), 569–572.

133 Dietrich, C., Bagatolli, L.A., Volovyk, Z.N., Thompson, N.L., Levi, M., Jacobson, K., and Gratton, E. (2001) Lipid rafts reconstituted in model membranes. *Biophysical Journal*, **80** (3), 1417–1428.

134 Evans, E. and Rawicz, W. (2094) 1990) Entropy-driven tension and bending elasticity in condensed-fluid membranes. *Physical Review Letters*, **64** (17), 2097.

135 Song, J.B. and Waugh, R.E. (1993) Bending rigidity of SOPC membranes containing cholesterol. *Biophysical Journal*, **64** (6), 1967–1970.

136 Meleard, P., Gerbeaud, C., Pott, T., FernandezPuente, L., Bivas, I., Mitov, M.D., Dufourcq, J., and Bothorel, P. (1997) Bending elasticities of model membranes: influences of temperature and sterol content. *Biophysical Journal*, **72** (6), 2616–2629.

137 Henriksen, J., Rowat, A.C., Brief, E., Hsueh, Y.W., Thewalt, J.L., Zuckermann, M.J., and Ipsen, J.H. (2006) Universal behavior of membranes with sterols. *Biophysical Journal*, **90** (5), 1639–1649.

138 Pan, J.J., Mills, T.T., Tristram-Nagle, S., and Nagle, J.F. (2008) Cholesterol perturbs lipid bilayers nonuniversally. *Physical Review Letters*, **100** (19), 198103.

139 Pan, J., Tristram-Nagle, S., Kucerka, N., and Nagle, J.F. (2008) Temperature dependence of structure, bending rigidity, and bilayer interactions of dioleoylphosphatidylcholine bilayers. *Biophysical Journal*, **94** (1), 117–124.

140 Mathai, J.C., Tristram-Nagle, S., Nagle, J.F., and Zeidel, M.L. (2008) Structural determinants of water permeability through the lipid membrane. *Journal of General Physiology*, **131** (1), 69–76.

141 Karatekin, E., Sandre, O., Guitouni, H., Borghi, N., Puech, P.H., and Brochard-Wyart, F. (2003) Cascades of transient pores in giant vesicles: line tension and transport. *Biophysical Journal*, **84** (3), 1734–1749.

142 Riske, K.A., Knorr, R.L., and Dimova, R. (2009) Bursting of charged multicomponent vesicles subjected to electric pulses. *Soft Matter*, **5**, 1983–1986.

143 Raffy, S. and Teissie, J. (1999) Control of lipid membrane stability by cholesterol content. *Biophysical Journal*, **76** (4), 2072–2080.

144 Genco, I., Gliozzi, A., Relini, A., Robello, M., and Scalas, E. (1993) Electroporation in symmetrical and asymmetric membranes. *Biochimica et Biophysica Acta*, **1149** (1), 10–18.

145 Isambert, H. (1998) Understanding the electroporation of cells and artificial bilayer membranes. *Physical Review Letters*, **80** (15), 3404–3407.

146 Betterton, M.D. and Brenner, M.P. (1999) Electrostatic edge instability of lipid membranes. *Physical Review Letters*, **82** (7), 1598–1601.

147 Kumaran, V. (2000) Instabilities due to charge-density-curvature coupling in charged membranes. *Physical Review Letters*, **85** (23), 4996–4999.

148 Schwister, K. and Deuticke, B. (1985) Formation and properties of aqueous leaks induced in human erythrocytes by electrical breakdown. *Biochimica et Biophysica Acta*, **816** (2), 332–348.

149 Kinosita, K. and Tsong, T.Y. (1977) Formation and resealing of pores of controlled sizes in human erythrocyte membrane. *Nature*, **268** (5619), 438–441.

150 Tekle, E., Astumian, R.D., and Chock, P.B. (1994) Selective and asymmetric molecular-transport across electroporated cell membranes. *Proceedings of the National Academy of Sciences of the USA*, **91** (24), 11512–11516.

151 Litster, J.D. (1975) Stability of lipid bilayers and red blood-cell membranes. *Physics Letters A*, **53** (3), 193–194.

152 Newman, C.M.H. and Bettinger, T. (2007) Gene therapy progress and prospects: ultrasound for gene transfer. *Gene Therapy*, **14** (6), 465–475.

153 Escoffre, J.M., Portet, T., Wasungu, L., Teissie, J., Dean, D., and Rols, M.P. (2009) What is (still not) known of the mechanism by which electroporation mediates gene transfer and expression in cells and tissues. *Molecular Biotechnology*, **41** (3), 286–295.

154 Chang, D.C. and Reese, T.S. (1990) Changes in membrane structure induced by electroporation as revealed by rapid-freezing electron microscopy. *Biophysical Journal*, **58** (1), 1–12.

155 Chang, D.C. (1992) Structure and dynamics of electric field-induced membrane pores as revealed by rapid-freezing electron microscopy, in *Guide to Electroporation and Electrofusion* (eds D.C. Chang, B.M. Chassy, J.A. Saunders, and A.E. Sowers), Academic Press, San Diego, pp. 9–27.

156 Puech, P.H., Borghi, N., Karatekin, E., and Brochard-Wyart, F. (2003) Line thermodynamics: adsorption at a membrane edge. *Physical Review Letters*, **90** (12), 128304.

157 Srividya, N. and Muralidharan, S. (2008) Determination of the line tension of giant vesicles from pore-closing dynamics. *Journal of Physical Chemistry B*, **112** (24), 7147–7152.

158 Portet, T., Dimova, R., Dean, D.S., and Rols, M.-P. (2010) Electric fields and giant vesicles. *Biophysical Journal*, **98** (3 Suppl. 1), 77a.

159 Ellens, H., Bentz, J., and Szoka, F.C. (1986) Fusion of phosphatidylethanolamine-containing liposomes and mechanism of the L_α-H_{II} phase-transition. *Biochemistry*, **25** (14), 4141–4147.

160 Lewis, R.N.A.H. and McElhaney, R.N. (1993) Calorimetric and spectroscopic studies of the polymorphic phase behavior of a homologous series of n-saturated 1,2-diacyl phosphatidylethanolamines. *Biophysical Journal*, **64** (4), 1081–1096.

161 Pink, D.A., McNeil, S., Quinn, B., and Zuckermann, M.J. (1998) A model of hydrogen bond formation in phosphatidylethanolamine bilayers. *Biochimica et Biophysica Acta*, **1368** (2), 289–305.

162 Kinosita, K. and Tsong, T.Y. (1977) Voltage-induced pore formation and hemolysis of human erythrocytes. *Biochimica et Biophysica Acta*, **471** (2), 227–242.

163 Chang, D.C., Chassy, B.M., Saunders, J.A., and Sowers, A.E. (eds) (1992) *Guide to Electroporation and Electrofusion*, Academic Press, San Diego.

164 Zimmermann, U. and Neil, G.A. (eds) (1996) *Electromanipulation of Cells*, CRC Press, Boca Raton.

165 Zimmermann, U. (1986) Electrical breakdown, electropermeabilization and electrofusion. *Reviews of Physiology Biochemistry and Pharmacology*, **105**, 175–256.

166 Llinas, R., Steinberg, I.Z., and Walton, K. (1981) Relationship between presynaptic calcium current and postsynaptic potential in squid giant synapse. *Biophysical Journal*, **33** (3), 323–351.

167 Lindau, M. and de Toledo, G.A. (2003) The fusion pore. *Biochimica et Biophysica Acta*, **1641** (2–3), 167–173.

168 Hafez, I., Kisler, K., Berberian, K., Dernick, G., Valero, V., Yong, M.G., Craighead, H.G., and Lindau, M. (2005) Electrochemical imaging of fusion pore openings by electrochemical detector arrays. *Proceedings of the National Academy of Sciences of the USA*, **102** (39), 13879–13884.

169 Shillcock, J.C. and Lipowsky, R. (2005) Tension-induced fusion of bilayer membranes and vesicles. *Nature Materials*, **4** (3), 225–228.

170 Veatch, S.L. and Keller, S.L. (2003) Separation of liquid phases in giant vesicles of ternary mixtures of phospholipids and cholesterol. *Biophysical Journal*, **85** (5), 3074–3083.

171 Kahya, N., Scherfeld, D., Bacia, K., Poolman, B., and Schwille, P. (2003) Probing lipid mobility of raft-exhibiting model membranes by fluorescence correlation spectroscopy. *Journal of Biological Chemistry*, **278** (30), 28109–28115.

172 Chiu, D.T., Wilson, C.F., Ryttsen, F., Stromberg, A., Farre, C., Karlsson, A., Nordholm, S., Gaggar, A., Modi, B.P., Moscho, A., Garza-Lopez, R.A., Orwar, O., and Zare, R.N. (1999) Chemical transformations in individual ultrasmall biomimetic containers. *Science*, **283** (5409), 1892–1895.

173 Fischer, A., Franco, A., and Oberholzer, T. (2002) Giant vesicles as microreactors

for enzymatic mRNA synthesis. *ChemBioChem*, **3** (5), 409–417.

174 Kulin, S., Kishore, R., Helmerson, K., and Locascio, L. (2003) Optical manipulation and fusion of liposomes as microreactors. *Langmuir*, **19** (20), 8206–8210.

175 Noireaux, V. and Libchaber, A. (2004) A vesicle bioreactor as a step toward an artificial cell assembly. *Proceedings of the National Academy of Sciences of the USA*, **101** (51), 17669–17674.

176 Yang, P., Lipowsky, R., and Dimova, R. (2009) Nanoparticle formation in giant vesicles: synthesis in biomimetic compartments. *Small*, **5** (18), 2033–2037.

177 Cohen, F.S., Akabas, M.H., and Finkelstein, A. (1982) Osmotic swelling of phospholipid-vesicles causes them to fuse with a planar phospholipid-bilayer membrane. *Science*, **217** (4558), 458–460.

178 Grafmuller, A., Shillcock, J., and Lipowsky, R. (2007) Pathway of membrane fusion with two tension-dependent energy barriers. *Physical Review Letters*, **98** (21), 218101.

179 Estes, D.J., Lopez, S.R., Fuller, A.O., and Mayer, M. (2006) Triggering and visualizing the aggregation and fusion of lipid membranes in microfluidic chambers. *Biophysical Journal*, **91** (1), 233–243.

180 Lentz, B.R. (2007) PEG as a tool to gain insight into membrane fusion. *European Biophysical Journal*, **36** (4–5), 315–326.

181 Kunishima, M., Tokaji, M., Matsuoka, K., Nishida, J., Kanamori, M., Hioki, K., and Tani, S. (2006) Spontaneous membrane fusion induced by chemical formation of ceramides in a lipid bilayer. *Journal of the American Chemical Society*, **128** (45), 14452–14453.

182 Jahn, R., Lang, T., and Sudhof, T.C. (2003) Membrane fusion. *Cell*, **112** (4), 519–533.

183 Bhattacharya, D. and Gupta, R.K. (2005) Nanotechnology and potential of microorganisms. *Critical Reviews in Biotechnology*, **25** (4), 199–204.

184 Mandal, D., Bolander, M.E., Mukhopadhyay, D., Sarkar, G., and Mukherjee, P. (2006) The use of microorganisms for the formation of metal nanoparticles and their application. *Applied Microbiology and Biotechnology*, **69** (5), 485–492.

185 Sanchez, C., Arribart, H., and Guille, M.M.G. (2005) Biomimetism and bioinspiration as tools for the design of innovative materials and systems. *Nature Materials*, **4** (4), 277–288.

186 Dickerson, M.B., Sandhage, K.H., and Naik, R.R. (2008) Protein- and peptide-directed syntheses of inorganic materials. *Chemical Reviews*, **108** (11), 4935–4978.

187 Mann, S., Hannington, J.P., and Williams, R.J.P. (1986) Phospholipid-vesicles as a model system for biomineralization. *Nature*, **324** (6097), 565–567.

188 Bhandarkar, S. and Bose, A. (1990) Synthesis of nanocomposite particles by intravesicular coprecipitation. *Journal of Colloid and Interface Science*, **139** (2), 541–550.

189 Khramov, M.I. and Parmon, V.N. (1993) Synthesis of ultrafine particles of transition-metal sulfides in the cavities of lipid vesicles and the light-stimulated transmembrane electron-transfer catalyzed by these particles. *Journal of Photochemistry and Photobiology A*, **71** (3), 279–284.

190 Korgel, B.A. and Monbouquette, H.G. (2000) Controlled synthesis of mixed core and layered (Zn,Cd)S and (Hg,Cd)S nanocrystals within phosphatidylcholine vesicles. *Langmuir*, **16** (8), 3588–3594.

191 Ishikawa, K., Sato, K., Shima, Y., Urabe, I., and Yomo, T. (2004) Expression of a cascading genetic network within liposomes. *FEBS Letters*, **576** (3), 387–390.

192 Kita, H., Matsuura, T., Sunami, T., Hosoda, K., Ichihashi, N., Tsukada, K., Urabe, I., and Yomo, T. (2008) Replication of genetic information with self-encoded replicase in liposomes. *ChemBioChem*, **9** (15), 2403–2410.

193 Hsin, T.M. and Yeung, E.S. (2007) Single-molecule reactions in liposomes. *Angeweldte Chemie International Edition*, **46** (42), 8032–8035.

194 Christensen, S.M. and Stamou, D. (2007) Surface-based lipid vesicle reactor systems: fabrication and applications. *Soft Matter*, **3** (7), 828–836.

8
Electroporation for Medical Use in Drug and Gene Electrotransfer

Julie Gehl

8.1
Introduction

Using electric pulses to manipulate cell membranes started in a community of physicists, actually with contributions from three different continents. The idea that dielectric breakdown of membranes could be used in medicine gained new life with the groundbreaking paper of Neumann *et al.* in 1982 [1], showing that DNA could be transferred to eukaryotic cells in a custom-built electroporation chamber. Eventually other researchers started to work with a short-circuited gel electrophoresis system in order to obtain electrical discharge which could be used.

As is known to molecular biologists, DNA electrotransfer to bacteria is a standard method, and a small pulse generator is everyday equipment in most laboratories. However, whereas the molecular biologist would not hesitate to let billions of bacteria succumb to side effects of high-field electroporation, this view is of course not shared by those working with mammalian cells–and doctors working in the clinic would have safety as the major priority [2].

In order to work with mammalian cells, pulse parameters need to be much better controlled, and this is efficiently done with a square wave electroporator [3]. Here, the pulse amplitude and duration may be individually controlled, allowing a much better adaptability to the target cells or tissues as well as to the molecule that is to be transferred. Another important stage of the development has been the availability of square wave pulse generators designed and approved for clinical use.

Alongside this technological development, important studies have been performed, enabling a much more detailed understanding of how cells and tissues respond to electric fields [4–6], and how the technologies may most optimally be used to transfer molecules or ablate tumor tissue [7–10]. This chapter is devoted to the medical applications of electroporation-based technologies, including how we may understand their scientific basis and exploitation.

Advances in Electrochemical Science and Engineering. Edited by Richard C. Alkire, Dieter M. Kolb, and Jacek Lipkowski

ISBN: 978-3-527-32885-7

8.2
A List of Definitions

Nomenclature tends to grow with the expansion of the field, and it would be useful to start this chapter with some definitions. These are of course debatable; indeed in the course of history scientific battles have been fought over single words and their use. The following is a list of terms and their meaning as used in this chapter.

Electroporation
Electroporation is use of electric fields to cause cell membrane permeabilization.

Electropermeabilization
This is the same as above. Electropermeabilization is technically more correct, as indeed what we know is that permeabilization structures are created, not necessarily that these have a particular shape, as the word "pore" would imply. However, as the word electroporation has taken precedence and is overwhelmingly used in the literature, this is also the term used in this chapter.

Electrotransfer
This term covers the use of electric pulses to assist transfer of molecules into cells. In principle a membrane would not necessarily need to be measurably porated or permeabilized, possibly only sufficiently destabilized or perturbed to allow electrophoretic transfer across it. Examples of uses are DNA electrotransfer, siRNA electrotransfer, etc.

Reversible Electroporation
This term describes permeabilization that may be reversed to a state where the cell membrane is resealed and the cell may continue its life cycle.

Irreversible Electroporation
This describes a state where a cell will reach a point of no return, a point where depletion of energy, ionic imbalance, and other factors will eventually lead to cell death.

Membrane Perturbation
A membrane may be temporarily destabilized to an extent where effects can occur that would not normally be possible–but the membrane is not measurably permeabilized in terms of molecules passing the membrane.

Pore

A large body of literature describes the occurrence of pores, of various sizes and lifetimes, as the result of electroporation. *In vivo* these pores may not necessarily be neatly rounded structures, but rather a mixture of shapes of permeabilization structures depending on the effect of transmembrane protein domains, cell shape, and surrounding matrix on the cell membrane response to the field, i.e. these permeabilization structures could either be circular in form, or have other forms depending on local factors.

Resealing Time

This is generally represented with τ, and defined as the point in time when two-thirds of the membrane permeabilization structures have resealed. Generally this is analyzed by measuring transport of molecules across a membrane.

Electrochemotherapy

Electrochemotherapy is the use of electric pulses to increase uptake of cytotoxic drugs. Drugs used in the clinic include bleomycin and cisplatin.

Voltage-to-Distance Ratio

An applied electric field is very often described in terms of units of $\mathrm{V\,cm^{-1}}$, indicating the applied voltage relative to the distance between electrodes measured in centimeters. This, in principle, does not describe the electric field in any detail since the actual field will vary depending on a number of factors in the electrode geometry. Direct comparison only works if it is the same set of electrodes being used in the same tissue.

8.3 How We Understand Permeabilization at the Cellular and Tissue Level

Before discussing the medical uses of electroporation, it is useful to look at how we may understand the concept of permeabilization at both the cellular and the tissue level. The simple explanation, which will definitely do for a quick introduction to the field, is shown in Figure 8.1. An electric field is applied that surpasses the capacitance of the membrane; dielectric breakdown of the membrane ensues, most prominently at the side of the cell facing the positive electrode since the cell has an intrinsic negative resting potential. Images of electroporation at the single-cell level may beautifully link theory to empirical findings (Figure 8.1).

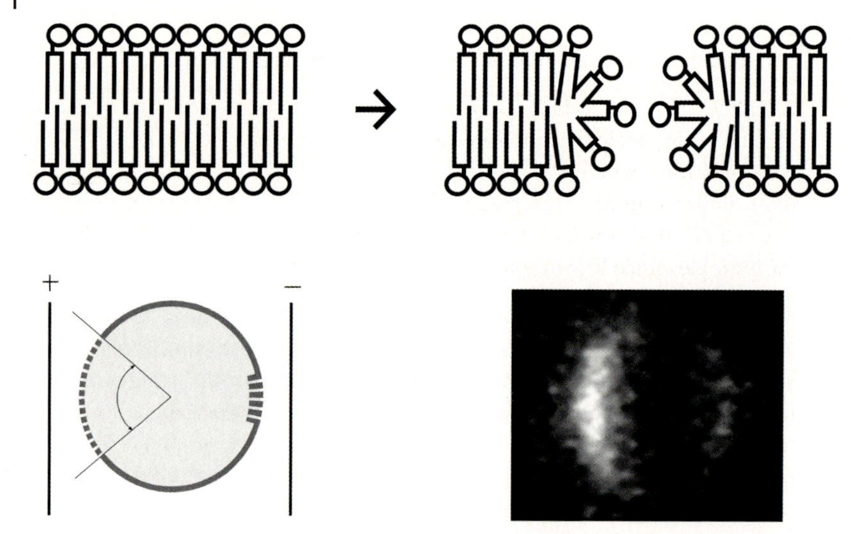

Figure 8.1 Schematic of cell electroporation. A basic understanding is that when the electric field exceeds the capacitance of the cell membrane, permeabilization structures will occur more prominently at the cell pole facing the positive electrode, due to the inherent negative resting potential of the cell. This is nicely supported by empirical data, the image at lower right showing propidium iodide uptake after pulsation, as described in [37, 38].

However, the situation gets a lot more complex when electroporation is performed in tissues.

1) Cells in tissues contain an abundance of proteins with transmembrane domains, meaning that deformation and response to the electric field differ from those of a lipid vesicle, or even a single cell after detachment.
2) Cells in tissues are rarely spherical, but rather have all types of shapes, from ellipsoid to square to stellate. This affects deformation in response to the field and the threshold for electroporation may depend on the direction of the field versus the orientation of the cell.
3) There are gap junctions, which may mean that one cell may be electrically connected to others, and thereby act as a much larger cell in the electric field [11, 12].
4) The influence of neighboring cells and structures [13] has to be taken into account. The field is likely to be inhomogeneous, and depends on the presence of neighboring structures such as high-conductive areas with fluid, or low-conductive areas on bone surfaces. Furthermore, there may be an influence

of permeabilization on neighboring cells, meaning that there is a dynamic development when one long or several shorter pulses are applied.

5) In tissues, the extracellular volume is small compared to the intracellular volume–completely the opposite of the situation when performing *in vitro* electroporation where the suspension medium volume by far exceeds the intracellular volume. This means that the threshold for electroporation is lowered [5]. It also means that the cell may tolerate permeabilization better, for example, since loss of ionic homeostasis may be less pronounced [6].

6) Furthermore, whereas a suspension medium may be viewed as a liquid, this would be an imprecise description of the extracellular volume. The extracellular volume is somewhere between a gel and a semistructured solid, containing extracellular matrix with complex glycoproteins. Indeed, the cell interior has also been described as a gel or molecular soup [14], so one could view tissue as gel-like or semistructured matter that contains lipid bilayer barriers in rich quantity.

This difference between the *in vitro* situation, which has of course served as a very important research object to understand basic points about electroporation, and tissue response to electric fields is important to highlight because proper attention must be paid to consequences of these differences. Schematic representations of this more complex situation are attempted in Figures 8.2 and 8.3.

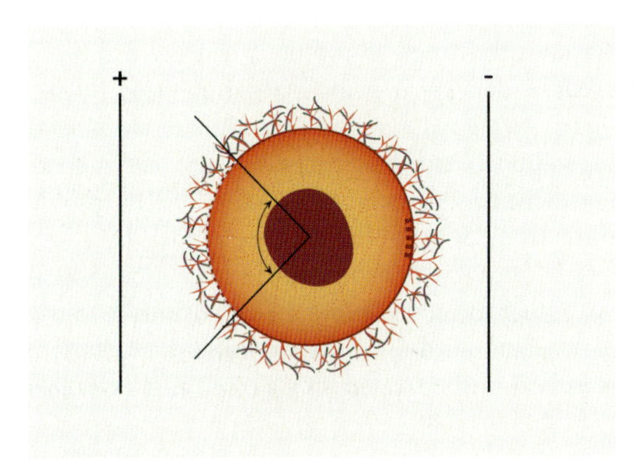

Figure 8.2 Cell electroporation: a more complex view. The cell membrane as shown in Figure 8.1 is of course a simplistic view of the situation. A more detailed (but far from complete) view would be including the cytoskeleton (dotted lines), which has also been shown to be involved in both formation and resealing of permeabilization structures. Also, transmembrane proteins are schematically depicted, indicating that the bilayer lipid membrane is more than just that, and also that molecules entering into the cell as a result of electrotransfer do so after negotiating both the extracellelular domains of charged proteins and the cell membrane itself.

Figure 8.3 Cell electroporation at the tissue level. Yet another layer of complexity is added when the cell is viewed as part of a tissue. The extracellular volume is not a liquid medium where molecules freely diffuse, but rather a dense mesh of structures and molecules interspersed between cells of varying sizes and shapes, and all linked with transmembrane proteins and intracellular structures. The distribution of molecules and the electric field distribution will necessarily be dynamic and also nonlinearly dynamic.

8.4 Basic Aspects of Electroporation that are of Particular Importance for Medical Use

8.4.1 Delivery of Drugs

There are basically two delivery routes for drugs: intravenous delivery and intratumoral delivery [10, 15]. Intravenous delivery is being extensively used for the chemotherapeutic agent bleomycin, with good success. Indeed there are several aspects that make this a favorable solution. First, bleomycin does not readily enter cells. Therefore, an intravenous delivery will have a favorable toxicity profile despite whole-body exposure to the drug. Second, the increase in toxicity is very marked once direct access to the cell cytosol is provided; thus even a limited local dose will exert a sufficient effect. Other drugs to be administered intravenously for delivery during electroporation would need to be able to fulfill this criterion, in order to have both sufficient effects in the treated area as well as a favorable toxicity profile.

Intratumoral delivery (or, in the case of drugs with other purposes, delivery to the local tissue) has the advantage of low whole-body exposure. On the other hand, the distribution of drug in the tissue will be uneven. Thus, washout to the systemic circulation is bound to occur. Washout may be delayed if local analgesic injections contain epinephrine, and also if the injection is timed so that cells are permeabilized immediately around the time of injection, as this will allow the permeabilized cells to take up drug at maximal concentration, and as the cells reseal, the maximal drug concentration could indeed be trapped inside the cells allowing sequestration of the drug intracellularly.

8.4.2
Delivery of DNA

DNA is a large and highly charged molecule with very limited diffusion capacity by itself. Injection in suspension may add a hydrostatic pressure that can aid the distribution in a tissue. However, there is also an advantage to this situation, which is that DNA also does not readily diffuse *out* of the injected area easily. Thus it has been shown that pulse delivery even minutes after DNA injection still leads to excellent levels of transgenic expression [16, 17].

Intravenous distribution of DNA has been tried without success. This is because DNA is degraded by DNAases and furthermore the whole-body distribution means that the concentration at the tissue level becomes very low. An intermediate solution is arterial injection, for example in the hepatic artery, and this may lead to an acceptable concentration at the tissue level, with resultant transgenic expression.

Certain tissues may allow distribution of a hydrophilic solution more easily than others, with muscle being an example of a tissue that allows distribution along fibers. In contrast, brain tissue is highly lipophilic in nature, giving a more focal distribution pattern for hydrophilic solutions.

DNA will move in a tissue as a function of electrophoretic forces on the polyanion. This has been measured, and is estimated to have importance only on a very small scale in terms of tissue distribution [18].

It has been shown that DNA must be administered *prior* to the application of an electric field, as DNA is unable to diffuse through cell membranes by itself [16]. DNA electrotransfer (Figure 8.4) works by steps involving membrane adsorption, membrane perturbation (by the electric field), electrophoretic movement of the DNA, and finally cellular processing of the internalized molecule, as nicely reviewed in [19].

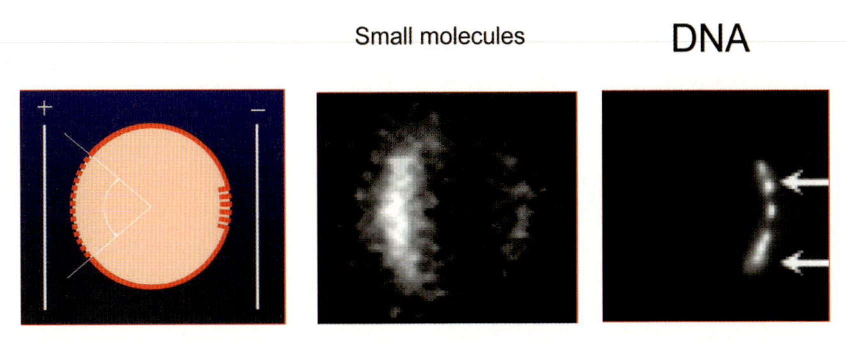

Figure 8.4 Schematic of DNA electrotransfer. Whereas small molecules such as used for chemotherapy diffuse through small permeabilization structures of cell membranes, this is not the case for the highly charged polyanion DNA. As can be seen on the right, DNA enters the cell moving towards the positive electrode driven by electrophoretic forces, as demonstrated in [40].

8.4.3 Delivery of Other Molecules

A number of other molecules may be delivered by electroporation: oligonucleotides [20], ions [21, 22], dyes [23], and radioactive tracers [24]—just to mention a few. For *in vivo* injection it is important to consider the following.

1) Washout of the molecule from tissue. Is it a very rapid process, for example, as seen for ^{51}Cr-EDTA which has an extremely rapid renal clearance? Or will the molecule remain in the target area much longer, as would be the case for proteins and larger nucleotides?
2) Charge of the molecule, which may affect uptake by electroporation at the cellular level, but charge may also lead to sequestration in the extracellular volume. An example of an oligonucleotide class (peptide nucleic acid) and how charge may play an important role are given in Joergensen *et al.* [20].

8.4.4 Delivery of Electric Pulses

Delivery of electric pulses is a science in itself, and a tremendous variety of pulse combinations is possible. Increasing the applied voltage will lead to increased permeabilization to the point of irreversible permeabilization. Increasing the number of pulses will also lead to increased permeabilization, but the correlation is only linear in the range of about 1–20 pulses. Usually the number of pulses chosen is 8–10, which gives a reasonable trade-off.

Pulse length is the third very important issue. Generally, short high-voltage pulses of, for example, 100 μs are used to augment uptake of small molecules, as it has been shown that the pulse amplitude defines the extent (and degree) of permeabilization, and so high-voltage pulses will lead to a large area of the membrane being permeabilized and thus open for diffusion of molecules. For DNA, typically long pulses are used, up to 400 ms long and with lower voltage. This allows an electrophoretic effect assisting DNA passage into the cytosol. Pulse characteristics used differ greatly, both as far as what is recommended for individual tissues and what is recommended from individual investigators.

8.4.5 End of the Permeabilized State

Resealing of cells after the end of the permeabilized state happens in part through energy-consuming processes and spontaneous resealing. Cells may succumb due to resealing not occurring before ionic imbalance and adenosine triphosphate loss bringing the life cycle to an end. The threshold for reversible and irreversible permeabilization may be well defined, although in a particular tissue the two phenomena may coexist due to variations in cell size and electrode geometry. Resealing is well described, and occurs in a time scale of the order of minutes.

8.4.6
The Vascular Lock

A particular phenomenon of great importance for the antitumor effect of electrochemotherapy is an antivascular effect. This effect has several elements:

1) A direct reflexory constriction at the level of the arterioles, lasting only a few minutes, in direct response to the electric pulses [25].
2) A permeabilizing effect on the tumor vessels leading to collapse of intratumoral capillaries [26].
3) A high interstitial pressure in the tumor combined with additional edema due to leakage of increased intracellular sodium after permeabilization, leading to further compression of the already collapsed tumor capillaries.
4) A direct toxic effect on endothelial cells as well as on tumor cells due to the internalization of bleomycin [27].

All these factors mean that electrochemotherapy is highly useful in the treatment of hemorrhaging tumors [28] (see also Figure 8.6).

8.5
How to Deliver Electric Pulses in Patient Treatment

8.5.1
Pulse Generators and Electrodes

For electrochemotherapy there is at this point only the Cliniporator equipment available (Carpi, IGEA, Italy). This generator allows simultaneous collection of voltage and current traces, and allows storage of treatment data, and is CE marked for clinical use. It works together with two types of needle electrodes as well as plate electrodes for cutaneous applications. It can also be used with endoscopic electrodes (Section 8.8.1), expandable soft tissue electrodes, and pin electrodes for bone metastases (Section 8.8.2). It can be used for gene therapy, as reviewed in [3].

8.5.2
Anesthesia

For smaller tumors, not too numerous and in areas where it may be applied, local anesthesia would be the optimal choice. Generally, an infiltration analgesia around the area treated gives good coverage, and the patient may be discharged immediately after the treatment procedure. It has been shown that the actual applied voltage is important for both the level of muscle contraction and the pain sensation, and therefore needle electrodes with, for example, a 4 mm gap may be preferred for the treatment with local anesthesia. Performed according to standard operating procedures, pain sensation should be minimal [15].

For larger tumors, multiple tumors, and tumors in areas where it may be difficult to apply sufficient local anesthesia (e.g., periost involvement in the scalp), general anesthesia is recommended. This may be handled by short-acting anesthetic agents such as propofol combined with short-acting opioids [15]. As pulses should be administered within approximately a 20 min window, the anesthesia can be short and the patient discharged on the same day [9].

8.6
Treatment and Post-treatment Management

Underlying muscle may be exited directly by the electric pulse, leading to a brief contraction. These contractions are limited to the muscle immediately beneath the treatment area and may be alleviated by lifting the lesion during treatment.

When using very high field strengths (e.g., as is used in the irreversible electroporation algorithm), interference with heart rhythm has been observed. This has been alleviated by the development of pulse generators avoiding the refractory phase of the heart cycle. However, for pulses used in electrochemotherapy no adverse events have been reported, despite use of electric pulses directly on the left chest wall.

After electrochemotherapy, necrosis/apoptosis will immediately ensue. If the tumor treated is superficially located, for example in the skin, an ulcerated area will form as a transition between the tumor and scar tissue. In particular in previously irradiated areas, this may take several weeks [9]. Standard wound care may be applied.

8.7
Clinical Results with Electrochemotherapy

8.7.1
Tumors Up to Three Centimeters in Size

The first clinical trial of electrochemotherapy was published in 1993 [29], with treatment of cutaneous metastases from head and neck cancer. This was followed by several studies of patients with malignant melanoma metastases and basal cell carcinoma [30, 31]. In the first study, intravenous or intraarterial bleomycin was used, and in the latter intratumoral injection of $5\,U\,ml^{-1}$ was used, for both the tumor and the surrounding area. In the first study, plate electrodes were used, where as in the latter a needle electrode device was used.

In a European Union framework program, four centers worked together on defining standard operating procedures for electrochemotherapy. Three different types of electrodes were used, local drug administration of either cisplatin or bleomycin (here in a $1\,U\,ml^{-1}$ concentration) or intravenous bleomycin was used, and either local or general anesthesia. In 41 patients, 171 tumor nodules were treated

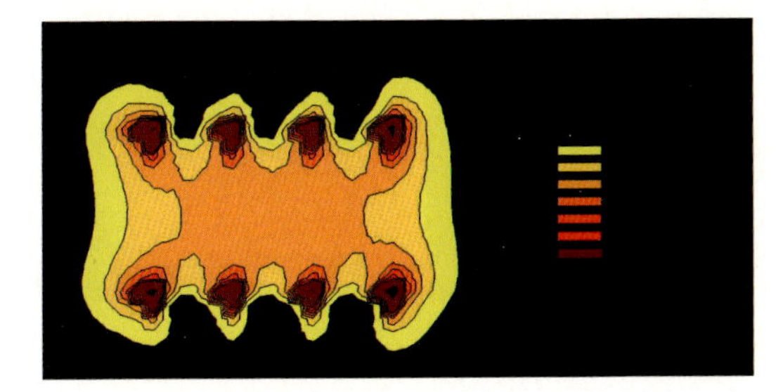

Figure 8.5 Example of electric field distribution. Assuming homogeneous conductivity (which is indeed an unlikely event), the field distribution between two arrays of opposing electrodes is shown. It is obvious that when speaking of applied voltage at the level of the electrodes, the actual field to which cells are exposed will depend on electrode geometry and conductivity. From [5].

and the result was that 73% of tumors were in complete remission, 11% in partial remission, and other lesions either no change or progression [10]. It was found that the three different administration types gave similar results, except in larger tumors where intravenous bleomycin seemed superior, likely associated with better distribution in the entire tumor volume. Furthermore, it was noted that there was a tissue-sparing effect in normal tissues (Figure 8.5). This is observed for both intravenous bleomycin and local injection of concentration $1\,U\,ml^{-1}$, whereas normal tissue necrosis may ensue with a concentration of $5\,U\,ml^{-1}$. Thus, with the high local concentration there may be a higher tumor response rate but at a cost of more local tissue damage. As electrochemotherapy can be repeated if necessary, it is this author's opinion that a good solution is to not use the higher concentration of bleomycin in order to obtain better preservation of normal tissues, while it will still be possible to retreat the approximately 25% of patients in need of better tumor control.

In 2006, Mir *et al.* [15] described standard operating procedures for electrochemotherapy of tumors up to 3 cm, and these are now frequently referred to as a standard of care. Several studies, most of them based on the 2006 standard operating procedures, have now been published, and confirm both the level of response rates and toxicity profile.

What is clear from the above mentioned studies is that electrochemotherapy works for a number of different tumor histologies. This may initially seem surprising, but is completely logical when taking the scientific background into account. First, the cell membrane is a unifying feature between all types of tumors, and although there may be slight variations in the threshold for permeabilization, if sufficient pulsing parameters are used all cells may be permeabilized. Furthermore, in the case of bleomycin, the increase in cytotoxicity is over 300-fold, which is far beyond what resistance mechanisms in cells may handle. This means (i) that all tumor histologies treated (planocullular carcinoma, adenocarcinoma,

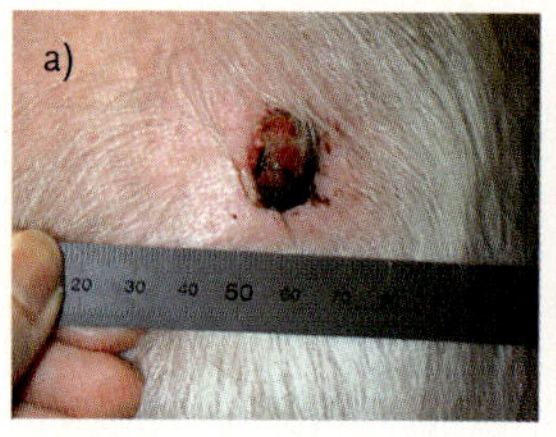

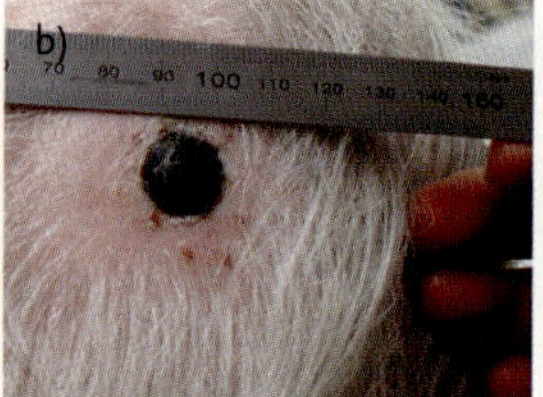

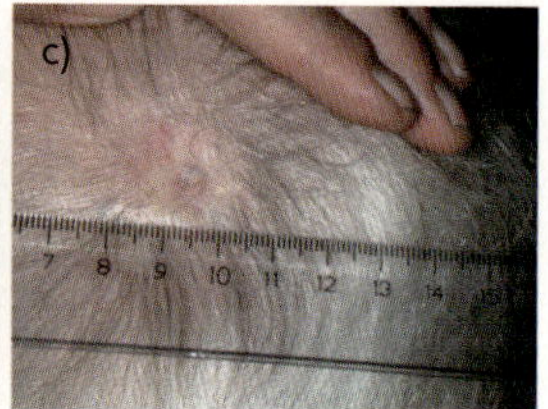

Figure 8.6 Melanoma treatment, showing a patient with ulcerated metastasis from malignant melanoma. (a) Before treatment and (b) after treatment with intravenous bleomycin and electroporation of the tumor area under general anesthesia. As can be seen, there are needle marks outside the treatment area in the normal tissue, indicating that tumor tissue is more responsive to the treatment (necrosis of lesion) than normal tissue (only needle marks present). (c) Six months follow-up showing complete resolution without recurrence. From [39].

transitocellular carcinoma, hypernephroma, malignant melanoma, basocellular carcinoma) have responded to electrochemotherapy, (ii) that tumors that generally do not show high response rates to chemotherapy (e.g., malignant melanoma; Figure 8.6) respond, and (iii) that tumors previously having progressed on chemotherapy may respond.

8.7.2
Larger Tumors

Whereas the initial studies were focused on assessing the response rates and safety profiles of smaller tumors, it was clear after the first results that also patients suffering from larger tumor recurrences would potentially benefit. One of the first case reports published was the treatment of a large chest wall recurrence of breast cancer, and this subsequently has prompted a phase II clinical investigation of the treatment of these tumors (www.clinicaltrials.gov). Initial results show impressive responses in a heavily pretreated patient population, with a favorable safety profile and patient-reported symptom alleviation.

Particular issues are relevant when treating large tumors. Thus healing time will be delayed as normal tissue needs to grow into the treatment site to replace necrotic tissue. This may be further delayed in previously irradiated tissues, where angiogenesis and wound healing is delayed.

8.8
Use in Internal Organs

After the excellent results on cutaneous tumors, an obvious question has been how to transfer these results to the treatment of internal tumors. At the time of writing, at least three different approaches are being pursued.

8.8.1 Endoscopic Use

An endoscopic applicator with a vacuum device has been developed. This enables endoscopic application and suction of a tumor into the device with subsequent application of drugs/genes and pulses. A study in veterinary clinics in spontaneous tumors in dogs has been carried out successfully (D. Soden, personal communication), and a phase I clinical trial has now opened for patients suffering from colorectal cancer (www.clinicaltrials.gov; NCT01172860).

8.8.2 Bone Metastases

A system with very stiff electrode pins has been developed for the treatment of bone metastases. In this system, a bone metastasis is defined, and pins positioned according to a treatment planning system, in and on all sides of the lesion. After pulse delivery the pins are removed and treatment completed. Extensive preclinical data have been obtained [32] and the first phase I clinical trial has commenced.

8.8.3 Brain Metastases, Brain Tumors, and Other Tumors in Soft Tissues

A novel electrode has been developed for the use of electrochemotherapy in soft tissues, the primary target being the brain (Figure 8.7). This electrode is expandable, so that an area larger than the insertion point may be treated. Preclinical data have shown results that are in essence comparable with clinical results from skin tumors, with a high level of complete responses [33]. A clinical trial has been opened (www.clinicaltrials.gov).

8.8.4 Liver Metastases

Finally, using the stiff electrode pins, a trial on the treatment of liver metastases has commenced (www.clinicaltrials.gov; NCT01264952) also using a treatment planning system.

8.9 Gene Electrotransfer

Gene transfer is a very exciting application of electroporation. The application is in principle very simple: DNA is injected in the target tissue, pulses are applied, and in this way gene therapy may be performed in a matter of a minute. Gene electrotransfer is nonviral, meaning better patient safety, reduced concerns for staff working safety, as well as markedly reduced cost of the procedure. The fact

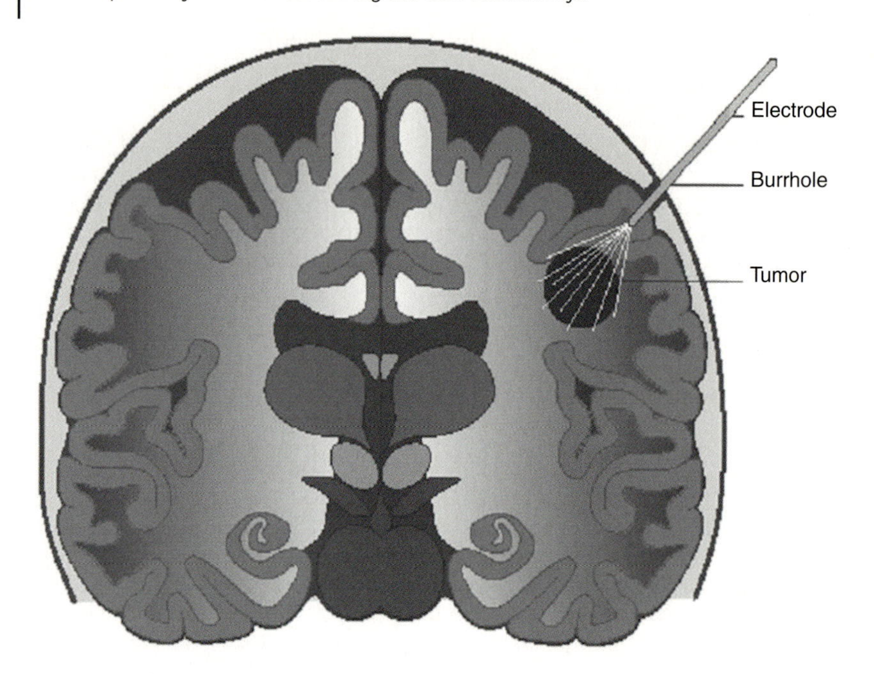

Figure 8.7 Brain electrode. Novel treatment strategies involve enabling the technology to work in internal tumors (see text). An example of this is an electrode developed for use in soft tissues, exemplified by tumors of the brain. Preclinical results using a prototype electrode are described in [33], and a clinical trial has commenced (www.clinicaltrials.gov).

that transgene expression only occurs in the particular region where injection and pulses are performed has an advantage in that the transfection area may be controlled in extent by the application in itself.

The transgenic expression is transient, as the DNA is not integrated in the host genome. This means that expression will depend on the longevity of the target cells:

1) Transfer to muscle cells, which have a very long lifespan, will lead to long-term secretion of the transgene, useful in protein deficiency syndromes.

2) Transfer to tumor cells will be short lived, and can be used for expression of proteins with tumoricidal effects, for example, antiangiogenic factors or cytokines.

3) Transfer to skin will lead to somewhat longer durations of transgenic expression (a few days to a few weeks), and this may be very useful for vaccinations, possibly also in the treatment of skin diseases.

Gene electrotransfer is a process involving several steps:

1) DNA must be approximated to the target cells. As DNA is large and highly charged, diffusion is slow. Once the DNA is in the vicinity of the target cell, it must adsorb to the cell membrane prior to transfer by electric pulses.

2) During the pulse application, DNA will interact with the cell membrane, after which DNA will either be internalized or trapped in the membrane (and then possibly later internalized).

3) Once in the cytosol, active transport to the nucleus occurs, and the microtubular system as well as the transporter dynein have been shown to play an important role in this process. Actin is also implicated in the early processes, and possibly further mechanisms for intracellular trafficking may coexist.

4) In order for the transgenic expression to take place, a number of factors are important. First, the cell must be in good condition, in order to express the transgene, and indeed it has been shown that optimal conditions for expression are at pulsing parameters where the cell is minimally perturbed, in terms of level of membrane permeabilization [6]. Second, choice of promoter will be important for transgenic expression, an important subject which falls outside the scope of this chapter. Third, the transgenic protein may or may not have immunogenic effects, and quenching may affect measurable levels of protein.

8.9.1 Gene Electrotransfer to Muscle

Muscle is clearly a very interesting target for gene electrotransfer. It is important for gene electrotransfer that the plasmid does not get integrated into the host genome. Muscle cells have a long lifespan, enabling long-term expression of the transgene. Thus, it has been shown that gene electrotransfer to muscle (Figure 8.8) enables highly efficient [16] expression for many months in rodent experiments. In humans it may actually be a life-long expression, yet to be shown. Naturally gene expression to muscle would be ideal for protein deficiency syndromes such as hemophilia or alpha-trypsin-1 deficiency. However, gene electrotransfer may also be used for expressing therapeutic molecules such as anticancer agents. Various inducible systems exist for experimental models, where expression of a transgene may be induced by addition of a certain molecule, for example doxycycline [34].

8.9.2 Gene Electrotransfer to Skin

Gene electrotransfer to skin leads to shorter term expression, from days to a few weeks [7, 35]. The skin consists of a number of different cell types, for example

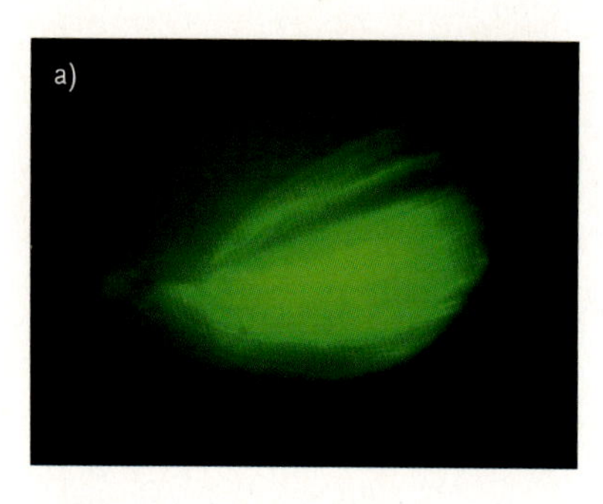

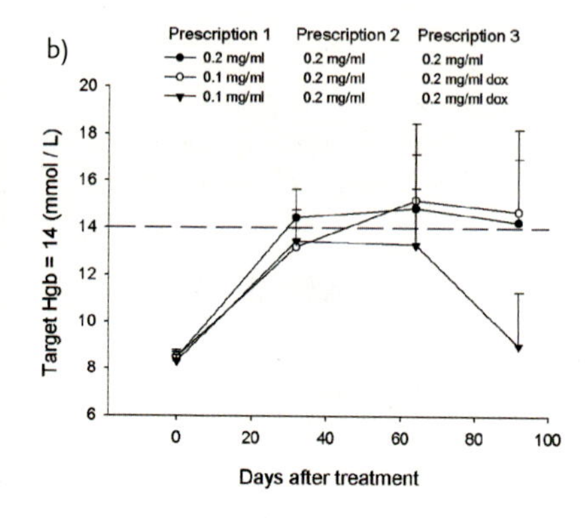

Figure 8.8 Gene electrotransfer to muscle. (a) Through simple injection of naked DNA intramuscularly followed be electric pulses, high-level and long-term expression may be obtained, as seen with transfection using the green fluorescent protein. (b) Transgenic expression of erythropoietin from a small muscle may in fact lead to therapeutic levels of the transgene, as evidenced by a dramatic increase in hemoglobin (Hgb) levels, erythropoietin stimulating the formation of new erythrocytes. From [34].

keratinocytes, Langerhans cells, and fibroblasts, and transfection levels may depend on depth of DNA injection and pulsing parameters. The skin has a number of different properties, and one very interesting aspect is its immunological properties. DNA vaccinations are likely to play an important role in the future, for several reasons, and in this aspect gene electrotransfer to skin may be very important as it allows good transfection rates of antigenic proteins in an immunologically active organ. Furthermore, the duration of expression of transgenic proteins would fit well with a vaccination strategy. Possible exploitations of gene electrotransfer to skin are represented schematically in Figure 8.9.

8.9.3
Gene Electrotransfer to Tumors

Gene electrotransfer to tumors may lead to high levels of an anticancer molecule directly at the tumor site. It has been shown in a number of preclinical studies how this may be utilized, and also the first clinical study on the use of gene electrotransfer to tumors was published in 2008 [36]. In this study, the plasmid coding for interleukin-12 (IL-12) was transferred to tumors in patients suffering from disseminated malignant melanoma. IL-12 is a cytokine known to have potent antitumor effects but some clinical studies have shown that systemic administration led to high levels of toxicity. By transferring the gene directly into the tumor, high local levels may be obtained while toxicity at the systemic level can be moderate. Indeed, in this study [36], significantly increased levels of IL-12 were shown in transfected tumors while overall toxicity was low. At the same time responses locally at the transfected site were observed, as well as distant responses, likely mediated by an immune response elicited from the treated areas.

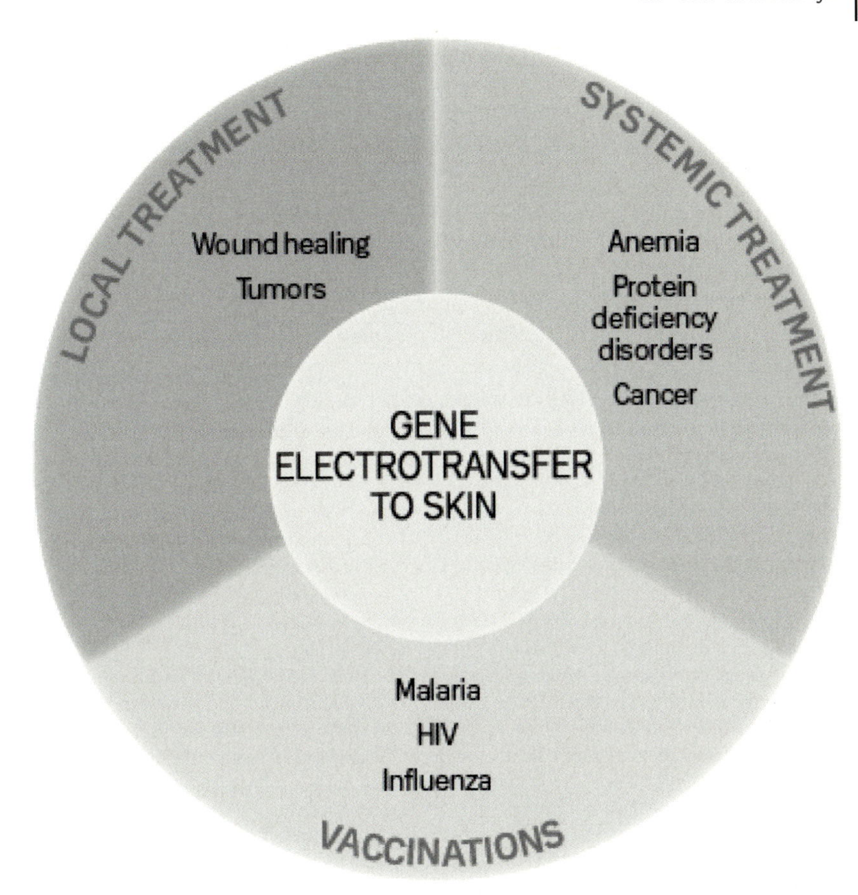

Figure 8.9 Gene electrotransfer to skin. Schematic of how gene electrotransfer to skin may be used in a variety of medical conditions, from vaccination to cancer treatment. Gene electrotransfer to skin gives rise to transgenic expression lasting a couple of weeks [7, 35].

Gene electrotransfer using other molecules may also be envisaged. For example, an ongoing trial is looking at the possibility to obtain responses by transfection of a plasmid coding for an antiangiogenic and antiproliferative molecule (www.clinicaltrials.gov; NCT01045915).

8.9.4 Gene Electrotransfer to Other Tissues

Gene electrotransfer to other tissues, for example kidney, testes and eyes, is extensively reviewed elsewhere and a detailed description falls beyond this chapter.

8.10 Conclusions

Drug and gene electrotransfer has a number of highly interesting perspectives. Clinical feasibility has been shown, and numerous novel applications are in development. Electrochemotherapy is estimated to become an important part of the armamentarium in the treatment of cancer of various histologies, and anatomical locations.

Gene therapy by electrotransfer has potential use in a number of different applications, from vaccinations to treatment of protein deficiencies such as hemophilia, or cancer treatment.

It will be very interesting to follow developments in the coming years, as this platform technology becomes increasingly used in a number of medical applications.

References

1 Neumann, E., Schaefer-Ridder, M., Wang, Y., and Hofschneider, P.H. (1982) Gene transfer into mouse lyoma cells by electroporation in high electric fields. *EMBO Journal*, **1**, 841–845.

2 Gehl, J. (2008) Electroporation for drug and gene delivery in the clinic: doctors go electric. *Methods in Molecular Biology*, **423**, 351–359.

3 Staal, L.G. and Gilbert, R. (2011) Generators and applicators: equipment for electroporation, in *Clinical Aspects of Electroporation* (eds S. Kee, J. Gehl, and E. Lee), Springer, New York, pp. 45–65.

4 Miklavcic, D., Beravs, K., Semrov, D., Cemazar, M., Demsar, F., and Sersa, G. (1998) The importance of electric field distribution for effective *in vivo* electroporation of tissues. *Biophysical Journal*, **74**, 2152–2158.

5 Gehl, J., Sørensen, T.H., Nielsen, K., Raskmark, P., Nielsen, S.L., Skovsgaard, T., *et al.* (1999) *In vivo* electroporation of skeletal muscle: threshold, efficacy and relation to electric field distribution. *Biochimica et Biophysica Acta*, **1428**, 233–240.

6 Hojman, P., Gissel, H., Andre, F., Cournil-Henrionnet, C., Eriksen, J., Gehl, J., and Mir, L.M. (2008) Physiological effects of high- and low-voltage pulse combinations for gene electrotransfer in muscle. *Human Gene Therapy*, **19**, 1249–1260.

7 Gothelf, A. and Gehl, J. (2010) Gene electrotransfer to skin: review of existing literature and clinical perspectives. *Current Gene Therapy*, **10**, 287–299.

8 Heller, L.C. and Heller, R. (2006) *In vivo* electroporation for gene therapy. *Human Gene Therapy*, **17**, 890–897.

9 Matthiessen, L.W., Chalmers, R.L., Sainsbury, D.C., Veeramani, S., Kessel, G., Humphreys, A.C., Bond, J., Muir, T. and Gehl, J. (2011) Management of cutaneous metastases using electrochemotherapy. *Acta Oncologica*, **50**, 621–629.

10 Marty, M., Sersa, G., Garbay, J.R., Gehl, J., Collins, C.G., Snoj, M. *et al.* (2006) Electrochemotherapy–an easy, highly effective and safe treatment of cutaneous and subcutaneous metastases: results of ESOPE (European Standard Operating Procedures of Electrochemotherapy) study. *European Journal of Cancer Supplement*, **4**, 3–13.

11 Fear, E. and Stuchly, M.A. (1998) Biological cells with gap junctions in low-frequency electric fields. *IEEE Transactions on Biomedical Engineering*, **45**, 856–866.

12 Andre, F., Gehl, J., Sersa, G., Preat, V., Hojman, P., Eriksen, J. *et al.* (2008)

Efficiency of high- and low-voltage pulse combinations for gene electrotransfer in muscle, liver, tumor, and skin. *Human Gene Therapy*, **19**, 1261–1271.

13 Gaylor, D.C., Prakah-Asante, K., and Lee, R.C. (1988) Significance of cell size and tissue structure in electrical trauma. *Journal of Theoretical Biology*, **133**, 223–237.

14 Pollack, G.H. (2011) *Cells, Gels and the Engines of Life*, Ebner and Sons, Seattle.

15 Mir, L.M., Gehl, J., Sersa, G., Collins, C.G., Garbay, J.R., Billard, V. *et al.* (2006) Standard operating procedures of the electrochemotherapy: instructions for the use of bleomycin or cisplatin administered either systemically or locally and electric pulses delivered by the Cliniporator™ by means of invasive or non-invasive electrodes. *European Journal of Cancer Supplement*, **4**, 14–25.

16 Mir, L.M., Bureau, M.F., Gehl, J., Rangara, R., Rouy, D., Caillaud, J.-M. *et al.* (1999) High efficiency gene transfer into skeletal muscle mediated by electric pulses. *Proceedings of the National Academy of Sciences of the USA*, **96**, 4262–4267.

17 Satkauskas, S., Bureau, M.F., Puc, M., Mahfoudi, A., Scherman, D., Miklavcic, D. *et al.* (2002) Mechanisms of *in vivo* DNA electrotransfer: respective contributions of cell electropermeabilization and DNA electrophoresis. *Molecular Therapy*, **5**, 133–140.

18 Zaharoff, D.A., Barr, R.C., Li, C.Y., and Yuan, F. (2002) Electromobility of plasmid DNA in tumor tissues during electric field-mediated gene delivery. *Gene Therapy*, **9**, 1286–1290.

19 Escoffre, J.M., Rols, M.P., and Dean, D.A. (2011) Electrotransfer of plasmid DNA, in *Clinical Aspects of Electroporation* (eds S. Kee, J. Gehl, and E. Lee), Springer, New York, pp. 145–157.

20 Joergensen, M., Agerholm-Larsen, B., Nielsen, P.E., and Gehl, J. (2011) Efficiency of cellular delivery of antisense peptide nucleic acid by electroporation depends on charge and electroporation geometry. *Oligonucleotides*, **21**, 29–37.

21 Gopal, R., Narkar, A.A., Mishra, K.P., Samuel, A.M., and Nair, N. (2003) Electroporation: a novel approach to enhance the radioiodine uptake in a human thyroid cancer cell line. *Applied Radiation and Isotopes*, **59**, 305–310.

22 Hojman, P., Spanggaard, I., Olsen, C.H., Gehl, J., and Gissel, H. (2011) Calcium electrotransfer for termination of transgene expression in muscle. *Human Gene Therapy*, **22**, 753–760.

23 Dinchuk, J.E., Kelley, K.A., and Callahan, G.N. (1992) Flow cytometric analysis of transport activity in lymphocytes electroporated with a fluorescent organic anion dye. *Journal of Immunological Methods*, **155**, 257–265.

24 Poddevin, B., Belehradek, J., Jr., and Mir, L.M. (1990) Stable [57Co]-bleomycin complex with a very high specific radioactivity for use at very low concentrations. *Biochemical and Biophysical Research Communications*, **173**, 259–264.

25 Gehl, J., Skovsgaard, T., and Mir, L.M. (2002) Vascular reactions to *in vivo* electroporation: characterization and consequences for drug and gene delivery. *Biochimica et Biophysica Acta*, **1569**, 51–58.

26 Sersa, G., Cemazar, M., Miklavcic, D., and Chaplin, D.J. (1999) Tumor blood flow modifying effect of electrochemotherapy with bleomycin. *Anticancer Research*, **19**, 4017–4022.

27 Cemazar, M., Parkins, C.S., Holder, A.L., Chaplin, D.J., Tozer, G.M., and Sersa, G. (2001) Electroporation of human microvascular endothelial cells: evidence for an anti-vascular mechanism of electrochemotherapy. *British Journal of Cancer*, **84**, 565–570.

28 Gehl, J. and Geertsen, P. (2000) Efficient palliation of hemorrhaging malignant melanoma skin metastases by electrochemotherapy. *Melanoma Research*, **10**, 585–589.

29 Belehradek, M., Domenge, C., Luboinski, B., Orlowski, S., Belehradek, J. Jr., and Mir, L.M. (1993) Electrochemotherapy, a new antitumor treatment. First clinical phase I–II trial. *Cancer*, **72**, 3694–3700.

30 Heller, R., Jaroszeski, M.J., Glass, L.F., Messina, J.L., Rapaport, D.P., DeConti, R.C. *et al.* (1996) Phase I/II trial for the

treatment of cutaneous and subcutaneous tumors using electrochemotherapy. *Cancer*, **77**, 964–971.

31 Glass, L.F., Fenske, N.A., Jaroszeski, M., Perrott, R., Harvey, D.T., Reintgen, D.S. *et al.* (1996) Bleomycin-mediated electrochemotherapy of basal cell carcinoma. *Journal of the American Academy of Dermatology*, **34**, 82–86.

32 Fini, M., Tschon, M., Ronchetti, M., Cavani, F., Bianchi, G., Mercuri, M. *et al.* (2010) Ablation of bone cells by electroporation. *Journal of Bone and Joint Surgery (British Volume)*, **92**, 1614–1620.

33 Agerholm-Larsen, B., Iversen, H.K., Ibsen, P., Moller, J.M., Mahmood, F., Jensen, K.S., and Gehl, J. (2011) Preclinical validation of electrochemotherapy as an effective treatment for brain tumors. *Cancer Research*, **71**, 3753–3762.

34 Hojman, P., Gissel, H., and Gehl, J. (2007) Sensitive and precise regulation of haemoglobin after gene transfer of erythropoietin to muscle tissue using electroporation. *Gene Theraoy*, **14**, 950–959.

35 Gothelf, A., Eriksen, J., Hojman, P., and Gehl, J. (2010) Duration and level of transgene expression after gene electrotransfer to skin in mice. *Gene Therapy*, **17**, 839–845.

36 Daud, A.I., DeConti, R.C., Andrews, S., Urbas, P., Riker, A.I., Sondak, V.K. *et al.* (2008) Phase I trial of interleukin-12 plasmid electroporation in patients with metastatic melanoma. *Journal of Clinical Oncology*, **26**, 5896–5903.

37 Gabriel, B. and Teissie, J. (1997) Direct observation in the millisecond time range of fluorescent molecule asymmetrical interaction with the electropermeabilized cell membrane. *Biophysical Journal*, **73**, 2630–2637.

38 Gehl, J. (2003) Electroporation: theory and methods, perspectives for drug delivery, gene therapy and research. *Acta Physiologica Scandinavica*, **177**, 437–447.

39 Gehl, J. (2005) Investigational treatment of cancer using electrochemotherapy, electrochemoimmunotherapy and electro-gene transfer. *Ugeskrift for Laeger*, **167**, 3156–3159.

40 Golzio, M., Teissie, J., and Rols, M.P. (2002) Direct visualization at the single-cell level of electrically mediated gene delivery. *Proceedings of the National Academy of Sciences of the USA*, **99**, 1292–1297.

Index

Advances in Electrochemical Science and Engineering. Edited by Richard C. Alkire, Dieter M. Kolb, and Jacek Lipkowski

ISBN: 978-3-527-32885-7

n